John Senior.

LIGHT TRANSMISSION OPTICS

Second Edition

DIETRICH MARCUSE

Member of Technical Staff
Bell Telephone Laboratories
Holmdel, New Jersey

Van Nostrand Reinhold Electrical/Computer Science and Engineering Series

VAN NOSTRAND REINHOLD COMPANY
New York / Cincinnati / Toronto / London / Melbourne

Library of Congress Catalog Card Number: 81-21804
ISBN: 0-442-26309-0

Manufactured in the United States of America

Published by Van Nostrand Reinhold Company Inc.
135 West 50th Street, New York, N.Y. 10020

Van Nostrand Reinhold Publishing
1410 Birchmount Road
Scarborough, Ontario M1P 2E7, Canada

Van Nostrand Reinhold Australia Pty. Ltd.
17 Queen Street
Mitcham, Victoria 3132, Australia

Van Nostrand Reinhold Company Limited
Molly Millars Lane
Wokingham, Berkshire, England

15 14 13 12 11 10 9 8 7 6 5 4 3 2 1

Library of Congress Cataloging in Publication Data

Marcuse, Dietrich, 1929-
    Light transmission optics.
    (Van Nostrand Reinhold electrical/computer science and engineering series)
    Includes index.
    1. Light—Transmission.   2. Optical wave guides.
3. Fiber optics.   I. Title.   II. Series.
QC389.M37   1982   535'.3   81-21804
ISBN 0-442-26309-0   AACR2

Van Nostrand Reinhold
Electrical/Computer Science and Engineering Series
Sanjit Mitra, Series Editor

PREFACE

A second edition of *Light Transmission Optics* became necessary when the first edition went out of print. Moreover, the many requests for copies of this book persuaded me that a second edition is needed.

Even though the field of optical fiber communications has grown tremendously in the 10 years since the appearance of *Light Transmission Optics*, most of the material contained in the first edition is still of interest today and is needed for an understanding of light propagation through free space, through optical fibers, and other types of optical waveguides. Even gas lenses, which are not currently under active consideration as waveguides for light communications systems, retain sufficient academic and historical interest to save them from being purged from this second edition. Except for correcting a few errors that had come to my attention, the contents of the first edition remain intact in the second edition.

Of the many new developments that have occurred since I wrote the manuscript of the first edition around 1970, the subject of multimode fibers seemed the most interesting. For this reason, I added chapter 11 which deals with several aspects of multimode fibers. In particular, I included the ray optics treatment of several refractive index profiles of special practical interest. Also included is a detailed derivation of the important WKB method and its application to light propagation in multimode fibers.

Chapter 12, another newcomer to this second edition, is devoted to problems of light dispersion in single-mode and multimode fibers. The theory of pulse propagation in single-mode fibers presented here includes the effects of first-and second-order dispersion. The treatment of dispersion in multimode fibers is based on geometrical optics and on the WKB theory.

Finally, chapter 8 was extended by inclusion of section 8.7, dealing with the approximation of the modes of single-mode fibers by Gaussian field distributions.

More could have been included in this second edition, but much of the material that does not appear in this book has already been treated by other authors and some of it is included in my book *Theory of Dielectric Optical Waveguides*, published in 1974 by Academic Press.

D. MARCUSE

CONTENTS

1

WAVE OPTICS

1.1 INTRODUCTION

Light is an electromagnetic phenomenon. Therefore optics should simply be a branch of electrodynamics. That optics is usually treated as a separate discipline has historical reasons, since light has been studied long before its electromagnetic character was understood. The one outstanding feature that sets optics apart from other branches of electromagnetic phenomena is the fact that we can detect light with our eyes while electromagnetic radiation at other than light frequencies goes largely unnoticed as long as no special detecting devices are being used.

A further advantage of light is its extremely short wavelength, which makes it possible to use methods of approximate analysis that cannot be employed for longer wavelength radiation. For this reason there are two different approaches to optics. Wave optics is directly based on Maxwell's equations to solve problems of light propagation, while ray optics uses the short wavelength of light to simplify many problems of light propagation. Ray optics is in many ways similar to the mechanics of point particles, while wave optics corresponds to the quantum theory of light rays. We shall discuss this relationship in detail in the chapter on geometrical optics.

In the present chapter, we introduce Maxwell's equations and derive the important wave equation. We then apply these basic equations to the solution of a few representative and important examples. The choice of applications of wave optics presented in this chapter is based on the desire of introducing only as much of the traditional wave optics as is required for the understanding of later chapters. Wave optics is an enormously large and well-developed field and is covered in many textbooks.[1] It appears therefore that we do not need to strive for exhaustive coverage but can limit the discussion to the essential properties of light reflection and refraction at the interface between two different dielectric media and, in the next chapter, to the fundamental problems of wave diffraction. The treatment in this first chapter is largely conventional, and the reader with a good background in wave propagation phenomena may want to do no more than glance at the treatment presented here.

1.2 MAXWELL'S EQUATIONS

The specific form which Maxwell's equations take is explained and justified in every textbook on electromagnetism (for particularly clear presentations, see references 2 and 3), so that we need only state them here. The electric field intensity vector **E** and the electric displacement vector **D** are related to the magnetic field intensity vector **H** and the magnetic flux density vector **B** by the following equation.*

$$\mathbf{V} \times \mathbf{H} = \frac{\partial \mathbf{D}}{\partial t} \tag{1.2-1}$$

and

$$\mathbf{V} \times \mathbf{E} = -\frac{\partial \mathbf{B}}{\partial t} \tag{1.2-2}$$

We have omitted the current term in the first of Maxwell's equations (1.2-1), since it is rarely advantageous to describe the generation of light by means of currents.

The two electric vectors are related to each other. In the general case, this relation can be quite complicated, being of a tensorial or even nonlinear nature. However, for many cases of practical interest, we can assume a simple linear relationship

$$\mathbf{D} = \varepsilon \mathbf{E} \tag{1.2-3}$$

which holds for linear, isotropic media. The constant ε is the dielectric permittivity. The ratio $\varepsilon/\varepsilon_0$ (ε_0 = the vacuum value of the permittivity) is called dielectric constant. The connection between **H** and **B** is similarly given by the equation

$$\mathbf{B} = \mu \mathbf{H} \tag{1.2-4}$$

The constant μ is known as the magnetic permeability; in nonmagnetic materials, its value is very nearly equal to the vacuum constant μ_0.

---

* The operator Δ is a vector whose components are given by

$$\left[\frac{\partial}{\partial x}, \frac{\partial}{\partial y}, \frac{\partial}{\partial z}\right].$$

In the absence of electric charges, the vector **D** satisfies the relation

$$\nabla \cdot \mathbf{D} = 0 \tag{1.2-5}$$

while the magnetic induction vector always obeys the additional relation

$$\nabla \cdot \mathbf{B} = 0 \tag{1.2-6}$$

Equations (1.2-1) through (1.2-6) completely describe the electromagnetic field in linear, isotropic media in the absence of currents and space charges.

A quantity of great importance is the power flow density vector

$$\mathbf{S} = \mathbf{E} \times \mathbf{H} \tag{1.2-7}$$

which is also known as the Poynting vector. It describes the flow of electromagnetic power in space. To obtain the power which flows through a surface A with an outward-directed normal unit vector **n** at every point, we have to evaluate the surface integral

$$P = \int_A \mathbf{S} \cdot \mathbf{n} \, dA \tag{1.2-8}$$

In many instances, it is permissible to think of light rays as the lines in space along which the electromagnetic energy of a narrow pencil of light propagates.

We will often have occasion to deal with monochromatic fields oscillating at one definite frequency f. It is then convenient to use a complex notation such that the components of the electric or magnetic fields can be expressed by equations of the following general form

$$F(x, y, z, t) = Re\{G(x, y, z)e^{i\omega t}\} \tag{1.2-9}$$

the radian frequency ω is defined by

$$\omega = 2\pi f \tag{1.2-10}$$

$G(x, y, z)$ can be a complex function of the real coordinate variables x, y, and z. The symbol $Re\{\}$ indicates that the real part of the expression in brackets is to be taken. Even though the expression must be interpreted as written in (1.2-9), the symbol $Re\{\}$ is always omitted from the equations, leaving it understood that only the real part of the quantity has physical significance. We thus write simply

$$F(x, y, z, t) = G(x, y, z)e^{i\omega t} \tag{1.2-11}$$

With the help of this complex notation, the Poynting vector (1.2-7) can be written in the form

$$\overline{\mathbf{S}} = \frac{1}{2}[\mathbf{E} \times \mathbf{H}^*] \tag{1.2-12}$$

The asterisk indicates complex conjugation. The factor 1/2 that appears in

(1.2-12) is necessary to account for the time average of the Poynting vector that is indicated by the overbar. The real part of (1.2-12) is the physical, time-averaged power flow vector.

1.3 THE WAVE EQUATION

Maxwell's equations can be modified in many ways to yield derived equations which may be more suitable for certain applications. For example, let us substitute (1.2-4) into (1.2-2) and take the curl of this latter equation. We obtain

$$\mathbf{V} \times (\mathbf{V} \times \mathbf{E}) = - \mu \frac{\partial}{\partial t} (\mathbf{V} \times \mathbf{H}) \tag{1.3-1}$$

In deriving Equation (1.3-1) we assumed that μ is a constant independent of the space coordinates. Substitution of (1.2-1) and (1.2-3) into (1.3-1) yields an equation which depends only on the vector $\mathbf{E}$ itself

$$\mathbf{V} \times (\mathbf{V} \times \mathbf{E}) + \varepsilon\mu \frac{\partial^2 \mathbf{E}}{\partial t^2} = 0 \tag{1.3-2}$$

It is worth noting that this equation holds even if ε varies in space. The $\mathbf{V} \times \mathbf{V} \times$ operator is not very easy to use, so that it is advantageous to introduce the vector identity

$$\mathbf{V} \times (\mathbf{V} \times \mathbf{E}) = \mathbf{V}(\mathbf{V} \cdot \mathbf{E}) - \nabla^2 \mathbf{E} \tag{1.3-3}$$

which holds if we use a cartesian coordinate system. Utilizing (1.2-3) and (1.2-5) enables us to rewrite equation (1.3-2) in the following way

$$\nabla^2 \mathbf{E} + \mathbf{V} \left[\mathbf{E} \cdot \frac{\mathbf{V}\varepsilon}{\varepsilon} \right] = \varepsilon\mu \frac{\partial^2 \mathbf{E}}{\partial t^2} \tag{1.3-4}$$

In the special case that ε is constant in space, the gradient of ε vanishes, and Equation (1.3-4) assumes the form of the wave equation

$$\nabla^2 \mathbf{E} = \varepsilon\mu \frac{\partial^2 \mathbf{E}}{\partial t^2} \tag{1.3-5}$$

The wave equation (1.3-5) holds for each component of the electric field vector—that is, each of its components satisfy the scalar wave equation

$$\nabla^2 \psi = \frac{1}{v^2} \frac{\partial^2 \psi}{\partial t^2} \tag{1.3-6}$$

with

$$v = (\varepsilon\mu)^{-1/2} \qquad (1.3\text{-}7)$$

having the physical significance of the velocity of light in the medium with dielectric constant $\varepsilon/\varepsilon_0$.

The wave equation (1.3-6) is approximately satisfied by each component of the electric field vector, even when ε varies in space, provided that its variation is slight over the distance of the light wavelength. We shall return to this point later.

The significance of the wave equation can easily be appreciated if we consider that every function of the general form

$$\psi = f\left[t - \frac{1}{v}\mathbf{n}\cdot\mathbf{r}\right] \qquad (1.3\text{-}8)$$

is a solution of this equation provided the second derivative of f exists. The components of the vector $\mathbf{r}$ are the coordinates of the point at which the field is being observed; $\mathbf{n}$ is a unit vector. For (1.3-8) to be a solution of the wave equation (1.3-6), the velocity v must be independent of frequency. We shall return to this point in a moment.

The solution (1.3-8) of the wave equation represents plane waves in a space which is homogeneously filled with a medium with dielectric constant $\varepsilon/\varepsilon_0$. To see that the function of (1.3-8) describes plane waves, it is necessary to consider a certain fixed value of its argument

$$u = t - \frac{1}{v}\mathbf{n}\cdot\mathbf{r} \qquad (1.3\text{-}9)$$

For any given value of u, the function has the corresponding fixed value $f(u)$. The value $u = \text{const.}$ is realized for a fixed value of the time t on a plane defined by the relation $\mathbf{n}\cdot\mathbf{r} = \text{const.}$ The vector $\mathbf{n}$ is directed perpendicular to the plane. The same value of the function is therefore to be found over an infinite plane in space. To watch how this particular value of the function behaves as the time t advances, we allow t to change by Δt and the vector $\mathbf{r}$ by $\Delta\mathbf{r}$ in such a way that u remains constant. The relation between the increment of time and the change of the position vector which keeps u unchanged is given by

$$\mathbf{n}\cdot\Delta\mathbf{r} = v\Delta t \qquad (1.3\text{-}10)$$

The endpoint of $\Delta\mathbf{r}$ lies again on a plane. The vector $\mathbf{n}$ is apparently the normal to both the original and the displaced plane. The plane described by (1.3-9) has moved in the direction of $\mathbf{n}$ by an amount $v\Delta t$ during the time interval Δt. This shows that the plane moves with the velocity v through space. Equation (1.3-8) describes, therefore, a plane wave disturbance moving with the velocity v. The form of the function $f(u)$ is arbitrary.

Changing the sign of v in (1.3-8) results in another solution of the wave equation. It can easily be seen that the function

$$\psi = f\left[t + \frac{1}{v}\mathbf{n}\cdot\mathbf{r}\right]$$

represents a plane wave traveling in the direction of $-\mathbf{n}$.

Solutions of the wave equation that are of particular importance are plane waves which vary sinusoidally with time at any point in space. Such a plane wave can be represented in the form

$$g = A\cos\left[\omega t - \frac{\omega}{v}\mathbf{n}\cdot\mathbf{r}\right]$$

The frequency of oscillation is

$$f = \frac{\omega}{2\pi}$$

ω is called the radian frequency. Because of its importance, we shall sometimes simply call ω by the name of frequency, leaving the fact understood that it is actually 2π times the frequency of oscillation. It is convenient to introduce the vector

$$\mathbf{k} = \frac{\omega}{v}\mathbf{n} \tag{1.3-11}$$

and write

$$g = A\cos(\omega t - \mathbf{k}\cdot\mathbf{r}) \tag{1.3-12}$$

We call $\mathbf{k}$ the propagation vector. If we let $\mathbf{r}$ advance an increment

$$\Delta\mathbf{r} = \lambda\mathbf{n}$$

requiring that the function (1.3-12) changes through a full cycle as $\mathbf{r}$ advances to $\mathbf{r} + \Delta\mathbf{r}$, we find the relation

$$k\lambda = 2\pi$$

or

$$k = \frac{2\pi}{\lambda} = \frac{\omega}{v} = \omega\sqrt{\varepsilon\mu} \tag{1.3-13}$$

with

$$k = |\mathbf{k}| \tag{1.3-14}$$

being the magnitude of the propagation vector. The right-hand part of Equation (1.3-13) follows from (1.3-11) and (1.3-7).

The great physical importance of waves of the form (1.3-12) stems from the fact that in most media, with the exception of vacuum, v is not a constant but depends on the frequency

$$v = v(\omega)$$

Sinusoidal waves of different frequencies travel with different phase velocities. This phenomenon is called dispersion. The general form (1.3-8) of a plane wave of arbitrary shape is therefore applicable only in vacuum. In other media it may be useful as a reasonable approximation when dispersion is not too pronounced. We can easily predict the way in which a general disturbance travels through a dispersive medium. Let us assume that we know that at a certain plane there exists a disturbance of the form $f(t)$. In order to find out how this disturbance will travel through the dispersive medium, we must decompose the arbitrary function f into a superposition of sinusoidal oscillations. This is accomplished by the Fourier integral transformation of the function $f(t)$. Each harmonic oscillation propagates through the medium as a plane wave (plane only since we were careful to assume a plane disturbance to begin with) according to (1.3-12). Positioning our coordinate system for convenience, so that the wave travels in the z direction, we can express the shape of the general plane wave, which at $z = 0$ assumes the form $f(t)$, by the Fourier integral

$$f(z, t) = \frac{1}{\pi} \int_0^\infty h(\omega)\cos[\omega t - kz + \theta(\omega)]d\omega \qquad (1.3\text{-}15)$$

Introducing the complex function

$$\phi(\omega) = h(\omega)e^{i\theta(\omega)} \qquad (1.3\text{-}16)$$

and extending the definition of the phase and amplitude to negative frequencies by the definitions (k also changes sign)

$$\theta(-\omega) = -\theta(\omega) \qquad (1.3\text{-}17)$$

and

$$h(-\omega) = h(\omega) \qquad (1.3\text{-}18)$$

we can rewrite the real representation (1.3-15) to take the form of the complex Fourier integral

$$f(z, t) = \frac{1}{2\pi} \int_{-\infty}^\infty \phi(\omega)e^{i(\omega t - kz)} d\omega \qquad (1.3\text{-}19)$$

The amplitude function $\phi(\omega)$ is determined by the known shape of the wave at $z = 0$

$$\phi(\omega) = \int_{-\infty}^\infty f(0, t)e^{-i\omega t} dt \qquad (1.3\text{-}20)$$

Equation (1.3-19) is not a solution of the wave equation when v depends on frequency, because the wave equation makes physical sense only when there is either no dispersion—that is, when v does not depend on frequency —or when the function $f(z, t)$ can be described by a very narrow frequency spectrum. The phase velocity v in (1.3-6) is meaningless for a general function of the type (1.3-19). However, (1.3-19) correctly describes the propagation of a general plane wave through a dispersive medium.

It is easy to extend the concept of fields traveling in dispersive media to disturbances other than plane waves. To do this, we introduce again the plane sinusoidal wave

$$g(x, y, z, t) = \phi(k_x, k_y, \omega)\exp[i(\omega t - \mathbf{k} \cdot \mathbf{r})] \qquad (1.3\text{-}21)$$

traveling in the direction of the vector $\mathbf{k}$. A complex notation is used. The real part of (1.3-21) describes the physical plane wave. The amplitude factor ϕ may depend on the independent variables k_x, k_y, and ω. Since the magnitude k of $\mathbf{k}$ must obey the relation (1.3-13), we can express k_z in terms of the independent variables

$$k_z = \sqrt{\left[\frac{\omega}{v}\right]^2 - k_x^2 - k_y^2} \qquad (1.3\text{-}22)$$

The z component (or any of the other components if we happen to choose the z component as an independent variable) of $\mathbf{k}$ can become imaginary if the expression under the square root sign becomes negative. In this case we no longer have a plane wave but encounter an evanescent wave. Evanescent waves are also allowed as solutions of the wave equation. Using a superposition of waves of the form (1.3-21) with all possible frequencies as well as all possible directions, we can construct the most general wave propagating in a dispersive medium

$$f(x, y, z, t) = \frac{1}{(2\pi)^3} \int_{-\infty}^{\infty} d\omega \int_{-\infty}^{\infty} dk_x \int_{-\infty}^{\infty} dk_y \, \phi(k_x, k_y, \omega)$$
$$\cdot \exp[i\,(\omega t - k_x x - k_y y - k_z z)] \qquad (1.3\text{-}23)$$

The z component of the propagation vector must be obtained from (1.3-22). Using a real notation, we can also write

$$f(x, y, z, t) = \frac{1}{4\pi^3} \int_{0}^{\infty} d\omega \int_{-\infty}^{\infty} dk_x \int_{-\infty}^{\infty} dk_y \, |\phi(k_x, k_y, \omega)|$$
$$\cdot \cos(\omega t - k_x x - k_y y - k_z z + \theta) \qquad (1.3\text{-}24)$$

This integral representation of the most general wave consists not only of sinusoidal plane waves traveling in all possible directions and at all possible frequencies but also of evanescent waves. For purely sinusoidally time-varying solutions of the wave equation, the integration over ω can be omitted from (1.3-24).

The existence of evanescent waves in free space may be surprising since one usually thinks of evanescent waves as being a feature of waveguides that are operated below their cutoff frequency. The evanescent waves occurring in our discussion are closely related to the phenomenon of total internal reflection of a wave trying to enter from a medium with high dielectric constant into another medium whose dielectric constant is lower. To understand this phenomenon, consider the sinusoidal plane wave of (1.3-21). It travels in space with a wavelength λ given by (1.3-13) in a direction determined by the components of the **k** vector. To simplify matters, we assume that the wave travels in the x-z plane so that $k_y = 0$. Now let us turn the **k** vector more and more into the direction of the x-axis. In the limit $k_z = 0$ the wave travels parallel with the x-axis. Its sinusoidal variation in space in x direction has then the spatial period λ. However, the mathematical apparatus allows us to make $k_x > 2\pi/\lambda$. In physical terms, this means that we are forcing the field to vary with a spatial period that is shorter than λ. It is indeed possible to do this, but the field reacts to our effort by contracting in z direction. It is not possible to force the field into spatial oscillations which are more rapid than its free space wavelength at the operating frequency and still have it extend throughout all of free space. This shows that evanescent waves occur whenever we impose spatial variations on the field which are more rapid than is consistent with the free space propagation of a sinusoidal plane wave. We shall see that this is what happens in total internal reflection.

Our discussion so far has centered on the wave equation (1.3-6). However, we obtained the wave equation as a special case of (1.3-4). It is necessary to investigate under what conditions the wave equation is at least a good approximation to (1.3-4), since this latter equation is far more difficult to handle and not very useful for actual calculations. It is fortunate that in most applications encountered in optics we can use the simple wave equation even though the condition for its validity—that is, the vanishing of the second term in (1.3-4)—is not strictly satisfied.

The order of magnitude of the terms occurring in (1.3-4) is dominated by the first term on the left-hand side and by the term on its right-hand side, which are of equal order. The following analysis applies only to order of magnitude estimates and must not be taken to be precise. The term on the right of (1.3-4) is of the order of

$$\varepsilon\mu \frac{\partial^2 E}{\partial t^2} = \omega^2 \varepsilon\mu E = \frac{\omega^2}{v^2}E = \left[\frac{2\pi}{\lambda}\right]^2 E \qquad (1.3-25)$$

Replacing the ∇ operator by a derivative with respect to some direction S in space, we can write to order of magnitude

$$\nabla\left(\mathbf{E}\cdot\frac{\nabla\varepsilon}{\varepsilon}\right) \approx \frac{\partial}{\partial S}\left(\mathbf{E}\cdot\frac{\nabla\varepsilon}{\varepsilon}\right) \approx \frac{2\pi}{\lambda}\mathbf{E}\cdot\frac{\nabla\varepsilon}{\varepsilon} + \mathbf{E}\cdot\frac{\partial}{\partial S}\left(\frac{\nabla\varepsilon}{\varepsilon}\right) \qquad (1.3\text{-}26)$$

We are interested in the case that the second term on the right of (1.3-26) is much smaller than the first term. Indicating the order of magnitude by placing the expression in brackets, we obtain as a comparison of the two terms of (1.3-4) in question

$$R = \frac{\left[\nabla\left(\mathbf{E}\cdot\dfrac{\nabla\varepsilon}{\varepsilon}\right)\right]}{\left[\varepsilon\mu\dfrac{\partial^2 E}{\partial t^2}\right]} \approx \frac{\dfrac{2\pi}{\lambda}\dfrac{\nabla\varepsilon}{\varepsilon}}{\left(\dfrac{2\pi}{\lambda}\right)^2} = \frac{1}{2\pi}\lambda\frac{\nabla\varepsilon}{\varepsilon} \approx \frac{1}{2\pi}\lambda\frac{\varepsilon_2-\varepsilon_1}{\varepsilon\Delta S} \qquad (1.3\text{-}27)$$

We have indicated the order of magnitude of the gradient of ε by the ratio of the difference $\varepsilon_2 - \varepsilon_1$ of the dielectric constant of two closely spaced points divided by their distance ΔS. As a final step we take $\Delta S = \lambda$ and obtain

$$R = \frac{1}{2\pi}\frac{\varepsilon_2-\varepsilon_1}{\varepsilon} \qquad (1.3\text{-}28)$$

If we want to neglect the second term on the left of (1.3-4), we must require $R \ll 1$. As (1.3-28) indicates, this means that the relative change of ε over the distance of one wave-length must be less than unity. This condition is often satisfied in inhomogeneous optical media. We shall find in our discussion of optical waveguides with inhomogeneous but continuous dielectric constants that $R \ll 1$ is true, so that we can solve the wave equation rather than the far more difficult equation (1.3-4). The only place where (1.3-28) is likely to be equal to or even larger than unity is at the interface between two regions of different dielectric constant. Most optical instruments consist of regions of uniform dielectric constant such as air into which other regions of different but again uniform dielectric constant are embedded. At the interface between glass lenses and air, for example, (1.3-28) is likely to be large. However, even in these cases we need solve only the wave equation, since it holds everywhere except at the interfaces. We handle such situations by solving the wave equation in the various homogeneous regions and joining these solutions by means of boundary conditions. The boundary conditions are the subject of a later section. We conclude this section by checking if we were justified in neglecting the second term on the right of (1.3-26). The order of magnitude of the neglected term in relation to the remaining term is

$$\frac{\left[\mathbf{E}\cdot\dfrac{\partial}{\partial S}\left(\dfrac{\nabla\varepsilon}{\varepsilon}\right)\right]}{\left[\dfrac{2\pi}{\lambda}\mathbf{E}\cdot\dfrac{\nabla\varepsilon}{\varepsilon}\right]} \approx \frac{1}{2\pi}\lambda\frac{\left(\dfrac{|\nabla\varepsilon|}{\varepsilon}\right)_2-\left(\dfrac{|\nabla\varepsilon|}{\varepsilon}\right)_1}{\dfrac{|\nabla\varepsilon|}{\varepsilon}\nabla S} \qquad (1.3\text{-}29)$$

We need not really require that (1.3-29) is much smaller than unity. The argument for neglecting the second term in (1.3-4) is equally valid if the two terms in (1.3-26) are of equal order of magnitude. This means that we need to require only that the change of the gradient of ε over the distance of one wavelength is about equal to or less than the gradient itself. This is likely to be true under those conditions which also make $R \ll 1$.

These arguments demonstrate that the wave equation can be used even if ε is not constant but varies in space provided that its variation is slight over the distance of one optical wavelength. With the exception of interfaces between two different dielectric media, this condition is almost always satisfied. The propagation of light in inhomogeneous media can therefore be studied by solving the wave equation. The difference between the wave equation and the more accurate equation (1.3-4) is negligible in most cases of practical interest.

1.4 OPTICAL SYSTEMS OF CYLINDRICAL SYMMETRY

The general solution (1.3-24) of the wave equation was constructed for the case in which ε and μ are constant in space. We shall always assume that μ is constant. However, when ε is a function of the space coordinates, the superposition of plane waves (1.3-24) is no longer a solution of the wave equation. It is still possible to construct general solutions from simpler solutions. This approach is known as the method of normal modes. Each plane wave in the last section can be regarded as a mode of the structure. The term "mode" is not always easy to define. A suitable definition may be to regard a mode as an eigensolution of Maxwell's equations belonging to a particular eigenvalue and satisfying all the boundary conditions of the problem. The plane waves used in the last section satisfy all of these requirements. Writing

$$\mathbf{E} = A\mathbf{e} \exp[i(\omega t - \mathbf{k} \cdot \mathbf{r})] \qquad (1.4\text{-}1)$$

$$\mathbf{H} = B\mathbf{h} \exp[i(\omega t - \mathbf{k} \cdot \mathbf{r})] \qquad (1.4\text{-}2)$$

with complex coefficients A and B and with $\mathbf{e}$ and $\mathbf{h}$ being unit vectors, we satisfy Maxwell's equations when the following set of equations (resulting from substitution of (1.4-1) and (1.4-2) in (1.2-1) and (1.2-2)) are satisfied

$$-i(\mathbf{k} \times \mathbf{h})B = i\omega\varepsilon A\mathbf{e} \qquad (1.4\text{-}3)$$

$$-i(\mathbf{k} \times \mathbf{e})A = -i\omega\mu B\mathbf{h} \qquad (1.4\text{-}4)$$

Using the unit vector $\mathbf{n}$ of (1.3-11), we find that (1.4-3) and (1.4-4) are satisfied when the following relations exist

$$\mathbf{n} \cdot \mathbf{e} = 0 \qquad (1.4\text{-}5)$$

$$\mathbf{h} = \mathbf{n} \times \mathbf{e} \qquad (1.4\text{-}6)$$

$$B = \sqrt{\frac{\varepsilon}{\mu}}\, A \qquad (1.4\text{-}7)$$

Equation (1.4-5) indicates that we are dealing with a transverse wave. Since there are no boundary conditions in this case, the plane waves (1.4-1) and (1.4-2) qualify as modes under our definition. Incidentally, each component of the vectors $\mathbf{E}$ and $\mathbf{H}$ satisfies the wave equation, and, by forming the superposition corresponding to (1.3-23) for each component, we obtain the most general solution of Maxwell's equations in a homogeneous medium. However, when ε is not constant in space, (1.4-1) and (1.4-2) are solutions neither of the wave equation nor of Maxwell's equations.

We discuss the solution of Maxwell's equations in an inhomogeneous medium for a special case of particular interest to us. It is often true that ε is independent of one space coordinate. Positioning the coordinate system suitably, we can assume that the dielectric constant does not depend on z.

We can then try to find mode solutions of the following form

$$\mathbf{E} = \mathbf{E}_0(x,y) \exp[i(\omega t - \beta z)] \qquad (1.4\text{-}8)$$

$$\mathbf{H} = \mathbf{H}_0(x,y) \exp[i(\omega t - \beta z)] \qquad (1.4\text{-}9)$$

Substitution of these equations into (1.2-1) and (1.2-2) leads, with the help of (1.2-3) and (1.2-4), to the following set of equations

$$\frac{\partial H_z}{\partial y} + i\beta H_y = i\omega\varepsilon E_x \qquad (1.4\text{-}10)$$

$$-i\beta H_x - \frac{\partial H_z}{\partial x} = i\omega\varepsilon E_y \qquad (1.4\text{-}11)$$

$$\frac{\partial H_y}{\partial x} - \frac{\partial H_x}{\partial y} = i\omega\varepsilon E_z \qquad (1.4\text{-}12)$$

$$\frac{\partial E_z}{\partial y} + i\beta E_y = -i\omega\mu H_x \qquad (1.4\text{-}13)$$

$$i\beta E_x + \frac{\partial E_z}{\partial x} = i\omega\mu H_y \qquad (1.4\text{-}14)$$

$$\frac{\partial E_y}{\partial x} - \frac{\partial E_x}{\partial y} = -i\omega\mu H_z \qquad (1.4\text{-}15)$$

With the help of (1.4-10), (1.4-11), (1.4-13), and (1.4-14), we can express the transverse field components in terms of E_z and H_z

$$E_x = -\frac{i}{\kappa^2}\left(\beta \frac{\partial E_z}{\partial x} + \omega\mu \frac{\partial H_z}{\partial y}\right) \qquad (1.4\text{-}16)$$

$$E_y = -\frac{i}{\kappa^2}\left(\beta\frac{\partial E_z}{\partial y} - \omega\mu\frac{\partial H_z}{\partial x}\right) \tag{1.4-17}$$

$$H_x = -\frac{i}{\kappa^2}\left(\beta\frac{\partial H_z}{\partial x} - \omega\varepsilon\frac{\partial E_z}{\partial y}\right) \tag{1.4-18}$$

$$H_y = -\frac{i}{\kappa^2}\left(\beta\frac{\partial H_z}{\partial y} + \omega\varepsilon\frac{\partial E_z}{\partial x}\right) \tag{1.4-19}$$

with

$$\kappa^2 = k^2 - \beta^2 \tag{1.4-20}$$

and

$$k^2 = \omega^2\varepsilon\mu \tag{1.4-21}$$

Similar equations in cylindrical polar coordinates can be found in section 8.2. Equations (1.4-16) through (1.4-19) are exact. Replacing H_x and H_y in (1.4-12) by (1.4-18) and (1.4-19) leads to

$$\frac{\partial^2 E_z}{\partial x^2} + \frac{\partial^2 E_z}{\partial y^2} + \kappa^2 E_z = 0 \tag{1.4-22}$$

Equation (1.4-15) similarly yields

$$\frac{\partial^2 H_z}{\partial x^2} + \frac{\partial^2 H_z}{\partial y^2} + \kappa^2 H_z = 0 \tag{1.4-23}$$

Equations (1.4-22) and (1.4-23) are exact only if ε is constant. It is remarkable that each of them contains either only E_z or H_z. The longitudinal **E** and **H** components are uncoupled and can be chosen arbitrarily as long as they satisfy the equations (1.4-22) and (1.4-23). In most general problems, coupling of the two longitudinal field components is required by the boundary conditions. If the boundary conditions do not achieve coupling of these components, it is possible to obtain mode solutions with either $E_z = 0$ or $H_z = 0$. These modes are called, respectively, transverse electric or TE modes and transverse magnetic or TM modes.

In the general case that ε depends on the x and y coordinate, (1.4-22) and (1.4-23) are no longer exact. However, they are still good approximations if the relative variation of ε is much less than unity over the region of one wavelength. For optical fields with their extremely short wavelengths, these approximations are usually very good and simplify the calculations considerably.

The propagation constant β has so far remained undetermined. It is the eigenvalue of the eigenvalue problem discussed earlier. Its value or rather its possible values are determined by the boundary conditions of the problem. General solutions of Maxwell's equations are obtained by superpositions of the set of modes.

1.5 BOUNDARY CONDITIONS

As mentioned repeatedly, Maxwell's equations do not determine the electromagnetic field completely. Out of the infinite possibilities of solutions of Maxwell's equations, we must select those which satisfy also the boundary conditions of the particular problem at hand.

In problems of inhomogeneous structures with no discontinuous regions of the dielectric constant, the only boundary condition is usually the requirement that the field vanish at infinity and be finite everywhere in space. We shall encounter problems of this type. The requirement of vanishing values of the field variables at infinity leads to guided modes. Guided modes have the property that the field is confined by the optical structure and does not lose power by radiation.

The most common type of boundary condition occurs when there are discontinuities in the dielectric constant. Such discontinuities exist, for example, if a glass lens is brought into the path of a light beam. We are interested only in boundary conditions for time-varying fields (which are different from those of static fields). The desired boundary conditions are obtained from Maxwell's equations in integral form. These equations can be derived by integrating (1.2-1) and (1.2-2) over an arbitrary area. Application of Stokes integral theorem leads to the following set of integral relations

$$\oint_S \mathbf{H} \cdot d\mathbf{S} = \int_A \frac{\partial \mathbf{D}}{\partial t} \cdot \mathbf{n} dA \qquad (1.5\text{-}1)$$

$$\oint_S \mathbf{E} \cdot d\mathbf{S} = -\int_A \frac{\partial \mathbf{B}}{\partial t} \cdot \mathbf{n} dA \qquad (1.5\text{-}2)$$

The unit vector $\mathbf{n}$ is directed perpendicular to the surface element dA, while $d\mathbf{S}$ points in a direction tangential to the curve S, which encloses the surface A.

Since the shape and position of the closed curve S are arbitrary, we are allowed to position it as shown in Figure 1.5.1. The closed integration path

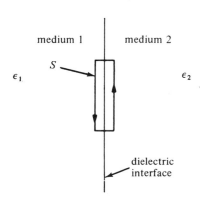

Figure 1.5.1 Illustration of the integration path for the derivation of boundary conditions. S indicates the closed curve, ε_1 and ε_2 are the dielectric permittivities of the two media.

along S is placed infinitesimally close to the boundary between two dielectric media with different values of ε. One part of the curve is located in medium 1, and the other in medium 2. The sections of the path running parallel to the boundary are considered to be much shorter than the wavelength of the radiation but still very long compared to the path sections that cross the boundary to close the integration path. In fact we shrink these latter path sections to zero length. The area enclosed by the curve thus becomes vanishingly small, and the right-hand sides of (1.5-1) and (1.5-2) can be set equal to zero. The values of $\mathbf{E}$ and $\mathbf{H}$ on the short but finite path section running entirely in one or the other medium can be taken as constant, so that we are allowed to write

$$\{(H_t)_1 - (H_t)_2\}\Delta S = 0$$

$$\{(E_t)_1 - (E_t)_2\}\Delta S = 0$$

The subscript t indicates that the components tangential to the boundary appear in these equations. The length of the infinitesimal integration path section in either medium is indicated by ΔS. The desired boundary conditions for the tangential field components follow immediately

$$(H_t)_1 = (H_t)_2 \qquad\qquad (1.5\text{-}3)$$

$$(E_t)_1 = (E_t)_2 \qquad\qquad (1.5\text{-}4)$$

The physical meaning of (1.5-3) and (1.5-4) is that the tangential components of the $\mathbf{E}$ field as well as of the $\mathbf{H}$ field must be continuous at the boundary. These boundary conditions, together with those at infinity discussed earlier, are used to select those solutions of Maxwell's equations which fit the particular physical problem under consideration. There are boundary conditions for the normal components of the vectors $\mathbf{D}$ and $\mathbf{B}$ which follow from the divergence relations (1.2-5) and (1.2-6). They state that the components of these vectors directed normal to the interface of the two media must also be continuous. However, the divergence relations can be obtained from Maxwell's equations by taking the divergence of (1.2-1) and (1.2-2). Since the divergence of the curl of a vector vanishes, we obtain immediately that the time derivative of the divergence of $\mathbf{D}$ and $\mathbf{B}$ must be zero. However, the time derivatives of these time-varying quantities can vanish only if the quantities themselves vanish. In the dynamic (time-varying) case, the divergence relations are a consequence of Maxwell's equations and are not independent conditions to be imposed separately. The boundary conditions for the normal components of $\mathbf{D}$ and $\mathbf{B}$ that follow from the divergence relations are likewise not independent but are already contained in Maxwell's equations. The boundary conditions for the normal components of $\mathbf{D}$ and $\mathbf{B}$ are therefore automatically satisfied if the relations (1.5-3) and (1.5-4) hold. Only for static magnetic and electric fields are the boundary conditions for the normal components independent of those for the tangential components.

Equations (1.5-3) and (1.5-4) are the only boundary conditions at discontinuities in the dielectric media which we need to consider.

There is another boundary condition which is often used in the idealized case of a perfect metallic conductor. This boundary condition requires that the electric field component tangential to the surface of the perfect conductor must vanish. In this case, no additional boundary condition for the **H** component is required. In optics, the notion of a perfect conductor is not very useful, since it cannot be realized to any satisfactory degree of approximation by existing metals. A metal must be treated as a dielectric material with complex values of its dielectric constant. In this case, the boundary conditions stated above are applicable and describe the situation completely.

1.6 REFLECTION AND REFRACTION AT A DIELECTRIC INTERFACE

We are now ready to solve one of the simplest but most important problems of wave propagation. Let us assume that a plane wave with propagation vector $\mathbf{k}_1$ is incident from a homogeneous medium with dielectric constant $\varepsilon_1/\varepsilon_0$ on the plane interface between this medium and another one with dielectric constant $\varepsilon_2/\varepsilon_0$. It is our purpose to study the phenomena of reflection and refraction of the wave as it crosses the interface.

The geometry of the problem is sketched in Figure 1.6.1. The coordinate system is located so that $k_y = 0$. We shall consider two different cases. Let

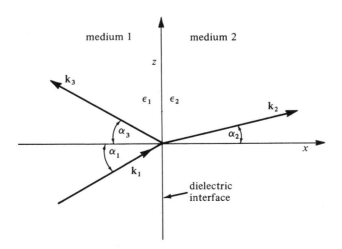

Figure 1.6.1   Reflection and refraction of a plane wave at the interface between two dielectric media. The propagation vectors of the incident, transmitted and reflected waves are indicated by $\mathbf{k}_1$, $\mathbf{k}_2$, and $\mathbf{k}_3$.

us first assume that the electric field is polarized in y direction, which means that the electric field vector is directed parallel to the plane of the interface. Later we shall assume that the E vector lies in the x-z plane. In either case, we will have to consider an incident and a reflected wave in medium 1 and a transmitted wave in medium 2. The equations for a plane wave traveling in the arbitrary direction $\mathbf{k}$ were already given in Section 1.4. All that remains to be done is to superimpose two plane waves in medium 1, allow for a plane wave in medium 2, and adjust the directions of the propagation vectors and the amplitudes of the waves so as to satisfy the boundary conditions (1.5-3) and (1.5-4).

In the case to be considered first, there is only an E_y component of the electric field, so that we have in medium 1

$$E_y = Ae^{-i\mathbf{k}_1 \cdot \mathbf{r}} + Be^{-i\mathbf{k}_3 \cdot \mathbf{r}} \tag{1.6-1}$$

and in medium 2

$$E_y = Ce^{-i\mathbf{k}_2 \cdot \mathbf{r}} \tag{1.6-2}$$

The factor exp $(i\omega t)$ is omitted for simplicity. The polarization of the $\mathbf{H}$ field follows from (1.4-6), so that we obtain in medium 1 (remember, only the y component of $\mathbf{e}$ exists)

$$H_x = -An_{1z}\sqrt{\frac{\varepsilon_1}{\mu}}e^{-i\mathbf{k}_1 \cdot \mathbf{r}} - Bn_{3z}\sqrt{\frac{\varepsilon_1}{\mu}}e^{-i\mathbf{k}_3 \cdot \mathbf{r}} \tag{1.6-3}$$

$$H_z = An_{1x}\sqrt{\frac{\varepsilon_1}{\mu}}e^{-i\mathbf{k}_1 \cdot \mathbf{r}} + Bn_{3x}\sqrt{\frac{\varepsilon_1}{\mu}}e^{-i\mathbf{k}_3 \cdot \mathbf{r}} \tag{1.6-4}$$

and in medium 2

$$H_x = -Cn_{2z}\sqrt{\frac{\varepsilon_2}{\mu}}e^{-i\mathbf{k}_2 \cdot \mathbf{r}} \tag{1.6-5}$$

$$H_z = Cn_{2x}\sqrt{\frac{\varepsilon_2}{\mu}}e^{-i\mathbf{k}_2 \cdot \mathbf{r}} \tag{1.6-6}$$

The relation between the vectors $\mathbf{n}$ and $\mathbf{k}$ is given by (1.3-11). The tangential field components E_y and H_z must now be made to satisfy the boundary conditions (1.5-3) and (1.5-4). We obtain the following equations at $x = 0$

$$Ae^{-ik_{1z}z} + Be^{-ik_{3z}z} = Ce^{-ik_{2z}z} \tag{1.6-7}$$

$$An_{1x}\sqrt{\varepsilon_1}e^{-ik_{1z}z} + Bn_{3x}\sqrt{\varepsilon_1}e^{-ik_{3z}z} = Cn_{2x}\sqrt{\varepsilon_2}e^{-ik_{2z}z} \tag{1.6-8}$$

These equations must hold for all values of z; in other words, the z-dependent term must cancel from the equation. This is possible only if

$$k_{1z} = k_{2z} = k_{3z} \tag{1.6-9}$$

The k_x components and consequently the component n_x of the vector **n** follow from (1.3-13) and (1.3-14)

$$k_{1x} = \sqrt{\omega^2 \varepsilon_1 \mu - k_{1z}} = n_{1x}k \qquad (1.6\text{-}10)$$

$$k_{2x} = \sqrt{\omega^2 \varepsilon_2 \mu - k_{2z}} = n_{2x}k \qquad (1.6\text{-}11)$$

$$k_{3x} = -k_{1x} = n_{3x}k \qquad (1.6\text{-}12)$$

The component k_{1z} can be chosen arbitrarily and determines the direction of the incident plane wave. The negative sign of (1.6-12) does not follow from (1.3-13) but is required to avoid a contradiction between (1.6-7) and (1.6-8). If $k_{3x} = \omega\sqrt{\varepsilon_1 \mu}\, n_{3x}$ had the same sign as k_{1x}, the left-hand sides of (1.6-7) and (1.6-8) would be proportional and a contradiction would result between their right-hand sides. Equations (1.6-7) and (1.6-8) can now be used to determine the amplitudes B and C relative to the arbitrary amplitude A of the incident plane wave. We obtain

$$-B + C = A$$

$$B + \frac{k_{2x}}{k_{1x}} C = A$$

This equation system has the solution

$$C = \frac{2A}{1 + \dfrac{k_{2x}}{k_{1x}}} \qquad (1.6\text{-}13)$$

$$B = \frac{\left(1 - \dfrac{k_{2x}}{k_{1x}}\right) A}{1 + \dfrac{k_{2x}}{k_{1x}}} \qquad (1.6\text{-}14)$$

with

$$\frac{k_{2x}}{k_{1x}} = \frac{n_{2x}}{n_{1x}} \sqrt{\frac{\varepsilon_2}{\varepsilon_1}} = \frac{\sqrt{\omega^2 \varepsilon_2 \mu - k_{1z}^2}}{\sqrt{\omega^2 \varepsilon_1 \mu - k_{1z}^2}} \qquad (1.6\text{-}15)$$

Equations (1.6-10) through (1.6-15) solve the problem completely. It remains to examine this solution and to determine the physical properties of the reflected and transmitted waves.

Since the magnitudes of the vectors $\mathbf{k}_1$ and $\mathbf{k}_3$ as well as their k_z components are identical, it is apparent from (1.6-12) and Figure 1.6.1 that

$$\alpha_1 = \alpha_3 \qquad (1.6\text{-}16)$$

The incident and reflected waves make equal angles with the normal to the reflecting surface. This is the law of reflection. Denoting the magnitude of the vectors $\mathbf{k}_1$ and $\mathbf{k}_2$ by k_1 and k_2, we obtain from (1.3-13) the following relation

$$\frac{k_2}{k_1} = \sqrt{\frac{\varepsilon_2}{\varepsilon_1}} \qquad (1.6\text{-}17)$$

The sine of the angle α_i is, according to Figure 1.6.1

$$\sin \alpha_i = \frac{k_{iz}}{k_i} \qquad i = 1, 2, \text{ or } 3 \qquad (1.6\text{-}18)$$

Using (1.6-9), (1.6-17), and (1.6-18), we can immediately write

$$\frac{\sin \alpha_1}{\sin \alpha_2} = \frac{n_2}{n_1} \qquad (1.6\text{-}19)$$

The term on the right-hand side is the ratio of the indices of refraction of the two media. The index of refraction is by definition

$$n_i = \sqrt{\frac{\varepsilon_i}{\varepsilon_0}} \qquad (1.6\text{-}20)$$

The very important Equation (1.6-19) is known as Snell's law. This simple law is all one needs to trace the trajectory of light rays through a sequence of homogeneous media. It is therefore possible to analyze the image-forming properties of lens systems by ray tracing with the help of Snell's law. The law of refraction of light rays indicates that a ray is broken toward the direction normal to the interface when it enters from one dielectric medium into another with higher dielectric constant

Figure 1.6.1 was drawn under the assumption that medium 2 has a higher index of refraction than medium 1. However, let us assume for the moment that $\varepsilon_1 > \varepsilon_2$ and therefore $n_1 > n_2$. We obtain in this case from (1.6-19)

$$\sin \alpha_2 = \frac{n_1}{n_2} \sin \alpha_1 > \sin \alpha_1 \qquad (1.6\text{-}21)$$

As the angle α_1 increases, $\sin \alpha_2$ approaches unity before α_1 reaches $90°$. In that limit the transmitted plane wave enters medium 2 parallel to the boundary between the two media. An infinitesimal increase of α_1 causes $\sin \alpha_2 > 1$, which cannot be satisfied for any real value of α_2. The physical interpretation of this phenomenon is that the transmitted wave no longer crosses the interface, so that only a reflected wave remains. This effect is called total internal reflection. However, our analysis does not break down in this limit, and we can study the behavior of the plane waves in more detail. Equations (1.6-9) and (1.6-18) tell us that

$$k_{2z} = k_{1z} > k_2 = \omega \sqrt{\varepsilon_2 \mu} \qquad (1.6\text{-}22)$$

so that, according to (1.6-11), k_{2x} assumes imaginary values. The x dependence of the wave in medium 2 is now given by*

$$e^{-|k_{2x}|x}$$

indicating that we have no longer a plane traveling wave in medium 2 but an evanescent wave whose field amplitude decays exponentially with increasing values of x.

Total internal reflection provides an example for the occurrence of evanescent waves in free space, which we discussed in section 1.3. Total internal reflection happens whenever the angle of the incident wave, approaching an interface between two dielectric media from the denser medium (the one with the higher index of refraction), exceeds a certain critical value given by the equation

$$\sin \alpha_{1c} = \frac{n_2}{n_1} \tag{1.6-23}$$

Equation (1.6-23) determines the critical angle for total internal reflection. The reason that we encounter an exponentially decaying wave in medium 2 if α_1 exceeds the critical angle is that the field in that medium is excited with a wavelength which is too short for waves propagating in the medium with the lower index of refraction (see the discussion in section 1.3).

The field amplitudes of the transmitted and reflected waves are given by (1.6-13) and (1.6-14). Equation (1.6-15) shows that the ration k_{2x}/k_{1x} is always positive.† As a consequence, C/A is always positive regardless of whether the waves enter from a dense medium into one with a low index of refraction or vice versa. The sign of B, the amplitude of the reflected wave, depends on the nature of the two media at the interface. When the wave is incident from medium 1 onto a medium with a higher index of refraction, (1.6-15) indicates that k_{2x}/k_{1x} is larger than unity, and consequently B is negative. Reflection from a denser medium changes the sign of the electric field vector. In the opposite case, if the light wave is reflected from a medium with a lower index of refraction compared to the medium from which it entered, B is positive, indicating that the **E** vector does not change sign on reflection.

It is interesting to study the amount of power which is reflected and transmitted at the interface. Since we are using a complex notation, we use (1.2-12).

$$\mathbf{S} = \frac{1}{2} Re\{\mathbf{E} \times \mathbf{H}^*\} \tag{1.6-24}$$

---

* The square root of (1.6-11) can be given a plus or a minus sign. We choose $k_{2x} = -i|k_{2x}|$ since the opposite choice would lead to a growing wave violating the boundary condition of finite (or vanishing) field intensity at infinity.
† We exclude total internal reflection for the moment.

In our special case, with E_y being the only nonvanishing component of the **E** vector, we obtain for the x component of **S**

$$P = |S_x| = \frac{1}{2}|Re(E_y H_z^*)| \tag{1.6-25}$$

The x component of the Poynting vector represents the power which flows perpendicular to the interface. The power S_z flowing parallel to the interface is not incident on the interface. We consider the power carried by the incident, the reflected, and the transmitted waves. The power of the incident wave is obtained by substituting the part containing the factor A of (1.6-1) and (1.6-4) into (1.6-25), resulting in

$$P_i = \frac{1}{2} n_{1x} \sqrt{\frac{\varepsilon_1}{\mu}} |A|^2 \tag{1.6-26}$$

The power of the reflected wave follows similarly from the parts of the field components containing the B factor

$$P_r = \frac{1}{2} n_{1x} \sqrt{\frac{\varepsilon_1}{\mu}} |B|^2 \tag{1.6-27}$$

The power carried by the transmitted wave is finally given by

$$P_t = \frac{1}{2} n_{2x} \sqrt{\frac{\varepsilon_2}{\mu}} |C|^2 \tag{1.6-28}$$

We can define reflection and transmission coefficients by the following relations

$$R = \frac{P_r}{P_i} = \frac{|B|^2}{|A|^2} \tag{1.6-29}$$

$$T = \frac{P_t}{P_i} = \frac{k_{2x}}{k_{1x}} \frac{|C|^2}{|A|^2} \tag{1.6-30}$$

In order to express the reflection and transmission coefficients in terms of the angle of the incident wave, we write (1.6-15) with the help of (1.6-18) in the following form

$$\frac{k_{2x}}{k_{1x}} = \frac{n_2 \cos \alpha_2}{n_1 \cos \alpha_1}$$

Replacing α_2 with the help of Snell's law (1.6-19) allows us to write

$$\frac{k_{2x}}{k_{1x}} = \frac{\sqrt{n_2^2 - n_1^2 \sin^2 \alpha_1}}{n_1 \cos \alpha_1} \tag{1.6-31}$$

The ratio of transmitted to incident power can now be obtained from (1.6-13), (1.6-30), and (1.6-31)

$$T_E = \frac{4n_1\cos\alpha_1 \sqrt{n_2^2 - n_1^2\sin^2 \alpha_1}}{(n_1 \cos \alpha_1 + \sqrt{n_2^2 - n_1^2 \sin^2 \alpha_1})^2} \tag{1.6-32}$$

The ratio of reflected to incident power is obtained similarly

$$R_E = \frac{(n_1\cos \alpha_1 - \sqrt{n_2^2 - n_1^2 \sin^2\alpha_1})^2}{(n_1\cos \alpha_1 + \sqrt{n_2^2 - n_1^2 \sin^2\alpha_1})^2} \tag{1.6-33}$$

These relations for T and R hold only as long as the square root appearing in (1.6-32) and (1.6-33) has a real value. For imaginary values of the square root, we have to examine our derivation more carefully. The square root becomes imaginary when the angle α_i exceeds its critical value as given by (1.6-23). The result is total internal reflection; no propagating wave exists in medium 2. The component k_{2x} of the propagation constant of the wave in medium 2 becomes imaginary, so that the component n_{2x} of the unit vector entering (1.6-6) is also imaginary, causing the expression (1.6-25) to vanish if applied to the transmitted wave. The transmission coefficient is

$$T = 0$$

in case of total internal reflection. The reflection coefficient assumes the value

$$R = 1$$

in case of total internal reflection. This becomes apparent when we consider that, in that case, k_{2x}/k_{1x} is a purely imaginary quantity, so that the absolute square value of B as given by (1.6-14) becomes $|A|^2$. In the general case, where a reflected as well as transmitted wave exists, it follows from (1.6-32) and (1.6-33) that

$$T + R = 1 \tag{1.6-34}$$

indicating that the total energy is conserved by our process.

The transmission and reflection coefficients assume particularly simple forms if the incident wave approaches the interface between the two dielectric media at right angles. Taking $\alpha_1 = 0$, we obtain from (1.6-32)

$$T_E = \frac{4n_1n_2}{(n_1 + n_2)^2} \tag{1.6-35}$$

and from (1.6-33) it follows that

$$R_E = \frac{(n_1 - n_2)^2}{(n_1 + n_2)^2} \tag{1.6-36}$$

The other extreme of grazing incidence must be discussed separately for the two possible cases. If $n_1 > n_2$, we have total internal reflection, and all the power is reflected for angles of incidence larger than the critical angle. In the other case $n_1 < n_2$, there is no total internal reflection, but in the limit of a wave that is incident parallel to the wall, $\alpha_1 = 90°$, we obtain from (1.6-32) and (1.6-33)

$$T_E = 0$$

and

$$R_E = 1$$

At grazing incidence, all the power is reflected from a dielectric interface regardless of whether the wave is incident on an optically denser or less dense medium.

Our discussion so far has concerned itself only with the special case that the E vector of the incident as well as the reflected and transmitted wave is directed parallel to the dielectric interface. The fact that the reflected as well as the transmitted waves also have their E vectors directed parallel to the E vector of the incident wave is not a special case but is a consequence of Maxwell's equations and the boundary conditions. Solutions of boundary value problems are unique. The fact that we obtained a solution of this particular kind means that no other solution can exist.

Next, we concern ourselves with a plane wave being incident on the plane interface between two dielectric media, having its E vector directed in the plane defined by the propagation vector and the direction normal to the surface. The transmission and reflection properties of plane waves crossing the interface between two dielectric media are polarization-dependent, so that it is necessary to study a wave polarized at right angles to the polarization studied above. We assume in medium 1 a superposition of incident and reflected waves of the form

$$E_x = A e_{1x} e^{-i\mathbf{k}_1 \cdot \mathbf{r}} + B e_{3x} e^{-i\mathbf{k}_3 \cdot \mathbf{r}} \tag{1.6-37}$$

$$E_z = A e_{1z} e^{-i\mathbf{k}_1 \cdot \mathbf{r}} + B e_{3z} e^{-i\mathbf{k}_3 \cdot \mathbf{r}} \tag{1.6-38}$$

$$H_y = A \sqrt{\frac{\varepsilon_1}{\mu}} e^{-i\mathbf{k}_1 \cdot \mathbf{r}} + B \sqrt{\frac{\varepsilon_1}{\mu}} e^{-i\mathbf{k}_3 \cdot \mathbf{r}} \tag{1.6-39}$$

In medium 2, the transmitted wave is of the general form

$$E_x = C e_{2x} e^{-i\mathbf{k}_2 \cdot \mathbf{r}} \tag{1.6-40}$$

$$E_z = C e_{2z} e^{-i\mathbf{k}_2 \cdot \mathbf{r}} \tag{1.6-41}$$

$$H_y = C \sqrt{\frac{\varepsilon_2}{\mu}} e^{-i\mathbf{k}_2 \cdot \mathbf{r}} \tag{1.6-42}$$

Equations (1.6-37) through (1.6-42) are solutions of Maxwell's equations. The constants A, B, and C as well as the directions of the propagation vectors must be determined from the boundary conditions (1.5-3) and (1.5-4). The boundary conditions at $x = 0$ lead to the equations (remember, $k_y = 0$)

$$Ae_{1z}e^{-ik_{1z}z} + Be_{3z}e^{-ik_{3z}z} = Ce_{2z}e^{-ik_{2z}z} \tag{1.6-43}$$

$$Ae^{-ik_{1z}z} + Be^{-ik_{3z}z} = \sqrt{\frac{\varepsilon_2}{\varepsilon_1}}\,Ce^{-ik_{2z}z} \tag{1.6-44}$$

The requirement that the z dependence must vanish from these equations leads again to the condition (1.6-9). Since the relations between the components of the propagation vectors are again exactly the same as in the case studied above, we find that Snell's law (1.6-19) holds also for this polarization. The directions of the reflected and transmitted waves are in this case exactly as they were in the case studied previously and are thus found to be independent of polarization. This holds also for the condition for total internal reflection in passing from a denser to a less dense medium.

It remains to determine the ratios C/A and B/A, which are obtained from the equations

$$Ae_{1z} + Be_{3z} = Ce_{2z} \tag{1.6-45}$$

$$A + B = \frac{n_2}{n_1}\,C \tag{1.6-46}$$

The solutions of these equations are

$$C = \frac{e_{1z} - e_{3z}}{e_{2z} - \dfrac{n_2}{n_1}\,e_{3z}}\,A \tag{1.6-47}$$

$$B = \frac{\dfrac{n_2}{n_1}\,e_{1z} - e_{2z}}{e_{2z} - \dfrac{n_2}{n_1}\,e_{3z}}\,A \tag{1.6-48}$$

The components of the $\mathbf{e}$ vectors can be expressed in terms of the components of the $\mathbf{k}$ vectors by the use of (1.4-5), which can also be written

$$\mathbf{k}_i \cdot \mathbf{e}_i = 0 \qquad i = 1, 2, 3 \tag{1.6-49}$$

Since our choice of polarization makes $e_y = 0$, we obtain

$$\frac{e_{iz}}{e_{ix}} = -\frac{k_{ix}}{k_{iz}} \tag{1.6-50}$$

This equation in conjunction with the normalization condition

$$e_{ix}^2 + e_{iz}^2 = 1 \tag{1.6-51}$$

enables us to determine the e_z components

$$e_{1z} = \frac{k_{1x}}{k_1} \tag{1.6-52}$$

$$e_{2z} = \frac{k_{2x}}{k_2} \tag{1.6-53}$$

$$e_{3z} = -e_{1z} \tag{1.6-54}$$

The negative sign in this last equation is required to avoid a contradiction between (1.6-45) and (1.6-46). Finally, using (1.6-17) and (1.6-20), we obtain the solutions

$$C = \frac{2A}{\dfrac{n_2}{n_1} + \dfrac{n_1}{n_2}\dfrac{k_{2x}}{k_{1x}}} \tag{1.6-55}$$

$$B = \frac{\dfrac{n_2}{n_1} - \dfrac{n_1}{n_2}\dfrac{k_{2x}}{k_{1x}}}{\dfrac{n_2}{n_1} + \dfrac{n_1}{n_2}\dfrac{k_{2x}}{k_{1x}}} A \tag{1.6-56}$$

The power transmission and reflection coefficients are obtained from (1.6-29) and (1.6-30) with the help of (1.6-31)*

$$T_H = \frac{4n_1 \cos\alpha_1 \sqrt{n_2^2 - n_1^2 \sin^2\alpha_1}}{\left(n_2 \cos\alpha_1 + \dfrac{n_1}{n_2}\sqrt{n_2^2 - n_1^2 \sin^2\alpha_1}\right)^2} \tag{1.6-57}$$

$$R_H = \frac{\left(n_2 \cos\alpha_1 - \dfrac{n_1}{n_2}\sqrt{n_2^2 - n_1^2 \sin^2\alpha_1}\right)^2}{\left(n_2 \cos\alpha_1 + \dfrac{n_1}{n_2}\sqrt{n_2^2 - n_1^2 \sin^2\alpha_1}\right)^2} \tag{1.6-58}$$

These expressions are valid only for real values of the square root. For imaginary values, we obtain again the relations $T = 0$ and $R = 1$. It can easily be seen that the conservation of power relation (1.6-34) holds also for R_H and T_H given by (1.6-57) and (1.6-58).

The reflection and transmission coefficients look rather similar in the two cases we have studied. However, there is one important difference. It is easy to show that R_E of (1.6-33) cannot vanish if $n_1 \neq n_2$. The reflection

---

* It is easy to show that (1.6-29) and (1.6-30) hold also in this case.

coefficient R_H for the wave whose electric vector is directed in the plane of the **k** vector and the normal to the surface vanishes for

$$\sin \alpha_1 = \frac{n_2}{\sqrt{n_1^2 + n_2^2}} \qquad (1.6\text{-}59)$$

as can easily be seen from (1.6-58). The angle defined by (1.6-59) is known as Brewster's angle. A wave polarized in the plane of incidence (the plane of incidence is defined by the **k** vector and the normal to the reflecting surface) traverses the interface without suffering reflection provided it is incident at the Brewster angle. The Brewster angle formula can be written more elegantly

$$\tan \alpha_{1b} = \frac{n_2}{n_1} \qquad (1.6\text{-}60)$$

With the help of Snell's law (1.6-19), we can express the Brewster angle of the refracted wave by the equation

$$\tan \alpha_{2b} = \frac{n_1}{n_2} \qquad (1.6\text{-}61)$$

A comparison of (1.6-60) and (1.6-61) shows that if the incident wave assumes the angle α_2 and if we change the index of refraction in medium 1 to n_2 and that in medium 2 to n_1, the wave is again incident at the Brewster angle. In other words, if the direction of a wave which is incident at the Brewster angle is reversed, it retraces its path and leaves again without reflection. A wave entering a piece of glass at the Brewster angle leaves the glass without reflection if its two faces are parallel.

The directions of the reflected wave (if it were to exist) and the transmitted wave stand in a remarkable relationship to each other if the wave is incident at the Brewster angle. To see this relationship, we multiply the left-hand side of (1.6-60) with the left-hand side of (1.6-61) and also multiply their right-hand sides, obtaining

$$\tan \alpha_{1b} \tan \alpha_{2b} = 1$$

This relationship can be expressed as

$$\cos \alpha_{1b} \cos \alpha_{2b} - \sin \alpha_{1b} \sin \alpha_{2b} = \cos (\alpha_{1b} + \alpha_{2b}) = 0 \qquad (1.6\text{-}62)$$

The sum of the angles of the incident and transmitted wave is $90°$

$$\alpha_{1b} + \alpha_{2b} = 90° \qquad (1.6\text{-}63)$$

provided that the wave is incident at the Brewster angle. It is true that the characteristic of the Brewster angle is the lack of a reflected wave. However, the law of reflection formally holds, so that the reflected wave, if it existed, would leave under an angle $\alpha_{3b} = \alpha_{1b}$, so that we also have

$$\alpha_{3b} + \alpha_{2b} = 90° \qquad (1.6\text{-}64)$$

It can be seen from Figure 1.6.1 that if this relationship holds, the direction of the transmitted wave and that of the reflected wave would be at right angles.

The relationship between the directions of the incident, transmitted, and (if there were one) reflected wave is shown in Figure 1.6.2. There is a plausible explanation for the physics of the Brewster angle, which can be understood with the help of Figure 1.6.2. To understand this explanation, it is important to realize that the change in phase velocity, which causes the phenomena of reflection and refraction, occurs because the field interacts with the bound electrons in material media. The electrons are forced to oscillate at the frequency of the light wave. The oscillating electrons, in turn, radiate new light waves. The superposition of the incident light wave and the wave created by the oscillating electrons causes the reflected and refracted waves to appear. Inside the material there is only the refracted wave, which itself stems from the combined action of the incident wave and the forced oscillation and consequent radiation of the electrons of the material. Since the electromagnetic radiation consists of transverse waves (see [1.4-5]), the forced motion of the electrons under the influence of the refracted wave is perpendicular to the direction of propagation of this wave. The wave that is incident with its electric vector polarized in the plane of the **k** vector and the normal direction to the dielectric interface causes the electrons inside

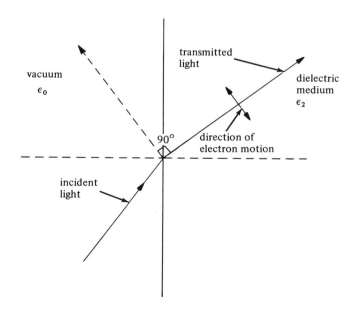

Figure 1.6.2 Illustration of the mechanism that explains the Brewster angle.

the material to oscillate in the same plane. Figure 1.6.2 shows a little double arrow, which is intended to indicate the direction of the back and forth movement of the forced electron motion. It is apparent from the discussion following (1.6-64) that the electrons move parallel to the direction of the would-be reflected wave. The theory of electric dipole radiation[2] teaches that electrons do not radiate in the direction of their motion. The reflected wave, which would be caused by the reradiation of the oscillating electrons, can therefore not materialize.

The phenomena of ordinary reflection and refraction, the Brewster angle, and total internal reflection are summarized in Figures 1.6.3 and 1.6.4. These figures are graphic representations of the transmission and reflection coefficients of (1.6-32), (1.6-33), (1.6-57), and (1.6-58). Figure 1.6.3 shows the reflection and transmission coefficients for both polarizations in the case that the wave enters the denser medium 2 from medium 1 with $n_1 < n_2$. In this case, there is no total internal reflection. However, the vanishing of the reflected wave at the Brewster angle is apparent for the curve of R_H. For the ratio $n_2/n_1 = 1.5$, for which Figure 1.6.3 is drawn, the Brewster angle is 56°. This figure shows also how the reflection increases as the wave

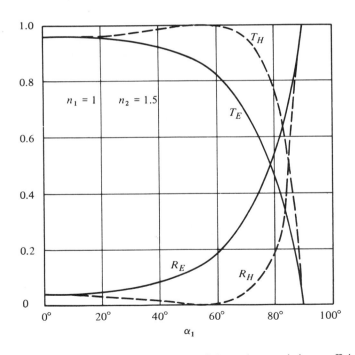

Figure 1.6.3 The reflection coefficients R_E and R_H and transmission coefficients T_E and T_H as functions of the angle of incidence α_1 in medium 1. The point $R_H = 0$ defines Brewster's angle. Medium 1 ($n_1 = 1$) is less dense than medium 2 ($n_2 = 1.5$).

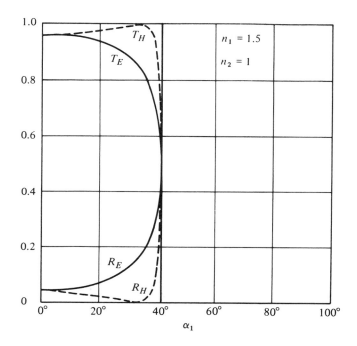

Figure 1.6.4 Reflection and transmission coefficients for light incidence from the denser medium 1 ($n_1 = 1.5$, $n_2 = 1$) as a function of the angle of incidence α_1. The Brewster angle ($R_H = 0$) and total internal reflection are clearly discernible.

is incident under increasingly grazing angles. The region where R is reasonably close to unity is extremely close to the 90° angle.

Figure 1.6.4 holds for the case that n_1 and n_2 have changed their places. The wave is now entering a medium with a lower index of refraction. In this case, there is total internal reflection. The total internal reflection angle for $n_1/n_2 = 1.5$ is equal to 41.8°.

The simple example of a plane wave crossing the interface between two dielectric media contains a great deal of important and useful physics. Further examples of wave propagation in dielectric media will be presented in other chapters.

CHAPTER

2

DIFFRACTION THEORY

2.1 INTRODUCTION

Diffraction is an important feature of wave propagation phenomena. If we consider light from the point of view of geometrical optics, we would expect sharp, well-defined shadows to appear behind the edge of an opaque obstacle. Wave optics, however, tells us that no sharp shadows can exist, since some of the light does reach into the geometrical shadow region. Another important application of diffraction theory is the study of the far field divergence angle of a collimated light beam. If a plane wave impinges on an aperture in an opaque screen, the light behind the screen begins to spread out and, far from the aperture, fills a region which is defined by the divergence angle of the spreading light beam. The divergence angle depends on the width of the aperture relative to the light wavelength.

A knowledge of diffraction phenomena is essential for the understanding of the lens waveguide to be discussed in Chapter 5. In the present chapter, we derive the necessary mathematical tools for the description of diffraction and apply them to a few representative examples. The most elementary examples are diffraction by a single slit and by a hole in an opaque screen. We have also included Bragg diffraction in this chapter. Often times this phenomenon is referred to as Bragg reflection. We choose to call it Bragg diffraction since its application is to three-dimensional diffraction gratings. Bragg diffraction can also be regarded as a special case of light scattering and, as such, is an important example of the general concept of scattered light. The treatment of Bragg diffraction by the method of coupled waves presented in the last section of this chapter is an example of coupled wave phenomena, which will be discussed in a different context in later chapters.

The subject of diffraction gratings and Bragg diffraction is of great importance for the understanding of holography. Even though a discussion of holography is outside the scope of this book, it may be useful for the reader to have this application in mind when reading this chapter.

2.2 THE KIRCHHOFF-HUYGENS DIFFRACTION INTEGRAL

The most striking features of wave propagation as opposed to particle motion are the phenomena of interference and diffraction. Diffraction is responsible for the occurrence of light in shadow regions which should not be reached if the propagation of light were strictly governed by geometrical optics. Whereas the polarization of the light wave played an important part in the reflection and refraction of light waves, its influence is not particularly important for the description of diffraction problems. This is particularly true if the wavelength of light is considerably shorter than any of the linear dimensions of the obstacles which the light wave encounters. For this reason it is sufficient to treat diffraction problems with the help of the scalar wave equation (1.3-6), ignoring the vector character of the light wave. Each component of the electromagnetic field is a solution of the wave equation, so that the scalar function ψ may represent any of the electric or magnetic components of the wave.

We limit our discussion to monocromatic radiation—that is, radiation at a certain frequency. More general forms of radiation can always be described by the superposition of all its sinusoidal components. Using the complex notation, we assume in this and the following sections that the time dependence of the wave is of the form

$$e^{i\omega t} \tag{2.2-1}$$

omitting the exponential factor (2.2-1) from all equations. A time dependence of the form (2.2-1) has the effect that differentiation with respect to time can be replaced by multiplication with the factor $i\omega$. The wave equation (1.3-6) assumes the form

$$\nabla^2\psi + k^2\psi = 0 \tag{2.2-2}$$

with k given by (1.3-13). Following Courant's example,[4] we call (2.2-2) the reduced wave equation. A wave equation must be a partial differential equation containing second derivatives with respect to time as well as with respect to spatial coordinates. Since the time derivative is missing from (2.2-2), the name reduced wave equation seems appropriate. Equation (2.2-2) is also known as Helmholtz equation

The solution of the reduced wave equation is substantially simplified by the use of the Dirac delta function.[5] The reader is assumed to be familiar with its properties. Mathematicians frown on the use of the delta function because of its highly singular behavior. However, it is a valuable tool in physics and engineering, and simplifies many calculations. Since it usually leads to the correct results, its use seems sufficiently justified.

Simultaneously with (2.2-2) we consider the equation.

$$\nabla^2 G + k^2 G = \delta(\mathbf{r} - \mathbf{r}')$$ (2.2-3)

The vectors in the argument of the delta function are a shorthand symbol for the following definition

$$\delta(\mathbf{r} - \mathbf{r}') = \delta(x - x')\delta(y - y')\delta(z - z')$$ (2.2-4)

The function G depends on two sets of variables

$$G = G(x, y, z, x', y', z')$$ (2.2-5)

G is known as Green's function of the problem.[6]

The solution of the reduced wave equation (2.2-2) can now be obtained as follows. We multiply (2.2-2) by G and (2.2-3) by ψ and subtract one equation from the other. After integration over a certain volume in space, we obtain

$$\psi(x', y', z') = \int_V (\psi \nabla^2 G - G \nabla^2 \psi) \, dV$$ (2.2-6)

The ψ function with the argument x', y', z', resulted from the integration of the product of the delta function and ψ. The integrand can be rewritten

$$\psi \nabla^2 G - G \nabla^2 \psi = \nabla \cdot (\psi \nabla G - G \nabla \psi)$$ (2.2-7)

Use of the divergence theorem[3] converts the volume integral to a surface integral so that we obtain as the solution of the reduced wave equation

$$\psi(x', y', z') = \int_S \left(\psi \frac{\partial G}{\partial n} - G \frac{\partial \psi}{\partial n} \right) dS$$ (2.2-8)

The integration extends over a closed surface S, and n is the direction of the outward normal of this surface, as shown in Figure 2.2.1. The point x', y', z', at which ψ is to be taken is located inside the closed surface S. If we knew the Green's function G, we could compute the values of ψ anywhere inside the closed surface from its values on S. It seems that not too much has been gained by the derivation of (2.2-8). However, in a homogeneous medium the Green's function is not hard to find, and knowledge of ψ on the surface S, or at least partial knowledge, exists in typical diffraction problems.

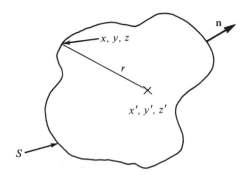

Figure 2.2.1 This figure illustrates the geometry of the integral representation (2.2-8). S is a surface enclosing the observation point. The unit vector normal to the surface is **n**.

The Green's function is given by

$$G = -\frac{1}{4\pi}\frac{e^{-ikr}}{r} \tag{2.2-9}$$

with

$$r = \{(x'-x)^2 + (y'-y)^2 + (z'-z)^2\}^{1/2} \tag{2.2-10}$$

It is easy to show that (2.2-9) is a solution of (2.2-3). We form the first partial derivative of G with respect to x

$$-4\pi\frac{\partial G}{\partial x} = -\left(ik + \frac{1}{r}\right)\frac{x-x'}{r}\frac{e^{-ikr}}{r} \tag{2.2-11}$$

Differentiating this expression a second time yields

$$-4\pi\frac{\partial^2 G}{\partial x^2} = \left\{\left(\frac{3ik}{r} + \frac{3}{r^2} - k^2\right)\frac{(x-x')^2}{r^3} - \frac{ik}{r^2} - \frac{1}{r^3}\right\}e^{-ikr} \tag{2.2-12}$$

The second derivatives with respect to y and z are of exactly the same form, with $x - x'$ replaced by $y - y'$ and $z - z'$, respectively. We therefore immediately obtain

$$\nabla^2 G + k^2 G = 0 \qquad \text{for } r \neq 0 \tag{2.2-13}$$

At $r = 0$, G has a singularity, so that the expressions (2.2-11) and (2.2-12) do not apply at that point. To study the nature of the singularity of the derivatives, we form the integral

$$\int_V (\nabla^2 G + k^2 G)dV = I \tag{2.2-14}$$

According to (2.2-13), the integrand vanishes everywhere with the exception of the point $r = 0$. We define the value of the integral over an infinitesimal volume v that includes the point $r = 0$ by the following relation

$$I = \lim_{v \to 0} \int_v (\nabla^2 G + k^2 G) dV = \lim_{\rho \to 0} \int_s \nabla G \cdot \mathbf{n} ds + \lim_{v \to 0} k^2 \int_v G \, dV \quad (2.2\text{-}15)$$

The surface s is an infinitesimal sphere around the point $r = 0$ with radius ρ and outward normal $\mathbf{n}$. The second term on the right-hand side can be evaluated with the help of (2.2-9)

$$\lim_{v \to 0} \int_v G \, dV = -\lim_{\rho \to 0} \int_0^\rho r \, dr = 0 \quad (2.2\text{-}16)$$

The first term on the right-hand side of (2.2-15) can be evaluated just as easily

$$\lim_{\rho \to 0} \int_s \nabla G \cdot \mathbf{n} ds = \lim_{\rho \to 0} \rho^2 \int_0^\pi d\theta \int_0^{2\pi} \left(\frac{\partial G}{\partial r} \right)_{r=\rho} \sin \theta \, d\varphi$$

$$= \lim_{\rho \to 0} \left\{ -\frac{\rho^2}{4\pi} \left[4\pi \left(-ik - \frac{1}{\rho} \right) \frac{e^{-ik\rho}}{\rho} \right] \right\} = 1 \quad (2.2\text{-}17)$$

Combining (2.2-16) and (2.2-17), we obtain

$$I = 1 \quad (2.2\text{-}18)$$

This is the result that we should obtain according to the defining equation for the Green's function. Our calculation shows that the Green's function (2.2-9) does indeed satisfy (2.2-3). The left-hand side of (2.2-3) vanishes everywhere (see [2.2-13]) except at $r = 0$. Its singularity at that point must be a delta function because the integral over an infinitesimally small volume including the singularity results in unity.

Substitution of (2.2-9) into (2.2-8) yields the desired solution of the wave equation

$$\psi(x', y', z',) = \frac{1}{4\pi} \int_s \left\{ \frac{\partial \psi}{\partial n} \frac{e^{-ikr}}{r} - \psi \frac{\partial}{\partial n} \left(\frac{e^{-ikr}}{r} \right) \right\} dS \quad (2.2\text{-}19)$$

In (2.2-19), n is the direction of the outward normal—that is, the direction pointing toward the outside of the volume that includes the point x', y', z'.

The Kirchhoff-Huygens diffraction integral (2.2-19) is very useful to solve a large number of diffraction problems. However, it should be borne in mind that there are alternate ways of solving diffraction problems. For example, there are other Green's functions that can be used in (2.2-8). It is possible to choose Green's functions that are solutions of (2.2-3) but satisfy certain boundary conditions on the surfaces S.[7] The Green's function can be chosen so that it vanishes on S. This special Green's function causes the term with $\partial\psi/\partial n$ to vanish from (2.2-8). This simplifies the problem, because it is no longer necessary to know both the function as well as its derivative over the entire surface S; knowledge of ψ on S is sufficient in this case.

However, this "simplification" is achieved at the expense of requiring a far more difficult Green's function than the simple function of (2.2-9). In fact, the Green's function that satisfies the boundary condition $G = 0$ on S can be found only for the simplest geometries. Another alternative is to find a Green's function that satisfies the boundary condition $\partial G/\partial n = 0$ on S. It is then necessary to know $\partial \psi/\partial n$ on S in order to evaluate the integral in (2.2-8). Finding this Green's function is of course no easier than finding the other one.

Instead of using solutions of the wave equation of the form (2.2-8) with an appropriate Green's function, we can use the method of expending the field in terms of plane waves outlined in Section 1.3 to solve diffraction problems. As an example, we solve in Section 2.3 a simple diffraction problem by two methods: by using the Kirchhoff-Huygens diffraction integral (2.2-19), and by using the solution formed as a superposition of plane waves as given by (1.3-23).

Certain approximations of the diffraction integral (2.2-19) are often useful for practical calculations.

A first approximation that is almost always valid for optical problems utilizes the short wavelength of light. In most problems of practical interest, the point x', y', z', at which the function ψ is being calculated is many light wavelengths removed from S, so that the condition

$$k \gg \frac{1}{r} \tag{2.2-20}$$

holds for all values of r occurring in the integral (2.2-19). It is then possible to neglect the derivative of $1/r$ compared to the derivative of $\exp(-ikr)$, so that we can write approximately

$$\psi(x', y', z') = \frac{1}{4\pi} \int_s \left(\frac{\partial \psi}{\partial n} + ik \frac{\partial r}{\partial n} \psi\right) \frac{e^{-ikr}}{r} \, dS \tag{2.2-21}$$

The next approximation is concerned with obtaining a simpler expression for r than (2.2-10). An approximation for r is possible in cases that involve apertures in diffracting screens that are small compared to the distance from the screen at which the field is being investigated. In those cases, the conditions

$$|x' - x| \ll |z' - z| \qquad \text{and} \qquad |y' - y| \ll |z' - z| \tag{2.2-22}$$

are satisfied. It is then convenient to use a Taylor expansion of r

$$r = z' - z + \frac{1}{2} \frac{(x' - x)^2 + (y' - y)^2}{z' - z} + \ldots \tag{2.2-23}$$

It is very important to realize that it is not sufficient that (2.2-22) be satisfied. Since r appears in the exponent of (2.2-19) only in the combination kr

(with kr being a large number) it is the product of k times the neglected terms in (2.2-23) that must be small compared to unity. Considerable errors occur if this point is missed.

Finally, let us assume that the surface S is an opaque, plane screen with a hole in it. Incident on this screen from the left is a plane wave. This is a diffraction problem of the simplest kind, but because of its simplicity it yields valuable insight into the properties of light diffraction. We shall discuss the physics of this problem in the following sections. In this section we are concerned only with the various approximate forms of the diffraction integral. The geometry of the diffraction screen with aperture is shown in Figure 2.2.2.

According to our geometry, we have on the screen

$$\frac{\partial}{\partial n} = -\frac{\partial}{\partial z} \tag{2.2-24}$$

so that we obtain with the help of (1.3-21) (taking $g = \psi$)

$$\frac{\partial \psi}{\partial n} = -\frac{\partial \psi}{\partial z} = ik_z\psi = ik\psi \cos \gamma \tag{2.2-25}$$

The derivative of r becomes

$$\frac{\partial r}{\partial n} = -\frac{\partial r}{\partial z} = \frac{z' - z}{r} = \cos \alpha \tag{2.2-26}$$

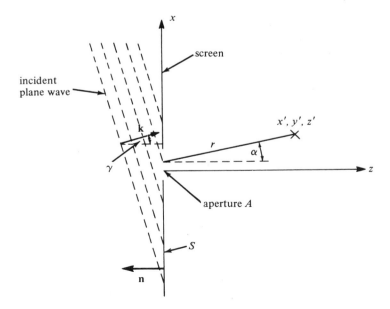

Figure 2.2.2   Diffraction of a plane wave by a slit in an infinite opaque screen.

Using (2.2-25) and (2.2-26), we obtain from (2.2-21)

$$\psi(x', y', z') = \frac{ik}{4\pi} \int_A (\cos \gamma + \cos \alpha) \psi(x, y, z) \frac{e^{-ikr}}{r} \, dS \qquad (2.2\text{-}27)$$

The integral over S includes the screen shown in Figure 2.2.2 and an infinite half sphere enclosing the volume to the right of the screen. We regard the incident wave as part of a very long but finite wave train. The finite wave train does not reach the infinite half sphere in finite time, so that the integral over its surface vanishes. It appears that the expression (2.2-27) is as precise as (2.2-21), from which it was derived. However, we have changed the range of integration from S to A, indicating that the integration is to be extended only over the aperture A in the screen. The step from (2.2-21) to (2.2-27) implies that the field ψ and its derivative $\partial\psi/\partial n$ vanish on the opaque screen. This assumption may seem logical at first glance. However, it can be proven that the solution of the wave equation must vanish everywhere if it and its first derivative vanish over any finite interval.[4] That means that, strictly speaking, our assumption of vanishing field plus vanishing first derivative on the opaque screen violates the principle that ψ must be a solution of the wave equation. In other words, this assumption is inconsistent with the fact that (2.2-19) was derived as a solution of the wave equation. In fact, if we were to calculate values for ψ from (2.2-19) with the assumption that ψ and its normal derivative vanish on the opaque screen, we would find that ψ, as obtained from this equation, does not vanish on the screen, contrary to our assumption. This is a serious mathematical problem. In order to be able to use the diffraction integral, we would have to know the values of ψ and its normal derivative not only over the aperture but also over the opaque screen. So, in order to obtain the desired solution, we would already have to know the function over areas where we really do not know anything at all. It is fortunate that in spite of this mathematical dilemma (2.2-27) yields results which agree very closely with experimental observations. The violation of mathematical principles is apparently not serious enough to prevent us from gaining useful results. The values of ψ and its normal derivative are very small on the screen, and are observable only over a region immediately adjacent to the aperture. The small error that is introduced by assuming that the function vanishes exactly is insufficient to cause real trouble. The diffraction integral in its form (2.2-27) is widely used with great success to solve optical diffraction and microwave antenna problems. Strictly speaking, it is not true that the field in the aperture is the same as the unperturbed field of the incident wave in the absence of the aperture, an assumption that is also implicit in (2.2-27). However, this objection is of a similar nature to the one concerning the vanishing of ψ on the screen.

In many instances the aperture in the screen is sufficiently small and the field point x', y', z' is sufficiently close to the axis that we can approximate

$$\cos \alpha \approx 1 \qquad (2.2\text{-}28)$$

If, finally, the plane wave is incident perpendicular, or at least nearly perpendicular, to the surface of the aperture, we can also set

$$\cos \gamma = 1 \qquad (2.2\text{-}29)$$

and obtain

$$\psi(x', y', z') = \frac{i}{\lambda} \int_A \psi(x, y, z) \frac{e^{-ikr}}{r} \, dS \qquad (2.2\text{-}30)$$

The approximation (2.2-30) is not only of practical but also of great historical interest. C. Huygens conjectured in 1690 that the propagation of light waves can be described by an integral of this form. To understand the interpretation of (2.2-30) in terms of Huygens principle, let us begin by examining the wave of the form (2.2-9) more closely. We have seen in Section 1.3 that a wave of the form (1.3-21) is a sinusoidal plane wave progressing with velocity $v = \frac{\omega}{k}$ in the direction of the vector $\mathbf{k}$. A similar argument shows that the wave of the form (2.2-9) spreads evenly on a sphere propagating with the velocity $v = \frac{\omega}{k}$ in a radial direction away from the origin $r = 0$. At $r = 0$, the wave amplitude is infinite, which indicates the location of the source generating the spherical wave. The wave amplitude of this spherical wave is not constant but decreases steadily as $1/r$.

The physical interpretation of (2.2-30) is now easy. Each point of the aperture can be interpreted as the source of a spherical wave. The strength of the source is determined by the amplitude of the incident wave at the particular point of the aperture. The field at the point x', y', z' is formed by the superposition of all the spherical waves arriving at this point. This interpretation is intuitively logical and was proposed by C. Huygens without the benefit of the wave equation. Our derivation shows also that Huygens principle is only an approximation. The exact form of the diffraction integral (2.2-19) is really quite different. The most striking feature of this principle—namely, the fact that the disturbances originating at the aperture propagate as spherical waves from each element of the aperture—is indeed correct. The form (2.2-27) is already an approximation, but even it is different from Huygens principle. It shows that it is not sufficient simply to assume that the incident wave acts as a source of spherical waves with a source strength proportional to the amplitude of the incident wave at each aperture point. It is necessary to attach to the source strength weight factors which take care of the directionality of the (fictitious) sources.

Returning to (2.2-27), we obtain, finally, useful approximations of the diffraction integral by substitution of (2.2-23). We repeat the warning that, in order to be able to use this approximation, one must check if the product of the third order term (which was neglected in [2.2-23]) with k is indeed much less than unity. If this condition is satisfied, we obtain the approximation

$$\psi(x', y', z') = \frac{i}{2\lambda} e^{-ikz'} \int_A (\cos \gamma + \cos \alpha) \frac{\psi}{z' - z}$$
$$\cdot \exp\left[ik\left(z - \frac{(x' - x)^2 + (y' - y)^2}{2(z' - z)}\right)\right] dS \quad (2.2\text{-}31)$$

In the spirit of the approximation (2.2-23), r in the denominator was replaced by $z' - z$. This form of the diffraction integral is actually most often used in practice. There are usually two regions which are being distinguished. When we are considering the field sufficiently close to the aperture that the term $x^2 + y^2$ (appearing implicitly in the exponent of the exponential function) must be considered, we speak of Fresnel diffraction. If this term is negligible, we speak of Fraunhofer diffraction. Fraunhofer diffraction considers the far field of the diffracted radiation. Fresnel diffraction concerns itself also with regions sufficiently far from the aperture that the approximation (2.2-31) is valid but not quite as far away as is required for Fraunhofer diffraction to apply. These names are based on the historical development of optics. A further discussion of the Fresnel approximation can be found in Section 4.6.

It is sometimes sufficient to study a two-dimensional problem instead of the three-dimensional problem discussed so far. Two-dimensional problems have the advantage of greater simplicity, and are often still sufficient to gain an understanding of a certain diffraction problem. The development of the diffraction integrals presented so far, and in particular of the important approximation (2.2-31), was based on a decomposition of the incident field into a continuous array of sources of spherical waves. Spherical waves occur naturally in three-dimensional problems. For a two-dimensional problem, cylindrical waves are the natural equivalent of spherical waves. The following derivation of diffraction integrals for two-dimensional problems is not commonly found in optical textbooks.

It is typical of a two-dimensional problem that all field variables, as well as the geometry in which the fields propagate, are independent of one of the space variables. The diffraction situation described by Figure 2.2.2. becomes two-dimensional when we can assume that none of the variables of interest depends on the y coordinate. Symbolically this can be expressed by the equation

$$\frac{\partial}{\partial y} = 0 \quad (2.2\text{-}32)$$

We refer to a problem as two-dimensional whenever (2.2-32) applies. In two dimensions, the reduced wave equation takes the form

$$\frac{\partial^2 \psi}{\partial x^2} + \frac{\partial^2 \psi}{\partial z^2} + k^2 \psi = 0 \qquad (2.2\text{-}33)$$

We proceed in exactly the same way as before and consider, in addition to the reduced wave equation (2.2-33), the inhomogeneous equation

$$\frac{\partial^2 G}{\partial x^2} + \frac{\partial^2 G}{\partial z^2} + k^2 G = \delta(z - z')\,\delta(x - x') \qquad (2.2\text{-}34)$$

Multiplying (2.2-33) with G and (2.2-34) with ψ and subtracting and integrating the resulting equation over an area in the $x - z$ plane yields the equation

$$\psi(x', z') = \int_C \left(\psi \frac{\partial G}{\partial n} - G \frac{\partial \psi}{\partial n} \right) dC \qquad (2.2\text{-}35)$$

The two-dimensional divergence theorem was used to convert the integral over the area to an integral over the closed curve encircling the area. The direction n is the outward normal of the curve C.

The Green's function can most easily be found by converting (2.2-34) to polar coordinates. The polar axis is placed in y direction, the distance from the polar axis is called r, and the angle between the radius r and the z-axis is called α. The equation (2.2-34) can then be written

$$\frac{\partial^2 G}{\partial r^2} + \frac{1}{r}\frac{\partial G}{\partial r} + \frac{1}{r^2}\frac{\partial^2 G}{\partial \alpha^2} + k^2 G = \delta(z - z')\delta(x - x') \qquad (2.2\text{-}36)$$

We are interested in solutions that are symmetrical in the $x - z$ plane—in other words, independent of α. The reason for this restriction becomes apparent when we study the more general solutions with an α dependence. These functions can be written as the product of a function that depends only on r times a function that depends only on α. The α dependent function is either $\cos(n\alpha)$ or $\sin(n\alpha)$. The constant n must be an integer, since we require that the function return to the same value as α changes from zero to 2π. When we now determine the singularity that appears after substitution in (2.2-36), we find that the integral over the left-hand side vanishes for all functions except those for which $n = 0$. In other words, only the functions which are independent of α have the correct singularity.

There is an additional restriction. We must require that the Green's function represent waves traveling away from the source point x', z'. There are also solutions representing waves that come in from infinity and converge to the point x', z'. These waves must be excluded, since it is apparent on physical grounds that the field must be composed of contributions stemming from a source region at a finite location. In our diffraction problems, we must expect that the field at x', z' is generated from (fictitious) sources located at the aperture. Waves converging from infinite distances to points on the aperture do not satisfy this requirement. This limitation on the admissible

Green's functions was first pointed out by A. Sommerfeld, and is called Sommerfeld's radiation condition.

Returning to Equation (2.2-36), we see that we can write it more simply

$$\frac{\partial^2 G}{\partial r^2} + \frac{1}{r}\frac{\partial G}{\partial r} + k^2 G = \delta(x - x')\delta(z - z') \qquad (2.2\text{-}37)$$

The left-hand side of this equation has the typical form of Bessel's differential equation. The solutions of (2.2-37) must therefore be cylinder functions. There are cylinder functions of various kinds. Since (2.2-37) is a second order differential equation we know that there exist two linearly independent solutions. The Bessel and Neumann functions are two independent solutions of this kind.[6,8] However, neither of these solutions is desirable in our case. The Bessel function does not have the singularity required by the right-hand side of (2.2-37). The Neumann function does have the necessary singular behavior, but it represents standing waves and thus violates Sommerfeld's radiation condition. A superposition of these two functions could be made to satisfy both requirements. Such a superposition exists ready made in form of the Hankel function. There are two Hankel functions. The Hankel function of the first kind represents waves coming in from infinity (with our choice of the time factor $\exp[i\omega t]$). The Hankel function of the second kind is just right for our purposes. It has the required singularity, and represents waves that travel outward from the origin of the coordinate system in which the function is represented. The proper Green's function is therefore

$$G(x, x', z, z') = \frac{i}{4} H_o^{(2)}(kr) \qquad (2.2\text{-}38)$$

It is left to the reader to prove that G satisfies (2.2-37), in particular that the singularity leads to the proper form of its right-hand side.

The solution of the two-dimensional diffraction problem can now be written in its exact form

$$\psi(x', z') = \frac{i}{4}\int_C \left(\psi\,\frac{\partial H_o^{(2)}}{\partial n} - H_o^{(2)}\,\frac{\partial\psi}{\partial n}\right) dC \qquad (2.2\text{-}39)$$

Just as in the three-dimensional case, it is convenient to use suitable approximations of this exact equation for actual calculations. As a first approximation, we use the fact that, owing to the short wavelength of light, the argument of the Hankel function is always very large. This allows us to use the asymptotic approximation for Hankel functions of large argument[8]

$$H_o^{(2)}(kr) = \sqrt{\frac{2}{\pi kr}}\, e^{-i(kr - \pi/4)} \qquad (2.2\text{-}40)$$

With the help of (2.2-40), Equation (2.2-39) assumes a form very similar to that obtained in the three-dimensional case (2.2-19)

$$\psi(x', z') = \frac{1}{\sqrt{8\pi k}} e^{-i\pi/4} \int_C \left\{ \frac{\partial \psi}{\partial n} \frac{e^{-ikr}}{\sqrt{r}} - \psi \frac{\partial}{\partial n} \left(\frac{e^{-ikr}}{\sqrt{r}} \right) \right\} dC \quad (2.2\text{-}41)$$

with

$$r = \sqrt{(z' - z)^2 + (x' - x)^2} \quad (2.2\text{-}42)$$

The difference between (2.2-19) and (2.2-41) is the occurrence of cylindrical waves in our two-dimensional equation. The amplitudes of these waves decay as $r^{-1/2}$ instead of r^{-1}, the decay with distance of the spherical waves.

We can immediately proceed to the approximation of (2.2-41) that corresponds to (2.2-31). Under conditions similar to those explained earlier, we find the approximation

$$\psi(x', z') = \frac{1}{2\sqrt{\lambda}} e^{i\pi/4} e^{-ikz'} \int_C \frac{\cos \gamma + \cos \alpha}{\sqrt{z' - z}} \psi(x, z)$$
$$\cdot \exp\left[ik \left(z - \frac{(x' - x)^2}{2(z' - z)} \right) \right] dC \quad (2.2\text{-}43)$$

Equation (2.2-43) does not really look much simpler than its three-dimensional equivalent (2.2-31). The simplification consists in the fact that the two-dimensional equation involves only a simple integration over the curve C. In case of a plane aperture, this integration extends simply over x. The integral of the three-dimensional problem is extended over the surface of the aperture which, in the case of a plane aperture, results in a double integral over x and y. This simplification can be a great help for analytical solutions. It is even more important, however, if a numerical solution on a computer is all that can be achieved. The time saving that results from having to perform only one integration, as compared to two, can be very substantial indeed. We shall have occasion to discuss the numerical integration of (2.2-43) for the investigation of a complicated wave transmission problem (see Section 5.8). This problem can be solved only on a computer. A good feeling for the behavior of the waves can be obtained by simplifying the problem to two dimensions. This simplification makes a computer solution economically feasible.

2.3 DIFFRACTION BY A SLIT IN AN OPAQUE SCREEN

As a first application of our diffraction integrals, we now proceed to solve the problem of wave propagation that occurs when a plane wave is forced to pass through a slit in an opaque screen. The problem of a rectangular aperture in an opaque screen is not much more complicated than the problem of the slit, but since not much additional information can be gained by solving the rectangular aperture problem, we concentrate on the slit.

It is instructive to solve this problem in two ways. We mentioned earlier that the diffraction integral method is not the only way to solve diffraction problems. To illustrate this point, we start by solving this problem by applying the solution of the wave equation in the form (1.3-23). For our present purpose, this equation can be simplified. Since we are interested in the problem of scattering of a plane, monochromatic wave, we can drop the integration over ω. This is not as arbitrary a procedure as might appear at first glance. A monochromatic wave consists of one precise frequency. Its spectrum must therefore appear as one sharp line. This can mathematically be expressed by taking

$$\phi(k_x, k_y, \omega) = 2\pi\bar{\phi}(k_x, k_y)\delta(\omega - \omega_o) \tag{2.3-1}$$

Substitution of (2.3-1) removes the ω integration from (1.3-23).

Our slit problem can be illustrated by Figure 2.2.2. The slit is given the width d and it is positioned so that it extends in the y direction. The incident wave is supposed to have its propagation vector directed in the $x - z$ plane so that $k_y = 0$. The geometry of the problem is such that in the absence of a k_y component no waves propagating in y direction will be generated at the slit. The scattered field consequently has no y dependence. Our problem is thus two-dimensional in the sense of the definition presented in the previous section. This insight into the problem allows us to simplify the integral (1.3-23) by writing

$$\phi(k_x, k_y, \omega) = (2\pi)^2\varphi'(k_x)\delta(\omega - \omega_o)\delta(k_y) \tag{2.3-2}$$

This form of the amplitude function ϕ transforms (1.3-23) into the following form

$$\psi(x', z', t) = \frac{1}{2\pi}\int_{-\infty}^{\infty} \varphi'(k_x) e^{-i(k_x x' + k_z z')}dk_x \tag{2.3-3}$$

The time dependence $\exp(i\omega_o t)$ has been omitted for simplicity. To simplify our problem further, we assume that the incident wave impinges perpendicularly on the aperture. The incident plane wave can therefore be expressed as

$$\psi(x, z) = \psi_o e^{-ikz} \tag{2.3-4}$$

We assume that the field in the slit can be represented with sufficient accuracy simply by truncating the function (2.3-4). By locating the slit at $z = 0$, the truncated function in the slit is

$$\psi(x, z) = \begin{cases} \psi_o & |x| < \dfrac{d}{2} \\[2mm] 0 & |x| > \dfrac{d}{2} \end{cases} \tag{2.3-5}$$

The amplitude function φ' can be obtained simply as the Fourier transform of (2.3-5)

$$\varphi'(k_x) = \int_{-d/2}^{d/2} \psi_o e^{ik_x x} dx = 2\psi_o \frac{\sin k_x \dfrac{d}{2}}{k_x} \tag{2.3-6}$$

The diffracted field behind the screen can now be expressed by the integral

$$\psi(x', z') = \frac{\psi_o}{\pi} \int_{-\infty}^{\infty} \frac{\sin k_x \dfrac{d}{2}}{k_x} \exp\left[-i(k_x x' + \sqrt{k^2 - k_x^2}\, z')\right] dk_x \tag{2.3-7}$$

The k_z component was obtained from (1.3-22) and (1.3-13) with $k_y = 0$.

The integral (2.3-7) cannot be solved exactly. We have to investigate the possibility of making some reasonable approximation in order to be able to find an analytical solution of our problem. Only if $d = \infty$ is a solution immediately apparent. Dirac's delta function can be represented in the following form[12]

$$\delta(k_x) = \lim_{d \to \infty} \frac{1}{\pi} \frac{\sin k_x \dfrac{d}{2}}{k_x} \tag{2.3-8}$$

Equation (2.3-8) allows us immediately to solve the integral (2.3-7) in the limit $d \to \infty$. The solution is

$$\psi(x', z') = \psi_o e^{-ikz'} \tag{2.3-9}$$

This solution comes as no great surprise, since it simply proves that the incident plane wave passes through an infinitely wide aperture undisturbed.

An approximation for the case of finite slit width is obtainable if we are content to study the diffracted field only at great distances behind the screen. To obtain an approximation for the far field, we use the method of stationary phase.[9] The method of stationary phase is extremely useful to obtain approximate solutions to far field diffraction patterns for a wide variety of applications. We will have occasion to use it repeatedly in this book. The method is based on the following argument. Let x' and z', or at least one of the two variables, be very large. As a consequence, the argument of the exponential function in (2.3-7) has a large value, and the exponential function varies very rapidly with varying k_x. A rapidly varying oscillatory function has the tendency to cancel out any contribution from a function which appears multiplied with it as the integrand of an integral. Therefore, if the function that multiplies the exponential factor varies comparatively slowly with respect to the variations of the exponential function, the result of the integration will be vanishingly small. Now let us assume that there is a point on the k_x axis at which the derivative of the argument of the exponential function vanishes.

Such a maximum or minimum point of the argument defines a small region in which the argument is stationary—in other words, it does not vary as k_x varies. Such a region makes a large contribution to the integral since the rapid oscillations cease for a certain region of the k_x axis and no cancellation takes place. The method of stationary phase consists in finding the point or points where the argument of the rapidly oscillating function has a stationary value. The argument is expanded in a Taylor series around this point, and the expansion is carried to the second order term. The first order term vanishes of course by the definition of the stationary point. The function multiplying the rapidly oscillating function is taken out of the integral with its argument taken at the stationary point. The method of stationary phase acts like a delta function in that the integral is proportional to the value of the function multiplying the oscillatory function with its argument taken at the stationary point.

Before we apply this method to the integral (2.3-7), we rewrite it in such a way that the exponential functions contained in its integrand become explicitly apparent

$$\psi(x', z') = \frac{\psi_o}{2i\pi} \int_{-\infty}^{\infty} \left(\frac{1}{k_x} e^{-i\theta_-} - \frac{1}{k_x} e^{-i\theta_+} \right) dk_x \tag{2.3-10}$$

with

$$\theta_- = \left(x' - \frac{d}{2} \right) k_x + \sqrt{k^2 - k_x^2}\, z' \tag{2.3-11}$$

and

$$\theta_+ = \left(x' + \frac{d}{2} \right) k_x + \sqrt{k^2 - k_x^2}\, z' \tag{2.3-12}$$

To find the stationary points of the phase θ_-, we form its derivative and set it equal to zero

$$\frac{\partial \theta_-}{\partial k_x} = x' - \frac{d}{2} - \frac{k_x}{\sqrt{k^2 - k_x^2}} z' = 0 \tag{2.3-13}$$

The stationary point is therefore

$$k_x^{(-)} = \frac{\left(x' - \dfrac{d}{2} \right)}{\sqrt{z'^2 + \left(x' - \dfrac{d}{2} \right)^2}} k \tag{2.3-14}$$

The negative sign of the square root must be rejected since it does not lead to a value of k_x that satisfies (2.3-13). The stationary point that belongs to θ_+ follows from (2.3-14) by changing the sign of d. To be able to expand the phases in Taylor series to the second power of $(k_x - k_x^{(-)})$, we need the second derivative of θ. Differentiating (2.3-13) once more, we obtain

$$\frac{\partial^2 \theta_-}{\partial k_x^2} = -\frac{k^2}{(k^2 - k_x^2)^{3/2}} z' \qquad (2.3\text{-}15)$$

The Taylor series expansion of θ_-

$$\theta_- = (\theta_-)_{k_x = k_x^{(-)}} + \frac{1}{2}\left(\frac{\partial^2 \theta_-}{\partial x^2}\right)_{k_x = k_x^{(-)}} (k_x - k_x^{(-)})^2 \qquad (2.3\text{-}16)$$

becomes with the help of (2.3-14)

$$\theta_- = \sqrt{z'^2 + \left(x' - \frac{d}{2}\right)^2}\, k - \frac{\left(z'^2 + \left(x' - \frac{d}{2}\right)^2\right)^{3/2}}{2kz'^2} u^2 \qquad (2.3\text{-}17)$$

with

$$u = k_x - k_x^{(-)} \qquad (2.3\text{-}18)$$

Our approximation holds only for the far field. It is therefore reasonable to use the fact that

$$z' \gg d \qquad (2.3\text{-}19)$$

to obtain the following approximations

$$\sqrt{z'^2 + \left(x' - \frac{d}{2}\right)^2} = r\left(1 - \frac{x'd}{2r^2}\right) \qquad (2.3\text{-}20)$$

and

$$\left(z'^2 + \left(x' - \frac{d}{2}\right)^2\right)^{3/2} = r^3\left(1 - \frac{3}{2}\frac{x'd}{r^2}\right) \qquad (2.3\text{-}21)$$

with

$$r^2 = x'^2 + z'^2 \qquad (2.3\text{-}22)$$

The method of stationary phase works by taking the function that multiplies the rapidly varying exponential function out of the integral. In our case, the factor $1/k_x$ appearing in (2.3-10) can be taken out of the integral and can be replaced by $1/k_x^{(-)}$ and $1/k_x^{(+)}$ respectively. The integral is extended only over the function $\exp(iau^2)$ with the result

$$\int_{-\infty}^{\infty} e^{iau^2}\, du = (1 + i)\sqrt{\frac{\pi}{2a}} = \sqrt{\frac{\pi}{a}}\, e^{i\pi/4} \qquad (2.3\text{-}23)$$

We have thus accumulated all the necessary facts and approximations to evaluate the integral (2.3-10). After performing the necessary substitutions, we obtain* (remember, the terms pertaining to θ_+ are obtained by replacing d with $-d$ in all equations)

$$\psi(x', z') = \sqrt{\frac{2}{\pi}}\, \psi_o\, e^{i\pi/4}\, \frac{e^{-ikr}}{\sqrt{kr}}\, \frac{\sin \pi \dfrac{d}{\lambda} \alpha}{\alpha} \tag{2.3-24}$$

The small angle

$$\alpha \approx \tan \alpha = \frac{x'}{z'} \approx \frac{x'}{r} \tag{2.3-25}$$

is the direction at which the field point x', z' appears as seen from the slit.

The method of stationary phase is not only a convenient mathematical approximation but it also makes physical sense. The point $k_x^{(\mp)}$ is the region of the k_x axis where most of the contribution to the integral comes from. Since the integral represents a superposition of plane waves, (2.3-14) tells us which portion of all the plane waves used in the superposition makes the most significant contribution to the diffraction process. The ratio $k_x^{(-)}/k$ defines the sine of the angle subtended by the direction of wave propagation and the z-axis. In the spirit of our far field approximation, we can write (2.3-14)

$$\frac{k_x^{(-)}}{k} \approx \frac{x'}{\sqrt{z'^2 + x'^2}} = \sin \alpha \approx \alpha \tag{2.3-26}$$

The right-hand side of (2.3-26) was obtained with the help of (2.3-25). The direction of the plane waves contributing most to the diffraction process is thus found to be identical to the direction at which the field point x', z' appears as seen from the aperture. This is an interesting result, for the plane waves traveling at other angles are not headed for the point of observation. The actual diffracted field is of course obtained as the superposition of infinitely many plane waves. The method of stationary phase points to the fact that only those waves in the immediate neighborhood (in k space) of the waves headed for the point of observation do really contribute to the field that is found there.

The diffracted field (2.3-24) has the shape of a cylindrical wave decaying as $r^{-1/2}$ with distance. This wave is modulated by the factor

$$\frac{\sin \pi \dfrac{d}{\lambda} \alpha}{\alpha} \tag{2.3-27}$$

The factor (2.3-27) as shown in Figure 2.3.1 has the familiar shape of the

---

* Terms containing d were neglected except in the exponent of the exponential function.

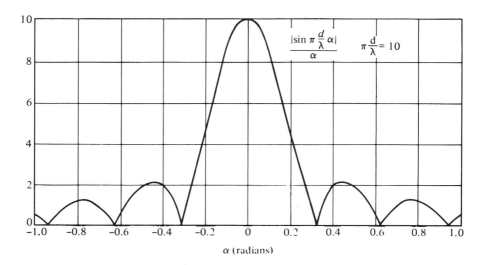

Figure 2.3.1   Diffraction pattern produced by a plane wave passing through a slit in an opaque screen.

(sin x)/x function. Its maximum appears at $\alpha = 0$. The function is symmetric with α, and decreases as $|\alpha|$ increases. The width of the main lobe of (2.3-27) is determined by the factor $\pi d/\lambda$. Let us define an angular width for the radiation lobe by the angle $\alpha = \Delta\alpha$ that points to the first zero of the function (2.3-27). Since this point appears when the argument of the sine function assumes the value π, we find

$$\Delta\alpha = \frac{\lambda}{d} \qquad (2.3\text{-}28)$$

This is an important formula. It tells us that the half width $\Delta\alpha$ of the main radiation lobe decreases with decreasing wavelength λ of the radiation and that it also decreases with increasing width d of the slit. As light passes a narrow slit, the field penetrating through the slit spreads out more when the slit is narrower. It also spreads out more when for a fixed width of the slit the wavelength is increased. Equation (2.3-28) is important for the design of directional antennas, for which our result is directly applicable. We shall see in Section 3.6 that (2.3-28) can be interpreted as a consequence of Heisenberg's famous uncertainty principle.[10]

It remains for us to show that the result (2.3-24) can also be obtained with the help of the diffraction integral (2.2-43). Since we are dealing with a far field pattern of diffracted radiation caused by a wave that is incident normally to the surface of the slit, we can make several simplifying approximations prior to applying (2.2-43).

We set

$$\cos \gamma = \cos \alpha = 1 \qquad (2.3\text{-}29)$$

because the wave is incident normally ($\gamma = 0$) and because we also assumed small values for α in our approximation (2.3-25). Finally, we neglect the term x^2 in the exponent of the exponential function under the integral (2.2-43). With these additional approximations, we can write (2.2-43) in the form

$$\psi(x', z') = \frac{e^{i\pi/4}}{\sqrt{\lambda}} e^{-ikz'} \frac{\psi_o}{\sqrt{z'}} e^{-ik(x'^2/2z')} \int_{-d/2}^{d/2} e^{ik(x'/z')x} \, dx \qquad (2.3\text{-}30)$$

The fact that x^2 was neglected in the exponent of the integrand identifies (2.3-30) as an example of Fraunhofer diffraction. It is implicit in the approximation (2.3-25) that $x' \ll z'$, so that we can replace z' by r everywhere except in the exponent of the exponential function, where the small number x'/z' appears multiplied by the large number k. Using the approximation

$$r = z' + \frac{x'^2}{2z'} \qquad (2.3\text{-}31)$$

for the term in the exponent in front of the integral and carrying out the simple integration, we can write (2.3-30) in its final form.

$$\psi(x', z') = \frac{2e^{i\pi/4}}{k\sqrt{\lambda}} \psi_o \frac{e^{-ikr}}{\sqrt{r}} \frac{\sin \pi \frac{d}{\lambda} \alpha}{\alpha} \qquad (2.3\text{-}32)$$

The approximation (2.3-25) was used once more. With $k = 2\pi/\lambda$, it is easily seen that (2.3-32) is equal to (2.3-24). To the approximation used in their derivation, both approaches used to derive the diffracted field of a narrow slit lead to the same result. Once (2.2-43) is available as a starting point, the method of the diffraction integral is far simpler than the method of the plane wave radiation modes.

2.4 DIFFRACTION BY A CIRCULAR APERTURE

The diffraction problem of the previous section could be treated with the help of the two-dimensional diffraction theory. In this section we discuss an example that can be treated with the use of the three-dimensional diffraction integral (2.2-31).

We consider the field of a plane wave that is incident perpendicularly to the surface of a circular aperture in an opaque screen. Using the approximations (2.3-29) and neglecting the terms x^2 and y^2 in the exponent of the exponential function, we use (2.2-31) in the following approximate form

$$\psi(x', y', z') = \frac{i}{\lambda z'} \psi_o e^{-ik[z' + (x'^2 + y'^2)/2z']} \int_A e^{ik(x' x + y'y)/z'} dS \quad (2.4\text{-}1)$$

The screen was assumed to be located at $z = 0$. To evaluate the integral, we introduce a polar coordinate system that is located with its polar axis parallel to the z' − axis. The coordinates x', y' and x, y can now be expressed as follows

$$x' = \rho \cos \beta' \qquad y' = \rho \sin \beta' \quad (2.4\text{-}2)$$

and

$$x = \sigma \cos \beta \qquad y = \sigma \sin \beta \quad (2.4\text{-}3)$$

The combination in which these coordinates appear in the integrand of (2.4-1) assumes the form

$$x'x + y'y = \rho\sigma \cos (\beta - \beta')$$

Placing the axis, from which the angles are measured, in a convenient way as to make $\beta' = 0$, we obtain simply

$$x'x + y'y = \rho\sigma \cos \beta \quad (2.4\text{-}4)$$

To simplify the notation, we introduce r_o as the radius that results from (2.2-23) with $z = x = y = 0$. We thus obtain from (2.4-1)

$$\psi(x', y', z') = \frac{i}{\lambda z'} \psi_o e^{-ikr_o} \int_0^a \sigma \, d\sigma \int_0^{2\pi} e^{ik(\rho/z')\sigma\cos\beta} \, d\beta \quad (2.4\text{-}5)$$

The β integral can be expressed in terms of the Bessel function of zero order, whose integral representation is[11]

$$J_o(z) = \frac{1}{2\pi} \int_0^{2\pi} e^{iz \cos \beta} \, d\beta \quad (2.4\text{-}6)$$

Using a well-known relation[11] between J_o and the Bessel function of first order J_1

$$\int z J_o(z) \, dz = z J_1(z) \quad (2.4\text{-}7)$$

we obtain from (2.4-5)

$$\psi(x', y', z') = i \frac{a}{\rho} \psi_o e^{-ikr_o} J_1\left(k \frac{\rho}{z'} a\right) \quad (2.4\text{-}8)$$

In analogy to (2.3-25), we introduce the angle α at which the field point x', y', z' appears as seen from the aperture

$$\alpha \approx \frac{\rho}{z'} \approx \frac{\rho}{r_o} \tag{2.4-9}$$

and thus obtain finally

$$\psi(x', y', z') = ia\psi_o \frac{e^{-ikr_o}}{r_o} \frac{1}{\alpha} J_1\left(2\pi \frac{a}{\lambda}\alpha\right) \tag{2.4-10}$$

The far field pattern of the field behind the circular aperture (Figure 2.4.1) looks quite similar to the field behind the slit (Figure 2.3.1) if both are plotted as functions of the angle α. Of course the field behind the slit extends parallel to the slit and shows diffraction fringes only in x direction. The diffraction fringes of the field behind the circular aperture consist of concentric rings. The half width of the main lobe can again be defined by the angle at which the field assumes its first null. The first zero of the Bessel function $J_1(z)$ appears at[11] $z = 3.832$, so that we get in analogy to (2.3-28)

$$\Delta\alpha = \frac{3.832}{2\pi} \frac{\lambda}{a} = 1.22 \frac{\lambda}{2a} \tag{2.4-11}$$

This formula is very similar to (2.3-28). If we consider that the hole diameter $2a$ corresponds to the slit width d, we find that the half angle $\Delta\alpha$ of the

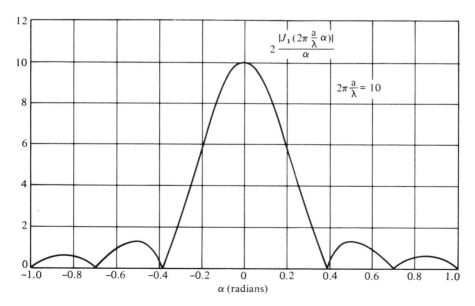

Figure 2.4.1 Diffraction pattern produced by a plane wave passing through a circular aperture in an opaque screen.

main radiation lobe of the diffraction field behind the circular aperture is very nearly the same as the half width of the main lobe of the diffraction field of the slit.

2.5 DIFFRACTION GRATINGS

The diffraction problems discussed in the previous two sections gave us valuable insight into the physics of light diffraction and helped us to develop useful approximations to solve far field problems. However, the slit and circular iris did not by themselves present very useful devices. The problem to be discussed in this section is of more practical importance.

Diffraction gratings are devices which are used to analyze the frequency content of multicolored light. A simple example of a diffraction grating is shown in Figure 2.5.1. An opaque screen is provided with a large number of slits of width d. These slits are parallel to each other and are spaced precisely periodically a distance a from each other. The total number of slits is $N + 1$ filling a region of the screen of length D, so that we have the relation

$$D = a(N + 1) \qquad (2.5\text{-}1)$$

To simplify future expressions, we assume that $N/2$ is an integer. The **k** vectors shown in Figure 2.5.1 represent the propagation vector of the

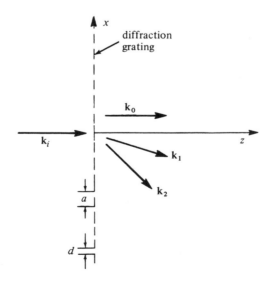

Figure 2.5.1 Illustration of a diffraction grating consisting of many thin slits in an opaque screen.

incident plane wave $\mathbf{k}_i$, which is assumed to be directed perpendicularly to the surface of the screen. The vector $\mathbf{k}_o$ represents a portion of the incident plane wave which continues to propagate in the original direction. The vectors $\mathbf{k}_1$ and $\mathbf{k}_2$ are added to indicate the scattered light. The only (but important) difference between this problem and the problem of the slit treated in Section 2.3 is the number of existing slits. In order to solve the problem of the diffraction grating, we must add all the scattered light components of each slit to obtain the total field that is found behind the diffraction grating.

Using (2.3-31), we replace the distance from the center of each slit to the field point x', z' by

$$r_n = z' + \frac{(x' - x_n)^2}{2z'} = r_o - \frac{x'x_n}{z'} + \frac{x_n^2}{2z'} \tag{2.5-2}$$

with

$$r_o = z' + \frac{x'^2}{2z'} \tag{2.5-3}$$

We solve the problem of the diffraction grating in the Fraunhofer approximation just as we did the diffraction problems in the previous two sections. This implies that we neglect the term with x_n^2 in (2.5-2). The coordinate $x_n = na$ determines the center of the nth slit of the grating. Even though the grating has very many slits, its total extension is supposed to be small in comparison to the distance at which the diffracted field is observed. We obtain the far field of the diffraction grating from (2.3-32) and (2.5-2) in the Fraunhofer approximation

$$\psi(x', z') = \frac{2e^{i\pi/4}}{k\sqrt{\lambda}} \psi_o \frac{e^{-ikr_o}}{\sqrt{r_o}} \frac{\sin\frac{\pi d}{\lambda}\alpha}{\alpha} \sum_{n=-N/2}^{N/2} e^{ik\alpha na} \tag{2.5-4}$$

Equation (2.3-25) was used to replace the ratio of x' over z' by the angle α at which the field point appears as seen from the center of the diffraction grating.

The sum in (2.5-4) can easily be evaluated with the result

$$\sum_{n=-N/2}^{N/2} e^{ik\alpha na} = e^{-ik\alpha(N/2)a} \frac{e^{ik\alpha(N+1)a} - 1}{e^{ik\alpha a} - 1}$$

$$= \frac{\sin\frac{1}{2}k\alpha(N+1)a}{\sin\frac{1}{2}k\alpha a} \tag{2.5-5}$$

Using $k = 2\pi/\lambda$ and (2.5-1), we obtain from (2.5-4)

$$\psi(x', z') = \frac{\sqrt{\lambda}}{\pi} \psi_o \, e^{i\pi/4} \frac{e^{-ikr_o}}{\sqrt{r_o}} \frac{\sin \dfrac{\pi d}{\lambda} \alpha \sin \pi \dfrac{D}{\lambda} \alpha}{\alpha \quad \sin \pi \dfrac{a}{\lambda} \alpha} \tag{2.5-6}$$

The far field of the diffraction grating consists of two essential terms. The first is the response of each individual slit as seen from (2.3-32), while the second term describes the function of the diffraction grating. The diffraction grating term

$$F(\alpha) = \frac{\sin \pi \dfrac{D}{\lambda} \alpha}{\sin \pi \dfrac{a}{\lambda} \alpha} \tag{2.5-7}$$

is schematically shown in Figure 2.5.2.

Each peak belongs to a different grating order. As long as the sine function in the denominator remains reasonably large the expression $F(\alpha)$ oscillates very rapidly with a frequency (in α-space) of $D/(2\lambda)$. Every time the sine function in the denominator vanishes at

$$\pi \frac{a}{\lambda} \alpha = m\pi \tag{2.5-8}$$

(m integer) it is apparent from (2.5-5) that the sine function in the numerator of (2.5-7) also vanishes. At the points (2.5-8) the function assumes the large value

$$F_{max} = \frac{D}{a} = N + 1 \tag{2.5-9}$$

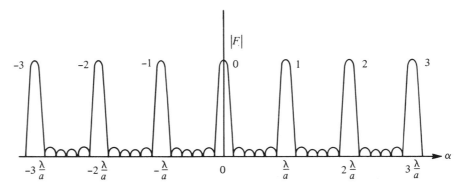

Figure 2.5.2 Illustration of the Function $|F(\alpha)|$ of Equation (2.5-7) that determines the diffraction pattern of a diffraction grating.

The peak response of the diffraction grating is proportional to the number of slits that make up the grating. To obtain the true grating response as a function of α, Figure 2.5.2 must be visualized as being multiplied by the response function of each individual slit shown in Figure 2.3.1. The angular half width of the grating lobes can be defined similar to the width of the single slit as the point at which the function F reaches its first null after passing through a maximum. The angular half width of the grating lobes is therefore given by

$$\Delta\alpha = \frac{\lambda}{D} = \frac{1}{N+1}\frac{\lambda}{a} \qquad (2.5\text{-}10)$$

The half width of the grating lobes is thus seen to be inversely proportional to the number of slits in the grating.

The lobe at $\alpha = 0$ is called the zero order lobe. It represents the field that passes the grating essentially undiffracted. The mth order grating lobe is obtained from (2.5-8). It corresponds to a wave traveling away from the grating at an angle

$$\alpha_m = m\,\frac{\lambda}{a} \qquad (2.5\text{-}11)$$

The relative half width of the mth grating order (lobe) is therefore

$$\frac{\Delta\alpha}{\alpha_m} = \frac{1}{m(N+1)} \qquad (2.5\text{-}12)$$

The far field of the grating consists of a number of nearly plane waves each traveling in a different direction. Each grating order can be made to appear as a line on a screen by focusing the far field pattern with the help of a lens. Equation (2.5-11) shows that the direction of each grating lobe depends on the wavelength of the radiation. The lens consequently focuses each wavelength of light contained in the incident plane wave on a different spot on the screen. This feature of diffraction gratings makes them useful for the spectral analysis of light. The spectral resolution of the diffraction grating can easily be obtained from (2.5-10) and (2.5-11). Suppose we assume that two spectral lines are still distinguishable if the peak of one falls on the first zero next to the peak of the other. This is called the Rayleigh criterion for the resolution of optical instruments. Our prescription means that the change of α_m for a certain change of the wavelength $\Delta\lambda$ must equal $\Delta\alpha$ of (2.5-10). This requirement can immediately be expressed in terms of the relative change in wavelength that the diffraction grating is capable of resolving

$$\frac{\Delta\lambda}{\lambda} = \frac{1}{mN} \qquad (2.5\text{-}13)$$

The number 1 was neglected relative to the large number N. It is apparent that the resolving power of the grating improves not only with increasing number of slits but also with the order m of the grating response. The resolving power of the instrument is increased if it is possible to work with a higher order grating lobe.

The equation (2.5-11) for the angle at which a certain grating lobe appears can be understood in very simple terms. Figure 2.5.3 shows two slits of a diffraction grating. Also shown are the propagation directions of two plane waves emerging from each slit. This plane wave concept of the diffracted waves is borrowed from the plane wave expansion of arbitrary fields (1.3-23). A grating lobe can result only if the two waves emerging from each slit interfere constructively. This means that the points B and C shown in the figure must correspond to points of equal phase. This requirement is met when the distance AB is equal to an integer multiple of the wavelength

$$AB = m\lambda \tag{2.5-14}$$

or

$$a \sin \alpha = m\lambda \tag{2.5-15}$$

In the small angle approximation used for its derivation, (2.5-11) is equal to (2.5-15).

The diffraction grating of Figure 2.5.1 can be described by a transmission response function T of the following form

$$T = \begin{cases} 1 & x_n - \dfrac{d}{2} < x < x_n + \dfrac{d}{2} \\ 0 & x_n - \left(a - \dfrac{d}{2}\right) < x < x_n - \dfrac{d}{2} \end{cases} \tag{2.5-16}$$

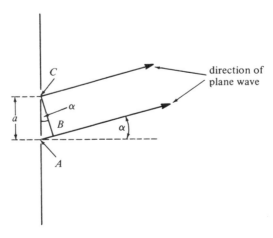

Figure 2.5.3   This figure aids in the intuitive explanation of a diffraction grating.

Equation (2.5-16) means that $T = 0$ on the screen portion of the grating but $T = 1$ in the slits. Even though we did not directly derive it that way, (2.5-4) can be obtained by substituting

$$\psi(x, z) = T\psi_o \qquad (2.5\text{-}17)$$

in the integrand of (2.2-43).

A diffraction grating results not only when the transmission function of the screen assumes the special form (2.5-16) but for almost any periodic transmission function. An example of particular interest for holography is a grating whose transmission function is given as a simple sinusoidal expression.

$$T = 1 + t \cos 2\pi \frac{x}{a} \qquad (2.5\text{-}18)$$

Substitution of (2.3-29), (2.5-17), and 2.5-18) into (2.2-43) results in

$$\psi(x', z') = \frac{\psi_o}{\sqrt{\lambda}} \frac{e^{-ikr_o}}{\sqrt{r_o}} e^{i\pi/4} \int_{-(N/2a)}^{(N/2a)} \left(1 + t \cos 2\pi \frac{x}{a}\right) e^{ik\alpha x} dx \qquad (2.5\text{-}19)$$

The usual approximations that apply to the case of Fraunhofer diffraction have been made, and (2.3-25) has been used in the exponent of the exponential function appearing in the integrand. The integral can be calculated most easily by expressing the cosine function in terms of exponential functions

$$\psi(x', z') = \frac{\psi_o}{\sqrt{\lambda}} e^{i\pi/4} \frac{e^{-ikr_o}}{\sqrt{r_o}} \left\{ 2 \frac{\sin k\alpha a \dfrac{N}{2}}{k\alpha} + t \frac{\sin\left(k\alpha a + 2\pi\right)\dfrac{N}{2}}{k\alpha + \dfrac{2\pi}{a}} \right.$$

$$\left. + t \frac{\sin(k\alpha a - 2\pi)\dfrac{N}{2}}{k\alpha - \dfrac{2\pi}{a}} \right\} \qquad (2.5\text{-}20)$$

The grating response of this sinusoidal grating is significantly different from the response of the slit grating (2.5-6). Whereas the slit grating had infinitely many grating lobes, the diffraction pattern (2.5-20) has only three lobes. The zero order lobe corresponds to the first term in the bracket. This term is simply the diffraction pattern of a slit whose size equals the size of the entire grating. This lobe is very narrow, and peaks at $\alpha = 0$. Each of the remaining terms in the bracket contributes one side lobe. The significant difference between the two gratings is the occurrence of the factor $\sin \pi \dfrac{a}{\lambda} \alpha$

in the denominator of the expression for the slit grating. This factor is absent from the expression for the sinusoidal grating. The denominators of the sinusoidal grating vanish only at one particular point, giving rise to a grating lobe as they do so. But they do not vanish periodically, as does the denominator of the slit grating. This example shows that it is possible to design diffraction gratings that have only one side lobe on each side of the zero order lobe.

We conclude the discussion of diffraction gratings with an analysis of a phase grating. The two gratings discussed so far modulated the amplitude of the incident plane wave by a periodic transmission function. It is also possible to obtain a diffraction grating by modulating the phase instead of the amplitude of the incident plane wave. This is possible by placing a clear transparent object in the path of the wave. The phase of the transmitted wave depends on the optical path length of the light passing a certain portion of the transparent object. Periodically changing the thickness of a glass plate results in a periodic phase modulation of the incident plane wave. Another way of creating a phase grating is by means of a sound wave in a liquid. In our complex notation, we can express the phase modulation simply by a phase transmission factor

$$T = e^{i(1 + t \cos 2\pi(x/a))} \tag{2.5-21}$$

Substitution of this expression into (2.5-17) and into (2.2-43) results

$$\psi(x', z') = \frac{\psi_o}{\sqrt{\lambda}} e^{i\pi/4} \frac{e^{-ikr_o}}{\sqrt{r_o}} \int_{-D/2}^{D/2} \exp\left[i\left(1 + t \cos 2\pi \frac{x}{a} + k\alpha x\right)\right] dx \tag{2.5-22}$$

The integral in this expression is considerably harder to evaluate than were the integrals encountered previously. To get a feeling for the behavior of this phase grating, let us begin with an approximate solution. We assume $t \ll 1$ and neglect terms of second and higher order in t. This allows us to obtain the following approximation

$$\psi(x', z') = \frac{\psi_o}{\sqrt{\lambda}} e^{i\pi/4} \frac{e^{-ikr_o}}{\sqrt{r_o}} e^{i} \int_{-D/2}^{D/2} \left(1 + it \cos 2\pi \frac{x}{a}\right) e^{ik\alpha x} dx \tag{2.5-23}$$

Except for a phase factor in front of the integral and for the fact that i times t, instead of t, appears in the equation, (2.5-23) is identical to the expression (2.5-19) for the sinusoidal amplitude grating. Expression (2.5-20) provides immediately the solution to our phase grating problem when we replace t by it and multiply the whole expression with the unimportant phase factor e^i. We therefore see that for small values of t there are again only three grating lobes: the zero order lobe, corresponding to the slit with the width of the entire grating; and two first order side lobes on either side of the slit. However, this result is not exact. We could keep higher order terms in the expansion of the transmission function (2.5-21) and would find that additional grating orders appear. According to (2.5-20), we see that the first order grating lobes appear at

$$\frac{ak\alpha}{2\pi} = \pm 1$$

or in terms of the light wavelength

$$\alpha\frac{a}{\lambda} = \pm 1 \tag{2.5-24}$$

The general grating orders occur for

$$\alpha\frac{a}{\lambda} = m \tag{2.5-25}$$

where m is any positive or negative integer or zero. To prove this assertion and to obtain an expression which allows us to determine the peak amplitude at each grating order and the width of the grating lobes, we set

$$v = \alpha\frac{a}{\lambda} = -m + \varepsilon \tag{2.5-26}$$

with integer m and

$$\varepsilon \ll 1 \tag{2.5-27}$$

Introducing the new variable

$$2\pi\frac{x}{a} = \theta \tag{2.5-28}$$

we obtain from (2.5-22)

$$\psi(x', z') = a\frac{\psi_o}{\sqrt{\lambda}}e^{i\pi/4}\frac{e^{-ikr_o}}{\sqrt{r_o}}\frac{e^i}{2\pi}\int_{-\pi(D/a)}^{\pi(D/a)}e^{i(v\theta + t\cos\theta)}d\theta \tag{2.5-29}$$

We subdivide the integration range in sections of length 2π. Using (2.5-1), we obtain

$$I = \int_{-\pi(D/a)}^{\pi(D/a)}e^{i(v\theta + t\cos\theta)}d\theta = \sum_{n=-N/2}^{N/2}\int_{(2n-1)\pi}^{(2n+1)\pi}e^{i(v\theta + t\cos\theta)}d\theta \tag{2.5-30}$$

Using (2.5-26) and (2.5-27), we can write approximately

$$I = \sum_{n=-N/2}^{N/2}e^{i2\pi n\varepsilon}\int_{(2n-1)\pi}^{(2n+1)\pi}e^{i(-m\theta + t\cos\theta)}d\theta \tag{2.5-31}$$

Because of the assumed smallness of ε, the function

$$e^{i\varepsilon\theta}$$

is very nearly constant throughout the integration range, and can therefore, be taken out of the integral. The integrand remaining in (2.5-31) is a periodic function with the period 2π. The integration range of an integral over a full period of a periodic function can be shifted without changing the value of the integral. In fact, we need only replace θ by $\theta' + 2n\pi$ to obtain

$$I = \sum_{n=-N/2}^{N/2} e^{i2n\pi\varepsilon} \int_{-\pi}^{\pi} e^{i(-m\theta' + t\cos\theta')} d\theta' \tag{2.5-32}$$

The integral in this expression is a Bessel function. One of the many integral representations of Bessel functions is[11]

$$J_n(z) = \frac{1}{2\pi} \int_{-\pi}^{\pi} e^{i(-n\theta + z\sin\theta)} d\theta \tag{2.5-33}$$

Introducing

$$\theta = \theta' + \frac{\pi}{2}$$

and shifting the integration range over the periodic function to make it again symmetrical, we obtain

$$J_n(z) = \frac{e^{-in(\pi/2)}}{2\pi} \int_{-\pi}^{\pi} e^{i(-n\theta + z\cos\theta)} d\theta \tag{2.5-34}$$

Equation (2.5-34) allows us to write (2.5-32) as follows

$$I = 2\pi e^{im(\pi/2)} J_m(t) \sum_{n=-N/2}^{N/2} e^{i2n\pi\varepsilon} \tag{2.5-35}$$

The summation was already carried out in (2.5-5), so that we obtain from (2.5-29)

$$\psi(x', z') = a \frac{\psi_0}{\sqrt{\lambda}} e^{i[(m+1/2)\pi/2 + 1]} \frac{e^{-ikr_o}}{\sqrt{r_o}} J_m(t) \frac{\sin\pi\varepsilon(N+1)}{\pi\varepsilon} \tag{2.5-36}$$

We replaced $\sin(\pi\varepsilon)$ by $\pi\varepsilon$.

Equation (2.5-36) shows clearly that the amplitude of the scattered radiation becomes large only when $\varepsilon = 0$. At that point we have

$$\lim_{\varepsilon \to 0} \frac{\sin\pi\varepsilon(N+1)}{\pi\varepsilon} = N + 1 \tag{2.5-37}$$

According to (2.5-26), the main grating lobes do occur for integer values of $\alpha\frac{a}{\lambda}$, as was asserted earlier. The phase grating has again infinitely many grating lobes that decay in amplitude as the Bessel function of mth order.

$$J_m(t) \tag{2.5-38}$$

Table 2.5.1 shows several values of $J_m(t)$.

TABLE 2.5.1   Sample values of $J_m(t)$ for several values of the order number m and the argument t of the Bessel function.

m	$t=$ 0.2	0.4	0.6	0.8	1.0
0	$9.90 \ 10^{-1}$	$9.60 \ 10^{-1}$	$9.12 \ 10^{-1}$	$8.46 \ 10^{-1}$	$7.65 \ 10^{-1}$
1	$9.95 \ 10^{-2}$	$1.96 \ 10^{-1}$	$2.87 \ 10^{-1}$	$3.69 \ 10^{-1}$	$4.40 \ 10^{-1}$
2	$4.98 \ 10^{-3}$	$1.97 \ 10^{-2}$	$4.37 \ 10^{-2}$	$7.58 \ 10^{-2}$	$1.15 \ 10^{-2}$
3	$1.66 \ 10^{-4}$	$1.32 \ 10^{-3}$	$4.40 \ 10^{-3}$	$1.02 \ 10^{-2}$	$1.96 \ 10^{-2}$
4	$4.16 \ 10^{-6}$	$6.61 \ 10^{-5}$	$3.31 \ 10^{-4}$	$1.03 \ 10^{-3}$	$2.48 \ 10^{-3}$
5	$8.32 \ 10^{-8}$	$2.65 \ 10^{-6}$	$1.99 \ 10^{-5}$	$8.31 \ 10^{-5}$	$2.50 \ 10^{-4}$
6	$1.39 \ 10^{-9}$	$8.84 \ 10^{-8}$	$1.00 \ 10^{-6}$	$5.56 \ 10^{-6}$	$2.09 \ 10^{-5}$

The amplitudes of the various grating lobes not only decrease with increasing order m, according to (2.5-38), but they also depend on the depth of modulation t. For $t \ll 1$, our approximation (2.5-23) holds for $m = \pm 1$.

The reader has probably noted the very close correspondence between the angular response of diffraction gratings and the frequency response of modulated radio waves. The case of simple sinusoidal amplitude modulation corresponds to the case of the grating with sinusoidal transmission function. The case of the phase grating corresponds to sinusoidal phase (or frequency) modulation of a radio wave. The simple slit grating, finally, corresponds to pulse amplitude modulation of a high-frequency carrier wave. This correspondence of deflection angle with frequency response of time-varying processes is not limited to diffraction gratings. It finds even more useful applications in the transformation properties of lenses, which will be discussed in a later chapter. The correspondence of deflection angle with the frequency of time-varying processes is caused by the fact that in the Fraunhofer regime of diffraction the diffraction integral assumes the form of a Fourier integral, as can immediately be seen from (2.5-19) or (2.5-22). The Fourier transform properties of diffraction processes, in particular with the insertion of lenses into the light beam, will be discussed in the chapter on lenses.

2.6 BRAGG DIFFRACTION: PERTURBATION THEORY

In the previous section, we studied two-dimensional diffraction gratings with zero thickness. In this section, we study thick phase gratings.

Let us assume that the index of refraction of an extended volume of a dielectric medium varies according to the law.

$$n = n_o + \eta \cos(\boldsymbol{\beta} \cdot \mathbf{r} + \varphi) \qquad (2.6\text{-}1)$$

The refractive index is sinusoidally space-modulated. The regions of constant index of refraction lie in planes perpendicular to the vector $\boldsymbol{\beta}$. The spacing

D between two successive maxima (or any other fixed value of the index) is given by

$$D = \frac{2\pi}{\beta} \qquad (2.6\text{-}2)$$

with β being the magnitude of the vector $\boldsymbol{\beta}$.

The index distribution of (2.6-1) can be realized by a sound wave in a dielectric medium. The compressions and rarefactions occurring in a sound field cause index changes of that kind. The index changes in a sound wave would travel through the medium at the speed of sound. This wave motion is absent from (2.6-1). We may regard this equation as a snapshot of a three-dimensional sound field. However, permanent index distributions of the kind discussed here can be created photographically by recording three-dimensional interference patterns in a photographic emulsion. After the emulsion is developed and the silver atoms are bleached away, an index distribution of the form (2.6-1) remains in the gelatine. Such permanent index distributions are of course stationary and do not travel through the medium. They confirm closely to our mathematical model.

The index distribution (2.6-1) creates a three-dimensional phase grating in the dielectric material. This grating has properties similar to the two-dimensional phase grating studied in the previous section. In fact, if the medium in which the grating exists is not too thick and if the incident wave enters the medium perpendicularly while the $\boldsymbol{\beta}$ vector is directed parallel to the surface of the medium, a good approximation to a two-dimensional grating results.

The geometry of our general problem is sketched in Figure 2.6.1. The

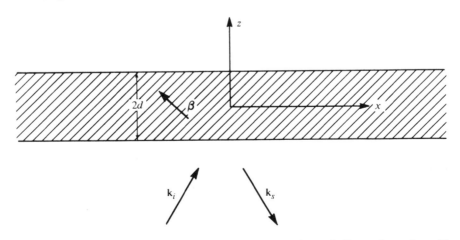

Figure 2.6.1 Schematic of Bragg diffraction. The hatched area indicates the region with periodic variation of the refractive index. The propagation vectors of the incident and scattered plane waves are indicated by $\mathbf{k}_i$ and $\mathbf{k}_s$, while β is a vector normal to the planes of constant values of the refractive index.

slanted lines indicate lines of constant index of refraction. The vector $\boldsymbol{\beta}$ as well as the vector $\mathbf{k}_i$ of an incident plane wave and the vector $\mathbf{k}_s$ of a scattered plane wave are also shown in the figure.

We solve this scattering problem by perturbation theory. We assume that

$$\eta \ll 1 \tag{2.6-3}$$

and write the square of the refractive index

$$n^2 \approx n_o^2 + 2\eta n_o \cos(\boldsymbol{\beta} \cdot \mathbf{r} + \varphi) = n_o^2 + \Delta\varepsilon \tag{2.6-4}$$

The reduced wave equation (2.2-2) has the form*

$$\mathbf{V}^2\psi + n^2 k_o^2 \psi = 0 \tag{2.6-5}$$

The factor k_o is the free space propagation constant

$$k_o = \omega\sqrt{\varepsilon_o\mu_o} \tag{2.6-6}$$

We postulate the existence of an incident plane wave

$$\psi_i = Ae^{-i\mathbf{k}_i \cdot \mathbf{r}} \tag{2.6-7}$$

with $k_i = n_o k_o$. The region outside the volume with the sinusoidally modulated index distribution is assumed to have the average value of the index of refraction n_o to avoid unnecessary complications resulting from reflection and refraction of waves at the interface between the two media. We know the effects of a plane dielectric interface from our discussion in Section 1.6 and need not burden ourselves with this well-known additional effect.

In addition to the incident wave, we expect to obtain a scattered wave, which we express as a Fourier integral

$$\psi_s = \int B(\mathbf{k}_s)e^{-i\mathbf{k}_s \cdot \mathbf{r}} d^3\mathbf{k}_s \tag{2.6-8}$$

The notation $d^3\mathbf{k}_s$ is an abbreviation for

$$d^3\mathbf{k}_s = dk_{sx}dk_{sy}dk_{sz} \tag{2.6-9}$$

The argument $\mathbf{k}_s$ of the function B shows in abbreviated form its dependence on the three variables k_{sx}, k_{sy}, k_{sz}. The integral is extended over all three variables from $-\infty$ to $+\infty$. The expansion (2.6-8) is different from the solution of the wave equation (1.3-23). Here we allow all three components of the vector $\mathbf{k}_s$ to vary independently through their entire range of integration. Any restriction that the wave equation may introduce must be expressed by the functional form of B.

The total field is the sum of the incident and reflected wave

$$\psi = \psi_i + \psi_s \tag{2.6-10}$$

---

* The scalar wave theory that is used here applies directly to a wave field that satisfies the condition $\mathbf{E} \cdot \boldsymbol{\beta} = 0$. For other polarizations the scalar theory does not necessarily yield the same results as the vector theory.

We substitute (2.6-4), (2.6-7), (2.6-8), and (2.6-10) into the reduced wave equation (2.6-5)

$$(n_o^2 k_o^2 - k_i^2)A e^{-i\mathbf{k}_i \cdot \mathbf{r}} + \Delta\varepsilon k_o^2 A e^{-i\mathbf{k}_i \cdot \mathbf{r}} + \int (n_o^2 k_o^2 - k_s^2)B(\mathbf{k}_s)e^{-i\mathbf{k}_s \cdot \mathbf{r}} d^3 k_s \qquad (2.6\text{-}11)$$
$$+ \int \Delta\varepsilon k_o^2 B(\mathbf{k}_s)e^{-i\mathbf{k}_s \cdot \mathbf{r}} d^3 k_s = 0$$

The incident wave is a solution of the undisturbed wave equation. The first term in (2.6-11) therefore cancels out. So far (2.6-11) is still exact. However, we wish to solve this problem by perturbation theory, and will settle for the first order approximation. Since both $\Delta\varepsilon$ and B are quantities that are small of first order, we can neglect their product and omit the last term in (2.6-11). We multiply the remaining terms with $\exp(i\mathbf{k}_s' \cdot \mathbf{r})$ and integrate over all space. The integral of the exponential function results in a delta function

$$\int e^{i(\mathbf{k}_s' - \mathbf{k}_s) \cdot \mathbf{r}} d^3 r = (2\pi)^3 \delta(k_{sx}' - k_{sx})\delta(k_{sy}' - k_{sy})\delta(k_{sz}' - k_{sz}) \qquad (2.6\text{-}12)$$
$$= (2\pi)^3 \delta(\mathbf{k}_s' - \mathbf{k}_s)$$

The last line of this equation is an abbreviated notation for the three dimensional delta function.

Using the procedure just outlined, we obtain from (2.6-11)

$$B(\mathbf{k}_s) = \frac{\eta k_o^2 A n_o}{(2\pi)^3 (k_s^2 - n_o^2 k_o^2)}$$

$$\cdot \int_V (e^{i[(\mathbf{k}_s + \boldsymbol{\beta} - \mathbf{k}_i) \cdot \mathbf{r} + \phi]} + e^{i[(\mathbf{k}_s - \boldsymbol{\beta} - \mathbf{k}_i) \cdot \mathbf{r} - \phi]}) d^3 r \qquad (2.6\text{-}13)$$

The value $\Delta\varepsilon$ defined by Equation (2.6-4) was substituted. Since $\Delta\varepsilon$ vanishes outside the shaded volume shown in Figure 2.6.1, the integration in (2.6-13) extends only over the interaction region with volume V. The integral extends over oscillatory functions, and does not contribute appreciably unless the arguments of the exponential functions vanish. This consideration shows that we obtain appreciable values only for

$$\mathbf{k}_s = \mathbf{k}_i \pm \boldsymbol{\beta} \qquad (2.6\text{-}14)$$

This important relation is known as the Bragg condition, The interaction volume is supposed to extend to infinity in the x and y directions. The x and y integrations lead therefore to delta functions, so that we can write (2.6-13) in integrated form

$$B(\mathbf{k}_s) = \frac{\eta n_o k_o^2 A}{\pi(k_s^2 - n_o^2 k_o^2)}$$

$$\left\{ \delta(k_{sx} + \beta_x - k_{ix})\delta(k_{sy} + \beta_y - k_{iy})\frac{\sin(k_{sz} + \beta_z - k_{iz})d}{k_{sz} + \beta_z - k_{iz}}e^{i\phi} \right.$$

$$\left. + \delta(k_{sx} - \beta_x - k_{ix})\delta(k_{sy} - \beta_y - k_{iy})\frac{-\sin(k_{sz} - \beta_z k_{iz})d}{k_{sz} - \beta_z - k_{iz}}e^{-i\phi} \right\} \qquad (2.6\text{-}15)$$

The scattered wave is thus obtained from (2.6-8) and (2.6-15)

$$
\psi_s = \eta n_o k_o^2 A \frac{1}{\pi} \Bigg[e^{i\phi} e^{-i(k_{sx}^- x + k_{sy}^- y)} \int_{-\infty}^{\infty} \frac{e^{-ik_{sz}z}}{k_s^{(-)2} - n_o^2 k_o^2}
$$

$$
\cdot \frac{\sin(k_{sz} + \beta_z - k_{iz})d}{k_{sz} + \beta_z - k_{iz}} dk_{sz}
$$

$$
+ e^{-i\phi} e^{-i(k_{sx}^+ x + k_{sy}^+ y)} \int_{-\infty}^{\infty} \frac{e^{-ik_{sz}z}}{k_s^{(+)2} - n_o^2 k_o^2}
$$

$$
\cdot \frac{\sin(k_{sz} - \beta_z - k_{iz})d}{k_{sz} - \beta_z - k_{iz}} dk_{sz} \Bigg] \qquad (2.6\text{-}16)
$$

The superscripts $+$ and $-$ that are attached to the components of $\mathbf{k}_s$ indicate that the values corresponding to the $+$ or $-$ sign in (2.6-14) must be substituted. The x and y components of $\mathbf{k}_s$ are thus determined by (2.6-14), while its z component is still arbitrary.

The integrals in (2.6-16) can easily be solved by contour integration. The integration path follows the real k_{sz}-axis from $-\infty$ to $+\infty$. Each integral has two poles lying on the path of integration. However, these poles can be moved off the real axis into the complex plane by a physical argument. We have always assumed that the index of refraction is a real quantity. This, however, is only an approximation. All real materials attenuate any electromagnetic wave traveling in them. We can drop the assumption of a perfectly lossless dielectric medium to our advantage. Assume that a plane wave is traveling in a perfect medium along the z-axis. With our choice of time-dependence (2.2-1), we can express this plane wave by the equation

$$
\psi = \psi_o e^{-in_o k_o z} \qquad (2.6\text{-}17)
$$

In a lossy dielectric material, the wave will be attenuated. We can describe an attenuated wave by choosing the complex dielectric constant

$$
n_o = n_r - in \qquad (2.6\text{-}18)
$$

For positive values of n_r and n_i, (2.6-17) describes a plane wave traveling along the positive z-axis and decaying exponentially according to the law

$$
\psi = e^{-n_i k_o z} \psi_o e^{-in_r k_o z} \qquad (2.6\text{-}19)
$$

This discussion shows us how to modify Equation (2.6-16) to allow for a slightly lossy dielectric. At the end of our calculation, we allow the imaginary part of the dielectric constant to return to zero.

Using n_o of (2.6-18) in (2.6-16) moves the poles of the integrand off the real axis. The integration path no longer passes through any poles, so that the integrals become well behaved and are easy to solve. Omitting the superscripts $+$ and $-$ for the moment from the k's, we can solve both integrals simultaneously. The critical factor is the denominator

$$D = k_s^2 - n_o^2 k_o^2 = k_{sz}^2 - (n_o^2 k_o^2 - k_{sx}^2 - k_{sy}^2) = (k_{sz} - k'_{sz})(k_{sz} + k'_{sz}) \qquad (2.6\text{-}20)$$

with

$$k'_{sz} = \sqrt{n_o^2 k_o^2 - k_{sx}^2 - k_{sy}^2} \qquad (2.6\text{-}21)$$

We assume that $n_i \ll 1$, and can write to a good approximation

$$k'_{sz} = \sqrt{n_r^2 k_o^2 - k_{sx}^2 - k_{sy}^2} - \frac{i n_i n_r k_o^2}{\sqrt{n_r^2 k_o^2 - k_{sx}^2 - k_{sy}^2}} \qquad (2.6\text{-}22)$$

We can take n_i as small as we like. In the limit $n_i \rightarrow 0$, (2.6-22) becomes exact. The integration path and the position of the poles are indicated by Figure 2.6.2. The dotted curves are supposed to indicate two infinitely large semicircles that close the integration path along the real axis.

We are interested only in values of z outside the three-dimensional diffraction grating. It is sufficient, therefore, if we limit ourselves to the case

$$|z| > d \qquad (2.6\text{-}23)$$

In the region indicated by (2.6-23), the exponential factor $\exp(-ik_{sz}z)$ determines the convergence or divergence behavior of the integrand. It is necessary to distinguish two different cases. We start by assuming that z is positive.

$$z > d \qquad (2.6\text{-}24)$$

In that case, the exponential factor vanishes on the curve C_2 shown in Figure 2.6.2. Adding the integration path along C_2 does not change the value of the

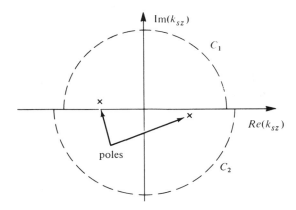

Figure 2.6.2   Integration path in the complex k_{sz} plane.

integral. Since there are no other poles than the one indicated, we can use Cauchy's integral theorem to contract the path of integration to an infinitesimal circle around the pole and obtain the value of the integral from the residue. For negative values of z

$$z < -d \qquad (2.6\text{-}25)$$

The integration path along C_1 gives no contribution for similar reasons. Carrying out the integrals in (2.6-16) by contour integration leads to the result

$$\psi_s = -i\eta n_o k_o^2 A \left\{ \frac{e^{i\phi}}{k_{sz}} e^{-i\mathbf{k}_s^- \cdot \mathbf{r}} \frac{\sin(k_{sz} + \beta_z - k_{iz})d}{k_{sz} + \beta_z - k_{iz}} \right.$$
$$\left. + \frac{e^{-i\phi}}{k_{sz}} e^{-i\mathbf{k}_s^+ \cdot \mathbf{r}} \frac{\sin(k_{sz} - \beta_z - k_{iz})d}{k_{sz} - \beta_z - k_{iz}} \right\} \qquad (2.6\text{-}26)$$

The value of k_{sz} appearing in (2.6-26) is given by (2.6-21). The prime was dropped for convenience. Equation (2.6-26) is the solution of the Bragg diffraction problem to first order of perturbation theory.

Equation (2.6-26) reveals several interesting properties of Bragg scattering that we now proceed to explore.

The Bragg condition stated in (2.6-14) is not precisely satisfied for the z component of this vector equation. This fact becomes apparent by looking at the sine factors in (2.6-26). The relation for the x and y components of (2.6-14) is exact because we assumed that the grating extended to infinity in x and y direction. By letting $d \to \infty$, we obtain delta functions for the sine factors (see [2.3-8]), and the z component of the Bragg condition also becomes precise. For finite values of d, there is some slight leeway to the precision with which the z component of (2.6-14) needs to be satisfied. The sine function terms drop off, as shown by Figure 2.3.1. The peak grows ever taller as d increases and reaches infinity in the limit $d \to \infty$. The width of the peak shrinks as its height grows. In most cases, we have $d \gg \lambda$, and the z component of the Bragg conditions is satisfied to a very good approximation.

However, we have found that, in addition to the Bragg condition, Equation (2.6-21) must also be satisfied. This equation can be rewritten as

$$n_o^2 k_o^2 = k_s^2 \qquad (2.6\text{-}27)$$

This relation is important but hardly surprising. It simply states that the scattered wave propagates with the same velocity as the incident wave and that both propagate with the phase velocity of plane waves in the medium with refractive index n_o. Equation (2.6-27) and the x and y components of (2.6-14) determine all the components of $\mathbf{k}_s$.

The scattered wave is a plane wave. Its amplitude is determined by the accuracy to which the component of (2.6-14) is satisfied via the sine functions appearing in (2.6-26). If the Bragg condition (2.6-14) is satisfied even for the z components, the scattered wave reaches its peak amplitude.

$$\psi_s = - i\eta n_o \frac{k_o^2 d}{k_{sz}} A e^{-i\mathbf{k}^\pm \cdot \mathbf{r}} e^{\mp i\phi} \qquad (2.6\text{-}28)$$

This simple equation shows the factors that determine the amplitude of the scattered wave. It is of course proportional to the amplitude of the incident wave. It is also proportional to the index change η and to the thickness d of the three-dimensional diffraction grating. An interesting feature is its dependence on the inverse value of k_{sz}. As the z component of the propagation vector decreases, the amplitude grows. We can understand this behavior when we realize that the wave travels a longer distance inside the grating when its propagation vector is inclined parallel to the grating boundary. The longer the scattered wave remains inside of the grating, the more it interacts with the incident wave, so that more power can be transferred from one wave to the other. In the limit $k_{sz} = 0$, our perturbation theory breaks down. The perturbation theory does not allow for a decrease of the amplitude of the incident wave. The exact theory of Section 2.7 accounts for the power loss from the incident wave by describing its decreasing amplitude.

The Bragg condition (2.6-14) and Equation (2.6-27) show how the directions of the incident and scattered wave are related. Both signs appearing in the equation are allowed, so that we obtain the two diagrams, Figures 2.6.3 and 2.6.4. Both figures represent isosceles triangles. The $\boldsymbol{\beta}$ vector is directed perpendicular to the planes of constant index of refraction. The incident

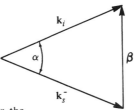

Figure 2.6.3   Illustration of the Bragg condition for the case that the minus sign in (2.6-14) is applicable.

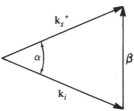

Figure 2.6.4   Illustration of the Bragg condition for the case that the plus sign in (2.6-14) is applicable.

wave is scattered off these planes. The two figures simply describe whether the incident wave impinges on the diffraction grating from above or below. The remarkable thing about Bragg diffraction is that scattering takes place just as light reflection does from a plane dielectric interface. If scattering takes place at all, it occurs specularly—according to the law of light reflection from mirrors. However, the condition of specular reflection is not at all sufficient for a scattered wave to appear. Scattering occurs only if the incident wave impinges on the reflecting planes of constant index under the angle $\frac{\alpha}{2}$.

This angle can easily be obtained by looking at the figures. It is obtained from the equation

$$\sin\frac{\alpha}{2} = \frac{\beta}{2k_i} = \frac{\lambda}{2D} \qquad (2.6\text{-}29)$$

This equation is also known as the Bragg condition. Equation (2.6-2) was used to obtain its right-hand side. We have indicated before that the Bragg condition is satisfied rigorously only when the thickness of the diffraction grating becomes infinitely large. For finite grating thickness, the Bragg condition is only approximately satisfied. We can use this fact to rederive the grating lobes of the two-dimensional phase grating from the laws of Bragg scattering. Let us assume that d is comparatively small. The Bragg condition is then not satisfied precisely. We assume that the β vector is located parallel to the side of the diffraction grating with the incident wave impinging on the grating perpendicularly, as shown in Figure 2.6.5. The Bragg condition holding now only for the x and y but not for the z component is shown in Figure 2.6.6. The length of the vectors $\mathbf{k}_s$ and $\mathbf{k}_i$ must of course still be

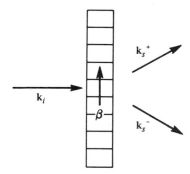

Figure 2.6.5   Thin Bragg diffraction grating at normal incidence.

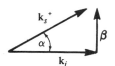

Figure 2.6.6   For a thin Bragg diffraction grating, the Bragg condition need not be satisfied exactly.

the same, but the vector diagram does not quite match up. For small angles, (2.6-29) still holds approximately, so that we obtain

$$\alpha = \frac{\lambda}{D} \qquad (2.6\text{-}30)$$

This equation is identical with the first order grating lobes obtained in Section 2.5, Equation (2.5-25). The quantity D was called a in Section 2.5.

The comparison of (2.6-30) with (2.5-25) brings out an important general point. Our derivation of the laws of Bragg diffraction based on first order perturbation theory allowed us to obtain only the first order of interference of the three-dimensional diffraction grating. This situation is completely analogous to the first order approximation of the theory of the two-dimensional phase grating, whose approximate theory led to the first order grating lobes (2.5-24). An exact theory of Bragg diffraction shows that there are also higher orders of interference, so that the exact Bragg condition must be written

$$\mathbf{k}_s = \mathbf{k}_i \pm m\boldsymbol{\beta} \qquad (2.6\text{-}31)$$

The integer m is the order of interference of the diffraction grating. The requirement of specular reflection still holds. The only difference is that there are now more possible directions for the incident beam to produce a grating response. However, in most practical applications, such as holograms or light scattering from sound waves in liquids, the change of refractive index η is so slight that only the first grating order can be strongly observed.

Since the theory of Bragg diffraction applies to light scattered by sound waves in liquids and solids, it is not surprising that we can obtain our results from a quantum theory of light scattering by phonons. This scattering process is also known as Brillouin scattering.[10] The quantum theory leads to Equation (2.6-31) by the requirement that the momentum is conserved between the interacting photons and phonons. The process of Brillouin scattering is a special case of Raman scattering.[10] Either the incident light photon emits a phonon and reradiates the remaining energy as another photon or a phonon is absorbed as the incident light photon is converted to the scattered photon. The momentum balance for scattering involving the emission or absorption of m phonons leads to Equation (2.6-31).

The requirement of conservation of energy leads to the equation

$$\omega_s = \omega_i \pm m\omega_{ph} \qquad (2.6\text{-}32)$$

with ω_s, ω_i, and ω_{ph} being the radian frequencies of the scattered photon, the incident photon and the phonon. The frequency shift that occurs for light scattering from sound waves comes about as a result of the Doppler effect. Quantum mechanically, it is a consequence of conservation of energy between the participating particles.

Our discussion of Bragg diffraction so far was carried out under the assumption of a three-dimensional phase grating. However, we can immediately extend the range of validity of our analysis to gratings with space-variable transmission loss instead of varying refractive index. We have seen that a complex refractive index of the form (2.6-18) describes a lossy medium. By using

$$n = n_o + i\xi \cos \boldsymbol{\beta} \cdot \mathbf{r} \qquad (2.6\text{-}33)$$

instead of (2.6-1), we are describing a three-dimensional grating with periodically varying transmission loss. The fact that (2.6-33) seems to describe a medium with alternating regions of loss and gain can easily be adjusted by providing a fixed amount of constant loss by proper choice of a slightly imaginary part attached yo n_o.

All our equations that we derived so far hold immediately for the case of the grating with variable loss if we replace everywhere

$$\eta = i\xi \qquad (2.6\text{-}34)$$

Another useful extension of our theory results by generalizing the shape of the index distribution from a simple sinusoid to an arbitrary but only slightly varying form. We use

$$n = n_o + \sum_i \eta_i \cos (\boldsymbol{\beta}_i \cdot \mathbf{r} + \phi_i) \qquad (2.6\text{-}35)$$

instead of (2.6-1) or (2.6-33). Because of the assumption

$$\eta_i \ll 1 \qquad (2.6\text{-}36)$$

there is no mixing of terms when we form n^2. The scattered wave resulting from this general index distribution is obtained immediately by attaching indices to η and φ appearing in (2.6-26) and summing the resulting expression. However, we must keep in mind that the x and y components of the Bragg condition (2.6-14) must be satisfied, so that the sum must be limited to those terms for which this is the case.

In its extended form, we can apply our theory to multilayered dielectrics that are often used in mirrors or filters. We simply expand the actual index distribution in a Fourier series. For any given incident wave, no more than one term in the series, can satisfy the Bragg condition. We obtain immediately the required layer thickness for any desired angle of incidence from (2.6-14). The amplitude of the reflected wave, however, can be obtained with reasonable accuracy only for mirrors with low reflectivity because of the limitations inherent in the perturbation theory. The most common case involves mirrors to be used for light incident normally to the stratification of the medium. The most efficient design is one utilizing the fundamental component in the series expansion (2.6-35). Under these conditions only the z components of the vectors in (2.6-14) exist, and these are equal (except for sign) to the magnitude of the vectors. We thus obtain the optimum condition for a periodic multilayered mirror from (2.6-14).

$$-k_s = k_i - \beta \tag{2.6-37}$$

Using (2.6-2) and $k_s = k_i = 2\pi/\lambda$, we obtain the required period length D for maximum reflection from the multilayered periodic structure

$$D = \frac{\lambda}{2} \tag{2.6-38}$$

2.7 BRAGG DIFFRACTION: COUPLED WAVE THEORY*

The treatment of Bragg diffraction in the last section yielded most of the important results. It allowed us to study the grating lobes of thin diffraction gratings and also yielded the important Bragg condition in the general vector form (2.6-14) and in the familiar form (2.6-29). The only shortcoming of the perturbation theory is its inability to predict the amplitudes of the incident and scattered waves in case of strong coupling or long interaction length.

The present section completes the discussion of Bragg diffraction by providing a coupled wave theory for the interaction between the incident wave and the diffracted wave.

Our starting points are Equation (2.6-4) and the reduced wave equation. We use the coordinate system of Figure 2.6.1 so that the planes $z = \pm d$ coincide with the boundaries of the region of the slab with periodic index variations. Using our knowledge gained from the perturbation theory that there is a definite diffracted plane wave, we assume that the field is of the form

$$\psi = A(z)e^{-ik_i \cdot \mathbf{r}} + B(z)e^{-ik_s \cdot \mathbf{r}} \tag{2.7-1}$$

The amplitude A of the incident wave and the amplitude B of the diffracted wave are both considered to be functions of z. The wave interaction is not strong enough to change the amplitudes very much over the distance of one wavelength. This important property of moderately strong coupling allows us to neglect the second derivatives of the amplitudes compared to the remaining terms. We shall use this device repeatedly in this book. It is an important mathematical approximation that leads to first order coupled differential equations for the wave amplitudes.

In order to be able to handle Bragg diffraction in media with periodic loss variations, we need to assume that the average value n_o of the refractive

---

* The coupled wave theory of Bragg diffraction was first derived by H. Kogelnik.[13]

index contains a loss term. We thus use (2.6-18) with the additional assumption

$$n_i \ll n_r \tag{2.7-2}$$

we also assume

$$|\eta| \ll n_r \tag{2.7-3}$$

and write (2.6-4) in the following approximate way

$$n^2 \approx n_r^2 - 2in_r n_i + 2\eta n_r \cos \boldsymbol{\beta} \cdot \mathbf{r} \tag{2.7-4}$$

The assumption (2.7-2) implies that the loss of the medium is weak enough so that the wave amplitude does not change very much over the distance of one wavelength. The relation (2.7-4) means likewise that the loss modulation or the change in refractive index is slight. These assumptions are always justified in cases of practical interest. If the losses of the medium are so high that the approximation (2.7-2) or (2.7-3) no longer holds, the incident wave is absorbed so strongly that Bragg diffraction cannot be observed.

Neglecting the second derivative of the amplitude, we obtain the approximation

$$\nabla^2(Ae^{-i\mathbf{k}_i \cdot \mathbf{r}}) = -\left(k_i^2 A + 2ik_{iz}\frac{\partial A}{\partial z}\right)e^{-i\mathbf{k}_i \cdot \mathbf{r}} \tag{2.7-5}$$

k_{iz} is the z component of the vector $\mathbf{k}_i$. The magnitude of the propagation constants is given by

$$k_i^2 = k_s^2 = n_r^2 k_o^2 \tag{2.7-6}$$

With the help of the equations and conditions stated in this section, we obtain by substitution of (2.7-1) into the reduced wave equation (2.6-5)

$$\left[-2ik_{iz}\frac{\partial A}{\partial z} - 2in_r n_i k_o^2 A + \eta n_r k_o^2 Be^{i(\mathbf{k}_i - \mathbf{k}_s - \boldsymbol{\beta}) \cdot \mathbf{r}}\right]e^{-i\mathbf{k}_i \cdot \mathbf{r}}$$

$$+ \left[-2ik_{sz}\frac{\partial B}{\partial z} - 2in_r n_i k_o^2 B + \eta n_r k_o^2 Ae^{i(\mathbf{k}_s - \mathbf{k}_i + \boldsymbol{\beta}) \cdot \mathbf{r}}\right]e^{-i\mathbf{k}_s \cdot \mathbf{r}}$$

$$+ \eta n_r k_o^2[Ae^{-i(\mathbf{k}_i + \boldsymbol{\beta}) \cdot \mathbf{r}} + Be^{-i(\mathbf{k}_s - \boldsymbol{\beta}) \cdot \mathbf{r}}] = 0 \tag{2.7-7}$$

We use our knowledge that the scattered and incident wave propagation vectors must satisfy the Bragg condition (2.6-14). The exponential factors inside the first two brackets are thus equal to unity. We can now follow the same procedure that was used to solve (2.6-11). We multiply (2.7-7) with $\exp(i\mathbf{k}_i \cdot \mathbf{r})$ and integrate the resulting equation over the entire x-y plane. In z direction, we let the integral extend over many periods of the oscillation but keep the integration interval short compared to the distance over which

$A(z)$ and $B(z)$ change appreciably. Except for the first term, all other terms in (2.7-7) vanish approximately and we obtain

$$\frac{\partial A}{\partial z} + \frac{n_r n_i k_o^2}{k_{iz}} A = \frac{\eta n_r k_o^2}{2ik_{iz}} B \qquad (2.7\text{-}8)$$

In a similar manner, we obtain by multiplication with $\exp(i\mathbf{k}_s \cdot \mathbf{r})$ and integration

$$\frac{\partial B}{\partial z} + \frac{n_r n_i k_o^2}{k_{sz}} B = \frac{\eta n_r k_o^2}{2ik_{sz}} A \qquad (2.7\text{-}9)$$

In the absence of the periodic index variation, $\eta = 0$, the equations (2.7-8) and (2.7-9) are uncoupled. Equation (2.7-8) has the solution

$$A = A_o e^{-\alpha_i(z+d)} \qquad (2.7\text{-}10)$$

with

$$\alpha_i = \frac{n_r n_i k_o^2}{k_{iz}} \qquad (2.7\text{-}11)$$

Equation (2.7-9) has the solution ($\eta = 0$)

$$B = B_o e^{-\alpha_s(z+d)} \qquad (2.7\text{-}12)$$

with

$$\alpha_s = \frac{n_r n_i k_o^2}{k_{sz}} \qquad (2.7\text{-}13)$$

In the absence of coupling, both waves travel inside the lossy medium with decreasing amplitudes. It is noteworthy that the amplitude decrease is not uniform in the direction of wave propagation $\mathbf{k}$, but the amplitudes are uniform along the planes $z = \text{const}$. If a plane wave travels inside a homogeneous lossy medium, its amplitude decreases with the factor $\exp\left(-\alpha\frac{\mathbf{k}}{|\mathbf{k}|}\cdot\mathbf{r}\right)$. Our situation is different, however. We have assumed that a plane wave is incident from a lossless medium and enters a lossy medium at the plane $z = -d$. The amplitude A_o of this wave is constant outside the lossy medium, and it decreases as a function of z inside the lossy medium. At $z = -d$, we have the boundary condition that the amplitude $A(z)$ must equal the amplitude of the incident wave A_o. The wave amplitude (2.7-10) satisfies this boundary condition, and is thus the correct solution of our problem. There is no reflection or refraction at the interface because of our assumption that the loss is sufficiently small to justify the condition (2.7-2) that allows us to neglect the second derivative of $A(z)$ with respect to z.

Introducing the abbreviation

$$\kappa = \frac{1}{2}\eta n_r k_o \tag{2.7-14}$$

we can write the system of coupled equations in the form

$$\frac{\partial A}{\partial z} + \alpha_i A = -i\frac{k_o}{k_{iz}}\kappa B \tag{2.7-15}$$

and

$$\frac{\partial B}{\partial z} + \alpha_s B = -i\frac{k_o}{k_{sz}}\kappa A \tag{2.7-16}$$

Differentiation of (2.7-15) allows us to eliminate B, with the result

$$\frac{\partial^2 A}{\partial z^2} + (\alpha_i + \alpha_s)\frac{\partial A}{\partial z} + \left(\alpha_i\alpha_s + \frac{k_o^2\kappa^2}{k_{iz}k_{sz}}\right)A = 0 \tag{2.7-17}$$

Because α_i, α_s, and κ are all small, we now must keep the second derivative of A. We use the trial solution

$$A = e^{\sigma z} \tag{2.7-18}$$

and obtain from (2.7-17)

$$\sigma^2 + (\alpha_i + \alpha_s)\sigma + \left(\alpha_i\alpha_s + \frac{k_o^2\kappa^2}{k_{iz}k_{sz}}\right) = 0 \tag{2.7-19}$$

This quadratic equation has the solution

$$\sigma_\pm = -\frac{\alpha_i + \alpha_s}{2} \pm \frac{1}{2}\sqrt{(\alpha_i - \alpha_s)^2 - 4\frac{k_o^2\kappa^2}{k_{iz}k_{sz}}} \tag{2.7-20}$$

We choose to call the expression with the positive sign σ_+ and with the negative sign σ_-. The most general solution of the differential equation (2.7-17) is thus

$$A = ae^{\sigma_+(z+d)} + be^{\sigma_-(z+d)} \tag{2.7-21}$$

From (2.7-15) we obtain

$$B = i\frac{k_{iz}}{k_o\kappa}\left[(\sigma_+ + \alpha_i)ae^{\sigma_+(z+d)} + (\sigma_- + \alpha_i)be^{\sigma_-(z+d)}\right] \tag{2.7-22}$$

We have now found the general solution of our problem of Bragg diffraction. It remains to study this solution to obtain some information about several cases of special interest.

There are two possible ways in which Bragg diffraction can be observed. Depending on the orientation of the β vector, the Bragg condition (2.6-14) is satisfied either for a reflected scattered wave or for a transmitted scattered wave. We thus must consider the case either that the scattered wave emerges on the same side of the scattering material from which the primary wave was incident or that it emerges on the opposite side.

First, we study the case of a reflected scattered wave. In this case, there is no scattered wave at the opposite side of the scattering layer of thickness $2d$ (Figure 2.6.1). The initial condition thus requires $B = 0$ at $z = d$ and $A = A_o$ at $z = -d$. These two conditions allow us to determine the amplitude coefficients a and b, with the result

$$a = \frac{(\sigma_- + \alpha_i)e^{-(\sigma_+ - \sigma_-)d}}{\sigma_- e^{-(\sigma_+ - \sigma_-)d} - \sigma_+ e^{(\sigma_+ - \sigma_-)d} - 2\alpha_i \sinh(\sigma_+ - \sigma_-)d} A_o \quad (2.7\text{-}23)$$

and

$$b = \frac{-(\sigma_+ + \alpha_i)e^{(\sigma_+ - \sigma_-)d}}{\sigma_- e^{-(\sigma_+ - \sigma_-)d} - \sigma_+ e^{(\sigma_+ - \sigma_-)d} - 2\alpha_i \sinh(\sigma_+ - \sigma_-)d} A_o \quad (2.7\text{-}24)$$

In the opposite case (transmission case), when the scattered wave emerges on the opposite side from the incident wave, we have the condition $B = 0$ and $A = A_o$ at $z = -d_o$. The amplitude coefficients become

$$a = \frac{\sigma_- + \alpha_i}{\sigma_- - \sigma_+} A_o \quad (2.7\text{-}25)$$

and

$$b = -\frac{\sigma_+ + \alpha_i}{\sigma_- - \sigma_+} A_o \quad (2.7\text{-}26)$$

The discussion of Bragg diffraction must next consider the case of a true phase grating and also the case that the grating is produced by periodic loss striations. In both cases we must distinguish between forward and backward scattering.

If the periodic changes of the refractive index are real (phase grating), we also assume that the background loss is negligible and set

$$\alpha_i = \alpha_s = 0 \quad (2.7\text{-}27)$$

From (2.7-20), we obtain in that case

$$\sigma_\pm = \pm i\rho \quad \text{(transmission case)} \quad (2.7\text{-}28)$$

with

$$\rho = \frac{k_o \kappa}{\sqrt{k_{iz} k_{sz}}} \quad (2.7\text{-}29)$$

For the transmission case (the scattered wave appears on the opposite side of the slab of scattering material), we obtain from (2.7-21), (2.7-22), (2.7-25), and (2.7-26)

$$A(z) = A_o \cos \rho(z+d) \quad \text{for} \quad -d \le z \le d \tag{2.7-30}$$

$$B(z) = -i \sqrt{\frac{k_{iz}}{k_{sz}}} A_o \sin \rho(z+d) \quad \text{for} \quad -d \le z \le d \tag{2.7-31}$$

This simple result shows clearly how the two waves exchange their power periodically. If

$$2\rho d = (2n+1)\frac{\pi}{2} \quad \text{with} \quad n = 0,1,2, \ldots \tag{2.7-32}$$

all the power of the incident wave is converted to the scattered wave. We thus have a result that could not have been obtained from the perturbation theory: Bragg diffraction can be 100 percent efficient. The amount of power conversion depends on the coupling strength and the thickness of the slab of material. If $2\rho d \ll 1$ we obtain the result that

$$B(d) = -2id \sqrt{\frac{k_{iz}}{k_{sz}}} \rho A_o = -i\frac{n_r \eta k_o^2}{k_{sz}} A_o d \tag{2.7-33}$$

The amplitude at $z = d$ thus agrees with the result (2.6-28) obtained from the perturbation theory.

In the reflection case (the scattered wave appears on the same side as the incident wave), the incident and scattered waves travel in opposite directions. This means that the z components of their propagation vectors have opposite signs, so that

$$k_{iz}k_{sz} < 0 \tag{2.7-34}$$

The parameter ρ of (2.7-29) is now imaginary, and it is more convenient to write

$$\sigma_\pm = \pm|\rho| \quad \text{(reflection case)} \tag{2.7-35}$$

The field amplitudes follow again from (2.7-21) and (2.7-22), but, in this case, with the help of (2.7-23) and (2.7-24)

$$A(z) = \frac{\cosh|\rho|(z-d)}{\cosh 2|\rho|d} A_o \quad \text{for} \quad -d \le z \le d \tag{2.7-36}$$

and

$$B(z) = i \sqrt{\left|\frac{k_{iz}}{k_{sz}}\right|} \frac{\sinh|\rho|(z-d)}{\cosh 2|\rho|d} A_o \quad -d \le z \le d \tag{2.7-37}$$

For $2|\rho|d \ll 1$, we obtain again from (2.7-37) for $B(-d)$ the result (2.6-28) of perturbation theory. There is a very interesting difference between the transmission and reflection cases. In the transmission case, the energy can oscillate between the two waves. Complete conversion can occur if the thickness of the scattering material is properly chosen. In the reflection case, on the other hand, there is no oscillation of the energy between the two waves. Complete energy conversion from the incident to the scattered wave can occur if the scattering medium is sufficiently thick. An increase in the thickness of the slab has no further effect. The adjustment of the thickness of the scattering material is not at all critical in this case. As long as the material is thick enough, complete conversion will always result. A multilayered dielectric mirror is an example for this case. In the transmission case, it is necessary to adjust the thickness of the material carefully in order to obtain complete conversion.

Next, we consider the case of a scattering medium with periodic loss variations (amplitude grating). We can now no longer assume that the loss constants of the waves vanish, since the imaginary part of the refractive index (2.6-1) must not become positive because there can be no gain in the medium. We must require

$$n_i \geq |\eta| \tag{2.7-38}$$

with η now being an imaginary quantity, so that

$$\kappa^2 < 0 \tag{2.7-39}$$

In the transmission case, we have

$$k_{iz}k_{sz} > 0 \tag{2.7-40}$$

so that

$$\sigma_+ = -\alpha + \gamma \tag{2.7-41}$$

and

$$\sigma_- = -\alpha - \gamma \tag{2.7-42}$$

with

$$\alpha = \frac{1}{2}(\alpha_i + \alpha_s) \tag{2.7-43}$$

and

$$\gamma = \frac{1}{2}\sqrt{(\alpha_i - \alpha_s)^2 + 4\left|\frac{k_o^2 \kappa^2}{k_{iz}k_{sz}}\right|} \tag{2.7-44}$$

we obtain the wave amplitudes for the transmission case in a lossy periodic medium (for $-d \leq z \leq d$)

$$A(z) = e^{-\alpha(z+d)} \left[\cosh \gamma(z+d) - \frac{\alpha_i - \alpha_s}{2\gamma} \sinh \gamma(z+d) \right] A_o \qquad (2.7\text{-}45)$$

$$B(z) = \frac{ik_o \kappa^* A_o}{\gamma k_{sz}} e^{-\alpha(z+d)} \sinh \gamma(z+d) \qquad (2.7\text{-}46)$$

This result is much more complicated than the case of the lossless phase grating. Total exchange of energy is of course no longer possible, since both waves suffer loss as they travel through the scattering medium. In the case of forward scattering at normal incidence, $k_{iz} = k_{sz}$, we also have $\alpha_i = \alpha_s = \alpha$, so that we obtain from (2.7-44), (2.7-11), and (2.7-14)

$$\gamma = \frac{|\eta|}{2n_i} \alpha \qquad (2.7\text{-}47)$$

and the expressions for the wave amplitudes simplify to the following form (for $-d \le z \le d$)

$$A(z) = A_o e^{-\alpha(z+d)} \cosh \left[\frac{1}{2} \frac{|\eta|}{n_i} \alpha(z+d) \right] \qquad (2.7\text{-}48)$$

and

$$B(z) = A_o e^{-\alpha(z+d)} \sinh \left[\frac{1}{2} \frac{|\eta|}{n_i} \alpha(z+d) \right] \qquad (2.7\text{-}49)$$

It is interesting to observe that the oscillatory exchange of power that occurs in the transmission case of the phase grating does not occur for the lossy grating. The amplitude of the scattered wave at $z = d$ is

$$B(d) = A_o e^{-2\alpha d} \sinh \left(\frac{|\eta|}{n_i} \alpha d \right) \qquad (2.7\text{-}50)$$

The constants η and n_i must obey (2.7-38). The best possible condition for a lossy grating is therefore

$$|\eta| = n_i \qquad (2.7\text{-}51)$$

and the amplitude of the forward scattered wave is

$$B(d) = \frac{1}{2} A_o (e^{-\alpha d} - e^{-3\alpha d}) \qquad (2.7\text{-}52)$$

The maximum amplitude is obtained for

$$2\alpha d = \ln 3 \qquad (2.7\text{-}53)$$

The maximum forward scattered amplitude at this optimum thickness is

$$B_{max}(d) = \frac{A_o}{3^{3/2}} = 0.192 \, A_o \qquad (2.7\text{-}54)$$

The power carried by the scattered wave is proportional to $|B|^2$. A grating with periodic loss variation obeying the optimum condition (2.7-51) can convert no more than 3.7 percent of the power of the incident wave to forward scattered power. It does of course not make sense to operate a volume grating with direct forward scattering, since the incident and scattered waves would not be distinguishable. Our calculation served the purpose of establishing the optimum value of near forward scattering with a lossy grating. It is interesting to note that for the same value of $\alpha_i = \alpha$ and for the length of material given by (2.7-53) the amplitude of the incident wave after traversing the material with the same value of n_i but with $\eta = 0$ would be

$$A(d) = A_o e^{-2\alpha d} = \frac{1}{3} A_o \qquad (2.7\text{-}55)$$

Next, we consider the reflection case of the lossy grating.
In the reflection case, (2.7-34) holds, so that we obtain

$$\sigma_+ = -\alpha + \delta \qquad (2.7\text{-}56)$$

and

$$\sigma_- = -\alpha - \delta \qquad (2.7\text{-}57)$$

with

$$\delta = \frac{1}{2}\sqrt{(\alpha_i - \alpha_s)^2 - 4\left|\frac{k_o^2 \kappa^2}{k_{iz} k_{sz}}\right|} \qquad (2.7\text{-}58)$$

and with α given by (2.7-43). The wave amplitudes follow from (2.7-21) and (2.7-22) with the help of (2.7-23) and (2.7-24) (for $-d \leq z \leq d$)

$$A(z) = A_o e^{-\alpha(z+d)} \frac{\delta \cosh \delta(z-d) - \frac{1}{2}(\alpha_i - \alpha_s)\sinh \delta(z-d)}{\delta \cosh 2\delta d + \frac{1}{2}(\alpha_i - \alpha_s) \sinh 2\delta d} \qquad (2.7\text{-}59)$$

$$B(z) = -\left|\frac{k_o \kappa}{k_{sz}}\right| A_o e^{-\alpha(z+d)} \frac{\sinh \delta(z-d)}{\delta \cosh 2\delta d + \frac{1}{2}(\alpha_i - \alpha_s) \sinh 2\delta d} \qquad (2.7\text{-}60)$$

The scattered wave appears now at $z = -d$. Assuming back scattering at normal incidence, $k_{iz} = -k_{sz}$ results in $\alpha_s = -\alpha_i$, and consequently $\alpha = 0$. Using also the optimum condition (2.7-51), we obtain from (2.7-58)

$$\delta = \frac{\sqrt{3}}{2}\alpha_i \tag{2.7-61}$$

Using

$$\frac{1}{2}(\alpha_i - \alpha_s) = \alpha_i \tag{2.7-62}$$

we obtain from (2.7-60)

$$B(-d) = \frac{iA_o}{2 + \sqrt{3}\coth(\sqrt{3}\alpha_i d)} \tag{2.7-63}$$

There is now no optimum thickness. The maximum value of $B(-d)$ is obtained as $2d$ tends toward infinity

$$B(-d)_{max} = \frac{iA_o}{2 + \sqrt{3}} = 0.268\, iA_o \tag{2.7-64}$$

The optimum power conversion efficiency in the reflection case is 7.2 percent. The lossy grating can have an optimum conversion efficiency for back reflection that is twice as high as the optimum conversion efficiency for forward scattering. However, the lossless phase grating is far more efficient than any grating with periodic loss variations, since it is capable of 100 percent conversion efficiency.

For forward (or nearly forward) scattering the $\boldsymbol{\beta}$ vector must be directed parallel to the planes $z = $ const. That means the planes of constant loss or constant refractive index must be oriented (nearly) perpendicular to the planes $z = $ const. For backward scattering, $\boldsymbol{\beta}$ must be directed (nearly) perpendicular to the planes $z = $ const so that the index striations are parallel to the planes $z = $ const.

3

GEOMETRICAL OPTICS

3.1 INTRODUCTION

One of the outstanding features of visible light is its short wavelength. There is electromagnetic radiation with wavelengths shorter than that of visible light. Ultraviolet radiation, X rays, and gamma rays all have wavelengths shorter than that of visible light. The region of visibility of electromagnetic radiation ranges from approximately 0.4 to 0.7 μ. Wavelengths longer than those of visible light are called infrared, microwaves, and radio waves. Often the infrared and ultraviolet regions of the electromagnetic spectrum are also considered to belong into the domain of optics. However, no matter how we define the optical region of the electromagnetic spectrum, we are always dealing with radiation whose wavelength is much shorter than any dimension which we commonly encounter in our daily lives. Many optical instruments consist of apparatus whose linear dimensions are much larger than that of the wavelength passing through them. In all cases, where the wavelength of light is much shorter than the dimensions of any obstacle that the light encounters, it is possible to utilize the short wavelength of light to solve the problem of light propagation in an approximate way. The approximation that is valid for short wavelength of light is known as geometrical optics. It allows us to treat the light propagation problem in a way far simpler than would be possible by solving Maxwell's equations or the wave equation. Geometrical optics is applicable for problems where light diffraction can be neglected.

It may be prudent to point out that geometrical optics is not limited to light propagation problems. Ray optics, as geometrical optics is also called, can be applied to all phenomena that are described by the wave equation and that satisfy the additional requirement that the wavelength is short compared to the dimensions of the apparatus through which it passes. Geometrical optics thus applies to many problems of sound propagation. Geometrical optics is very similar to the classical mechanics of a point particle. In fact, the relation between wave optics and ray optics is analogous to the relation between wave mechanics and ordinary mechanics.

We start out by deriving the equations of geometrical optics from the wave equation. Next, we derive the same equations from Fermat's principle and develop the Hamiltonian formalism of geometrical optics. Liouville's theorem ordinarily found in books on statistical mechanics is included since it has an interesting and important application to optics. Also included in this chapter is a quantum theory of ray optics showing that quantization of the equations of ray optics leads to the wave equation. An interesting relationship between the relativistic mechanics of a point particle and relativistic quantum mechanics shows that the so-called paraxial approximation of ray optics is equivalent to nonrelativistic mechanics of a point particle.

The Hamiltonian formulation of ray optics and the quantum theory of light rays are not absolutely essential for the understanding of the rest of the book. Readers less interested in these subjects may skip Sections 3.5 and 3.6.

3.2 DERIVATION OF RAY OPTICS FROM THE WAVE EQUATION*

In Chapter 1 we have derived the wave equation (1.3-6) from Maxwell's equations. It was pointed out that the wave equation is satisfied by every component of the electric and magnetic field vectors in a homogeneous medium. However, even in inhomogeneous media, we can use the wave equation to a good approximation provided the index of refraction of the medium changes only very slightly over distances comparable with the wavelength of the radiation. The wave equation describes not only electromagnetic phenomena but also a variety of other wave propagation problems. To a linear approximation, sound propagation in liquids, gases, and solids can be described by the wave equation. Taking the wave equation as our starting point of geometrical optics makes it clear that we are about to derive an approximate theory that applies not only to light but also to sound propagation. The vector nature of light is lost in the geometrical optics description. Light is treated as a scalar process. In terms of the photon language, this means that we neglect the properties of the photon spin.

Before we actually begin, let us immediately limit ourselves to a time harmonic process assuming that all field variables have a time dependence,

---

* Excellent descriptions of ray optics can be found in references 1, 15, and 16.

which in complex notation can be expressed by $\exp(i\omega t)$, so that the wave equation can be replaced by the reduced wave equation (2.2-2)

$$\nabla^2\psi + k^2\psi = 0 \tag{3.2-1}$$

with k being expressed in any of the following ways

$$k = \frac{2\pi}{\lambda} = \omega\sqrt{\varepsilon\mu_o} \tag{3.2-2}$$

Let us hasten to add that the reduced wave equation is actually not as much of a restriction on the full wave equation (1.3-6) as might be suspected, because we have seen in Section 1.3 that in a dispersive medium the wave equation makes physical sense only for a narrow band of frequencies, so that the phase velocity v of the radiation can be defined. Only in free space does the wave equation apply to all frequencies. However, the free space case is the least interesting for the purposes of geometrical optics.

In order to obtain the desired approximation, we assume that the wave function ψ is expressed in the following form

$$\psi = \psi_o(x,y,z)e^{-ik_oS(x,y,z)} \text{ with } k_o = \omega\sqrt{\varepsilon_o\mu_o} \tag{3.2-3}$$

This form of the wave function has the advantage that it separates the rapid variations caused by the short wavelength of the radiation field in free space from the much slower variations of its amplitude. Both ψ_o and S are functions which can be assumed to be slowly varying with respect to the wavelength of the radiation. Substitution into the reduced wave equation results in the following expression

$$k_o^2\left(\frac{k^2}{k_o^2} - \nabla S\cdot\nabla S\right)\psi_o - ik_o(2\nabla S\cdot\nabla\psi_o + \psi_o\nabla^2 S) + \nabla^2\psi_o = 0 \tag{3.2-4}$$

The terms in (3.2-4) are grouped in decreasing order of magnitude. Because the wavelength λ is so short, k_o must be considered as a very large quantity. We may furthermore assume that both ψ_o and S are real quantities. This means that the term that is multiplied by ik_o must vanish separately from the remaining two terms. Some authors[14] use complex ψ_o and S, arriving at the notion of complex rays. We do not use this formalism in this book but assume, instead, that (3.2-3) represents the decomposition of the complex number ψ into amplitude ψ_o and phase k_oS. For short wavelength, the first term in (3.2-4) is much larger than the last term, so that we may write to a good approximation

$$(\nabla S)^2 = n^2 \tag{3.2-5}$$

The index of refraction is defined as usual by the relation

$$n^2 = \frac{k^2}{k_o^2} = \frac{\varepsilon}{\varepsilon_o} \tag{3.2-6}$$

Equation (3.2-5) is known as the eikonal equation. It determines the function S, which allows us to define the surfaces of constant phase via the equation

$$S(x, y, z) = \text{const} \qquad (3.2\text{-}7)$$

The surfaces of constant phase define the shape of the radiation field, so that the eikonal equation determines the wave propagation in the geometrical optics approximation.

As the alternate name "ray optics" implies, geometrical optics uses the notion of light rays to describe the propagation of the radiation field. The idea of light rays arises naturally if we watch how a thin pencil of light propagates in space. In most media, light is scattered continuously out of its path of propagation, making the trajectory of the light pencil visible. Everybody is familiar with the sight of a thin beam of light in a smoke-filled room. Intense beams of light become visible even in relatively clean air as sufficient power is scattered by dust particles to allow us to follow the trajectory of the beam. However, the notion of light rays has been generalized somewhat. It is intuitively obvious that if we place a second pencil of light close to an existing one, both will follow very nearly the same path if they started out parallel and in close proximity to each other. We can thus think of bundles of light rays tracing out the direction of propagation of a more extended light field. In terms of wave optics, we can think of a pencil of light as a plane wave that has somehow been truncated to fill only a very narrow region of space. To some approximation this can be done by passing an extended plane wave through a hole in an opaque screen. We have seen in Sections 2.3 and 2.4 that a plane wave does not continue in its narrow truncated form if it is passed through a very narrow hole. However, if the hole is large compared to the wavelength (an essential assumption for the entire field of geometrical optics), the light beam passing through the hole will spread only gradually by diffraction, and the notion of a truncated plane wave will remain approximately valid for some distance behind the hole. We know that the surfaces of constant phase are perpendicular to the direction of light propagation of a plane wave. We are thus led to define light rays as the orthogonal trajectories to the phase fronts of a light wave. If we know the surfaces of constant phase, we can construct the light rays by drawing lines perpendicular to the phase fronts. As the phase fronts curve in space, owing to changes in the index of refraction, so do the light rays.

It is desirable to be able to calculate the light rays directly without having to construct the phase fronts from the eikonal equation.

We use a fixed point of origin and draw a vector from there to all points on the light ray. If this vector $\mathbf{r}$ were known for all points along the ray, we would have a mathematical description of the light ray. Defining s as the distance measured along the light ray, we obtain the unit vector

$$\mathbf{u} = \frac{d\mathbf{r}}{ds} \qquad (3.2\text{-}8)$$

By definition, we require the unit vector **u**, which is tangential to the light ray, to be perpendicular to the phase fronts. From (3.2-7) we obtain a vector perpendicular to the phase fronts by taking the gradient of S

$$\mathbf{v} = \nabla S \tag{3.2-9}$$

The vectors **v** and **u** are required to be parallel. The magnitude of the vector **v** can be found from the eikonal equation (3.2-5)

$$|v| = n \tag{3.2-10}$$

This allows us immediately to formulate the relation

$$\mathbf{u} = \frac{\mathbf{v}}{n} \tag{3.2-11}$$

or in more detail

$$n\frac{d\mathbf{r}}{ds} = \nabla S \tag{3.2-12}$$

Differentiation with respect to s can be expressed as the product of the unit vector (3.2-8) with the ∇ operator

$$\frac{d}{ds} = \sum_i \frac{dx_i}{ds}\frac{\partial}{\partial x_i} = \frac{d\mathbf{r}}{ds}\cdot\nabla \tag{3.2-13}$$

In order to be able to derive the ray equation, we form the gradient of both sides of the eikonal equation (3.2-5)

$$2\nabla S\cdot\nabla\nabla S = 2n\nabla n \tag{3.2-14}$$

The product $\nabla\nabla$ defines a tensor operator. The relation (3.2-14) can be confirmed by using a vector component notation. Utilizing (3.2-12) and (3.2-13), we can rewrite (3.2-14) to obtain the following relation

$$\frac{d}{ds}\nabla S = \nabla n \tag{3.2-15}$$

Taking the derivative of (3.2-12) with respect to s and using (3.2-15) finally results in the desired ray equation

$$\frac{d}{ds}\left(n\frac{d\mathbf{r}}{ds}\right) = \nabla n \tag{3.2-16}$$

The ray equation and the eikonal equation are two alternate descriptions of geometrical optics. The ray equation is more convenient for determining the trajectory of light rays in inhomogeneous media. However, in its exact (exact in the framework of ray optics) form, Equation (3.2-16) is hard to solve. For many practical applications, we need an approximation of the ray equation. Many optical problems involve light rays that always travel very nearly parallel to the optical axis of the system. This does not mean that the rays may not depart considerably from this axis, but it does mean that the angle α which the light ray forms with the optical axis remains sufficiently small so that we can use the approximation

$$\left. \begin{array}{c} \tan \alpha \approx \sin \alpha \approx \alpha \\ \cos \alpha \approx 1 \end{array} \right\} \tag{3.2-17}$$

Making the optical axis of the system equal to the z-axis allows us to take

$$ds \approx dz \tag{3.2-18}$$

to the degree of accuracy implied by the approximation (3.2-17). The simplification resulting from (3.2-17) is known as the paraxial approximation. In this approximation the Equation (3.2-16) simplifies to the paraxial ray equation

$$\frac{d}{dz}\left(n \frac{d\mathbf{r}}{dz} \right) = \nabla n \tag{3.2-19}$$

The paraxial ray equation will be used in Chapter 7 to solve ray propagation problems in optical waveguides with parabolic index distribution.

The ray equation can be obtained from an important variational principle. Before turning to this alternate derivation, we investigate the boundary condition that light rays obey at the interface between two media with different dielectric constants.

3.3 BOUNDARY CONDITIONS FOR LIGHT RAYS

The boundary condition for light rays can be obtained from (3.2-12). Using a result of the theory of line integrals,[17] we obtain immediately

$$\oint \nabla S \cdot d\mathbf{l} = 0 \tag{3.3-1}$$

The integral is extended over a closed curve with $d\mathbf{l}$ being directed tangentially to the curve. Applying (3.3-1) to (3.2-12), we obtain the important relation

$$\oint n \frac{d\mathbf{r}}{ds} \cdot d\mathbf{l} = 0 \tag{3.3-2}$$

Equation (3.3-2) can be expressed in slightly different form as follows

$$\int_{P_1}^{P_2} n \frac{d\mathbf{r}}{ds} \cdot d\mathbf{l}_1 = \int_{P_1}^{P_2} n \frac{d\mathbf{r}}{ds} \cdot d\mathbf{l}_2 \tag{3.3-3}$$

Equation (3.3-3) states that the result of integrating $n \dfrac{d\mathbf{r}}{ds} \cdot \mathbf{t}$ (with $\mathbf{t}$ being the vector tangential to the curve and parallel to $d\mathbf{l}$) between two points P_1 and P_2 is independent of the path taken. The subscripts 1 and 2 on $d\mathbf{l}$ indicate integration along two different curves. This result can be interpreted in two different ways. First, let us consider a field of rays in a region of space. The relation (3.3-3) then states that if we integrate the projection of the vector $n\dfrac{d\mathbf{r}}{ds}$ onto the direction of the curve (along which the integral is extended) from point P_1 to point P_2, the result of this integration will be independent of the particular choice of the integration path. The vector $\dfrac{d\mathbf{r}}{ds}$ is of course directed tangentially to the direction of each light ray that is crossed by the integration curve. This is an interesting and surprising result. Another interpretation of (2.3-3) is a result that is even more important in ray optics. This time let us assume that the two points P_1 and P_2 are connected by two different light rays. We can then place the two integration curves so that each coincides with one of the light rays. The vectors $\dfrac{d\mathbf{r}}{ds}$ and $d\mathbf{l}$ are parallel in this case, and their product is simply the line element ds of the light ray. The relation (3.3-3) assumes the form

$$\int_{P_1}^{P_2} n\,ds_1 = \int_{P_1}^{P_2} n\,ds_2 \qquad (3.3\text{-}4)$$

The integral of n along a light ray is known as the optical path length. Equation (3.3-4) can then be explained as follows. Two points P_1 and P_2 can be connected by two different light rays only if the optical path length along each ray is the same.* This result is important for an image-forming system. An image is formed when each ray departing from an arbitrary point of an object meets all other rays from the same object point at some other point in space. The point at which all rays stemming from a given object point converge is called the image point. Equation (3.3-4) states that the optical path lengths along all rays connecting an object point with an image point are identical.

After this digression, we return to our stated objective of deriving boundary conditions for light rays. Assume that a field of light rays traverses the interface between two media with different index of refraction as shown in Figure 3.3.1. Also shown in this figure is a curve C that runs parallel to the interface in medium 1, crosses over to medium 2, and returns to its starting point by running parallel to the interface in medium 2. Using the curve C as the integration path of (3.3-2), we obtain

---

* This result holds only if the light field is continuous, so that there are infinitely many rays that smoothly fill the space between the two rays in question. It does not hold if one ray reaches P_2 via a mirror while the other ray takes a direct path.

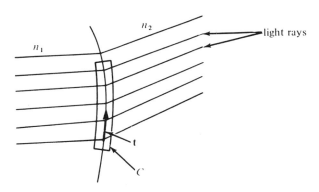

Figure 3.3.1 Illustration of the integration path for the derivation of boundary conditions for light rays. C is a closed curve running on either side of the interface between the two media; $\mathbf{t}$ is a unit vector in tangential direction to the interface.

$$n_1\left(\frac{d\mathbf{r}}{ds}\right)_1 \cdot \mathbf{t} = n_2\left(\frac{d\mathbf{r}}{ds}\right)_2 \cdot \mathbf{t} \tag{3.3-5}$$

The unit vector $\mathbf{t}$ is directed tangential to the interface. Equation (3.3-5) was obtained by making the path length parallel to the interface sufficiently short so that the vectors $\frac{d\mathbf{r}}{ds}$ can be considered as constant along these path sections. The sections of the integration path crossing the interface between the two media were chosen to be vanishingly short. The vectors $\frac{d\mathbf{r}}{ds}$ and $\mathbf{t}$ are both of length unity. Using the angles shown in Figure 3.3.2, we obtain immediately the desired boundary condition

$$n_1\sin \alpha_1 = n_2\sin \alpha_2 \tag{3.3-6}$$

Equation (3.3-6) has already been obtained in Section 1.6, Equation (1.6-19). We have thus rederived Snell's law by the methods of geometrical optics.

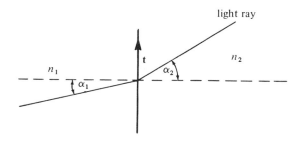

Figure 3.3.2 Illustration of the angles α_1 and α_2 of Snell's law.

This result is particularly gratifying, as we might have had reservations about applying our ray optics results right at the interface between two dielectric media. This reservation might have been based on the argument that the equations of ray optics follow from the wave equation, which itself is only an approximation in case of inhomogeneous media. The approximate nature of the wave equation is based on its derivation from Maxwell's equations. At the interface the index of refraction (or the dielectric constant) varies drastically, thus violating the condition $(\varepsilon_2 - \varepsilon_1)/\varepsilon \ll 1$ (derived in Section 1.3) that was required for the validity of the wave equation. To obtain (3.3-6) from our ray optics equations forces us to apply these offsprings of the wave equation exactly in the region where the applicability of the wave equation becomes questionable. However, our result (3.3-6) is confirmed by the exact derivation of (1.6-19) that was based on Maxwell's equations. Furthermore, it is possible to derive the eikonal equation directly from Maxwells' equations.[18]

It is possible to justify the validity of our procedure by the following argument. Let us regard the interface between the two media not as an abrupt discontinuity but as a region where the refractive index changes continuously from its value n_1 to n_2. This change is allowed to take place sufficiently rapidly so that it is accomplished over a distance that is short compared to other dimensions of interest but long compared to the wavelength of light. In this case, we are allowed to use the wave equation in the transition region. We obtain our result (3.3-6) as before by allowing the path sections that run parallel to the "interface" to be much longer than the sections of the integration path that connect the two regions of space. Since ray optics holds in the limit $\lambda \to 0$, the smooth transition region can be arbitrarily short and still be long compared to the wavelength.

3.4 FERMAT'S PRINCIPLE

We derived the eikonal equation and the ray equation in Section 3.2 from the reduced wave equation. In this section, an alternate approach will be taken. The equations of geometrical optics can be obtained from a variational principle. This approach has the advantage that it elucidates the close relation between ray optics and classical mechanics.

Most laws of physics can be derived from variational principles. The most famous variational principle of physics is Hamilton's principle of least action.[19] Fermat's principle is very similar in form to Hamilton's celebrated law. However, there is one general difference. Whereas Hamilton's principle is based on minimizing functions of time, Fermat's principle minimizes a function of a length coordinate. The similarity between classical mechanics and optics requires us, therefore, to replace time with this length coordinate. Except for this peculiarity, the analogy between ray optics and mechanics is close.

Fermat's principle is based on the concept of optical path length that we introduced in (3.3-4). It states that a light ray always chooses a trajectory that minimizes the optical path length. In rare cases, this minimum can actually be a maximum, as Luneberg[16] has pointed out. It is also important to understand that this path length minimization is not necessarily an absolute minimum. All that the principle requires is that any path in the immediate neighborhood be longer than the minimum trajectory. In mathematical terms, Fermat's principle assumes the form,

$$\int_{P_1}^{P_1} n(x, y, z)ds = \text{minimum} \tag{3.4-1}$$

Instead of path length, we can introduce the concept of transit time by dividing (3.4-1) by the constant c, the velocity of light in vacuum. The quantity c/n is the velocity of light in a medium with refractive index n. The expression under the integral sign is then the transit time that the light requires to travel the distance ds. Fermat's principle can therefore also be written

$$\int_{P_1}^{P_2} dt = \text{minimum} \tag{3.4-2}$$

The points P_1 and P_2 in (3.4-1) are two fixed points in space. The light ray must pass from P_1 to P_2 in the minimum time—that is, it must find the trajectory which takes it from P_1 to P_2 in the shortest optical path length. The solution to this problem in free space is trivial, since a straight line is the shortest connection between any two points, so that light will travel fastest by choosing a straight line path. If the points P_1 and P_2 are located in two different homogeneous dielectric media with a plane interface, Fermat's principle leads directly to Snell's law of refraction. This result can be obtained by direct calculation. That it is true follows from our treatment by showing that the eikonal and ray equations follow from Fermat's principle and the already established fact that Snell's law is a consequence of these equations of ray optics.

The solution of the variational problem (3.4-1) is accomplished more easily by a transformation to a new variable of integration. We use the definition of the line element

$$ds = \sqrt{dx^2 + dy^2 + dz^2} = \sqrt{1 + x'^2 + y'^2}\, dz \tag{3.4-3}$$

with

$$x' = \frac{dx}{dz}\,;\, y' = \frac{dy}{dz} \tag{3.4-4}$$

to express (3.4-1) in the form

$$\int_{P_1}^{P_2} L(x, y, x', y', z)dz = \text{minimum} \tag{3.4-5}$$

The function L is given by

$$L(x, y, x', y', z) = n(x, y, z)\sqrt{1 + x'^2 + y'^2} \qquad (3.4\text{-}6)$$

Equation (3.4-5) is of exactly the same form as Hamilton's principle of least action.[19] The only difference between the principle of Hamilton and that of Fermat is the replacement of the time coordinate t in Hamilton's principle by the length coordinate z in Fermat's equation. In classical mechanics, the function L is called the Lagrangian. The z coordinate is usually chosen to coincide with a preferred direction of the optical system known as the optical axis. Most optical systems have an axis of symmetry, which can be an axis of revolution.

The solution of the problem (3.4-5) is well known in variational calculus, and need not be derived here.[20] It is given by the Euler equations of the variational problem

$$\frac{d}{dz}\frac{\partial L}{\partial x'} - \frac{\partial L}{\partial x} = 0 \qquad (3.4\text{-}7)$$

$$\frac{d}{dz}\frac{\partial L}{\partial y'} - \frac{\partial L}{\partial y} = 0 \qquad (3.4\text{-}8)$$

Substitution of (3.4-6) results in

$$\frac{d}{dz}\frac{nx'}{\sqrt{1 + x'^2 + y'^2}} = \sqrt{1 + x'^2 + y'^2}\,\frac{\partial n}{\partial x} \qquad (3.4\text{-}9)$$

$$\frac{d}{dz}\frac{ny'}{\sqrt{1 + x'^2 + y'^2}} = \sqrt{1 + x'^2 + y'^2}\,\frac{\partial n}{\partial y} \qquad (3.4\text{-}10)$$

With the help of (3.4-3), these equations can be written in the more concise form

$$\frac{d}{ds}\left(n\frac{dx}{ds}\right) = \frac{\partial n}{\partial x} \qquad (3.4\text{-}11)$$

$$\frac{d}{ds}\left(n\frac{dy}{ds}\right) = \frac{\partial n}{\partial y} \qquad (3.4\text{-}12)$$

Comparison with (3.2-16) identifies these equations with the x and y components of the ray equation. In fact, our derivation from Fermat's principle indicates that the two equations (3.4-11) and (3.4-12) should be sufficient to determine the ray trajectory. This would indicate that the z component of the ray equation is redundant. We can show indeed that the corresponding equation for z can be obtained from the equations for x and y. To facilitate this proof, we begin by rewriting the relation (3.4-3)

$$\sqrt{1 + x'^2 + y'^2} = \frac{ds}{dz} = \frac{ds}{\sqrt{(ds)^2 - dx^2 - dy^2}} = \frac{1}{\sqrt{1 - \dot{x}^2 - \dot{y}^2}} \qquad (3.4\text{-}13)$$

with

$$\dot{x} = \frac{dx}{ds} \; ; \; \dot{y} = \frac{dy}{ds} \qquad (3.4\text{-}14)$$

Next, we form

$$\frac{d}{ds}\left(n\frac{dz}{ds}\right) = \frac{d}{ds}\left(n\sqrt{1 - \dot{x}^2 - \dot{y}^2}\right) = \frac{dn}{ds}\sqrt{1 - \dot{x}^2 - \dot{y}^2} - n\frac{\dot{x}\ddot{x} + \dot{y}\ddot{y}}{\sqrt{1 - \dot{x}^2 - \dot{y}^2}}$$

$$= \frac{(1 - \dot{x}^2 - \dot{y}^2)\dfrac{dn}{ds} - n(\dot{x}\ddot{x} + \dot{y}\ddot{y})}{\sqrt{1 - \dot{x}^2 - \dot{y}^2}} \qquad (3.4\text{-}15)$$

From (3.4-11), we obtain by multiplication with $\dot{x}$

$$n\dot{x}\ddot{x} = \dot{x}\frac{\partial n}{\partial x} - \dot{x}^2\frac{dn}{ds}$$

and from (3.4-12) follows similarly

$$n\dot{y}\ddot{y} = \dot{y}\frac{\partial n}{\partial y} - \dot{y}^2\frac{dn}{ds}$$

Utilizing these equations, we can simplify (3.4-15)

$$\frac{d}{ds}\left(n\frac{dz}{ds}\right) = \frac{\dfrac{dn}{ds} - \dot{x}\dfrac{\partial n}{\partial x} - \dot{y}\dfrac{\partial n}{\partial y}}{\sqrt{1 - \dot{x}^2 - \dot{y}^2}}$$

With the help of

$$\frac{dn}{ds} = \frac{\partial n}{\partial z}\frac{dz}{ds} + \frac{\partial n}{\partial x}\frac{dx}{ds} + \frac{\partial n}{\partial y}\frac{dy}{ds}$$

and using (3.4-13), we finally obtain

$$\frac{d}{ds}\left(n\frac{dz}{ds}\right) = \frac{\partial n}{\partial z} \qquad (3.4\text{-}16)$$

We have thus derived the z component of the vector ray equation (3.2-16), proving that not all three components of this equation are independent. The two equations (3.4-11) and (3.4-12) are sufficient for the description of ray trajectories.

3.5 HAMILTONIAN FORMULATION OF RAY OPTICS

The analogy between ray optics and mechanics becomes most striking if we express the equations of ray optics in Hamiltonian form. The reader is assumed to be familiar with Hamilton's equations of motion.[19]

The transition from the Euler equations (in mechanics, they are usually called the Lagrange equations) to Hamilton's equations is achieved in the following way. We begin by introducing the generalized momenta p_x and p_y, which are called the variables canonically conjugate to x and y. The generalized momenta are defined by equations

$$p_x = \frac{\partial L}{\partial x'} \tag{3.5-1}$$

$$p_y = \frac{\partial L}{\partial y'} \tag{3.5-2}$$

Next, we define the Hamiltonian by the relation

$$H(x, y, p_x, p_y) = p_x x' + p_y y' - L \tag{3.5-3}$$

The Lagrangian L is given by (3.4-6). The key element in this definition is the choice of independent variables for the Hamiltonian H. The Lagrangian was assumed to depend on x, y, x', and y', as indicated by (3.4-6). The independent variables of the Hamiltonian are by definition x, y, p_x, and p_y. This change of variables is achieved by replacing x' and y' with p_x and p_y with the help of (3.5-1) and (3.5-2). The different sets of independent variables make it apparent that p_x, p_y, x', and y' are treated as being independent of x and y, whereas x' and y' depend on p_x and p_y. We utilize this interdependence or mutual independence of these variables to form the derivative

$$\frac{\partial H}{\partial p_x} = x' + p_x \frac{\partial x'}{\partial p_x} + p_y \frac{\partial y'}{\partial p_x} - \frac{\partial L}{\partial x'} \frac{\partial x'}{\partial p_x} - \frac{\partial L}{\partial y'} \frac{\partial y'}{\partial p_x}$$

Using (3.5-1) and (3.5-2) and remembering (3.4-4), we obtain one of Hamilton's equations

$$\frac{dx}{dz} = \frac{\partial H}{\partial p_x} \tag{3.5-4}$$

By an equivalent calculation, we obtain also

$$\frac{dy}{dz} = \frac{\partial H}{\partial p_y} \tag{3.5-5}$$

The remaining two equations are found by forming from (3.5-3)

$$\frac{dH}{\partial x} = -\frac{\partial L}{\partial x} \tag{3.5-6}$$

Utilizing (3.4-7) and (3.5-1) allows us to write

$$\frac{dp_x}{dz} = -\frac{\partial H}{\partial x} \tag{3.5-7}$$

and by a similar calculation, we obtain the last of Hamilton's equations

$$\frac{dp_y}{dz} = -\frac{\partial H}{\partial y} \tag{3.5-8}$$

The final step to a complete formulation of ray optics in Hamiltonian form consists of expressing the Hamiltonian in terms of its proper variables. This is achieved by expressing x' and y' in terms of p_x and p_y by using (3.5.1) and (3.5-2). From (3.5-1), we obtain with the help of (3.4-6)

$$p_x = \frac{nx'}{\sqrt{1 + x'^2 + y'^2}} \tag{3.5-9}$$

and from (3.5-2), we find

$$p_y = \frac{ny'}{\sqrt{1 + x'^2 + y'^2}} \tag{3.5-10}$$

These equations can be solved for x' and y', with the result

$$x' = \frac{p_x}{\sqrt{n^2 - p_x^2 - p_y^2}} \tag{3.5-11}$$

and

$$y' = \frac{p_y}{\sqrt{n^2 - p_x^2 - p_y^2}} \tag{3.5-12}$$

We use these equations to replace x' and y' in (3.5-3) to obtain the Hamiltonian in its proper form

$$H = -\sqrt{n^2 - p_x^2 - p_y^2} \tag{3.5-13}$$

The Hamiltonian of ray optics has some resemblance to the relativistic energy of a point particle with rest mass m_o^2[1]

$$E = c\sqrt{m_o^2 c^2 + p_x^2 + p_y^2 + p_z^2} \tag{3.5-14}$$

However, the resemblance between the ray optics Hamiltonian and the Hamiltonian of point particles is more convincing in the "nonrelativistic" approximation. The analog to the mechanical, nonrelativistic case is the paraxial approximation that we have already encountered in Section 3.2 (See Equations [3.2-18] and [3.2-19]). The derivatives x' and y' describe the slope of the ray relative to the z axis. In the paraxial approximation, the ray slope is assumed to be very slight, so that we can assume

$$x' \ll 1 \text{ and } y' \ll 1 \tag{3.5-15}$$

The relationship between the coordinates and the momenta (3.5-9) and (3.5-10) requires that (3.5-15) also imply

$$p_x \ll n \text{ and } p_y \ll n \tag{3.5-16}$$

This assumption allows us to expand the square root in (3.5-13) to obtain the Hamiltonian of ray optics in the paraxial approximation

$$H = \frac{p_x^2 + p_y^2}{2n_o} - n \tag{3.5-17}$$

We also must express the index of refraction as the sum of a constant part n_o plus a small part Δn that varies in space

$$n = n_o - \Delta n \tag{3.5-18}$$

This assumption is necessary for the paraxial approximation to apply, because if the index of refraction varies drastically in space, we are likely to encounter rays with large slopes. The assumption

$$\Delta n \ll n_o \tag{3.5-19}$$

in conjunction with (3.5-16) allowed us to replace n by n_o in the p-dependent part of the Hamiltonian. The differential equations of ray optics do not apply at discontinuities of the dielectric constant where Δn may not be small. However, such discontinuities are treated in the usual manner by piecing together the solutions applying to the continuous sections by means of boundary conditions.

The paraxial Hamiltonian of geometrical optics has a very close correspondence to the nonrelativistic Hamiltonian of the mechanics of point particles.[19]

$$H = \frac{p_x^2 + p_y^2 + p_z^2}{2m} + V \tag{3.5-20}$$

The ray optics problem has one dimension less than does the corresponding problem of point particles. The particle potential V is replaced in a very logical way by the index of refraction of the optical medium. The difference in sign of the two potential terms is immaterial. We can actually use Δn of (3.5-18) as the potential of ray optics and get complete agreement even for the sign of the potential term. The additive constant n_o has no physical significance, since any potential is determined only up to an arbitrary constant. The Hamilton-Jacoby partial differential equation[19]

$$\frac{\partial S}{\partial t} + H\left(x, y, z, \frac{\partial S}{\partial x}, \frac{\partial S}{\partial y}, \frac{\partial S}{\partial z}\right) = 0 \tag{3.5-21}$$

plays an important part in Hamiltonian mechanics of point particles. The function S is introduced as the solution of the Hamilton-Jacoby partial differential equation (3.5-21). The derivatives of S with respect to the co-ordinates replace the momenta in the Hamiltonian function. True to the general correspondence between mechanics and ray optics, we must replace the time variable t by the space variable z and reduce the dimension of the problem by one. We thus obtain the Hamilton-Jacoby equation of ray optics

$$\frac{\partial S}{\partial z} = -H\left(x, y, \frac{\partial S}{\partial x}, \frac{\partial S}{\partial y}\right) \tag{3.5-22}$$

Squaring both sides of this equation and using the explicit expression (3.5-13) for the Hamiltonian leads to the Hamilton-Jacoby equation of ray optics

$$\left(\frac{\partial S}{\partial z}\right)^2 = n^2 - \left(\frac{\partial S}{\partial x}\right)^2 - \left(\frac{\partial S}{\partial y}\right)^2 \tag{3.5-23}$$

It is obvious that we have rederived the eikonal equation (3.2-5)

$$(\nabla S)^2 = n^2 \tag{3.5-24}$$

We have thus come full circle, arriving at the eikonal equation that formed the starting point for the derivation of the ray equation in Section 3.2. The derivation of geometrical optics from Fermat's principle is thus fully equivalent to the earlier derivation from the reduced wave equation. The various formulations of the laws of mechanics or ray optics from a variational principle are typical of variational problems of this kind. This interrelationship between the variational principle, the Lagrangian and Hamiltonian equations of motion, and Hamilton-Jakoby's partial differential equation is beautifully elucidated in Courant's classic book.[4]

We shall see in the next section that we can go even further and rederive the wave equation as the Klein-Gordon equation of relativistic quantum mechanics of ray optics.

Returning to Hamilton's differential equations (3.5-4), (3.5-5), (3.5-7), and (3.5-8), we show that these too are equivalent to the ray equation. Using (3.5-4) and (3.5-5), we obtain from the Hamiltonian (3.5-13)

$$\frac{dx}{dz} = \frac{p_x}{\sqrt{n^2 - p_x^2 - p_y^2}} \tag{3.5-25}$$

and

$$\frac{dy}{dz} = \frac{p_1}{\sqrt{n^2 - p_x^2 - p_y^2}} \tag{3.5-26}$$

These equations already appeared in (3.5-11) and (3.5-12). From (3.5-7) and (3.5-8), we obtain

$$\frac{dp_x}{dz} = \frac{n}{\sqrt{(n^2 - p_x^2 - p_y^2)}} \frac{\partial n}{\partial x} \qquad (3.5\text{-}27)$$

$$\frac{\partial p_y}{\partial z} = \frac{n}{\sqrt{n^2 - p_x^2 - p_y^2}} \frac{\partial n}{\partial y} \qquad (3.5\text{-}28)$$

From (3.5-25) and (3.5-26), we find

$$\frac{ds}{dz} = \sqrt{1 + x'^2 + y'^2} = \frac{n}{\sqrt{n^2 - p_x^2 - p_y^2}} \qquad (3.5\text{-}29)$$

This equation allows us to write (3.5-27) and (3.5-28) in the simpler form

$$\frac{dp_x}{ds} = \frac{\partial n}{\partial x} \qquad (3.5\text{-}30)$$

$$\frac{dp_y}{ds} = \frac{\partial n}{\partial y} \qquad (3.5\text{-}31)$$

Using (3.5-29) once more, we obtain from (3.5-25) and (3.5-26) the generalized momenta in the simple form

$$p_x = n \frac{dx}{ds} \qquad (3.5\text{-}32)$$

and

$$p_y = n \frac{dy}{ds} \qquad (3.5\text{-}33)$$

Substitution of (3.5-32) into (3.5-30) yields the x component of the ray equation (3.2-16) or (3.4-11). The y component follows from the other two equations.

The paraxial approximation of the ray equation follows similarly if we use the paraxial Hamiltonian in Hamilton's equations. From (3.5-4) and (3.5-17), we obtain

$$\frac{dx}{dz} = \frac{1}{n_o} p_x \qquad (3.5\text{-}34)$$

Equation (3.5-7) and (3.5-17) lead to the expression

$$\frac{dp_x}{dz} = \frac{\partial n}{\partial x} \qquad (3.5\text{-}35)$$

Combining (3.5-34) and (3.5-35) results in the x component of the paraxial ray equation

$$n_o \frac{d^2x}{dz^2} = \frac{\partial n}{\partial x} \qquad (3.5\text{-}36)$$

The corresponding y component is

$$n_o \frac{d^2 y}{dz^2} = \frac{\partial n}{\partial y} \qquad (3.5\text{-}37)$$

These equations appear here in a slightly more consistent form than their earlier version (3.2-19) because the index of refraction on the left side is now replaced by its average value n_o. This is necessary for a consistent paraxial approximation since the products of Δn of (3.5-18) with the derivatives of x and y are terms of second order that must be neglected in a consistent first order approximation.

Figures 3.5.1 and 3.5.2. show a section of a light ray and the angles that it defines with the x-z and y-z planes. With the definitions indicated in these figures, we can express the generalized momenta (3.5-32) and (3.5-33) by the angles α_x and α_y

$$p_x = n \sin \alpha_x \qquad (3.5\text{-}38)$$

$$p_y = n \sin \alpha_y \qquad (3.5\text{-}39)$$

If we think of the x-y plane as the interface between two media with different dielectric constants, and locate the coordinate system so that a ray passing through the interface from one medium into the other travels in the y-z plane, we can express Snell's law (3.3-6) in terms of the generalized momenta as follows

$$\left. \begin{array}{l} (p_x)_1 = (p_x)_2 = 0 \\[2mm] (p_y)_1 = (p_y)_2 \end{array} \right\} \qquad (3.5\text{-}40)$$

We see that the generalized momenta are conserved as the ray travels from

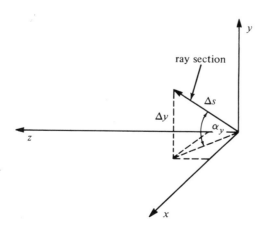

Figure 3.5.1   This diagram shows the definition of the angle α_y.

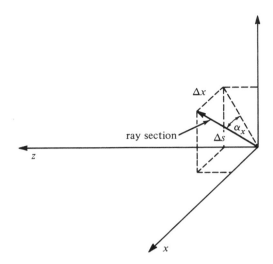

Figure 3.5.2   This diagram shows the definition of the angle α_x.

one medium to the other. This result follows more generally from (3.5-30) and (3.5-31), because, with our choice of the coordinate system, we have everywhere (including at the interface) $\dfrac{\partial n}{\partial x} = 0$, $\dfrac{\partial n}{\partial y} = 0$ so that p_x and p_y must be constant.

The generalized momentum of the ray must not be confused with photon momentum. The two have nothing in common. The photon momentum describes actual mechanical momentum that is carried by the photon, while the generalized momentum of the light ray was derived purely formally from the Hamiltonian formalism of ray optics. The generalized momentum of the light ray describes its slope with respect to the fixed coordinate system.

3.6 QUANTUM THEORY OF LIGHT RAYS*

Fermat's principle allowed us to obtain all the relevant equations of geometrical optics that we had already derived from the reduced wave equation. However, purely algebraic manipulations cannot succeed in rederiving the wave equation from Fermat's principle since we neglected certain terms to derive the equations of ray optics from wave optics. These neglected terms remind us that geometrical optics was only an approximation of wave optics and cannot describe optical phenomena as completely as wave optics does.

---

* This section is not essential for understanding the rest of the book. Readers with no knowledge of quantum mechanics are advised to skip it.

We show in this section that the wave equation can indeed be recovered from ray optics, not by purely algebraic transformations but, rather, by a drastic step—quantization. Applying the rules of quantum mechanics to ray optics results in a quantum theory of light rays.[42] This theory, however, turns out to be identical to the scalar wave theory described by the reduced wave equation. We pointed out in the previous section that the exact equations (exact in the framework of ray optics) of ray optics are equivalent to relativistic mechanics and that the paraxial equations are equivalent to nonrelativistic mechanics. This analogy carries over into the quantum theory of light rays. The reduced wave equation is thus obtained from the equivalent of the relativistic Klein-Gordon equation[22] of wave mechanics, while the ray optics equivalent of the usual nonrelativistic Schrödinger equation follows from the paraxial approximation. G. Eichmann[66] has been able to go even one step further. He developed a Dirac type theory of ray optics and showed that the Dirac equation of light rays is equivalent to the time independent Maxwell equations.

Quantization of a physical theory is accomplished by replacing all of the variables by operators. In wave mechanics, the coordinates retain their meaning as numbers but the canonically conjugate variables—the momenta—become differential operators[5,10]

$$p_x = - i\kappa \frac{\partial}{\partial x} \tag{3.6-1}$$

and

$$p_y = - i\kappa \frac{\partial}{\partial y} \tag{3.6-2}$$

We refrained from designating the constant appearing in (3.6-1) and (3.6-2) by the usual symbol $\hbar$, because our quantum theory is slightly different from the usual quantum mechanics of point particles. The time coordinate of mechanics is now replaced by the z coordinate. The unit "time" in Planck's constant must therefore also be replaced by a length coordinate, and $\hbar$ can no longer be interpreted as having the dimension of energy times time. Our constant κ, which takes the place of $\hbar$, must, instead, have the dimension of the Hamiltonian times length. The Hamiltonian (3.5-13) is dimensionless, so that κ must have the dimension of length.

The energy or Hamilton operator is expressed by a time derivative in ordinary quantum mechanics. We thus use the relation

$$H = i\kappa \frac{\partial}{\partial z} \tag{3.6-3}$$

using again the correspondence between the time variable and the length variable z.

It is customary in relativistic quantum mechanics to square the relation (3.6-3) instead of using square root expressions for the operators. We write therefore

$$H^2 = -\kappa^2 \frac{\partial^2}{\partial z^2} \qquad (3.6\text{-}4)$$

Applying this operator equation to a wave function results in the ray optics equivalent of the relativistic wave equation. This equation is known as the Klein-Gordon equation,[22] to distinguish it from the usual nonrelativistic Schrödinger equation. Using (3.5-13), (3.6-1), and (3.6-2), we obtain the Klein-Gordon equation of the quantum theory of ray optics by applying (3.6-4) to a wave function

$$n^2\psi + \kappa^2 \frac{\partial^2\psi}{\partial x^2} + \kappa^2 \frac{\partial^2\psi}{\partial y^2} = -\kappa^2 \frac{\partial^2\psi}{\partial z^2} \qquad (3.6\text{-}5)$$

Regrouping the equation results in

$$\nabla^2\psi + \frac{n^2}{\kappa^2}\psi = 0 \qquad (3.6\text{-}6)$$

This wave equation of the quantum theory of geometrical optics is obviously identical with the reduced wave equation (3.2-1), which we shall write for our present purposes in the form

$$\nabla^2\psi + \left(\frac{2\pi n}{\lambda_o}\right)^2\psi = 0 \qquad (3.6\text{-}7)$$

to show explicitly the occurrence of the free space wavelength λ_o. Comparison between the two equations allows us to determine the constant κ of our quantum theory, which so far was unknown. We find

$$\kappa = \frac{\lambda_o}{2\pi} \qquad (3.6\text{-}8)$$

This is a most satisfactory result. Our constant κ is the equivalent of the quantum mechanical constant $\hbar$. It is a well-known feature of quantum theory that its results coincide with classical mechanics in the limit $\hbar \to 0$. We can expect that our quantum theory of ray optics coincides with geometrical optics in the limit $\kappa \to 0$. However, we see now that this limit is identical with the limit $\lambda_o \to 0$. We know from our previous derivation of ray optics from wave optics that the ray optics equations become exact in the limit $\lambda_o \to 0$. It is most fitting that "Planck's constant" of the quantum theory of ray optics turns out to be the vacuum wavelength of light.

We can immediately use all the well-known results of quantum mechanics and apply them to the quantum theory of ray optics, which, as we have shown, is identical to the usual scalar wave theory of light or to any other phenomenon described by the wave equation. In fact, our discovery of the equivalence of wave optics with the quantum theory of rays benefits ray optics more than it does wave optics. Wave optics is well understood in its own right. The limit of validity of ray optics, however, is open to discussion, since ray optics is only an approximate theory. Since we now know that ray optics is the classical "mechanics" of wave optics, we can use our experience with classical mechanics and wave mechanics to guide us in our expectations as to the validity and applicability of ray optics. Quantum mechanics did not replace classical mechanics. The latter is still used as much as ever in those areas where it is known to apply. We know, for example, that the motion of electrons in electric and magnetic fields is described with high precision by the laws of classical mechanics. Only when the fields become very strong, as in the vicinity of a nucleus, do we need quantum mechanics to describe the motion of electrons correctly. For such strong fields, classical mechanics fails. We also know that classical mechanics cannot be applied in situations where an electron is forced to exhibit its wave nature. These analogies can help to guide us in our choice of using ray optics or wave optics for any given optical problem, Ehrenfest's theorem[22] helps to establish the limits of classical mechanics. We shall discuss it later in this section and draw conclusions as to the applicability of ray optics.

An immediate by-product of our quantum theory of ray optics is a paraxial approximation for the reduced wave equation. We have pointed out repeatedly that the paraxial approximation corresponds to nonrelativistic mechanics. Using the paraxial Hamiltonian (3.5-17) in the operator equation (3.6-3) allows us to obtain the ray optics equivalent of the nonrelativistic Schrödinger equation. This equation is also the paraxial approximation of the reduced wave equation. It has the form

$$-\frac{\lambda_o^2}{8\pi^2 n_o}\left(\frac{\partial^2\psi}{\partial x^2} + \frac{\partial^2\psi}{\partial y^2}\right) - n\psi = i\frac{\lambda_o}{2\pi}\frac{\partial\psi}{\partial z} \tag{3.6-9}$$

By regrouping terms, we obtain the paraxial wave equation

$$\frac{\partial^2\psi}{\partial x^2} + \frac{\partial^2\psi}{\partial y^2} + i\frac{4\pi}{\lambda_o}n_o\frac{\partial\psi}{\partial z} + \frac{8\pi^2}{\lambda_o^2}n_o n\psi = 0 \tag{3.6-10}$$

Equation (3.6-9) is of the form of the nonrelativistic Schrödinger equation.

Another important quantum mechanical equation is the energy eigenvalue equation

$$H\psi = E\psi \tag{3.6-11}$$

which can be written in a form more suitable for the relativistic Hamilton operator

$$H^2\psi = E^2\psi \tag{3.6-12}$$

From (3.5-17) and (3.6-11), we obtain for the paraxial version of the energy eigenvalue equation

$$-\frac{\lambda_o^2}{8\pi^2 n_o}\left(\frac{\partial^2\psi}{\partial x^2} + \frac{\partial^2\psi}{\partial y^2}\right) - n\psi = E\psi \qquad (3.6\text{-}13)$$

Using a wave function of the form (for the plus sign in the exponent compare [3.6-27])

$$\psi = \psi_o(x, y)e^{+i\beta z} \qquad (3.6\text{-}14)$$

with the propagation constant β, we find by comparing (3.6-9) with the energy eigenvalue equation (3.6-13) that the eigenvalue E is proportional to the propagation constant

$$E = -\frac{\lambda_o}{2\pi}\beta \qquad (3.6\text{-}15)$$

The same result can be obtained from (3.6-12). Equation (3.6-15) is equivalent to the quantum mechanical expression

$$E = \frac{h}{2\pi}\omega \qquad (3.6\text{-}16)$$

The difference in sign between (3.6-15) and (3.6-16) occurs only because (3.6-14) describes a wave traveling in negative z direction. Planck's constant h becomes λ_o; as we know, the frequency ω of ordinary quantum mechanics is replaced by the propagation constant in z direction, $-\beta$, of the quantum theory of rays. The "energy eigenvalue" equations, (3.6-11) and (3.6-12), are identical to the eigenvalue problem of finding the modes of an optical waveguide. We shall solve such a problem on a later chapter for an optical waveguide with parabolic index distribution.

The wave function ψ of the quantum mechanics of rays is the usual scalar wave function of wave optics. In our present theory, it assumes the additional interpretation as a probability amplitude. The wave function describes the state of a statistical ensemble of rays. Its absolute square value*

$$\bar{P} = |\psi(x, y, z)|^2 \qquad (3.6\text{-}17)$$

assumes the meaning of the probability density for finding a light ray inside the unit area in the x, y plane at the position z. The total probability of finding the light ray in a given area A is consequently defined as

$$P = \int_A |\psi|^2 dA \qquad (3.6\text{-}18)$$

---

* The probability interpretation works only for the paraxial ray theory. The wave function of the Klein-Gordon equation cannot be interpreted as a probability amplitude.[111]

The requirement that the light ray must cross every plane perpendicular to the z-axis yields the normalization condition

$$\int_{-\infty}^{\infty} \int_{-\infty}^{\infty} |\psi(x, y, z)|^2 dx dy = 1 \tag{3.6-19}$$

The wave function describes the probability of finding a ray in a field of light rays whose statistical state is characterized by ψ.

We can also transform the wave function to momentum space.[10] To be able to accomplish this transformation, we need to know the momentum eigenfunctions and eigenvalues. The eigenvalue problem for the momentum operator p_x of (3.6-1) has the form

$$- i\kappa \frac{\partial \psi_{px}}{\partial x} = p'_x \psi_{px} \tag{3.6-20}$$

The solution of this equation is the momentum eigenfunction belonging to the momentum eigenvalue p'_x.

$$\psi_{px} = \frac{1}{\sqrt{2\pi}} e^{i(p'_x/\kappa)x} e^{i\phi(z)} \tag{3.6-21}$$

The factor $(2\pi)^{-1/2}$ is required for the proper normalization of the momentum eigenfunction; the phase function ϕ is arbitrary. Since its eigenvalue spectrum is continuous, the eigenfunction must be normalized with respect to a delta function

$$\int_{-\infty}^{\infty} \psi_{p'_x}^* \psi_{p''_x} dx = \frac{1}{2\pi} \int_{-\infty}^{\infty} e^{i(1/\kappa)(p''_x - p'_x)x} dx = \delta(p''_x - p'_x) \tag{3.6-22}$$

To be able to give a physical interpretation to the momentum eigenfunction, we must choose its phase ϕ such that ψ_{px} becomes also a solution of the Klein-Gordon equation (3.6-5). Using the trial solution

$$\psi_{px} = \frac{1}{\sqrt{2\pi}} e^{i[(p'_x/\kappa)x + (p'_z/\kappa)z]} \tag{3.6-23}$$

we find that (3.6-23) solves the "relativistic" Schrödinger equation (3.6-5) when the following relation holds

$$p_x^2 + p_z^2 = n^2 \tag{3.6-24}$$

Our trial solution solves the "relativistic" Schrödinger equation only when the refractive index is a constant $n = n_o$. This result means that a momentum eigenfunction can be a physical state—that is, one that also satisfies the Schrödinger equation—only when the refractive index, n, is constant in space. If this requirement is satisfied, a momentum eigenstate exists and represents a plane wave traveling in the direction of the vector

$$\mathbf{p} = p_x \mathbf{e}_x + p_z \mathbf{e}_z \tag{3.6-25}$$

($\mathbf{e}_x$ and $\mathbf{e}_z$ are unit vectors in x and z directions). This result agrees with our physical interpretation of ray optics. A momentum eigenstate can exist as a physical state only when every measurement (that is performed to determine the value of the rays "momentum") results in the same accurate value. The rays "momentum" is its slope according to (3.5-38). A light ray can have a definite slope when it belongs to a plane wave. The rays are the orthogonal trajectories to the phase fronts of the accompanying wave. If the phase fronts of the wave are curved, the ray slope—its "momentum"—is different in different parts of space. A "momentum" eigenstate must therefore be a plane wave. With our choice of constants, (3.6-23) describes a plane wave traveling in the negative direction of the vector $\mathbf{p}$ if we use for its time dependence the factor

$$e^{i\omega t} \tag{3.6-26}$$

A wave function that is simultaneously an eigenfunction of the p_x and p_y operators is a plane wave in a more general direction

$$\psi = \frac{1}{2\pi} e^{i(p_x x + p_y y + p_z z)/\kappa} \tag{3.6-27}$$

This wave solves the "relativistic" Schrödinger equation (3.6-5) for constant $n = n_o$ provided that the following relation holds

$$p_x^2 + p_y^2 + p_z^2 = n_o^2 \tag{3.6-28}$$

Had we required the momentum eigenfunction to be a solution of the "nonrelativistic" Schrödinger equation (3.6-9), we would have obtained the relation

$$\frac{1}{2n_o}(p_x^2 + p_y^2) + p_z = n_o \tag{3.6-29}$$

This corresponds of course to the paraxial approximation of (3.6-28) with $p_z \approx n_o$, its paraxial value.

We are now able to perform the transformation of an arbitrary wave function ψ to momentum space. This transformation is accomplished by

$$\phi(p_x, p_y, z) = \frac{1}{2\pi} \int_{-\infty}^{\infty} \int_{-\infty}^{\infty} \psi(x, y, z) e^{-(i/\kappa)(p_x x + p_y y)} dx dy \tag{3.6-30}$$

The probability of finding a ray with "momentum" p_x, p_y in the interval $dp_x dp_y$ is given by

$$dP = |\phi(p_x, p_y)|^2 \, dp_x dp_y \tag{3.6-31}$$

The inverse transformation of (3.6-30) presents a problem, since the range of physically possible values of p_x and p_y is only from $-n$ to n. However, κ is a very small quantity, so that the range of p_x/κ is $-2\pi n/\lambda_o$ to $+2\pi n/\lambda_o$. For very small values of λ_o, the integration range is very large. One can assume that the function $\phi(p_x, p_y)$ becomes vanishingly small close to the end of the integration range, since the probability of finding rays traveling perpendicular to the axis must be very nearly zero in most practical cases. We do not make an appreciable error by taking $\phi = 0$ outside the physical range of integration and thus extend the integration range from $-\infty$ to $+\infty$. The transformation (3.6-30) can thus be viewed as a Fourier integral transform with the inverse transformation

$$\psi(x, y) = \frac{1}{2\pi} \int_{-\infty}^{\infty} \int_{-\infty}^{\infty} \phi(p_x, p_y) e^{(i/\kappa)(p_x x + p_y y)} d\left(\frac{p_x}{\kappa}\right) d\left(\frac{p_y}{\kappa}\right) \quad (3.6\text{-}32)$$

The interpretation of the wave function as a probability amplitude allows us immediately to define the expectation values[10,22] of all the operators appearing in the quantum theory of rays. For any operator A of the theory, we can define the expectation value by the equation

$$\langle A \rangle = \int_{-\infty}^{\infty} \int_{-\infty}^{\infty} \psi^*(x, y) A \psi(x, y) dx dy \quad (3.6\text{-}33)$$

The expectation value can be expressed in momentum space by the relation

$$\langle A \rangle = \int_{-\infty}^{\infty} \int_{-\infty}^{\infty} \phi^*(p_x, p_y) A_p \phi(p_x, p_y) d\left(\frac{p_x}{\kappa}\right) d\left(\frac{p_y}{\kappa}\right) \quad (3.6\text{-}34)$$

The operator A_p is the momentum space equivalent of A[10]. We are now ready to discuss Ehrenfest's theorem. We remarked earlier that this theorem provides a bridge between classical and quantum theory. Ehrenfest's theorem in relativistic quantum mechanics assumes an awkward and not very instructive form. Most commonly it is used in nonrelativistic quantum mechanics. For this reason we shall limit ourselves to the paraxial approximation and derive Ehrenfest's theorem from the "nonrelativistic" Schrödinger equation (3.6-9), which, in operator notation, assumes the form

$$H\psi = i\kappa \frac{\partial \psi}{\partial z} \quad (3.6\text{-}35)$$

with H of (3.5-17). Ehrenfest's theorem involves the derivatives of the expectation values of the coordinates x and y and the "momenta" p_x and p_y. We therefore take the derivative of the expectation value of the x coordinate

$$\frac{d\langle x \rangle}{dz} = \frac{d}{dz} \int_{-\infty}^{\infty} \int_{-\infty}^{\infty} \psi^* x \psi \, dx dy = \int_{-\infty}^{\infty} \int_{-\infty}^{\infty} \left(\frac{d\psi^*}{dz} x\psi + \psi^* x \frac{d\psi}{dz} \right) dx dy$$

$$(3.6\text{-}36)$$

The derivatives of the wave function can be eliminated with the help of the Schrödinger equation (3.6-35)

$$\frac{d\langle x \rangle}{dz} = \frac{i}{\kappa} \int_{-\infty}^{\infty} \int_{-\infty}^{\infty} [(H\psi^*)x\psi - \psi^* x H\psi] dx dy$$

The following redistribution of terms can be made for any Hermitian operator[22]

$$\frac{d\langle x \rangle}{dz} = \frac{i}{\kappa} \int_{-\infty}^{\infty} \int_{-\infty}^{\infty} \psi^* (Hx - xH)\psi dx dy \qquad (3.6\text{-}37)$$

In our special case, it can be justified by performing partial integrations using the specific form of the Hamiltonian that results by substitution of (3.6-1) and (3.6-2) into (3.5-17). The operator expression in parenthesis can easily be evaluated. We begin with the commutators

$$xp_x - p_x x = i\kappa \qquad (3.6\text{-}38)$$

and

$$xp_y - p_y x = 0 \qquad (3.6\text{-}39)$$

These expressions are well-known and form the cornerstone of quantum mechanics.[10] They can be verified by letting the expressions (3.6-38) and (3.6-39) operate on an arbitrary wave function using the explicit expressions (3.6-1) and (3.6-2) for p_x and p_y. Repeated application of the same commutators results in

$$xp_x^2 - p_x^2 x = p_x xp_x + i\kappa p_x - p_x xp_x + i\kappa p_x = 2i\kappa p_x \qquad (3.6\text{-}40)$$

$$xp_y^2 - p_y^2 x = 0 \qquad (3.6\text{-}41)$$

Using these commutation relations, we immediately obtain the desired result

$$Hx - xH = -\left[\frac{1}{2n_o} \left(xp_x^2 - p_x^2 x + xp_y^2 - p_y^2 x \right) + xn - nx \right] = -\frac{i\kappa}{n_o} p_x$$
$$(3.6\text{-}42)$$

Equation (3.6-37) now assumes the form of one of the equations of Ehrenfest's theorem

$$\frac{d\langle x \rangle}{dz} = \frac{1}{n_o} \int_{-\infty}^{\infty} \int_{-\infty}^{\infty} \psi^* p_x \psi dx dy \qquad (3.6\text{-}43)$$

or, using the definition of expectation values of operators (3.6-33), we can write it in the form

$$\frac{d\langle x \rangle}{dz} = \frac{1}{n_o} \langle p_x \rangle \qquad (3.6\text{-}44)$$

A similar expression can be derived for the y component

$$\frac{d\langle y \rangle}{dz} = \frac{1}{n_o} \langle p_y \rangle \qquad (3.6\text{-}45)$$

We also need the derivative of the expectation value of p_x

$$\frac{d\langle p_x\rangle}{dz} = \int_{-\infty}^{\infty}\int_{-\infty}^{\infty}\left(\frac{d\psi^*}{dz}p_x\psi + \psi^*p_x\frac{d\psi}{dz}\right)dxdy$$

$$= \frac{i}{\kappa}\int_{-\infty}^{\infty}\int_{-\infty}^{\infty}\psi^*(Hp_x - p_xH)\psi dxdy \qquad (3.6\text{-}46)$$

To evaluate this commutator requires the commutation relation

$$p_xf(x) - f(x)p_x = -i\kappa\frac{\partial f(x)}{\partial x} \qquad (3.6\text{-}47)$$

This commutation relation is immediately verified by letting the expression (3.6-47) operate on an arbitrary wave function and by using (3.6-1). With the help of this commutation relation and the fact that p_x commutes with itself and p_y, we obtain from (3.6-46).

$$\frac{d\langle p_x\rangle}{dz} = \left\langle\frac{\partial n}{\partial x}\right\rangle \qquad (3.6\text{-}48)$$

A similar relation holds for the y component

$$\frac{d\langle p_y\rangle}{dz} = \left\langle\frac{\partial n}{\partial y}\right\rangle \qquad (3.6\text{-}49)$$

Equations (3.6-44), (3.6-45), (3.6-48), and (3.6-49) are known as Ehrengest's theorem. These equations are very similar to the paraxial ray equations (3.5-34) and (3.5-35). In fact, we can combine equations (3.6-44) and (3.6-48) into an equation that looks almost like the paraxial ray equation (3.5-36)

$$n_o\frac{d^2\langle x\rangle}{dz^2} = \left\langle\frac{\partial n(x, y)}{\partial x}\right\rangle \qquad (3.6\text{-}50)$$

A similar expression holds of course for the y component. If equation (3.6-50) would read

$$n_o\frac{d^2\langle x\rangle}{dz^2} = \frac{\partial}{\partial\langle x\rangle}n(\langle x\rangle, \langle y\rangle) \qquad (3.6\text{-}51)$$

we could conclude that the expectation value of the x coordinate of a light ray moves itself like such a ray. However, equation (3.6-50) is just sufficiently different that such a statement cannot be made generally. All we can say is that if $\partial n(\langle x\rangle, \langle y\rangle)/\partial\langle x\rangle$ is sufficiently similar to $\langle\partial n(x, y)/\partial x\rangle$, the expectation value of a ray moves similarly to a real ray. This is an extremely important theorem, for it says that the center of gravity of the light field (which is the same as the expectation value $\langle x\rangle$) moves approximately like a light ray. This comparison between the rays of ray optics and the center of gravity of the light field is the key point to the interpretation of ray optics. We use ray optics only to obtain a simplified and approximate description of the trajectory of light beams. It would be helpful if we were assured that the ray equation describes the trajectory of the center of gravity of the actual light beam. In that case, ray optics would tell us exactly where the bulk of the light field is going. However, the comparison between the actually derived equation (3.6-50) and the desired equation (3.6-51) shows us that the ray theory does not necessarily predict the motion of the center of gravity of light beams. To explore in more detail under what conditions the ray equation predicts the motion of the center of gravity of the light field, we expand the index of refraction into a power series

$$n = n_o + n_1 x + n_2 y + n_3 x^2 + n_4 xy + n_5 y^2 + n_6 x^3 + n_7 x^2 y$$
$$+ n_8 xy^2 + n_9 y^3 + \cdots \quad (3.6\text{-}52)$$

The derivative of this index distribution is

$$\frac{\partial n}{\partial x} = n_1 + 2n_3 x + n_4 y + 3n_6 x^2 + 2n_7 xy + n_8 y^2 + \cdots \quad (3.6\text{-}53)$$

If only the terms of first and second order existed in (3.6-52), we would have exactly

$$\left\langle \frac{\partial}{\partial x} n(x, y) \right\rangle = \frac{\partial}{\partial \langle x \rangle} n(\langle x \rangle, \langle y \rangle) \quad (3.6\text{-}54)$$

It is only in this case that we can state with precision that the center of gravity of a light beam moves in accordance with ray optics. Index distributions that are described by an expansion of the form

$$n = n_o + n_1 x + n_2 y + n_3 x^2 + n_4 xy + n_5 y^2 \quad (3.6\text{-}55)$$

are of course possible. In fact, optical fibers with parabolic index distribution have been made.[23,112] Such "square law media" are of interest as light waveguides, and will be discussed in Chapter 7. A square law medium is extremely pleasing because we know that the ray optics solutions describe exactly the motion of the center of gravity of every light beam traveling in it. In other more general media, the ray equations still give us solutions, but it is not clear how these mathematical rays are connected with the actual motion of the light field. In media with slowly and slightly varying refractive index, the index distribution can approximately be described by an expansion using only first and second order terms. Only to the approximation afforded by the expansion (3.6-55) can we say that the ray describes the motion of the center of gravity of the light field. For very general index distributions, we can say little about the connection between the ray optics solutions and the actual trajectory of the light beam.

In conclusion, we take a look at the uncertainty principle as it applies to our quantum theory of light rays. It is shown in every book on quantum theory[10,22] that the commutation relations (3.6-38) (and the corresponding relation for the y components) lead to the following uncertainty relations

$$\Delta x \Delta p_x \geq \frac{1}{2} \kappa = \frac{\lambda_o}{4\pi} \tag{3.6-56}$$

and

$$\Delta y \Delta p_y \geq \frac{\lambda_o}{4\pi} \tag{3.6-57}$$

The uncertainties of x and p_x etc. are defined by the relations

$$\Delta x = [\langle (x - \langle x \rangle)^2 \rangle]^{1/2} \tag{3.6-58}$$

and

$$\Delta p_x = [\langle (p_x - \langle p_x \rangle)^2 \rangle]^{1/2} \tag{3.6-59}$$

The expectation values are defined by (3.6-33). These relations state that if a light beam is prepared in such a way that x is known to a certain precision, then p_x can be measured only to the accuracy implied by (3.6-56). Let us consider two examples. First, we assume that the state of the ray is described by a plane wave of the form (3.6-23). This state is a " momentum " eigenstate, and assures us that every measurement of the ray's slope must result in the value p_x'. The position of the ray is totally unknown in this case, since an infinite plane wave extends throughout all of space and no particular ray position can be defined. Next, we consider a light beam that is passed through a very narrow slit. The position of the ray passing through the slit is known to within the width of the slit. The slope of the emerging ray, however, becomes less and less well defined as the width of the slit is reduced, because the light wave is diffracted more and more and spreads apart as the field travels away from the slit. If we assume that we can take $\Delta x = d$ to be the slit width, we would obtain from (3.6-56)

$$\Delta p_x \geq \frac{\lambda_o}{4\pi d} \tag{3.6-60}$$

or, using (3.5-38) for small values of α, we obtain the uncertainty of the ray angle for $n = 1$

$$\Delta \alpha \geq \frac{\lambda_o}{4\pi d} \tag{3.6-61}$$

Comparison with (2.3-28) shows us that our earlier diffraction theory resulted in an angular spread of approximately the same order of magnitude. Our present result is of course only an inequality, and does not contradict the result (2.3-28). We can expect, however, that it must be possible to choose more advantageous field distributions that cause the beam to spread less than a truncated plane wave with uniform amplitude distribution in the slit. Gaussian beams (Chapter 6) are such minimum uncertainty wave packets.[22]

3.7 LIOUVILLE'S THEOREM[24]

Liouville's theorem is important in statistical mechanics. Our discussion stressing the close similarity between ray optics and mechanics has prepared us not to be surprised to find an application for mechanical theorems in optics. Light rays are not usually treated by statistical theories. However, quantum mechanics is a statistical theory, so that wave optics, being the quantum theory of ray optics, amounts to a statistical treatment of light rays. Statistical ideas in a strictly classical sense can also be applied to advantage to describe the collective motion of bundles of rays. The great importance of Liouvilles' theorem in optics is based on its ability to provide guidance in situations where simple-minded intuition could easily lead to grave errors. An example of this kind will be presented in Chapter 4. It involves a generalization of the concept of a thin lens as a phase transformer. Liouville's theorem can be applied to test the physical realizability of mathematical assumptions that appear plausible intuitively but violate physical principles in a subtle way. Liouville's theorem is not customarily discussed in optics books. But it is a powerful tool and includes as special cases several well-known optical theorems.

The derivation of Liouville's theorem can be found in books on statistical mechanics.[25] We shall rederive it here to avoid forcing the reader into a lengthy digression into the field of statistical mechanics. The starting point for the formulation and the understanding of the theorem is the notion of phase space. We have seen that the Hamiltonian formulation of ray optics uses two sets of coordinates; the position of the ray x and y, and the ray's "momentum" p_x and p_y. Together these variables define a space of four dimensions (six in the classical mechanics of a single point particle). The physical state of a light ray can be described as a point in phase space. Each point in phase space provides information about the position and slope ("momentum") of the ray. A trajectory in phase space gives us the same amount of information that we would obtain by studying the position and slope of the ray as a function of the z coordinate. The extension of the description of rays from the study of a single ray to that of many rays and, finally, to a statistics of rays leads naturally to the consideration of volume elements in phase space. Consider a bundle of rays filling a certain area in real space and having a certain spread in directions. Each ray of the bundle occupies a point in phase space that is slightly different from the position of every other ray of the bundle. If the ray positions in real space and their slopes are limited in range, their representation points in phase space fill a certain finite volume. If we now follow the development of this ray bundle as it propagates through the optical system, we observe that the volume that was originally occupied by the rays changes its shape and moves through phase space. Instead of studying the history of every single ray in the bundle, it is easier to observe the motion of the volume of the representation points in phase space. Liouville's theorem provides information about the phase space volume of a bundle of rays. It can be expressed in several equivalent forms. In terms of phase space volume it states: *The volume in phase space, filled with representation points of a bundle of rays, remains constant.* This remarkable result is based on the ray equations. The rays themselves may move far apart, causing the phase space volume to deform and stretch out. The volume that was originally occupied by the rays in their starting position remains the same, however, even though we may no longer be able to recognize its original shape.

Liouville's theorem is based on the density ρ of representation points in phase space. The density is the number of representation points per unit of phase space volume. This density obeys the continuity equation

$$\frac{\partial \rho}{\partial z} + div(\rho \mathbf{v}) = 0 \tag{3.7-1}$$

Note that the time coordinate of mechanics is again replaced by z. The continuity equation states that the total number of points must be constant. Its proof is accomplished by noting that the change of the total number of points enclosed inside a fixed volume V_f of phase space must be accounted for by the motion of representation points into and out of that volume. Since the representation points move as functions of z, we have

$$\frac{\partial}{\partial z} \int_{V_f} \rho dV = -\int_S \rho \mathbf{v} \cdot \mathbf{n} dS \tag{3.7-2}$$

The integral on the left is the total number of points in the volume V_f, while the integral on the right accounts for the particle current with current density $\rho \mathbf{v}$ (particle velocity $\mathbf{v}$) that flows through the closed surface surrounding V_f. The unit vector $\mathbf{n}$ points in the direction of the outward normal. A negative current flow means, therefore, that the volume is gaining particles, accounting for the negative sign on the right-hand side. Application of the divergence theorem to the integral on the right-hand side leads to (3.7-1) when we observe that the volume V_f is arbitrary.

We have discussed this flow problem as if it takes place in ordinary three-dimensional space with z representing the time coordinate. The mathematical operations that were involved are, however, not limited to three-dimensional space, so that every statement applies to four-dimensional phase space. The divergence appearing in (3.7-1) is the generalization of the usual three-dimensional expression to the four dimensions of phase space. In component notation, we can write (3.7-1) as follows

$$\frac{\partial \rho}{\partial z} + \sum_{i=1}^{2} \left[\frac{\partial}{\partial x_i}\left(\rho \frac{dx_i}{dz}\right) + \frac{\partial}{\partial p_i}\left(\rho \frac{dp_i}{dz}\right) \right] = 0 \tag{3.7-3}$$

We use the notation $x_1 = x$, $x_2 = y$, $p_1 = p_x$, and $p_2 = p_y$. The coordinates and momenta represent actual rays that obey the laws of ray propagation. Using (3.5-4) through (3.8-8) allows us to write

$$\frac{\partial \rho}{\partial z} + \sum_{i=1}^{2} \left[\frac{\partial}{\partial x_i} \left(\rho \frac{\partial H}{\partial p_i} \right) - \frac{\partial}{\partial p_i} \left(\rho \frac{\partial H}{\partial x_i} \right) \right] = 0 \qquad (3.7\text{-}4)$$

It is useful to remember that it is at this point that the laws of ray optics were used. Carrying out the derivatives of the products allows us to cancel certain terms, so that we obtain

$$\frac{\partial \rho}{\partial z} + \sum_{i=1}^{2} \left(\frac{\partial \rho}{\partial x_i} \frac{\partial H}{\partial p_i} - \frac{\partial \rho}{\partial p_i} \frac{\partial H}{\partial x_i} \right) = 0 \qquad (3.7\text{-}5)$$

Using Hamilton's equations once more results in the expression

$$\frac{\partial \rho}{\partial z} + \sum_{i=1}^{2} \left(\frac{\partial \rho}{\partial x_i} \frac{dx_i}{dz} + \frac{\partial \rho}{\partial p_i} \frac{dp_i}{dz} \right) = 0 \qquad (3.7\text{-}6)$$

The density ρ is a function of x, y, p_x, and p_y. It is apparent, therefore, that the left-hand side is the total derivative of ρ with respect to z, and we obtain Liouville's theorem in its first form

$$\frac{d\rho}{dz} = 0 \qquad (3.7\text{-}7)$$

The density of representation points in phase space is independent of the z coordinate. We can use this result to derive the second form of Liouville's theorem. Let us enclose a certain number of representation points in a volume V and allow the boundaries of this volume to change in such a way that the total number of points contained inside does not vary. It is then by definition

$$\frac{d}{dz} \int_V \rho dV = 0 \qquad (3.7\text{-}8)$$

We can arrange the representation points any way we want. We choose the arrangement so that ρ is constant inside the volume V. This allows us to take ρ out of the integral and, using (3.7-7), also move it in front of the derivative sign. The result is the equation

$$\frac{dV}{dz} = 0 \qquad (3.7\text{-}9)$$

This is Liouville's theorem in its second form. It states that the volume enclosing a fixed number of representation points remains constant in z. This volume gets carried along by the representation points, but the numerical value of the volume does not change even though its shape may deform substantially.

We can express Liouville's theorem in a third form, which is most useful for our future purposes. To obtain this last version of our theorem requires a little introduction. Let us assume a ray at an input plane $z = z_i$ with x_i, y_i,

p_{xi}, p_{yi}. Following the ray from the initial point $z = z_i$ to an arbitrary point $z = z_f$ along its trajectory changes, in general, all the values of the coordinates. Let us designate the variables at an arbitrary point of the ray by x_f, y_f, p_{xf}, p_{yf}. Our ray is a member of a bundle of rays that fill a certain volume in phase space. The neighbors of our ray have coordinates whose initial values and final values are somewhat different from those of our original ray. The point to be made is that, as we vary the initial coordinates of the ray at the fixed plane $z = z_i$ its final corrdinates at the fixed plane $z = z_f$ vary too. In fact, the final coordinates are functions of the initial coordinates

$$x_f = x_f(x_i, y_i, p_{xi}, p_{yi}) \tag{3.7-10}$$

$$p_{xf} = p_{xf}(x_i, y_i, p_{xi}, p_{yi}) \tag{3.7-11}$$

Corresponding functional relations hold for the other two coordinates. The z coordinate is held constant, and need not be included in this functional relationship. Next, we need a result from the theory of volume integrals. If an integral, which is expressed in terms of a set of variables, is to be transformed to a new set of primed variables, the transformation is accomplished with the help of the Jacobian of the transformation.[26] This transformation property of volume integrals can be expressed by the equation

$$dV_f = \frac{\partial(x_f, y_f, p_{xf}, p_{yf})}{\partial(x_i, y_i, p_{xi}, p_{yi})} dV_i \tag{3.7-12}$$

We have immediately applied this theorem to our set of variables. The equations (3.7-10) and (3.7-11) and the remaining equations not explicitly shown constitute a transformation from one set of variables to a new one, so that the transformation theory of volume integrals applies. Liouville's theorem in its second form (3.7-8) states that the two volume elements must be the same, since the volume of the conglomerate of representation points in phase space does not change. We must have

$$dV_f = dV_i \tag{3.7-13}$$

The Jacobian appearing in (3.7-12) must consequently have the value unity. From the definition of the Jacobian, we obtain our third version of Liouville's theorem

$$\frac{\partial(x_f, y_f, p_{xf}, p_{yf})}{\partial(x_i, y_i, p_{xi}, p_{yi})} = \begin{vmatrix} \dfrac{\partial x_f}{\partial x_i} & \dfrac{\partial y_f}{\partial x_i} & \dfrac{\partial p_{xf}}{\partial x_i} & \dfrac{\partial p_{yf}}{\partial x_i} \\[2mm] \dfrac{\partial x_f}{\partial y_i} & \dfrac{\partial y_f}{\partial y_i} & \dfrac{\partial p_{xf}}{\partial y_i} & \dfrac{\partial p_{yf}}{\partial y_i} \\[2mm] \dfrac{\partial x_f}{\partial p_{xi}} & \dfrac{\partial y_f}{\partial p_{xi}} & \dfrac{\partial p_{xf}}{\partial p_{xi}} & \dfrac{\partial p_{yf}}{\partial p_{xi}} \\[2mm] \dfrac{\partial x_f}{\partial p_{yi}} & \dfrac{\partial y_f}{\partial p_{yi}} & \dfrac{\partial p_{xf}}{\partial p_{yi}} & \dfrac{\partial p_{yf}}{\partial p_{yi}} \end{vmatrix} = 1 \tag{3.7-14}$$

Liouville's theorem holds not only for continuous distributions of the refractive index but also for rays crossing the interface between media with different index of refraction and also for light reflection from curved mirrors. The verification of the validity of the theorem for light reflection is left as an exercise for the reader. The validity of Liouville's theorem for dielectric interfaces can be ascertained by the following argument. Let us assume that the abrupt index distribution of the interface is replaced by an arbitrary continuous distribution, as shown in Figure 3.7.1. The z-axis of the coordinate system was chosen to coincide with the normal to the (smoothed out) interface. The refractive index is constant in two regions, and is a continuous function $n(z)$ inside the transition region. With our choice of coordinate system and index distribution, we obtain from (3.5-30) and (3.5-31)

$$p_x = \text{const} \tag{3.7-15}$$

$$p_y = \text{const} \tag{3.7-16}$$

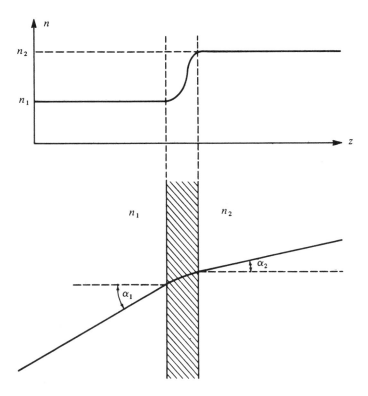

Figure 3.7.1   The abrupt index change at a dielectric interface has been smoothed out into a continuous distribution.

throughout all space. This result is completely independent of the shape of the function $n(z)$. Equations (3.5-32) and (3.5-33) can be augmented by the relation

$$p_z = n \frac{dz}{ds} \tag{3.7-17}$$

All three equations lead to the square of the length of the "momentum" vector

$$p_x^2 + p_y^2 + p_z^2 = n^2 \left[\left(\frac{dx}{ds} \right)^2 + \left(\frac{dy}{ds} \right)^2 + \left(\frac{dz}{ds} \right)^2 \right] = n^2 \tag{3.7-18}$$

Using the angle α_1, defining the ray slope in medium 1, and α_2 for the slope in region 2 (see the definition of these angles as shown in Figure 3.7.1), we obtain immediately Snell's law from (3.7-15) (we have assumed that the coordinate system is located so that $p_y = 0$)

$$(p_x)_1 = n_1 \sin \alpha_1 = (p_x)_2 = n_2 \sin \alpha_2 \tag{3.7-19}$$

This discussion shows that Snell's law holds not only for an abrupt change of the index at an interface but even for an arbitrary index distribution of arbitrary width. Since the index distribution does not enter into the argument at all, we can allow it to become as abrupt as we like and obtain the discontinuous interface as a legitimate limiting process. This discussion shows that we can assume every discontinuous change of the refractive index to be smoothed out, so that the ray equations and consequently Liouville's theorem hold. The ray trajectories in the smoothed-out medium are asymptotically identical to the trajectories in the discontinuous medium if the index variation is allowed to become more and more abrupt.

A second demonstration of the validity of Liouville's theorem uses an abrupt interface from the start. It is of course possible to verify the validity of (3.7-14) by direct calculation for any desired case. However, for arbitrary interfaces, such a direct calculation is extremely tedious. Our second demonstration shows how such a direct calculation can be simplified by a rotation of the coordinate system.

Consider the ray problem that is sketched in Figure 3.7.2. A plane interface is located at an arbitrary angle to the coordinate system. The problem is reduced to two dimensions for simplicity. It is far easier to trace rays through an interface that is located perpendicularly to the z-axis of the coordinate system. For this reason we introduce the coordinates shown by dotted lines in Figure 3.7.2. However, rotating the coordinate system presents certain difficulties reminiscent of the problem of Galilei versus the Lorentz transformation.[21] Liouville's theorem is critically based on the fact that the ray coordinates x, y, p_x, and p_y are all taken at the same value of the z coordinate. The rays at point P_1 and P_2 shown in Figure 3.7.2 satisfy this requirement if expressed in the original coordinate system x, z. However, in the rotated coordinate system, each point has a different value of z'. In order to be able to use Liouville's theorem, we must not simply refer the ray position to the coordinates x' and z' that result from the rotation alone but must use new coordinates by following each ray to its point of intersection with a plane that is perpendicular to the z'-axis. We are not allowed to simply rotate the coordinate system geometrically but must change the coordinates of each point to make certain that each ray is again referred to the same point z'. This procedure is analogous to the problem of simultaneity encountered in the theory of relativity. We see once more the relationship of ray optics with relativistic mechanics. Only in the paraxial (nonrelativistic) approximation, where all angles (inclusive of the angles of the rays with the normal to the interface) are small, would a simple rotation of the coordinate system be approximately correct. The phase space representation of the four rays at the points P_1 and P_2 as expressed in the old coordinate system is shown in Figure 3.7.3. The points indicated as 1 and 2 in Figure 3.7.3 correspond to P_1 in Figure 3.7.2, while the points 3 and 4 correspond to P_2. In the new coordinate system, the positions of the four rays must be expressed by the intersections of the dotted line (which runs through P_1) with the rays. The corresponding phase space representation is shown in Figure 3.7.4. The points have all moved up along the p_x'-axis, and the original square shows a slight trapezoidal deformation. Using an average slope angle α between the slopes of ray 3 and 4, we can express the relation between the length increment dx' in the new coordinate system with that of dx in the old one by simple geometry in the form

$$dx = (\cos \phi - \sin \phi \tan \alpha)dx' \tag{3.7-20}$$

The "momentum" in the old coordinate system follows from (3.5-38) or (3.7-19)

$$p_x = n_1 \sin \alpha \tag{3.7-21}$$

Its increment is given by

$$dp_x = n_1 \cos \alpha \, d\alpha \tag{3.7-22}$$

The momentum in the new coordinate system is

$$p_x' = n_1 \sin (\alpha + \phi) \tag{3.7-23}$$

with the increment

$$dp_x' = n_1 \cos (\alpha + \phi)d\alpha \tag{3.7-24}$$

The increment in the old coordinate system can now be expressed in terms of the increment in the new one

$$dp_x = \frac{\cos \alpha}{\cos (\alpha + \phi)} dp_x' \tag{3.7-25}$$

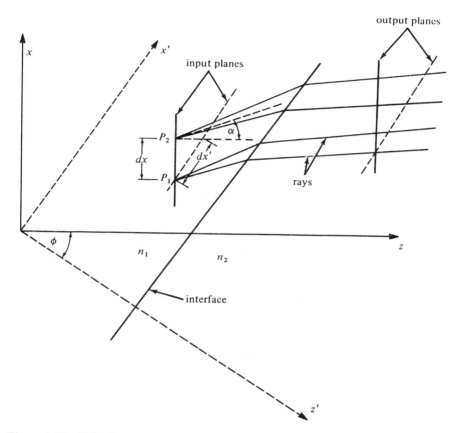

Figure 3.7.2 This diagram aids in the proof that Liouville's theorem holds for rays crossing a dielectric interface between two different media. A new coordinate system is introduced whose x' axis is parallel to the interface.

The volume elements in phase space expressed in the two coordinate systems are related by the expression

$$dxdp_x = \frac{\cos \alpha(\cos \phi - \tan \alpha \sin \phi)}{\cos(\alpha + \phi)} dx'dp_x' \qquad (3.7\text{-}26)$$

Or, using $dV = dxdp_x$ and an addition theorem of circular functions, we obtain the result

$$dV = dV' \qquad (3.7\text{-}27)$$

This calculation shows that the phase space volume is conserved by the transformation to the new coordinate system. This is a convenient result, for it shows that we are allowed to rotate the coordinate system to any convenient

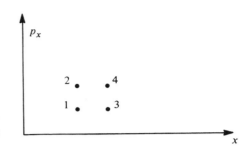

Figure 3.7.3   Phase space diagram for the rays that pass through the points P_1 and P_2 of Figure 3.7.2.

position when we try to confirm Liouville's theorem for rays that are either reflected by a mirror or, as in this case, refracted at a dielectric discontinuity. The result of the invariance of the phase space volume is independent of the value of the refractive index, and of course holds also when we rotate the coordinate system back to its original position after tracing the rays through the interface. This result also holds for the real three-dimensional case. Our limitation to a two-dimensional problem was motivated by convenience.

To prove Liouville's theorem for rays passing through an interface between two media with different index of refraction, we use a rotated coordinate system with its z'-axis perpendicular to the point at which the rays intersect the interface. The boundary condition for rays, (3.3-5), can be reexpressed with the help of (3.5-32) and (3.5-33) in the form

$$(\mathbf{p}_t)_1 = (\mathbf{p}_t)_2 \tag{3.7-28}$$

The tangential components of the $\mathbf{p}$ vector remain continuous on crossing the interface. With the choice of the rotated coordinate system, the tangential components are p_x' and p_y'. We know that Liouville's theorem holds along the ray trajectory in free space, so that we can use ray positions immediately adjacent to the interface to prove the validity of (3.7-14). Crossing the interface does not change the ray position. Since the tangential x', y' components of $\mathbf{p}$ are also the same on either side of the interface we find

$$dV_1 = dV_2 \tag{3.7-29}$$

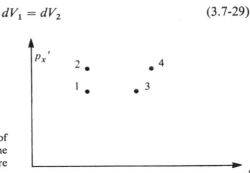

Figure 3.7.4   Phase space diagram of the rays that cross the dotted line passing through point P_1 of Figure 3.7.2.

so that Liouville's theorem holds for rays at index discontinuities. Once we know that we are allowed to rotate the coordinate system to our convenience the proof that Liouville's theorem holds at a discontinuous dielectric interface has become trivial. This method of verifying Liouville's theorem directly is most advantageous for checking its validity for reflection from arbitrary surfaces. We did convince ourselves by the limiting procedure presented earlier that the theorem must hold at dielectric interfaces. However, the case of reflection cannot be handled in the same manner, so that a proof by a direct method appears desirable. This proof for reflected rays is left to the reader as an exercise. The direct calculation becomes quite simple when we use the fact that we are allowed to rotate the coordinate system and refer the ray positions to a plane of constant z' values.

Next we use Liouville's theorem to prove Abbe's sine theorem[1] of optical imaging. This theorem is not important for us in the context of this book. Its derivation is included only as a demonstration of the usefulness of Liouville's theorem. The sine condition applies to image-forming systems. More specifically, it applies to imaging without aberrations. Let us assume that we want to determine the relations between ray angles and distances between neighboring image and object points. Figure 3.7.5 shows the geometry of the problem. The object point P_1 is located on the optical axis of the system. Its image point P_1' is also located on the optical axis. In fact, the optical axis is defined by this relationship. Because of our assumption of perfect optical imaging, every ray leaving P_1 under an arbitrary angle must pass through P_1'. The same is of course also true for every ray leaving the neighboring object point P_2. They must all pass through its image P_2'. Liouville's theorem in its two-dimensional form requires

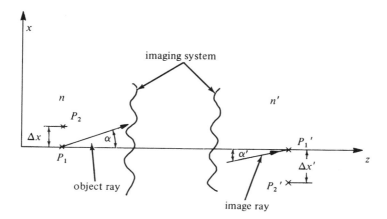

Figure 3.7.5 Illustration of the geometry of the optical imaging system. Object points P_1 and P_2 and image points P_1' and P_2' as well as the direction of two typical rays used in Abbe's sine theorem are shown.

$$\begin{vmatrix} \dfrac{\partial x'}{\partial x} & \dfrac{\partial p'_x}{\partial x} \\[2ex] \dfrac{\partial x'}{\partial p_x} & \dfrac{\partial p'_x}{\partial p_x} \end{vmatrix} = 1 \tag{3.7-30}$$

We assume that Δx and $\Delta x'$ are the vertical separations between the two adjacent object and image points. Because of the requirement of stigmatic (perfect) imaging, we must require that the ray position x' in the image plane is independent of the ray angle (its "momentum"). This means, in terms of the derivatives appearing in (3.7-30)

$$\frac{\partial x'}{\partial p_x} = 0 \tag{3.7-31}$$

Liouville's theorem (3.7-30) then requires that the following relation between neighboring object and image points and ray "momenta" must hold

$$\Delta x \Delta p_x = \Delta x' \Delta p'_x \tag{3.7-32}$$

Since the position of each pair of object and image points is independent of the angle at which a ray leaves the object point, we can integrate the relation with respect to p_x and obtain

$$\Delta x p_x = \Delta x' p'_x \tag{3.7-33}$$

Finally, by using (3.5-38), we obtain Abbe's sine condition

$$n \Delta x \sin \alpha = n' \Delta x' \sin \alpha' \tag{3.7-34}$$

This condition is usually derived without the benefit of Liouville's theorem.[1] Our derivation shows that the sine condition is only a special case of Liouville's theorem. The ratio $\Delta x'/\Delta x$ is the magnification of the imaging system. The sine theorem provides a relation between image magnification and the angles of the departing and arriving rays. It is based on the assumption of stigmatic imaging.

Finally, let us derive another relation that is important for our future use of light propagation through rather general optical transmission systems. It is convenient to use the idea of a thin optical transformer. This is a device that has (ideally) zero thickness but changes the angle of every ray passing through it as a function of its position in the transformer. Equivalently we can say that the thin optical transformer imparts a position-dependent phase shift to a wave passing through it. A sketch of the thin optical transformer is shown in Figure 3.7.6. The thin optical transformer may be a thin lens. The fact that it is shown positioned perpendicular to the z-axis is no restriction, since we have seen that we are allowed to rotate the coordinate system without affecting the validity of Liouville's theorem. It is again sufficient to apply Liouville's theorem in its two-dimensional form (3.7-30).

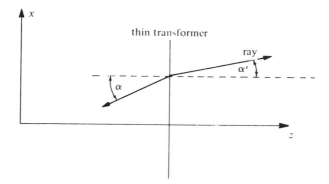

Figure 3.7.6   The thin optical transformer. The incident ray with angle α leaves the transformer with angle α'.

We choose the two points x and x' on the ray trajectory to lie immediately in front and behind the transformer. This assumption implies that $x = x'$ and that x' is independent of p_x. Liouville's theorem (3.7-30) results in

$$\frac{\partial p_x'}{\partial p_x} = 1 \tag{3.7-35}$$

We multiply (3.7-35) by dp_x and integrate. The integration introduces an arbitrary constant that may be a function of x

$$p_x' = p_x + F(x) \tag{3.7-36}$$

In terms of the angles, this equation becomes

$$\sin \alpha' = \sin \alpha + F(x) \tag{3.7-37}$$

The refractive index of the media in front and behind the transformer is assumed to be identical. The function $F(x)$ is arbitrary, and gives us considerable freedom of choice of thin optical transformers. However, the change of the ray angle that the transformer can accomplish is severely restricted by Liouville's theorem. Using the addition theorem of sine functions, we can express (3.7-37) in terms of the angle difference $\alpha' - \alpha$

$$\sin \frac{\alpha' - \alpha}{2} = \frac{F(x)}{2 \cos \frac{\alpha' + \alpha}{2}} \tag{3.7-38}$$

In the paraxial approximation, we assume that both α and α' are very small, so that we can use the approximate form of (3.7-38) to obtain the well-known thin lens formula

$$\alpha' - \alpha = F(x) \qquad\qquad (3.7\text{-}39)$$

In the paraxial approximation, we can think of a thin optical transformer (thin lens) as a device that changes the angle of a ray as a function of ray position on the transformer but is independent of the angle of incidence. However, the change in angle must depend on the angle of incidence if this angle is large. We are not allowed to assume that we can make a thin optical device that simply changes the ray angle independent of the angle of incidence. This is an important and surprising result. Being used to working with thin lenses at near normal incidence, one gets used to the idea that a thin lens does nothing else than break each ray by a fixed amount depending only on the point where the ray penetrates the lens but is independent of the angle of incidence. If this simple rule is applied to thin lenses that are tilted with respect to the axis or perhaps even warped to a more general form, grave errors can result if the consequences of Liouville's theorem are not taken into account. The ray trajectory through a thin lens is given by (3.7-38). The only freedom left to the experimenter is the choice of the function $F(x)$. The consequences of this discussion for the transmission of light beams through systems of warped lenses will be pursued in a later chapter.

An application of Liouville's theorem to the problem of the compression of a bundle of light rays by tapered pieces of dielectric material and by lenses can be found in reference 102.

CHAPTER

LENSES

4.1 INTRODUCTION

No other components of optical systems are of such extraordinary importance as lenses. The principal component of every conventional image-forming system, including the eyes of animals, is a lens. Lenses in man-made optical systems are usually made of glass. The purpose of the lens in an image-forming system is to alter the path of light rays in such a way that all rays that originate at an object point converge again at an image point. No lens can do this job perfectly for all the requirements that are usually imposed on image formation. In order to improve the performance of single lenses, complicated compound systems are usually employed in an attempt to reach an acceptable compromise between several conflicting requirements.

The construction of single and compound lenses is exhaustively covered in most books describing the imaging properties of optical instruments.[1,15] For this reason we shall not go into any detailed description of the many means of correcting lenses for the various aberrations. We shall limit the discussion to those properties of lenses that are important for the transmission of light through lens waveguides and to the Fourier transform properties of lenses that are important in optical data processing and optical analog computers. The principle of operation of these devices can be explained adequately in terms of thin lenses. A lens is called thin when the ray trajectory inside the lens need not be considered in detail. A thin lens acts as a device that simply changes the direction of light rays passing through it. In terms of wave optics, we can describe a thin lens as a phase transformer that changes the phase of the wave passing through it. The phase change depends on the part of the lens on which the wave is incident.

Our description of (thin) lenses will use ray optics as well as wave optics.

4.2 RAY OPTICS OF THIN LENSES

Let us define a thin lens as a device that deflects every light beam incident parallel to the optical axis in such a way that it crosses the optical axis at a fixed distance f after traversing the lens. Figure 4.2.1 shows a thin converging lens and its effect on two light rays for the two-dimensional case. The thin lens must obey the general formula (3.7-37). In order to satisfy the requirements that every ray incident parallel to the optical axis crosses the same axis at the distance f from the lens, the function $F(x)$ assumes the following form (note $\alpha = 0$)

$$F(x) = \sin \alpha' = \frac{x}{\sqrt{x^2 + f^2}} \qquad (4.2\text{-}1)$$

For rotationally symmetric lenses, we replace the distance x with the radial distance r at which the ray enters the lens

$$F(r) = \frac{r}{\sqrt{r^2 + f^2}} \qquad (4.2\text{-}2)$$

The general thin lens formula for a ray that impinges on the lens at an angle α (measured with respect to the normal to the plane of the thin lens) and leaves at an angle α' is given by (3.7-37) and (4.2-2)

$$\sin \alpha' = \sin \alpha + \frac{r}{\sqrt{r^2 + f^2}} \qquad (4.2\text{-}3)$$

In the paraxial approximation, the thin lens equation assumes the familiar form

$$\alpha' - \alpha = \frac{r}{f} \qquad (4.2\text{-}4)$$

If the focal length f is also a function of r, the thin lens is not distortion-free. Actual physical lenses are never entirely free of distortions. Even if we

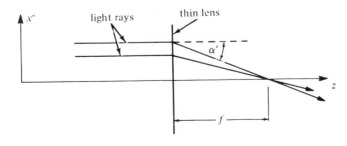

Figure 4.2.1   Schematic of an ideal thin lens. All horizontal input rays pass through the focal point after leaving the lens.

are allowed to use the thin lens approximation, we must expect that f is a function not only of r but also of frequency. As long as r remains small, it is usually possible to neglect the r dependence of f and apply the paraxial approximation with f being a constant. We use the paraxial formula (4.2-4) to describe the image-forming properties of the lens. It is apparent from (4.2-3) as well as (4.2-4) that a ray passing the lens on the optical axis $r = 0$ does not suffer any change in direction. The knowledge that a ray incident parallel to the optical axis passes the axis at the focal point f and that a ray through f leaves the lens traveling parallel to the optical axis and finally that a ray that passes the lens with $r = 0$ traverses the lens without change of direction is sufficient to determine the image of any object placed in front of the lens. Let us designate the distance of the object (from the lens) by a and the distance of the image (from the lens) by b. Figure 4.2.2 shows the construction that allows us to determine the image. Of the three rays shown in the figure, two are sufficient to determine the image point. From the ray that passes through the focus on the right of the lens, we obtain from simple geometry

$$\frac{h_o}{f} = \frac{h_o + h_i}{b} \qquad (4.2\text{-}5)$$

A similar relation is obtained for the ray passing the left focal point

$$\frac{h_i}{f} = \frac{h_o + h_i}{a} \qquad (4.2\text{-}6)$$

Adding the two equations results in an important equation that relates the object distance a and the image distance b to the focal length of the lens.

$$\frac{1}{f} = \frac{1}{a} + \frac{1}{b} \qquad (4.2\text{-}7)$$

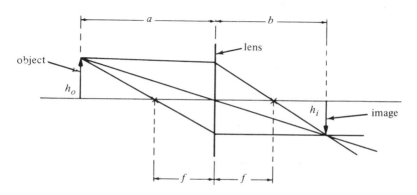

Figure 4.2.2   Image formation with the help of an ideal thin lens.

This equation shows that for an object at infinity we obtain $b = f$. As the object approaches the left focal point, b increases. At $a = f$, we find $b = \infty$. For $a < f$, b becomes negative. Instead of rejecting negative image distances, we interpret negative values of b as belonging to a so-called virtual image. The real object and its virtual image are shown in Figure 4.2.3.

An observer viewing the object through the lens interprets the rays reaching him on the right-hand side of the lens as coming from the virtual image shown in Figure 4.2.3. Figure 4.2.2 explains the principle of the photographic camera. If light-sensitive film is placed at the position b of the real object, its image is recorded on the film. Figure 4.2.3 is an explanation of the action of the magnifying glass. The viewer holds the lens so that the virtual image is formed at a convenient viewing distance b. The object must be located between the lens and the left focal point. By taking the ratio of (4.2-5) and (4.2-6), we obtain the image magnification

$$\frac{h_i}{h_o} = \frac{b}{a} \qquad (4.2\text{-}8)$$

We close this section with a brief comment on the resolution of optical instruments without getting deeper into the image-forming properties of optical systems. The precision with which an image point can be determined depends on the limitation on the precision of determining the position of a ray. The rays of ray optics are only an approximate description of optical phenomena. The exact description must be given in terms of wave optics. Since wave optics has been identified in Section 3.6 as the quantum theory of ray optics, we know that the wave function is the probability amplitude for finding a ray. The position and "momentum" of light rays are two quantities that cannot be determined with arbitrary precision. In order to be able to locate a light ray to an accuracy Δx in traverse direction, we must allow for a spread of the ray's "momentum." From the uncertainty principle (3.6-56), we obtain the limit on the accuracy of determining the position of the light ray or, alternately expressed, on the resolution of an image point

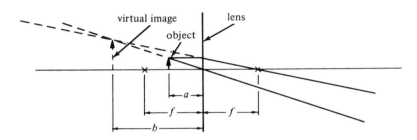

Figure 4.2.3   A virtual image is formed of an object that is located between the lens and its focal point.

$$\Delta x \geq \frac{\lambda}{4\pi \, \Delta p} \tag{4.2-9}$$

The "momentum" of the ray is related to its slope by (3.5-38). All rays originating at an object point converge on the corresponding image point, as shown in Figure 4.2.4. The uncertainty of the ray's momenta must be related to the maximum angle of the ray permitted by the lens aperture. Using

$$\Delta p = n \sin \alpha \tag{4.2-10}$$

as a rough estimate of the uncertainty of the ray's "momentum," we obtain an estimate for the resolution limit of the optical system

$$\Delta x \geq \frac{\lambda}{4\pi n \sin \alpha} \tag{4.2-11}$$

This formula agrees with Rayleigh's famous resolution limit at least to the order of magnitude.[1] We shall return to the subject of optical resolution in connection with the wave optics description of lenses.

4.3 WAVE OPTICS OF THIN LENSES

The understanding of wave optics properties of thin lenses does not require the use of Maxwell's theory. The polarization of light has an influence on its reflection properties from the front and back surfaces of lenses. However, if we are willing to disregard such reflection phenomena, we are justified to use the scalar wave equation to describe the wave optics of lenses. Furthermore, we limit the discussion again to the description of thin lenses. The additional complications that arise from thick lenses can best be treated

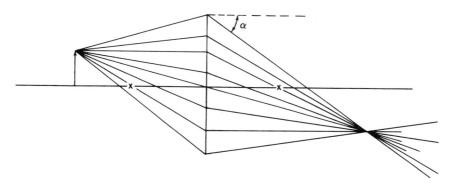

Figure 4.2.4   All rays from an object point converge at the image point of an ideal thin lens.

by means of ray optics. Such treatments can be found in most books on optics.[1,15]

The thin lens is a phase transformer. If we consider a wave field in front of and immediately behind the lens, we find that the phase of the wave has changed while its amplitude, for all practical purposes, remains the same. Let us call the input field immediately in front of the lens ψ_i and the output field immediately behind the lens ψ_o. Assuming, for simplicity, a wave field of rotational symmetry, we can describe the action of the lens by the equation

$$\psi_o(r) = e^{i\gamma(r)} \psi_i(r) \qquad (4.3\text{-}1)$$

r is the distance measured perpendicular from the optical axis. We require our thin lens to have the same properties as the lens described by (4.2-4) in the previous section. The fact that the rays are the orthogonal trajectories to the wave fronts allows us to translate Figure 4.2.1 into wave optics. The result is shown in Figure 4.3.1.

The dotted lines represent the wave fronts. The incident plane wave is changed to a spherical wave converging on the focal point at distance f behind the lens. The phase of the wave immediately after passing the lens is given by

$$\gamma(r) = -(\rho^2 - r^2)\,K \qquad (4.3\text{-}2)$$

The negative sign is required in order to obtain an advancing wave with the time dependence (2.2-1). Equation (4.3-2) can be justified as follows. At $r = \rho$, the radius of the lens, the phase shift is zero. It increases for decreasing values of r, and reaches its maximum at $r = 0$. This behavior becomes immediately obvious if we remember that converging lenses are thick in their center section on the optical axis and become thinner toward their

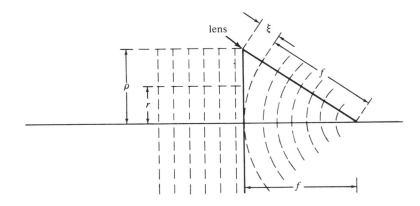

Figure 4.3.1   The thin lens acts as a phase transformer. An incident plane wave leaves the lens as a spherical wave converging on the focal point.

periphery. The thick portion presents a longer optical path length to the light field than does the thin portion on the rim. A longer path means more phase shift. The assumption of a strict square law dependence is an approximation that is valid for lenses that obey the relation

$$\rho \ll f \qquad (4.3\text{-}3)$$

The square law behavior becomes obvious if we keep in mind that the phase shift must be the same for r and $-r$. The square law is the lowest power that satisfies this requirement. The existence of fourth order terms would indicate that the lens has aberrations. These simple arguments allow us to determine the phase shift up to an unknown parameter K. This constant must describe the power of the lens. We determine this constant in the paraxial approximation. Since every wave front is a surface of constant phase, we know that the wave front at distance ξ from the rim of the lens (see Figure 4.3.1) has the phase $\gamma(0)$. We obtain the relation

$$\gamma(0) = -\rho^2 K = -k\xi \qquad (4.3\text{-}4)$$

The phase shift $k\xi$ corresponds to a wave traveling from the rim of the lens the distance ξ to the wave front that touches the thin lens at $r = 0$. The constant k is the propagation constant (1.3-13). The distance ξ can be obtained from the theorem of Pythagoras applied to the triangle shown in Figure 4.3.1

$$(f + \xi)^2 = f^2 + \rho^2 \qquad (4.3\text{-}5)$$

Neglecting ξ^2 leads to the solution

$$\xi = \frac{\rho^2}{2f} \qquad (4.3\text{-}6)$$

The constant K now follows from (4.3-4) and (4.3-6)

$$K = \frac{k}{2f} \qquad (4.3\text{-}7)$$

The phase shift factor of the thin lens is finally obtained from (4.3-2) and (4.3-7)

$$\gamma(r) = -\frac{k}{2f}(\rho^2 - r^2) \qquad (4.3\text{-}8)$$

The phase shift is all we need to calculate the transmission of waves through the thin lens. Let us begin by assuming a simple two-dimensional case. Following our usual practice, we call a problem two-dimensional if it is independent of one of the coordinates. Assuming no dependence on y, we consider the transmission of a wave field from a plane a distance a in front of the lens to a distance b behind the lens. In our two-dimensional problem we are dealing with the thin lens equivalent of a cylindrical lens. The geometry of our problem is shown in Figure 4.3.2.

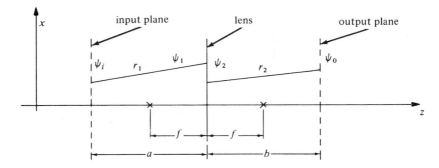

Figure 4.3.2 This diagram defines some of the quantities used for the wave optics description of image formation by a thin lens.

The transmission of the input field ψ_i at z_i to the position just in front of the lens is described by the diffraction integral (2.2-43)

$$\psi_1 = \frac{e^{i\pi/4}}{\sqrt{\lambda\,(z'-z_i)}}\, e^{-ik(z'-z_i)} \int_{-\infty}^{\infty} \psi_i\,(x_i, z_i)\, \exp\left[-ik\,\frac{(x'-x_i)^2}{2\,(z'-z_i)}\right] dx_i \quad (4.3-9)$$

In the spirit of the paraxial approximation, we replaced the cosine factors appearing in (2.2-43) by unity. The field describes the wave immediately to the left of the lens. Passing the lens transforms this field to (see [4.3-1] and [4.3-8])

$$\psi_2 = e^{-i(k/2f)(\rho^2 - x'^2)}\,\psi_1 \quad (4.3-10)$$

describing the wave immediately to the right of the lens. The field at the output plane is finally obtained by

$$\psi_o = \frac{e^{i\pi/4}}{\sqrt{\lambda\,(z_o-z')}}\, e^{-ik(z_o-z')} \int_{-\infty}^{\infty} \psi_2\,(x', z')\, \exp\left[-ik\,\frac{(x_o-x')^2}{2\,(z_o-z')}\right] dx' \quad (4.3-11)$$

When we substitute (4.3-9) into (4.3-10) and use this equation to replace ψ_2 in (4.3-11), we obtain a double integral. The integration not involving the input field ψ_i leaves us with the integral

$$I = \int_{-\infty}^{\infty} \exp\left[-\frac{i}{2}k\left(\frac{1}{a}+\frac{1}{b}-\frac{1}{f}\right)x'^2\right] \exp\left[ik\left(\frac{x_i}{a}+\frac{x_o}{b}\right)x'\right] dx' \quad (4.3-12)$$

The relations

$$z' - z_i = a \quad (4.3-13)$$

and

$$z_o - z' = b \quad (4.3-14)$$

were used to simplify the notation of (4.3-12). The integral is of the form

$$I = \int_{-\infty}^{\infty} e^{i(\alpha x^2 + \beta x)} \, dx = \sqrt{\frac{i\pi}{\alpha}} \, e^{-i(\beta^2/4\alpha)} \tag{4.3-15}$$

With

$$\alpha = -\frac{k}{2}\left(\frac{1}{a} + \frac{1}{b} - \frac{1}{f}\right) \tag{4.3-16}$$

and

$$\beta = k\left(\frac{x_i}{a} + \frac{x_o}{b}\right) \tag{4.3-17}$$

we obtain the relation between the input field and the output field as follows

$$\psi_o = \frac{\sqrt{\pi} \, e^{i(3/4)\pi}}{\lambda\sqrt{ab}} \, e^{-ik(z_o - z_i)} \, e^{-i(k/2f)\rho^2} \frac{1}{\sqrt{\alpha}} \int_{-\infty}^{\infty} \psi_i(x_i, z_i)$$

$$\cdot \exp\left\{-\frac{i}{2}k\left[\frac{x_i^2}{a} + \frac{x_o^2}{b} + \frac{k}{2\alpha}\left(\frac{x_i}{a} + \frac{x_o}{b}\right)^2\right]\right\} \, dx_i \tag{4.3-18}$$

In its full generality, this equation is quite complicated. There are, however, several interesting special cases that are worth investigating. Let us begin by considering the case

$$\alpha = 0 \tag{4.3-19}$$

Since α appears in the denominator of (4.3-18), we are apparently faced with a singularity. However, this case can be handled quite easily if we return to (4.3-15). For $\alpha = 0$ the x^2 term disappears from the integrand, and we are left with the representation of the delta function.

$$I = \int_{-\infty}^{\infty} e^{i\beta x} dx = 2\pi\delta(\beta) \tag{4.3-20}$$

We can combine (4.3-15) with (4.3-20) to obtain the result of the limiting process

$$\lim_{\alpha \to 0} \sqrt{\frac{i\pi}{\alpha}} \, e^{-i(\beta^2/4\alpha)} = 2\pi\delta(\beta) \tag{4.3-21}$$

We have thus found another representation of the delta function. Using our result in (4.3-18) results in

$$\psi_o = \frac{2\pi i}{\lambda\sqrt{ab}} \, e^{-ik(a+b)} \, e^{-i(k/2f)\rho^2} \int_{-\infty}^{\infty} \psi_i(x_i, z_i) \, e^{-i(k/2)(x_i^2/a + x_o^2/b)}$$

$$\delta\left[k\left(\frac{x_i}{a} + \frac{x_o}{b}\right)\right] \, dx_i \tag{4.3-22}$$

The integral can now be evaluated with the help of the delta function appearing in its integrand

$$\psi_o (x_o, z_o) = i \sqrt{\frac{a}{b}} \, e^{-ik(a+b)} \, e^{-i(k/2f)[\rho^2 + (a/b) \times \frac{2}{o}]} \psi_i \left(-\frac{a}{b} x_o, z_i \right) \qquad (4.3\text{-}23)$$

The relations

$$k = \frac{2\pi}{\lambda} \qquad (4.3\text{-}24)$$

and

$$\frac{1}{a} + \frac{1}{b} = \frac{1}{f} \qquad (4.3\text{-}25)$$

following from (4.3-16) with $\alpha = 0$ were used to simplify (4.3-23). Equation (4.3-23) is the law of image formation expressed in wave optics terms. The condition $\alpha = 0$ results in (4.3-25), which is the required relation between object distances, image distances, and the focal length of the lens. This relation appeared already as Equation (4.2-7) in our ray optics treatment of optical imaging by lenses. Equation (4.3-23) expresses the fact that the image field and the object field are identical except for an inversion of the image coordinates, expressed by the negative sign attached to the argument x_o. In addition to the image inversion, we see that a scale factor a/b has appeared, expressing the fact that the image is different in size from the object. The image magnification factor b/a was already encountered in our ray optics treatment of image formation (4.2-8). The phase factors attached to (4.3-23) are of no consequence for the image-forming property of the lens. Both Photographic film and the retina of the eye are insensitive to the phase of the light field. They respond to $|\psi|^2$ instead of ψ. The factor $\sqrt{a/b}$ shows that the light intensity of the image is reduced from the intensity of the object if the image is magnified, $b/a > 1$.

The reader has undoubtedly noticed that we have tacitly assumed that our lens is so large that its actual limits could be ignored and the integration could be extended from $-\infty$ to $+\infty$. This is obviously an approximation that is well justified if the amount of light that misses the lens is small. However, the finite size of the lens that was ignored in our calculation leads to diffraction effects and limits the achievable image resolution (see Section 4.6).

Returning to Equation (4.3-18), we continue our discussion of the wave optics of lenses by considering another special case. We assume that

$$a = b = f \qquad (4.3\text{-}26)$$

This means the object field and the image field are both located at the front and back focal planes of the lens. Substitution of (4.3-16) and (4.3-26) into (4.3-18) leads to the result

$$\psi_o(x_o, z_o) = \frac{e^{i\pi/4}}{\sqrt{\lambda f}} e^{-2ikf} e^{-i(k/2f)\rho^2} \int_{-\infty}^{\infty} \psi_i(x_i, z_i) e^{i(k/f)x_o x_i} dx_i \quad (4.3-27)$$

Except for an unimportant phase factor, Equation (4.3-27) establishes the fact that the output field in the back focal plane of the lens represents the Fourier transform of the input field located in the front focal plane. The lens is therefore capable of performing a Fourier transformation. This result is of great importance for optical data processing and spatial filtering. Since the Fourier transform of any input field located at the front focal plane is accessible in the back focal plane, it is possible to influence this "spectrum" by spatial filtering, phase shifting, or multiplication with other functions in order to alter the transmitted field in any desired fashion. We can make the Fourier transform relationship even more suggestive by introducing the new coordinates

$$u = \sqrt{\frac{k}{f}} x_i = \sqrt{\frac{2\pi}{\lambda f}} x_i \quad (4.3-28)$$

and

$$v = \sqrt{\frac{k}{f}} x_o \quad (4.3-29)$$

Introducing the functions

$$\psi_i(x_i, z_i) = f(u) \quad (4.3-30)$$

and

$$\psi_o(x_o, z_i) = F(v) \quad (4.3-31)$$

allows us to write (4.3-27) in the form

$$F(v) = \frac{1}{\sqrt{2\pi}} \int_{-\infty}^{\infty} f(u) e^{iuv} du \quad (4.3-32)$$

The phase factor was omitted since it can always be removed by displacing the input or output plane by no more than one optical wavelength. Such a minute "defocusing" has no effect on the quality of the optical images, but allows us to adjust the phase in such a way as to make it an exact multiple of 2π. The only limitation on the perfect realization of Fourier transformations by thin lenses is the presence of a finite lens aperture. Our derivation of (4.3-32) ignored the lens aperture. The integral (4.3-12) must, strictly speaking, be taken between finite limits. However, as long as the amount of light that falls outside the lens aperture is small, our approximation (4.3-32) is valid.

The transmission from front to back focal plane achieves a nearly perfect Fourier transformation. However, we obtain a Fourier transform relationship also by placing the input field in direct contact with the lens. Equation (4.3-18) cannot be used directly in this case, since we assumed that certain paraxial approximations could be made which would not be true for this extreme case. The result for this case follows, however, immediately from (4.3-10) and (4.3-11)

$$\psi_o = \frac{e^{i\pi/4}}{\sqrt{\lambda b}} e^{-ikb} e^{-(i/2)k(\rho^2/f + x_o^2/b)} \int_{-\infty}^{\infty} \psi_i(x', z') e^{(i/2)k[(1/f - 1/b)x'^2 + (2/b)x_o x']} dx'$$

(4.3-33)

Collecting the output field again in the back focal plane

$$b = f$$

(4.3-34)

and using the transformations (4.3-28) through (4.3-31) with x_i replaced by x' results in

$$F(v) = \frac{1}{\sqrt{2\pi}} e^{-ik(x_o^2/2f + z_o)} \int_{-\infty}^{\infty} f(u) e^{iuv} du$$

(4.3-35)

The uninteresting constant phases were again omitted, using the same argument that was advanced earlier. The Fourier transformation (4.3-35) is marred by the occurrence of the phase factor that is not constant but depends on the output coordinate x_o. However, this flaw can be removed by considering the output field not in the plane $z_o = f$ but on the curved surface.

$$\frac{x_o^2}{2f} + z_o = f + z'$$

Considering the output field on this parabolic surface instead of on the plane allows us again to regard the result of the field transmission through the lens from a plane directly to the left of the lens as a Fourier transformation.

So far we have limited the discussion to the two-dimensional case. It is not hard to extend it to the general case using the results of the two-dimensional discussion. In the three-dimensional case, we start with (2.2-31) instead of (2.2-43). The differences between these two equations are slight and can easily be taken into account. The most important difference is the occurrence of the product of an x-dependent and a y-dependent phase factor under the integration sign. This has the effect that, instead of the integral (4.3-12), we now encounter the product of this integral with another one that is obtained by interchanging all the x parameters with corresponding y parameters. Keeping this fact in mind and carefully inspecting the two equations (2.2-31) and (2.2-43) allows us to write the result corresponding to (4.3-18) for the two-dimensional case.

$$\psi_o = \frac{\pi e^{i(3/2)\pi}}{\lambda^2 ab} e^{-ik(z_o - z_i)} e^{-i(k/2f)\rho 2} \frac{1}{\alpha} \int_{-\infty}^{\infty} \int_{-\infty}^{\infty}$$

$$\cdot \left\{ \psi_i(x_i, y_i, z_i) \exp\left[-\frac{i}{2} k \left(\frac{x_i^2 + y_i^2}{a} + \frac{x_o^2 + y_o^2}{b} + \frac{k}{2\alpha} \left(\frac{x_i}{a} + \frac{x_o}{b} \right)^2 \right. \right. \right.$$

$$\left. \left. \left. + \frac{k}{2\alpha} \left(\frac{y_i}{a} + \frac{y_o}{b} \right)^2 \right) \right] \right\} dx_i dy_i \tag{4.3-36}$$

The conclusions to be drawn from (4.3-36) are almost identical to the discussion of the two-dimensional case. An inverted, magnified image of the input field results when the object and image distance are chosen so that $\alpha = 0$. If the input plane is located at $a = f$ and the output plane at $b = f$, we find again that the output field is the (two-dimensional) Fourier transform of the input field. Using (4.3-26) and the abbreviations

$$u_x = \sqrt{\frac{k}{f}} x_i \tag{4.3-37}$$

$$u_y = \sqrt{\frac{k}{f}} y_i \tag{4.3-38}$$

$$v_x = \sqrt{\frac{k}{f}} x_o \tag{4.3-39}$$

$$v_y = \sqrt{\frac{k}{f}} y_o \tag{4.3-40}$$

we obtain from (4.3-36)

$$F(v_x, v_y) = \frac{1}{2\pi} \int_{-\infty}^{\infty} \int_{-\infty}^{\infty} f(u_x, u_y) e^{i(u_x v_x + u_y v_y)} du_x du_y \tag{4.3-41}$$

We used (4.3-30) and (4.3-31) with a straightforward extension to two variables. A constant phase factor was omitted, using the argument advanced earlier.

A comparison of the Fourier transform relationship between the input and output fields (from the front focal plane or the plane directly in contact with the lens) transformed by a lens with Equations (2.2-31) and (2.2-43) makes it apparent that the field in the back focal plane of the lens corresponds to the far field diffraction in the Fraunhofer region. We remind the reader that the Fraunhofer region is defined by the fact that the quadratic terms in the argument of the exponential functions become negligibly small. The far field near the axis of the structure is also proportional to the Fourier transform of the input field without using a lens.

4.4 OPTICAL FOURIER TRANSFORM AND SPATIAL FILTERING[7]

The fact that lenses can be used to form the Fourier transform of two dimensional light distributions is of great importance for optical data processing. We have shown in Equations (4.3-32) and (4.3-41) that the Fourier transform of an object positioned in the front focal plane of the lens appears in its back focal plane. The Fourier transform of the input field distribution is thus easily accessible, and can be influenced to modify the field distribution in certain ways. The optical analog of frequency filtering of electrical signals can be performed without the need for elaborate electrical circuitry simply by placing apertures in the path of the light wave. More sophisticated spatial filtering techniques involve placing of attenuators or phase shifters in the back focal plane of the lens to alter the Fourier spectrum of the input field distribution.

Let us assume that we want to obtain the Fourier transform of a given function. We begin by preparing an optical transparency whose amplitude transmission coefficient represents the desired function. In principle, it is possible to consider transparencies with a given amplitude and phase characteristic. However, controlling the phase of a transparency requires extremely accurate control of the optical path length through the transparency to within a small fraction of the wavelength. This is a very exacting task that can usually not be realized. However, reproducing an optical field with a desired amplitude and phase is possible by using the methods of holography.[7]

For the purpose of the present discussion, we limit ourselves to the simpler two-dimensional case. Let us assume that we have prepared a transparency with an amplitude transmission function $t(x)$. Illumination of this transparency in the front focal plane of a cylindrical lens with a plane wave of light results in the following field distribution immediately to the right of the transparency

$$\psi_i(x_i) = At(x_i) = f(u) \tag{4.4-1}$$

The relation between x_i and u is given by (4.3-28). The constant A is the amplitude of the incident plane wave. After passing the transparency, this plane wave has its amplitude modulated by the amplitude transmission function $t(x)$ of the transparency. According to (4.3-32), we find the Fourier transform of the input field (4.4-1) in the back focal plane of the cylindrical lens

$$\psi_o(x_o) = F(v) = \frac{1}{\sqrt{2\pi}} \int_{-\infty}^{\infty} f(u)e^{iuv}du \tag{4.4-2}$$

Electrical engineers are accustomed to think of Fourier transformations of time functions. The Fourier transform of a time function is a function of frequency. The time coordinate is now replaced by the spatial coordinate

$$u = \sqrt{\frac{2\pi}{\lambda f}}\, x_i \tag{4.4-3}$$

while the radian frequency ω is replaced by another spatial coordinate

$$v = \sqrt{\frac{2\pi}{\lambda f}}\, x_o \tag{4.4-4}$$

The coordinate x_o in the output plane (back focal plane) of the cylindrical lens is proportional to the "frequency" space of the conventional Fourier transform of a time function.

This simple relationship between a function and its Fourier transformation is ideally suited to perform filtering operations on the Fourier transform in the "frequency" domain. However, since the frequency variable is here replaced by another space variable, such a filtering operation is known as spatial filtering.

Let us assume that we want to simulate the effect of a low-pass filter. In the case of a time-dependent electrical signal, we would have to transmit the corresponding voltage or current through an electrical network that allows all frequencies up to a cut-off frequency ω_c to pass while all frequencies above ω_c are rejected. The spatial filter that corresponds to this low pass is represented simply by a slit in an opaque screen. The slit extends from $x_o = -x_{oc}$ to $x = x_{oc}$, with x_{oc} given by

$$x_{oc} = \sqrt{\frac{\lambda f}{2\pi}}\, \omega_c \tag{4.4-5}$$

Light arriving at the back focal plane at $|x_o| > |x_{oc}|$ is intercepted by the opaque screen. A second lens can be used to perform another Fourier transformation and to display the function

$$\psi_t(x_t) = G(w) = \frac{1}{\sqrt{2\pi}} \int_{-\omega_c}^{\omega_c} \psi_o(x_o) e^{ivw} dv \tag{4.4-6}$$

with

$$w = \sqrt{\frac{k}{f}}\, x_t \tag{4.4-7}$$

The correct mathematical form of the inverse Fourier transformation is obtained by replacing w (and consequently x_t) with $-w$. This replacement causes the required negative sign to appear in the exponent of the exponential function. The sign change required to display the Fourier transformation and its proper inverse shows that the image produced by the succession of the two lenses (in the absence of the low-pass filter) appears inverted with respect to the input field. The spatial filtering operation just described is sketched in Figure 4.4-1.

The simple optical device shown in Figure 4.4.1 is an analog computer. It allows us to obtain the result of the low-pass filtering operation on a function $t(x)$. The resultant function is recorded on the photographic plate located in the back focal plane of the second lens. However, we must remember that the photographic plate is sensitive to field energy instead of amplitude. If we operate the plate in its linear range, we record the absolute square value of the field distribution and not the field distribution itself. The result of the analog computation performed by our optical system is obtained when we take the square root of the optical density recorded on the film. For the same reason, we find that the optical density of our input transparency represents the square of the desired function $t(x)$. The transmission coefficient $t(x)$ represented amplitude and not power transmission.

We must not forget that the Fourier transformation performed by the lenses is not exact but is only an approximation that is caused by the finite extent of the lenses. The integral in (4.4-2) is not truly extended over an infinite range but extends only over the area of the lens. The approximation can be quite good, however, if we use sufficiently large lenses to ensure that most of the light passes through the lens aperture.

If we place the photographic plate in the back focal plane of the first lens, we record the (square of the) Fourier transformation of the input function $t(x)$. The optical analog computer represents a Fourier analyzer in that configuration.

The spatial filtering operation is of course not limited to low-pass filters. We can use a high-pass filter with equal ease by replacing the slit with a stop

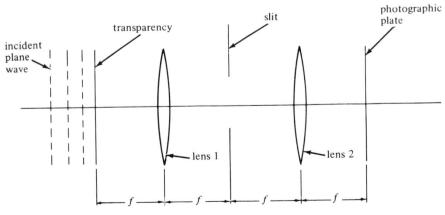

Figure 4.4.1   This optical analog computer acts as a low-pass filter. If the screen with the slit is replaced by a transparency, more general filtering operations can be performed.

that prevents light from passing in the region $|x_o| < |x_{oc}|$ but passes it for $|x_o| > |x_{oc}|$. More sophisticated filtering operations are possible by placing optical transparencies instead of opaque apertures in the path of the light beam at the back focal plane of the first lens. A particularly interesting example of an analog computation of this kind is the formation of the convolution of two functions.

Let us assume that we want to form the convolution of the function $t(x)$ with another function $g(x)$. We achieve this (at least in principle) by preparing a transparency with an amplitude transmission coefficient that is the Fourier transform of $g(x)$. The preparation of such a transparency is in general a very difficult task, so that only very simple functions can be represented in this way. The difficulties arise from the fact that the phase of a transparency cannot be accurately controlled because of random variations of the film thickness and variations of the refractive index of the material. The problem of random phase variations can be eased by placing the transparency inside a container with precisely controlled, optically flat glass walls. The optical path length of the transparency can be controlled to a reasonable degree by filling the glass vessel, which contains the transparency, with an index-matching oil. However, since the index of refraction of the transparency is not constant, it is not possible to equalize the random variations of the optical path length through the transparency perfectly. However, even this method of path length equalization does not solve our problem. The Fourier transform G of $g(x)$ is characterized by a phase as well as an amplitude function. It is impossible to prepare a transparency with a prescribed phase shift that varies in a given manner. This problem can be overcome by using holographic techniques that allow the storage of amplitude as well as phase information in a photographic film or plate. A discussion of spatial filters prepared by holographic techniques can be found in Goodman's book.[7]

For the present discussion, let us assume that we have a transparency whose amplitude transmission response is given by

$$G(v) = \frac{1}{\sqrt{2\pi}} \int_{-\infty}^{\infty} g\left(\sqrt{\frac{f}{k}}\, u\right) e^{iuv} du \qquad (4.4\text{-}8)$$

Placing this transparency in the back focal plane of the first lens shown in Figure 4.4.1 causes the field distribution $F(v)$—the Fourier transform of $t(x)$, according to (4.4-1) and (4.4-2)—to be multiplied by the amplitude transmission function $G(v)$. The field distribution immediately to the right of the transparency in the back focal plane of the first lens is now $F(v)G(v)$.

The second lens performs a second Fourier transformation on this function, so that the field in the back focal plane of the second lens is given by

$$H(w) = \frac{1}{\sqrt{2\pi}} \int_{-\infty}^{\infty} F(v)G(v) e^{ivw} dv \qquad (4.4\text{-}9)$$

Substitution of (4.4-8) results in

$$H(w) = \frac{1}{2\pi} \int_{-\infty}^{\infty} du\, g\left(\sqrt{\frac{f}{k}}\, u\right) \int_{-\infty}^{\infty} F(v) e^{i(u+w)v} dv \qquad (4.4\text{-}10)$$

The inverse of the Fourier integral (4.4-2) is given by

$$f(u) = \frac{1}{\sqrt{2\pi}} \int_{-\infty}^{\infty} F(v) e^{-iuv} dv \qquad (4.4\text{-}11)$$

Using this result, we can rewrite (4.4-10) in the following way

$$H(w) = \frac{1}{\sqrt{2\pi}} \int_{-\infty}^{\infty} g\left(\sqrt{\frac{f}{k}}\, u\right) f(-u-w) du \qquad (4.4\text{-}12)$$

Using the original spatial coordinate x_i of (4.4-3) and a corresponding expression for w

$$w = \sqrt{\frac{k}{k}}\, x_t \qquad (4.4\text{-}13)$$

we can write (4.4-12) in the form

$$H\left(\sqrt{\frac{k}{f}}\, x_t\right) = \frac{A}{\sqrt{\lambda f}} \int_{-\infty}^{\infty} g(x_i) t(-x_i - x_t) dx_i \qquad (4.4\text{-}14)$$

Equation (4.4-1) was used to express f in terms of the original function t. Finally, we introduce the function

$$\phi(-x_t) = H\left(\sqrt{\frac{k}{f}}\, x_t\right) \qquad (4.4\text{-}15)$$

and obtain as the field distribution in the back focal plane of the second lens

$$\phi(x_t) = \frac{A}{\sqrt{\lambda f}} \int_{-\infty}^{\infty} g(x_i) t(x_t - x_i) dx_i \qquad (4.4\text{-}16)$$

Note that the definition (4.4-15) of the function $\phi(x)$ implies an inversion of the coordinate direction in the output plane.

The integral appearing in (4.4-16) is known as the convolution of the functions g and t. Filtering the field distribution $t(x)$ with the Fourier transformation of the function $g(x)$ results in the computation of the convolution of the two functions by the analog computer shown in Figure 4.4.1. (Note that the slit shown in Figure 4.4.1 is now replaced by the transparency with amplitude transmittance G). A photographic plate in the output plane records the absolute square of the function $\phi(x)$ rather than the function itself.

The full three-dimensional version of the two-dimensional optical arrangements discussed so far leads to the twofold Fourier transformation of (4.3-41). Spatial filtering, using ordinary lenses instead of the cylindrical lenses considered so far, can be accomplished in complete analogy to the two-dimensional case. Use of ordinary lenses and twofold Fourier transform transparencies of functions of two variables leads to the twofold convolution of functions of two variables. Let the input transparency be given by $t(x, y)$ and the spatial filter be the Fourier transformation of the function $g(x, y)$

$$G(v_x, v_y) = \frac{1}{2\pi} \int_{-\infty}^{\infty} \int_{-\infty}^{\infty} g\left(\sqrt{\frac{f}{k}} u_x, \sqrt{\frac{f}{k}} u_y\right) e^{i(u_x v_x + u_y v_y)} du_x du_y \quad (4.4\text{-}17)$$

The output field distribution in the back focal plane of the second (ordinary) lens is then given by the expression

$$\phi(x_t, y_t) = \frac{A}{\lambda f} \int_{-\infty}^{\infty} \int_{-\infty}^{\infty} g(x_i, y_i) t(x_t - x_i, y_t - y_i) dx_i dy_i \quad (4.4\text{-}18)$$

The convolution of two functions or of a function with itself (autocorrelation) is a valuable operation. The autocorrelation procedure can be used to recover a signal that is buried in noise.

The optical analog computers discussed in this section require coherent light for their operation. The light output of a laser is coherent to a high degree, while light from thermal sources is incoherent. Light is called coherent if the field amplitudes at different points in space have a definite phase relationship to each other. The lack of a definite phase relationship of the field amplitudes at different points in space characterizes incoherent light. If partial correlations exist between the phases of the field amplitudes at different points in space, we speak of partially coherent light. In addition to coherence, we must require the light to be monochromatic (that is, consisting of only one frequency or a very narrow band of frequencies) if it is to be useful for the optical analog computer functions discussed in this section. The light of lasers is ideally suited for these purposes. However, it might be necessary to force the laser to operate at one of many possible frequencies and to force it into one definite transverse mode (see the discussion following Equation [6.6-26]) to ensure that its output is sufficiently monochromatic and spatially coherent. Ordinary light from a thermal light source or a gas discharge can be made monochromatic and coherent by filtering out a narrow band of frequencies and by passing it through a narrow pinhole. However, these procedures waste so much light power that the remaining partially coherent light has usually insufficient power to be of much use for spatial filtering experiments.

4.5 GAS LENSES

Glass lenses are ideally suited for image-forming optical instruments. Even though a simple glass lens made of only one type of glass has a considerable number of undesirable aberrations, it is possible to obtain excellent lenses by combining lenses made of different glasses.[1,15] However, there are applications for lenses other than image formation that require different properties of the lens. We shall discuss the problems of light transmission over large distances in detail in Chapter 5. A very promising light waveguide consists of a succession of lenses that are used to counteract the tendency of the light beam to spread apart by diffraction. If the light losses caused by scattering inside the glass and at the surface of the lens are ignored, such lens waveguides can be made with extremely low losses. In fact, using lenses that are large enough and taking care that the light beam does not depart appreciably from the axis of the waveguide allows us to design a lens waveguide with arbitrarily low losses. However, lenses are not ideal, so that we cannot ignore the light that is lost by scattering inside and at the surface of the lens. Even if we provide the lens surfaces with antireflection coatings, we must still expect to lose a fraction of 1 percent of the light at each lens surface. Most of this loss is caused by residual reflection and by random scattering from dust and other imperfections at the surface of the lens. The main cause of loss must therefore be attributed to the air-lens interface. If it were possible to design lenses that do not have interfaces with drastic variations of the index of refraction, the scattering and reflection loss problem would be eliminated.

Such low-loss lenses are indeed possible. The first lens of this type was demonstrated by D. W. Berreman,[27] who used a warm metallic helical structure inside a cool metal tube to heat the gas inside the tube and to cause it to circulate in convective flow. He could show that the density variation of the nonuniformly heated gas caused refractive index changes that had the property of focusing a light beam traveling near the axis of the tube. Berreman's lens, even though it is quite effective, is very hard to analyze theoretically. We shall therefore discuss a different type of gas lens that is equally effective and can be analyzed to predict its performance.

Since it is possible to design lenses that act by utilizing the refractive index change in a gas, we have eliminated the loss problem of glass lenses. In a gas, whose index of refraction is nonuniform because of nonuniform heating effects, there is no interface between regions of homogeneous materials with different indices of refraction. Reflection and surface scattering losses are thus completely eliminated. However, we must caution the reader not to expect that gas lenses are ideal in every respect. Gas lenses suffer from a series of shortcomings all their own. They usually have fairly strong aberrations—that is, their focal length depends on the position of the light beam with respect to the center of the lens.[28] Gas lenses cannot be made with large apertures, since gas flowing through large aperture tubes tends to break into turbulent flow. Finally, gas lenses are more costly to operate than a glass lens, since energy is required to maintain the temperature gradients essential for their operation.[29] However, the low loss properties of gas lenses are so impressive that it is likely that this outstanding feature may be sufficiently attractive to ensure their use in certain special applications. The low-loss properties of gas lenses have been dramatically demonstrated by A. C. Beck.[30] It is well known that helium-neon lasers cannot tolerate much loss inside their cavities. It is usually sufficient to insert a microscope cover slide into the laser beam inside its cavity to kill the oscillation. Glass lenses as focusing devices inside helium-neon laser cavities are quite troublesome. A. C. Beck demonstrated the operation of a helium neon laser with not only one but with up to seventy-eight gas lenses inside the laser cavity. The cavity with all seventy-eight lenses inside was 78 m long. The helium-neon laser with this extremely long cavity operated with a conventional laser tube of 30 cm length.

The gas lens that we are about to study is sketched in Figure 4.5.1.

The gas lens consists of a heated metal tube. A laminar flow of gas is maintained inside the tube. The gas enters the tube at room temperature from the left, and leaves the tube at a higher temperature at the right. Since the gas is heated at the walls of the tube, the heat penetrates radially into the gas and establishes a temperature gradient. At each cross section we find the coolest gas at the axis of the tube, while the gas heats up toward the wall.[31]

According to the ideal gas law, the volume V, pressure p, and temperature T of the gas are related by the following equation

$$pV = \mu RT \qquad (4.5\text{-}1)$$

The ideal gas constant R is

$$R = 8.315 \ W \sec \text{degree}^{-1} \text{mole}^{-1} \qquad (4.5\text{-}2)$$

The constant μ indicates the number of moles contained in the volume V. Let M be the weight of one mole of the gas. (One mole is the amount of gas

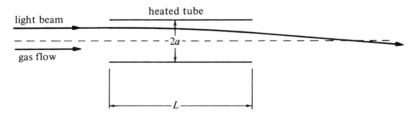

Figure 4.5.1 Schematic of a gas lens.

equal to its molecular weight in grams. For example, oxygen O_2 has the molecular weight 32. One mole of oxygen is therefore 32 grams of this gas.) This means that

$$m = \mu M \tag{4.5-3}$$

is the actual mass of the gas that is contained in the volume V. The density ρ of the gas is defined as its mass per unit volume

$$\rho = \frac{m}{V} \tag{4.5-4}$$

Using these definitions, we can rewrite the ideal gas law in the following form

$$\frac{p}{\rho} = \frac{R}{M} T \tag{4.5-5}$$

It is an experimentally observed fact that the refractive index of a gas depends on its density

$$n = 1 + (n_o - 1) \frac{\rho}{\rho_o} \tag{4.5-6}$$

In this equation, ρ_o represents some constant average density of the gas while n_o is the refractive index at that density.

The gas lens is operated at constant (atmospheric) pressure. This means that the gas density is inversely proportional to its temperature. Equation (4.5-6) can therefore be expressed as follows

$$n = 1 + (n_o - 1) \frac{T_o}{T} \tag{4.5-7}$$

Equation (4.5-7) shows that the refractive index of the gas decreases with increasing temperature. The refractive index of the gas is higher at the center of the tube, where the gas is cool, while the index decreases toward the wall of the gas tube. A light beam traveling inside the gas lens is deflected toward the region of higher index of refraction. The light beam bends, as shown in Figure 4.5.1, and leaves the gas lens with a steeper angle with respect to the optical axis than its input slope.

Solution of the Thermodynamic Problem

The detailed analysis of the gas lens requires some digression into thermodynamics. Since the relevant equations are easy to derive, their derivation will be given here.

Let us assume a certain volume at rest (with respect to the tube) in the moving fluid with nonuniform temperature. The heat content of the small volume is given by

$$c_p \int_V \rho T dV \qquad (4.5\text{-}8)$$

The constant c_p is the specific heat at constant pressure. Multiplication by the temperature T and the mass $dm = \rho dV$ gives the amount of heat contained in the element of mass dm; integration yields the amount of heat contained in the volume V. If the amount of heat contained in V is changing, it is because heat is flowing into the volume by heat conduction and because heat is transported by the flowing gas into the volume. The amount of heat that enters the volume per second by heat conduction is proportional to the gradient of the temperature T

$$k \int_A \nabla T \cdot \mathbf{dA} \qquad (4.5\text{-}9)$$

The vector $\mathbf{dA}$ is directed toward the outside of the closed surface A. This equation accounts for the influx of heat through the surface A of the volume. The constant k is the thermal conductivity. Finally, heat may be transported by the flow of hot gas into the volume. The net inflow of heat by this mechanism is

$$-c_p \int_A \rho T \mathbf{v} \cdot \mathbf{dA} \qquad (4.5\text{-}10)$$

The negative sign is required since $\mathbf{dA}$ is directed out of the volume. The combination $-\mathbf{v} \cdot \mathbf{dA}$ specifies a volume of material that enters the volume V per unit time. Multiplication by $c_p \rho T$ gives us the amount of heat stored in this volume. Combining all three equations allows us to express the balance of energy as follows

$$c_p \frac{\partial}{\partial t} \int_V \rho T dV = k \int_A \nabla T \cdot \mathbf{dA} - c_p \int_A \rho T \mathbf{v} \cdot \mathbf{dA} \qquad (4.5\text{-}11)$$

The change of energy in the volume per unit time on the left of the equation is brought about by heat influx (first term on the right) and by gas inflow (second term on the right). Application of the divergence theorem to the two surface integrals yields

$$\int_V \left\{ c_p \left[\frac{\partial}{\partial t} (\rho T) + \text{div} (\rho T \mathbf{v}) \right] - k \nabla^2 T \right\} dV = 0 \qquad (4.5\text{-}12)$$

Since the volume is totally arbitrary, we conclude that the integrand of the volume integral must vanish. We write this expression in the slightly different form

$$c_p \left[T \left(\frac{\partial \rho}{\partial t} + \text{div} (\rho \mathbf{v}) \right) + \rho \frac{\partial T}{\partial t} + \rho \mathbf{v} \cdot \nabla T \right] = k \nabla^2 T \qquad (4.5\text{-}13)$$

Finally, we use the continuity equation that was derived in section 3.7, Equation (3.7-1). We use this equation here for the case of three-dimensional space and replace the coordinate z appearing in (3.7-1) by the time t. This allows us to drop the first term on the right-hand side of (4.5-13), so that we finally obtain the partial differential equation of heat transfer in a flowing fluid

$$\frac{k}{c_p} \nabla^2 T = \rho \frac{\partial T}{\partial t} + \rho \mathbf{v} \cdot \nabla T \tag{4.5-14}$$

For the gas lens, we are interested only in the stationary state that establishes itself after the gas has flowed through the tube for a short time. The stationary state is mathematically indicated by

$$\frac{\partial T}{\partial t} = 0 \tag{4.5-15}$$

The remaining problem is still impossible to solve if we insist on treating the density ρ and the velocity v as variable quantities. An exact solution of the problem would force us to consider the dynamics of the fluid flow in the presence of a nonuniform heat distribution. This problem is far too difficult to handle. We settle, therefore, for an approximate solution, assuming that ρ is approximately constant throughout the tube and that the velocity profile conforms to the laminar, viscous flow through a tube at constant temperature. The velocity distribution of a viscous fluid in laminar flow at constant temperature is simply given by (a = radius of the tube)

$$\left.\begin{matrix} v_x = 0 \\ v_y = 0 \end{matrix}\right\} \tag{4.5-16}$$

$$v_z = v_o \left[1 - \left(\frac{r}{a}\right)^2 \right] \tag{4.5-17}$$

The flow problem in this simplified form is known as the Graetz problem. It is discussed in the book on heat transfer by M. Jakob.[32] The stationary problem in its approximate form follows from (4.5-14) through (4.5-17)

$$\alpha \left(\frac{\partial^2 T}{\partial r^2} + \frac{1}{r}\frac{\partial T}{\partial r} + \frac{\partial^2 T}{\partial z^2} \right) = v_o \left[1 - \left(\frac{r}{a}\right)^2 \right] \frac{\partial T}{\partial z} \tag{4.5-18}$$

with

$$\alpha = \frac{k}{\rho c_p} \tag{4.5-19}$$

We assumed that the temperature distribution is independent of the azimuth ϕ. This assumption is one more approximation, since the gravitational field of the earth will tend to distort the gas flow if the gas lens is oriented horizontally. This gravity effect leads to observable distortions of the lens, but, for the sake of simplicity, we ignore this gravitational distortion since its treatment makes the solution of our problem considerably harder.[33]

We solve the heat distribution problem in the gas lens by introducing the new variable

$$\Theta = \frac{T - T_w}{T_w - T_o} \tag{4.5-20}$$

T_w is the temperature of the tube wall, and T_o is the temperature of the gas entering the tube. The function Θ is a solution of the same differential equation (4.5-18), but it has a simpler boundary value. We must require that $T = T_w$ at $r = a$. Using the independent variable

$$u = \frac{r}{a} \tag{4.5-21}$$

we require that Θ assumes the value

$$\Theta(u) = 0 \qquad \text{at } u = 1 \tag{4.5-22}$$

For physical reasons, we expect our solution to be symmetrical in u, so that we require that Θ is a symmetrical function of u. The partial differential equation (4.5-18) can be solved by the trial solution

$$\Theta = A R(u) e^{-\beta^2 (\alpha/a^2 v_o) z} \tag{4.5-23}$$

Substituting (4.5-23) into (4.5-18) (with Θ replacing T), we obtain the ordinary differential equation

$$\frac{d^2 R}{du^2} + \frac{1}{u} \frac{dR}{du} + \beta^2 (1 - u^2) R = 0 \tag{4.5-24}$$

The term $\dfrac{\partial^2 T}{\partial z^2}$ was neglected in (4.5-18). This approximation is valid for sufficiently high values of v_o.

The solutions of the differential equation (4.5-24) are related to the so-called Whittaker functions. However, since these functions are not extensively tabulated, it is necessary to obtain their power series expansion.

Because of the assumption that R must be an even function of u, we can limit the power series to even orders of u

$$R = \sum_{v=0}^{\infty} C_{2v} u^{2v} \tag{4.5-25}$$

It is convenient to require of R that

$$R(0) = 1 \tag{4.5-26}$$

so that we obtain

$$C_o = 1 \tag{4.5-27}$$

Substitution of (4.5-25) into (4.5-24) using (4.5-27) leads to

$$C_2 = -\frac{1}{4}\beta^2 \qquad (4.5\text{-}28)$$

if we take $u = 0$. Comparison of coefficients of u^2 in the equation that results from substitution of (4.5-25) into (4.5-24) leads to the following recursion relation for the coefficients

$$C_{2\nu} = \frac{\beta^2}{(2\nu)^2}(C_{2\nu-4} - C_{2\nu-2}) \qquad \text{for } \nu \geq 2 \qquad (4.5\text{-}29)$$

If β were known, our problem would be solved. However, the eigenvalue β must be determined from the eigenvalue equation

$$R(1) = 0 \qquad (4.5\text{-}30)$$

that follows from the boundary condition (4.5-22).

The solution of this eigenvalue problem is possible only numerically with the help of a computer. The eigenvalue problem has an infinite number of solutions. The different solutions are identified by attaching subscripts to β and R. The solution of the actual physical problem is obtained as a superposition of all the eigensolutions. The function θ can therefore be written in the most general case

$$\Theta = \sum_{\mu=0}^{\infty} A_\mu R_\mu(u) e^{-(\alpha/a^2 v_0)\beta_\mu^2 z} \qquad (4.5\text{-}31)$$

The solution of the higher order eigenvalues β_μ runs into the following problem. The series expansion (4.5-25) does not converge very readily. The coefficients $C_{2\nu}$ grow to enormously large values before they begin to decrease. It was possible to compute solutions of the eigenvalue problem only up to β_8, even with the use of double precision, since the coefficients grow to values in the order of 10^{20} while the values of $R(u)$ do not exceed unity. For this reason it was necessary to use two series expansion for $R(u)$. The expansion (4.5-25) can be used for values $0 \leq u \leq 0.5$. The products of $C_{2\nu}u^{2\nu}$ stay within reasonable limits. For values of $u > 0.5$, it is necessary to introduce the variable

$$w = 1 - u \qquad (4.5\text{-}32)$$

and use the series expansion

$$R(u) = \sum_{\nu=0}^{\infty} D_\nu w^\nu \qquad (4.5\text{-}33)$$

The differential equation for R in terms of the variable w becomes

$$(1-w)\frac{d^2R}{dw^2} - \frac{dR}{dw} + \beta^2(2w - 3w^2 + w^3)R = 0 \qquad (4.5\text{-}34)$$

The condition (4.5-30) now reads

$$R = 0 \qquad \text{for } w = 0 \qquad (4.5\text{-}35)$$

and leads to

$$D_o = 0 \qquad (4.5\text{-}36)$$

Substitution of (4.5-33) into (4.5-34) leads to the recursion relations

$$\left. \begin{aligned} D_2 &= \frac{1}{2} D_1 \\[6pt] D_3 &= \frac{1}{3} D_1 \\[6pt] D_4 &= \left(\frac{1}{4} - \frac{1}{6} \beta^2 \right) D_1 \end{aligned} \right\} \qquad (4.5\text{-}37)$$

$$D_\nu = \frac{1}{\nu(\nu-1)} [(\nu-1)^2 D_{\nu-1} - \beta^2 (2D_{\nu-3} - 3D_{\nu-4} + D_{\nu-5})] \qquad (4.5\text{-}38)$$

The eigenvalue β and the coefficient D_1 must now be chosen such that the function and its first derivative match continuously with the series expansion (4.5-25). This formidable problem can be solved by a computer.[34] The result of the eigenvalues, the first derivative R' of $R(u)$ at $u = 1$ and the derivative $\dfrac{\partial R}{\partial \beta}$, at $u = 1$ and $\beta = \beta_\mu$, are listed in Table 4.5.1.

TABLE 4.5.1 This table shows solutions of the eigenvalue problem, the derivative of the function $R_\mu(u)$ at $u = 1$, and its derivative with respect to the eigenvalue β evaluated at $u = 1$ and $\beta = \beta_\mu$.

μ	β_μ	$R'_\mu(1)$	$\dfrac{\delta R_\mu}{\delta \beta}(\beta = \beta_\mu, u = 1)$
0	2.70436	−1.01430	−0.50090
1	6.67903	1.34924	0.37146
2	10.67338	−1.57232	−0.31826
3	14.6711	1.74600	0.28648
4	18.6699	−1.89090	−0.26449
5	22.6691	2.01647	0.24799
6	26.6686	−2.12814	−0.23491
7	30.6682	2.22038	0.22485
8	34.6679	−2.32214	−0.21548
9	38.6676	2.40274	0.20779
10	42.6667	−2.48992	−0.20108
11	46.6667	2.56223	0.19516
12	50.6668	−2.64962	−0.18988
13	54.6668	2.70216	0.18513
14	58.6668	−2.76421	−0.18083

The first three functions $R_\mu(u)$ are shown in Figure 4.5.2. In order to be able to determine the expansion coefficients A_μ in (4.5-31), we need the orthogonality relations of the functions $R_\nu(u)$. With the help of the differential equation (4.5-24), it is possible to prove that the following orthogonality relation holds

$$\int_0^1 u(1 - u^2)R_\mu(u)\, R_\nu(u)\,du = 0 \qquad \text{for } \mu \neq \nu \qquad (4.5\text{-}39)$$

The normalization of the R functions follows from the integral

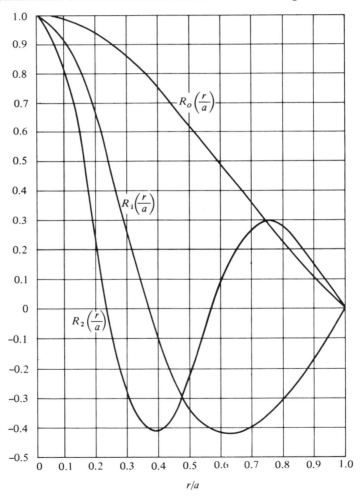

Figure 4.5.2   Plot of the first three functions $R_\mu(r/a)$. (From D. Marcuse and S. E. Miller, Analysis of a Tabular Gas Lens, *B.S.T.J.*, Vol. 43, No. 4, July 1964, pp. 1759–82. Copyright 1964, The American Telephone and Telegraph Co., reprinted by permission.)

$$\int_0^1 u(1 - u^2)R_\mu^2(u)du = \frac{1}{2\beta_\mu}\left[\frac{\partial R_\mu}{\partial \beta}\frac{\partial R_\mu}{\partial u}\right]_{\substack{u=1 \\ \beta=\beta_\mu}} \tag{4.5-40}$$

The expansion coefficients appearing in the series (4.5-31) can be obtained from the condition $T = T_o$ at $z = 0$. The input temperature of the gas T_o is constant in u. This input condition leads to

$$\Theta = -1 \qquad \text{at } z = 0 \tag{4.5-41}$$

From (4.5-41) and (4.5-31), we obtain the expansion coefficients with the help of the relations (4.5-39) and (4.5-40)

$$\frac{1}{2\beta_\mu}\left(\frac{\partial R_\mu}{\partial \beta}\frac{\partial R_\mu}{\partial u}\right)_{\substack{u=1 \\ \beta=\beta_\mu}} A_\mu = -\int_0^1 u(1 - u^2)R_\mu(u)du = \frac{1}{\beta_\mu^2}\left\{\frac{\partial R_\mu}{\partial u}\right\}_{u=1} \tag{4.5-42}$$

The solution of the integral was also obtained with the help of the differential equation (4.5-24). The expansion coefficient is therefore

$$A_\mu = \frac{2}{\beta_\mu\left\{\dfrac{\partial R_\mu}{\partial \beta}\right\}_{\substack{u=1 \\ \beta=\beta_\mu}}} \tag{4.5-43}$$

The derivatives of the R function with respect to the eigenvalue are listed in Table 4.5.1. The temperature distribution inside the gas lens is now obtained from (4.5-20), (4.5-31), and (4.5-43)

$$T = T_w + 2(T_w - T_o)\sum_{\mu=0}^\infty R_\mu(u)[\beta_\mu (\partial R_\mu/\partial\beta)]^{-1} e^{-(\alpha/a^2 v_o)\beta_\mu^2 z} \Big|_{\substack{u=1 \\ \beta=\beta_\mu}} \tag{4.5-44}$$

This series expansion converges very well for z values that are larger than zero. The entries in Table 4.5.1 show that the values of β_μ increase very rapidly with increasing values of μ. The exponential function in (4.5-44) soon becomes vanishingly small and can be neglected for all but the first few μ values. The convergence of the series is slow only at $z = 0$.

This discussion provides an approximate solution of the heat flow problem in the gas lens. The temperature distribution in the gas lens is shown in Figure 4.5.3 for several values of the normalized length along the tube. The excellent convergence of the series (4.5-44) becomes apparent when we realize that for the value 0.043 of the normalized length parameter $\alpha z/(a^2 v_o)$, the exponential factors in the series expansion are, in order of their appearance, 0.73, 0.147, 0.0075, 0.000096. This shows that the first three terms of the series are already a very good approximation to the actual temperature distribution.

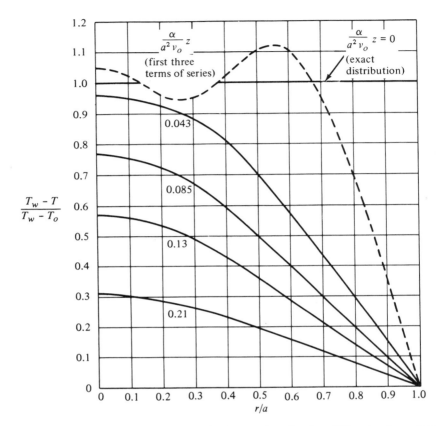

Figure 4.5.3 Temperature distribution in the gas lens for five different values of the normalized length parameter $\alpha z / a^2 v_0$. The dotted line is an approximation, using only the first three terms of the series expansion. (From D. Marcuse and S. E. Miller, Analysis of a Tabular Gas Lens, *B.S.T.J.*, Vol. 43, No. 4, July 1964, pp. 1759–82. Copyright 1964, The American Telephone and Telegraph Co., reprinted by permission.)

The normalization of the length coordinate used to label the different curves in Figure 4.5.3 is suggested by the factors appearing in the exponent of the exponential functions in (4.5-44).

Ray Tracing

After the temperature distribution inside the gas lens is established, we can determine its optical properties. We use the paraxial ray equation (3.2-19). Since the index of refraction of a gas is very nearly unity (for air at 20°C, $n = 1.000293$), we use on the left-hand side of (3.2-19) $n = 1$. Using polar coordinates r, ϕ, z and the fact that for rays that are incident parallel to the optical axis there is no ϕ dependence of the ray trajectory, we obtain to a good approximation

$$\frac{d^2r}{dz^2} = \frac{\partial n}{\partial r} \qquad (4.5\text{-}45)$$

The refractive index is given by (4.5-7). For moderate temperature differences, we can use the approximation

$$\frac{\partial n}{\partial r} = -(n_o - 1)\frac{T_o}{T^2}\frac{\partial T}{\partial r} \approx -(n_o - 1)\frac{1}{T_o}\frac{\partial T}{\partial r} \qquad (4.5\text{-}46)$$

The paraxial ray equation assumes the form

$$\frac{d^2r}{dz^2} = -(n_o - 1)\frac{1}{T_o}\frac{\partial T}{\partial r} \qquad (4.5\text{-}47)$$

Equations (4.5-47) and (4.5-44) can be used to obtain the ray trajectory of paraxial rays passing through the gas lens. Because of the complicated form of the temperature distribution, it is impossible to obtain an analytical solution of (4.5-47). We must resort to numerical solutions obtained with the aid of a computer to determine the lens properties of our device.[34]

Two quantities are useful to determine the performance of the gas lens. As in every lens, we need to know the focal length of the gas lens. However, because the physical structure of the gas lens is long, we need an additional parameter to characterize its performance. It turns out that the gas lens, even though it is quite long, can be approximately treated as a thin lens. However, this equivalent thin lens is not plane, but must be described by a distorted rotationally symmetric surface. The theory of thick lenses uses the concept of principal planes to describe their properties.[1,15] We use this concept for the gas lens, but, instead of principal planes, we must use principal surfaces for its description. The definition of the principal surfaces is shown in Figures 4.5.4 and 4.5.5.

The gas flow is directed from left to right. Figure 4.5.4 shows a light ray incident parallel to the optical axis from the left. Its trajectory inside the lens is not shown. Instead, the trajectory of the incident ray is continued, as indicated by the broken straight line. Likewise, we have continued the exit ray backward into the lens, as indicated by another straight line. The intersection of these two lines defines one point of the principal surface P_+. The set of points defined by all possible input rays incident parallel to the optical axis does not lie on a plane. The determination of this principal surface is one of the tasks of the numerical solution of (4.5-47). Another principal surface P_- is defined by Figure 4.5.5. Its definition is identical to that of P_+, with the only exception that the incident light ray enters the lens opposite to the direction of the gas stream parallel to the optical axis from

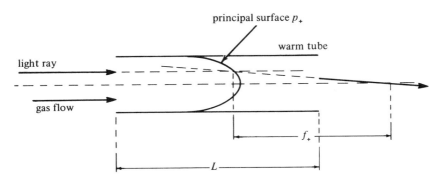

Figure 4.5.4 Definition of the principal surface of the gas lens. The direction of the light beam is parallel to the gas flow.

the right. There is no reason why the two surfaces P_+ and P_- should coincide. Their separation is an indication of the optical thickness of the gas lens. If, however, both surfaces should coincide, one principal surface would be sufficient to allow us to trace rays through the lens. In that case, we are justified in calling the lens optically thin. The thin lens in Section 4.2 could be drawn as a plane. The ray trajectories were obtained by breaking each ray at its point of intersection with the plane by an amount that was determined by the focusing power of the lens. The description of the gas lens can similarly be given by breaking each ray at its point of intersection with the principal surface. If there are two principal surfaces, only rays incident parallel to the optical axis could be traced unambiguously by breaking them at their corresponding principal surface. If the lens has only one principal surface, then each ray must be broken at this surface; ray tracing through the gas lens is no harder than ray tracing through an ordinary thin lens

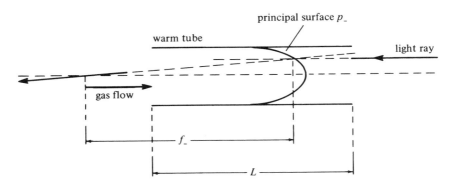

Figure 4.5.5 Definition of the second principal surface of the gas lens. Here the light beam is traveling in a direction opposite to the gas flow.

provided we know the shape of the principal surface. The power of the gas lens is again described in terms of a focal length. As indicated in Figures 4.5.4 and 4.5.5, we measure the focal length from the point at which the rays intercept the principal surfaces. If there are two distinctly different principal surfaces, we would not be surprised to find also that we need two different focal lengths f_+ and f_- to describe the power of the lens. Fortunately, the gas lens is practically a thin lens. Its two principal surfaces coincide very nearly, and so do the focal lengths f_+ and f_-. The numerical solution of the ray equation (4.5-47) provides us with the position and the slope of the rays at the point where they leave the lens. We designate the points of the principal surface P_+ by

$$z = P_+(r) \qquad (4.5\text{-}48)$$

and those of P_- by

$$z = P_-(r) \qquad (4.5\text{-}49)$$

The ray position at $z = 0$ is indicated by $r_\pm(0)$, and its position at $z = L$ by $r_\pm(L)$. Likewise, we indicate the ray slopes dr/dz by $r'_\pm(0)$ and by $r'_\pm(L)$. The focal length is then given by

$$f_+ = -\frac{r_+(0)}{r'_+(L)} \qquad (4.5\text{-}50)$$

and

$$f_- = \frac{r_-(L)}{r'_-(0)} \qquad (4.5\text{-}51)$$

The principal surfaces are obtained by

$$P_+ = L + \frac{r_+(0) - r_+(L)}{r'_+(L)} \qquad (4.5\text{-}52)$$

and

$$P_- = \frac{r_-(L) - r_-(0)}{r'_-(0)} \qquad (4.5\text{-}53)$$

The index $(+)$ refers to rays traveling to the right, while $(-)$ indicates rays traveling to the left. The plane $z = 0$ coincides with the gas input at the left of the tube. The choice of sign is dictated by the fact that $r'_+(L) < 0$ while $r'_-(0) > 0$.

In order to be able to present the data in terms of dimensionless quantities, we introduce several normalizing parameters. A useful quantity for this purpose is the parameter.*

---

* For air at 20°C we have: $\rho = 1.21 \times 10^{-3}$ g/cm^3, $k = 6.28 \times 10^{-5}$ cal/(cm sec degree), $c_p = 0.24$ cal/(g degree), $n-1 = 2.93 \times 10^{-4}$.

$$V = \frac{\alpha L}{a^2} = \frac{kL}{\rho c_p a^2} \qquad (4.5\text{-}54)$$

Since the exponents of the exponential functions appearing in (4.5-44) must be dimensionless quantities, it is clear that V has the dimension of velocity. It is important to remember that V is proportional to the length of the gas lens. It is a parameter that, aside from the length of the lens, is determined by the kind of gas that is used (its thermal conductivity k and specific heat c_p enter in) and by the cross section of the lens. Since the coordinate z enters (4.5-44) in the combination

$$s = \frac{\alpha}{a^2 v_o} z = \frac{V}{v_o} \frac{z}{L} \qquad (4.5\text{-}55)$$

it is convenient to express the ray equation in terms of this coordinate. Using also (4.5-21), we obtain from (4.5-47)

$$\frac{d^2 u}{ds^2} = - \left(\frac{v_o}{V}\right)^2 \left(\frac{L}{a}\right)^2 (n_o - 1) \frac{1}{T_o} \frac{\partial T}{\partial u} \qquad (4.5\text{-}56)$$

Equation (4.5-56) contains only dimensionless quantities, and is therefore particularly well suited for numerical evaluation. From (4.5-44) and (4.5-56), it is apparent that the constant

$$C = (n_o - 1) \frac{T_w - T_o}{T_o} \left(\frac{L}{a}\right)^2 \qquad (4.5\text{-}57)$$

is another useful dimensionless parameter for the evaluation of the lens. It combines information about the length and cross section of the lens with information about the index of refraction of the gas and the temperature of the heated tube.

The results of the numerical solution of our gas lens problem are shown in a number of figures. Figure 4.5.6 presents the focal length of the gas lens for rays very close to the axis of the lens ($r_+(0)/a = 0.1$) as a function of the normalized gas velocity. The most striking feature of these focal length curves is the existence of a minimum. The fact that there is an optimum gas velocity for which the focal length becomes shorter than for any other velocity can easily be explained. Let us begin by considering a gas lens with no gas flow at all, $v_o = 0$. In this case, the gas in the tube remains stationary and heats up to the temperature of the warm tube. With the gas at a uniform temperature, there is no lens action at all, $f = \infty$. As the gas begins to flow through the lens, there will be a temperature distribution at the input end of the tube. However, for very low gas velocities, the gas will warm up completely long before it reaches the end of the tube, and the resulting lens action remains weak. However, as we increase the gas velocity, the lens gets stronger. This trend explains the decreasing focal length to the

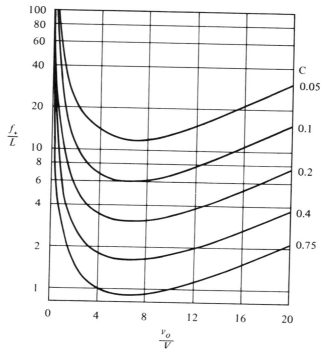

Figure 4.5.6 Normalized focal length, f_+/L, is plotted as a function of the normalized on-axis gas velocity v_o/V. (From Theory of Thermal Gradient Gas Lenses, *IEEE Trans. on MTT*, Vol. 13, No. 6, pp. 734–739, November 1965. Reproduced with permission.)

left of the minimum in Figure 4.5.6. As the gas velocity is increased more and more, there comes the moment when the gas begins to leave the tube before it has heated up very much. The lens begins to grow weaker; the focal length is increasing again. At infinite gas velocity (if this were possible with laminar flow), the gas would pass through the lens without changing its input temperature, and no gas lens action would be possible. We see that the focal length must be infinite at both extremes, for zero velocity as well as for infinite gas velocity. The C parameter corresponds to different temperatures of the gas tube or to gases with different indices of refraction or a combination of both conditions. For air blowing through a tube of 20 cm length and 0.3 cm radius that is heated to 50°C above the temperature of the entering gas, we obtain $C = 0.2$. The optimum flow velocity, $v_o/V = 6.4$, corresponds to an actual on-axis velocity of $v_o = 272$ cm/sec. (This on-axis velocity results in a gas flow of 0.04 liters/sec.) The optimum focal length is $f = 3L = 60$ cm. Even shorter focal lengths can be obtained if we use longer gas tubes, gases with higher index of refraction, and higher operating temperatures. Focal length values of about 20 cm have been realized even with air.[35]

The focusing power of the gas lens is rather remarkable. However, we must study the lens distortions to obtain an overall picture of the performance of this type of lens. Figures 4.5.7 and 4.5.8 show the dependence of the focal length on the position of the ray entering the tube. We see from these figures that the r dependence of the focal length is moderate for the optimum value of the flow velocity $v_o/V = 6.4$. At lower velocities, there is more aberration, as shown by the curves for $v_o/V = 2$ of Figure 4.5.7. The focal length distortion becomes very serious for high flow velocities. This is apparent from the curves in Figure 4.5.8. The focal length curves show also that the focal lengths f_+ (defined for a beam entering the lens in the direction of the gas stream) and f_- (for a ray traveling in opposite direction) are very nearly the same.

However, the focal length curves alone do not allow us to draw definite conclusions as to the overall distortions of the lens. We must remember

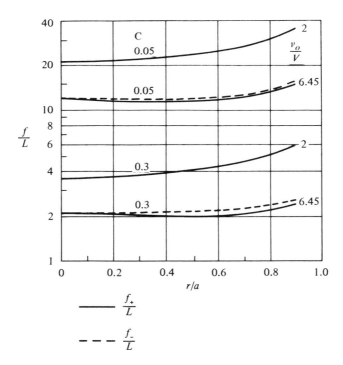

Figure 4.5.7 Focal length as a function of the entrance position r/a of the ray for $v_o/V = 2$ and $v_o/V = 6.45$. The solid line indicates f_+/L, while the broken line illustrates f_-/L. (From Theory of Thermal Gradient Gas Lenses, *IEEE Trans. on MTT*, Vol. 13, No. 6, pp. 734–739, November 1965. Reproduced with permission.)

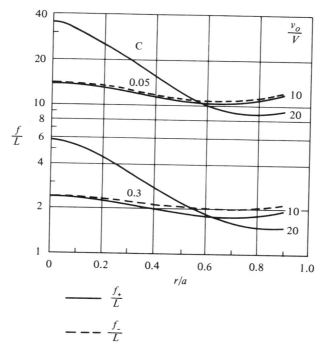

Figure 4.5.8   Focal length as a function of entrance position of the ray for $v_o/V = 10$ and $v_o/V = 20$. The solid and broken lines have the same meaning as in Figure 4.5.7. (From Theory of Thermal Gradient Gas Lenses, *IEEE Trans. on MTT*, Vol. 13, No. 6, pp. 734–739, November 1965. Reproduced with permission.)

that the focal length of the lens was measured with respect to the principal surfaces. Only after we have examined the shape of the principal surfaces do we get an overall impression of the lens distortions.

The shape of the principal surfaces is shown in Figures 4.5.9 and 4.5.10. The figures show that the two principal surfaces P_+ and P_- are nearly identical, supporting our claim that the gas lens acts like a thin lens. However, the distortion of the principal surface is rather severe. For flow velocities corresponding to the optimum of Figures 4.5.6, we find that the principal surface is located about midway in the lens, but that its extremes on-axis and close to the wall of the tube extend over nearly one-third of the length of the gas lens. For short lenses and long focal length, this actual distortion can be kept low. We see from a comparison of Figures 4.5.7 and 4.5.9 that the focal length distortion and the principal plane distortion have the tendency to cancel each other. Figure 4.5.7 shows that the focal length is longer for rays entering the lens closer to the wall. Figure 4.5.9 shows, on

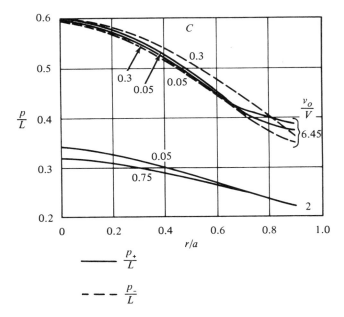

Figure 4.5.9   Shape of the principal surface for $v_o/V = 2$ and $v_o/V = 6.45$. The solid line indicates p_+/L, while the broken line illustrates p_-/L. (From Theory of Thermal Gradient Gas Lenses, *IEEE Trans. on MTT*, Vol. 13, No. 6, pp. 734–739, November 1965. Reproduced with permission.)

the other hand, that the principal surface is located closer to the input end of the tube. If we locate the focal point with respect to the input end of the tube at $z = 0$, we see that the two effects, increasing focal length and receding principal surface, have the tendency to cancel, so that the location of the focal point with respect to the gas tube remains more nearly the same. If it were important to design a gas lens with minimum distortions, this cancellation effect could be used to advantage.

In order to study the validity of the thin lens assumption, the following computer simulation was conducted.[36] Rays were traced through a succession of 100 gas lenses. The ray tracing was done in one case by solving the ray equation numerically to obtain the ray position at each lens. The second method consisted in using a thin lens equivalent, as shown in Figure 4.5.11, to simulate the gas lens. The shape of this equivalent lens was chosen to fit the principal surface of the gas lens, and the focal length that was assigned each point on the distorted thin lens was chosen to correspond to the focal length f_+, calculated by tracing parallel input rays through the lens. The actual ray tracing was done with the help of (3.7-37). The close agreement between the two methods is shown in Figures 4.5.12 and 4.5.13. The lenses are spaced in both figures approximately two to two and a half times their

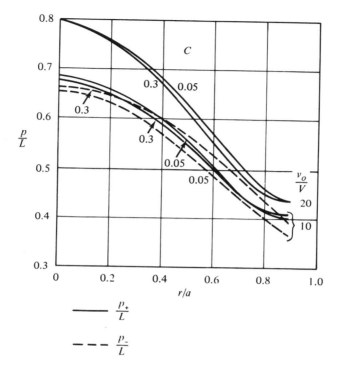

$$\frac{p_+}{L}$$

$$\frac{p_-}{L}$$

Figure 4.5.10   Shape of the principal surface for $v_o/V = 10$ and $v_o/V = 20$. The meaning of the solid and broken lines is the same as in Figure 4.5.9. (From Theory of Thermal Gradient Gas Lenses, *IEEE Trans. on MTT*, Vol. 13, No. 6, pp. 734–739, November 1965. Reproduced with permission.)

focal length. The lenses used in Figure 4.5.12 are approximately four times weaker than those used in Figure 4.5.13. The agreement between the two methods is better for the weaker lenses. But even for the stronger lenses, the agreement is close.

Gas lenses are quite efficient in their focusing properties and can be made with reasonably small distortions. However, compared to good, corrected glass lenses, their image-forming qualities remain poor. Their low-loss behavior makes them attractive in applications where light must be passed through many lenses in succession. We shall return to such applications in the next chapter.

4.6 RESOLUTION LIMIT OF IMAGE FORMATION

In Section 4.2, we derived an expression, Equation (4.2.11), for the resolution limit of image formation. That derivation was based on the uncertainty

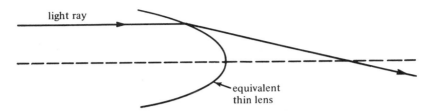

Figure 4.5.11   The equivalent thin lens of a gas lens is warped.

principle of the quantum theory of ray optics. In this section, we discuss the Rayleigh criterion for the resolution of optical instruments, and make some comments on a discussion that has been carried on in the literature and that has added considerable confusion to the subject. It has been claimed that the Rayleigh limit is only a convenient measure for the obtainable resolution of optical instruments but that it should be possible, at least in principle, to obtain resolution without limit.[7] This claim has been discussed in a very elegant paper by G. Toraldo di Francia,[37] who has shown that even though it should be possible to obtain infinite resolution on strictly mathematical

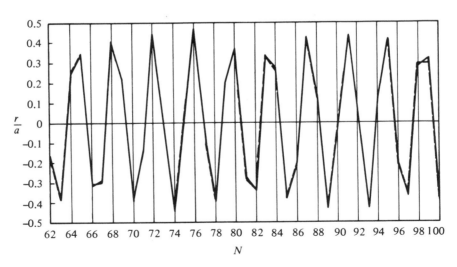

Figure 4.5.12   Comparison of ray trajectories through gas lenses and equivalent thin lenses. N = lens number, a = radius of gas lens, D = lens spacing, f_o = focal length for rays close to the optical axis, L = length of gas lens, D/a = 1200, D/f_o = 2.16, L/a = 50. (From D. Marcuse, Comparison Between a Gas Lens and its Equivalent Thin Lens, *B.S.T.J.*, Vol. 8, No. 8, October 1966, pp. 1339–44. Copyright 1966, The American Telephone and Telegraph Co., reprinted by permission.)

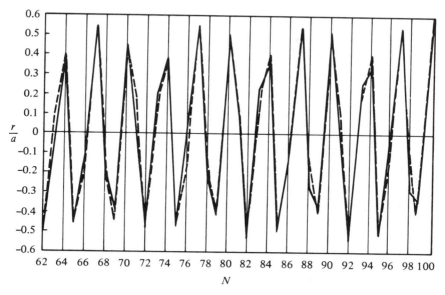

Figure 4.5.13   This figure is similar to Figure 4.5.12, with $D/a = 330$, $D/f_o = 2.74$, and $L/a = 50$. (From D. Marcuse, Comparison Between a Gas Lens and its Equivalent Thin Lens, *B.S.T.J.*, Vol. 8, No. 8, October 1966, pp. 1339–44. Copyright 1966, The American Telephone and Telegraph Co., reprinted by permission.)

grounds, practical realization of this mathematical ideal is nevertheless impossible.

We begin the discussion by presenting the classical argument by Lord Rayleigh. Following the procedure of Toraldo di Francia, we consider the image-forming system shown in Figure 4.6.1.

For simplicity, we consider the geometry of Figure 4.6.1 as a two-dimensional problem. The essential points about the resolution criterion can be obtained satisfactorily from the two-dimensional problem; the numerical factors involved are only very slightly different from those of the corresponding three-dimensional problem. Using the notation of Section 4.3, we obtain the Fourier transform of the object field $f(x)$ in the back focal plane of the first lens according to (4.3.32)

$$F(v) = \frac{1}{\sqrt{2\pi}} \int_{-u_o/2}^{u_o/2} f(u)e^{iuv} du \qquad (4.6-1)$$

We assumed that the object field exists over only a limited range, so that the integral can be taken from $-\frac{1}{2} u_o$ to $\frac{1}{2} u_o$. The second lens subjects the field in the Fourier transform plane to another Fourier transformation.

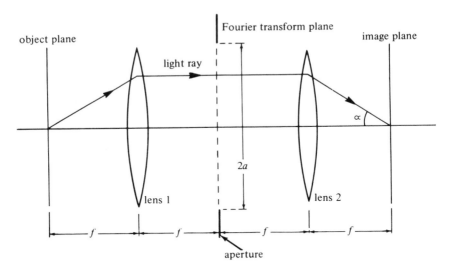

Figure 4.6.1 Schematic of the image forming system used for the discussion of the resolution limit of optical imaging systems.

However, we assume that there is an aperture in the Fourier transform plane limiting the region through which light beams can pass. The existence of such apertures is unavoidable. Even if no real aperture is mounted in the back focal plane of the first lens, there would be an effective aperture that is caused by the finite size of the two lenses. An optical system unlimited by apertures is impossible. Using the same formula once more and assuming that the aperture in the Fourier transform plane extends from $-v_o$ to v_o (in terms of the v variable), we obtain the field distribution $\bar{f}$ in the image plane

$$\bar{f}(w) = \frac{1}{\sqrt{2\pi}} \int_{-v_o}^{v_o} F(v)e^{ivw}dv \tag{4.6-2}$$

Substitution of (4.6-1) into (4.6-2) results in the following expression

$$\bar{f}(w) = \int_{-u_o/2}^{u_o/2} \frac{\sin v_o(u + w)}{\pi(u + w)} f(u)du \tag{4.6-3}$$

In order to be able to present Rayleigh's argument, we assume a "point" source (in our two-dimensional treatment, this corresponds of course to a line source) in the image plane

$$f(u) = \delta(u) \tag{4.6-4}$$

The "point" source is represented by a delta function that gives an infinite field contribution at the coordinate origin. The image of this "point" source follows from (4.6-3) and (4.6-4)

$$\bar{f}_\delta(w) = \frac{\sin v_o w}{\pi w} \qquad (4.6\text{-}5)$$

In order to study the limit in which two objects can still be resolved, we would need at least one more object point in the object plane. However, it is unnecessary to complicate the problem any further. Using the Rayleigh argument, we say that a second object point must not lie so close to the first point that the two images completely overlap. As a practical criterion, we assume that two objects can be recognized as being distinct if the maximum of the image field produced by one object falls on the first zero of the diffraction pattern that represents the image of the other point. Since the image and object sizes in our arrangement are the same (there is no magnification), we conclude that the image of the two objects can be resolved if the separation $\Delta u = \Delta w$ of the two objects is at least given by

$$v_o \Delta u \geq \pi \qquad (4.6\text{-}6)$$

In order to be able to use this information, we must transform the coordinates into the real space coordinates system, using the Equations (4.3-28) and (4.3-29).

$$\Delta u = \sqrt{\frac{2\pi}{\lambda f}} \, \Delta x \qquad (4.6\text{-}7)$$

and

$$v_o = \sqrt{\frac{2\pi}{\lambda f}} \, a \qquad (4.6\text{-}8)$$

The object separation is given in terms of the real x coordinate by Δx, while a is the half width of the aperture in the Fourier transform plane. Using (4.6-6), we obtain the Rayleigh criterion for the resolution limit of two object points

$$\Delta x \geq \frac{\lambda f}{2a} \qquad (4.6\text{-}9)$$

In terms of the steepest ray angle α that can reach the image point, we can write the Rayleigh criterion

$$\Delta x \geq \frac{\lambda}{2\alpha} \qquad (4.6\text{-}10)$$

Since the derivation was based on diffraction theory that was derived under the assumption of small angles, it is logically consistent to use the approximate relation $a/f = \alpha$. Except for a factor $(2\pi)^{-1}$, (4.6-10) agrees with the resolution criterion (4.2-11). Since the criteria for image resolution are not identical in the two cases, the agreement between these two formulas is as close as might be expected.

The Rayleigh criterion is a useful order of magnitude estimate of the possible resolution that can be achieved with image forming instruments. The factors ($1/2$ or $1/4\pi$) appearing in these expressions must not be taken too literally. What is important is the fact that the resolution limit is to order of magnitude given by $\lambda/\sin\alpha$ or by $\lambda f/a$.

In the last few years there have appeared articles in technical magazines disputing the fact that the Rayleigh limit constitutes an insurmountable barrier to possible image resolution. This view has most recently been expressed by Goodman in his book on Fourier optics.[7] The argument goes as follows. An object field of finite extent has a Fourier transform that is an analytic function. Even though only part of this Fourier transform can be utilized, because of the presence of the aperture (real or effective) in the Fourier transform plane, it must be possible to recover the entire function. An analytic function can be constructed completely if the function is known in a small neighborhood of a point. This fact can be understood if we realize that any analytic function can be expressed in a power series at least in a finite domain. Knowing the function and all its derivatives at one point allows us to extend it as far as the radius of convergence of the power series expansion. However, the process does not stop there. We can use the function so obtained to determine again the expansion coefficients from a knowledge of the function and its derivatives near the edge of the convergence limit and thus construct a new power series expansion that extends the function into an even larger domain. Proceeding in this way, it is possible to construct the entire analytic function from its values in a small initial area. This procedure is known as analytic continuation.[38]

The critics of the Rayleigh criterion claim that since the Fourier transform (an analytic function) of the image of finite extent is known over a finite area, it must be possible to construct the entire Fourier transform by analytic continuation and thus reconstruct a complete image of the object to any degree of accuracy unhampered by the Rayleigh limit.

This argument is mathematically correct. However, there are other correct statements of principle that cannot be realized in practice. Consider, for example, a signal that is transmitted by a distant radio transmitter. Disregarding the quantum nature of radiation, we can say that the signal of this transmitter must reach out into space as far as we want to go. Of course the amount of energy arriving at any given point in space becomes increasingly weaker as we increase the distance between transmitter and receiver. However, the solution of the problem with the help of Maxwell's theory assures us that, in principle, the signal reaches as far into space as we please. The only problem is, can we retrieve it. The practical limit to signal detection is the inevitable presence of noise. If we increase the distance between transmitter and receiver far enough, we get to a point where the signal power vanishes in thermal noise. Of course it is possible to recover signals out of the noise by special techniques. However, to state that we can bridge arbitrary distances with a transmitter of a given finite power and signal bandwidth because the signal is always there, at least in principle, is meaningless since the presence of noise puts definite limits on the rate at which we can recover the signal. The limits to the reception of signals in the presence of noise have been shown by Shannon.[39]

The statement that the Rayleigh limit has no bearing on the theoretical recoverability of the image of an object is of a similar nature. Toraldo di Francia has clarified this problem, and we shall discuss his argument.

Toraldo di Francia's argument is based on the properties of certain functions, scaled versions of prolate spheroidal functions, whose remarkable properties are discussed in a series of papers by Landau, Pollack, Slepian, and Sonnenblick.[40,41,62,63] The property of these $\psi_i(u)$ functions that concerns us here is the fact that they form a complete set of functions simultaneously over the infinite domain $-\infty \leq u \leq \infty$ as well as over a finite domain $-\frac{u_o}{2} \leq u \leq \frac{u_o}{2}$. This property is expressed by their orthogonality conditions. The first orthogonality condition extends over the infinite domain

$$\int_{-\infty}^{\infty} \psi_i(u)\psi_j(u)du = \delta_{ij} \qquad (4.6\text{-}11)$$

(δ_{ij} is the Kronecker δ symbol), while the orthogonality condition over the finite domain assumes the form

$$\int_{-u_o/2}^{u_o/2} \psi_i(u)\psi_j(u)du = \lambda_i\delta_{ij} \qquad (4.6\text{-}12)$$

The ψ functions are solutions of the integral equation

$$\lambda_i\psi_i(v) = \int_{-u_o/2}^{u_o/2} \frac{\sin v_o(v-u)}{\pi(v-u)} \psi_i(u)du \qquad (4.6\text{-}13)$$

The parameter λ_i is the eigenvalue of the integral equation (4.6-13). The eigenvalue λ_i as well as the functions $\psi_i(u)$ are dependent on a constant c, which is defined by the following equation

$$c = \frac{1}{2} v_o u_o \qquad (4.6\text{-}14)$$

The interval from $-\frac{u_o}{2}$ to $\frac{u_o}{2}$ over which the $\psi_i(u)$ functions are orthogonal determines the exact form of these functions via their dependence on the parameter c. For the following discussion, it is helpful to know what values to expect for c. Using (4.3-28) and (4.3-29), we can write c as

$$c = \frac{\pi}{\lambda f} aD \tag{4.6-15}$$

The object size D corresponds to u_o, and a is the half width of the aperture. We pointed out in Section 2.2 that the Fresnel approximation of diffraction theory holds if the neglected terms in the series expansion (2.2-23) after multiplication with $k = 2\pi/\lambda$ (the wave length λ must not be confused with the eigenvalue λ_i) are much smaller than unity. We must require

$$\frac{1}{8} k \frac{a^4}{f^3} = \frac{\pi}{4} \frac{a^4}{\lambda f^3} \ll 1 \tag{4.6-16}$$

in order to ensure that the Fresnel approximation is justified. With $f = 1\ m$ and $\lambda = 0.5\ \mu$, we need $a = 1.5$ cm to make the expression on the left of (4.6-16) approximately equal to 0.1. Using for simplicity $a = D$, we obtain from (4.6-15)

$$c = 1410 \tag{4.6-17}$$

The number c can become quite large even for relatively small apertures. Incidentally, c is a measure of the validity of the Fresnel approximation in diffraction problems. The Fresnel approximation works better for smaller values of c. The parameter c gives the order of magnitude of the second term in the expansion (2.2-23) (after multiplication with $k = 2\pi/\lambda$) which, to order of magnitude, can be written

$$N = \frac{1}{2} k \frac{a^2}{L} = \pi \frac{a^2}{\lambda L} \tag{4.6-18}$$

N is called the Fresnel number. We wrote L instead of f in (4.6-18) to indicate more generally the distance along the optical axis. The choice of the focal length f in (4.6-15) was suggested by our current problem as shown in Figure 4.6.1. For $a = D$ and $L = f$, the parameter c is identical with the Fresnel number N. For our example, we obtain $N = 1410$.

After this digression into the validity range of the Fresnel approximation, let us return to the discussion of the eigenvalues λ_i. Slepian and Sonnenblick[41] give an approximation for λ_i that is valid for large values of c and i

$$\lambda_v = \frac{1}{1 + e^{\pi b}} \tag{4.6-19a}$$

with

$$b = \frac{\frac{\pi}{2} v - c + \frac{\pi}{4}}{0.2886 + 2 \ln 2 + \frac{1}{2} \ln c} \tag{4.6-19b}$$

This equation shows that $\lambda_v \approx 1$ for $v \ll \dfrac{2}{\pi} c$ but $\lambda_v \ll 1$ for $v \gg \dfrac{2}{\pi} c$. The transition from the region where λ_v is essentially unity to the region where λ_v drops off to very small values appears at $v = v_c$, with

$$v_c = \frac{2}{\pi} c \qquad (4.6\text{-}20)$$

To show how dramatic the change of the eigenvalue is, let us look at an example. For $v = 890$, we get

$$\lambda_{890} = 0.96 \qquad (4.6\text{-}21)$$

For $v = 920$, we obtain

$$\lambda_{920} = 10^{-5} \qquad (4.6\text{-}22)$$

For smaller values of c, the transition is even more abrupt. Using the tables of Slepian and Sonnenblick, we find that for $c = 30$

$$\lambda_{19} = 0.356 \qquad (4.6\text{-}23)$$

but

$$\lambda_{25} = 4.74 \times 10^{-6} \qquad (4.6\text{-}24)$$

and

$$\lambda_{30} = 1.32 \times 10^{-11} \qquad (4.6\text{-}25)$$

We are now ready to discuss Toraldo di Francia's argument. Every function can be expanded in terms of the complete orthogonal set of ψ functions. The object field can be represented by an expansion

$$f(u) = \sum_{i=0}^{\infty} a_i \psi_i(u) \qquad (4.6\text{-}26)$$

with the expansion coefficient determined by

$$a_i = \int_{-\infty}^{\infty} f(u)\psi_i(u)du \qquad (4.6\text{-}27)$$

The output field (image field) can similarly be represented by

$$\tilde{f}(-w) = \sum_{i=0}^{\infty} A_i \psi_i(w) \qquad (4.6\text{-}28)$$

with

$$A_i = \int_{-\infty}^{\infty} \tilde{f}(-w)\psi_i(w)dw \qquad (4.6\text{-}29)$$

Because of the eigenvalue equation (4.6-13) for the ψ functions, we can immediately state the connection between the expansion coefficients a_i and A_i. Substitution of (4.6-26) into (4.6-3) and use of the eigenvalue equation (4.6-13) yields

$$\bar{f}(-w) = \sum_{i=0}^{\infty} a_i \lambda_i \psi_i(w) \tag{4.6-30}$$

so that the comparison of (4.6-28) and (4.6-30) allows us to state

$$A_i = \lambda_i a_i \tag{4.6-31}$$

In an image-forming system, the image field $\bar{f}(w)$ is observed, and the object field $f(u)$ is not known. The critics of the Rayleigh limit of image resolution claim that the image field can, in principle, be constructed from the object field no matter how small the aperture of the image forming system. That this statement is mathematically true can be seen from Equation (4.6-31). Since we know the image field, we know, in principle, all the coefficients A_i. From them we can obtain the expansion coefficients of the object field series (4.6-26)

$$a_i = \frac{A_i}{\lambda_i} \tag{4.6-32}$$

Substitution of (4.6-32) into (4.6-26) provides us with the precise image field regardless of the size of the aperture and the wavelength of light used in the image-forming system. This proves the fact that, mathematically speaking, there is no limitation caused by the finite aperture of the image forming system.

Let us now consider the practical problem and see whether there is hope that our mathematical result can be utilized. To obtain the result that the image can always be reconstructed, we need to assume that all the coefficients A_i can actually be measured with arbitrary precision. However, we know from our discussion of the eigenvalue λ_i that for $i \gg \frac{2}{\pi} c$ the eigenvalue becomes extremely small. This shows that the expansion coefficients of the image field must be much smaller than the corresponding expansion coefficients for the object field. We thus face the problem of measuring very small quantities with very high precision. Since, according to (4.6-32), the expansion coefficients a_i are obtained from A_i by multiplication with a very large number, it is obvious that the measuring error that occurred in the attempt to measure A_i is also magnified by the very large amount $1/\lambda_i$. The precise determination of the object fields requires us to measure the expansion coefficients of the image field with extreme precision. It is thus apparent that there is an analogy between the problem of detection of a radio signal originating at a distant transmitter and the problem of recovering the object field from the image field.

It is impossible to obtain the image without interference of some sort of noise. This noise might be thermal noise in a photodetector used to probe the image field, a television picture tube, or even a photographic emulsion. That there is noise in electrical detection is a well-known fact. Photographic emulsions contribute "noise" because of the finite grain size of the silver halid crystals after development. But even if we could make an ideal emulsion, we would not be able to reduce the grain size below the size of an individual molecule. This shows that noise in photographic emulsions is just as inevitable as noise in electrical circuits. As a practical matter, therefore, there is no hope at all of being able to reproduce the object field precisely from knowledge of the image field. Some information will always get lost, even though, in a strict mathematical sense, we should be able to construct the object field from the image field in spite of diffraction. This mathematical statement is of only limited use in the real world. Only a far more detailed study could show whether the resolution limit given by the Rayleigh criterion could be improved. The connection between our present discussion and the Rayleigh criterion is not a simple and straightforward one even though it should be possible to determine a practical resolution limit based on Equation (4.6-32). Our intent was to point out that arguments against the actual limitation predicted by the Rayleigh criterion are of very doubtful value if they are based on the observation that an analytical function can, in principle, be constructed from its values within a finite domain. Unless the influence of noise interfering with the detection process are included in the argument, not much can be gained from an argument based on an idealized mathematical fact.

CHAPTER
5

LENS WAVEGUIDES

5.1 INTRODUCTION

We have seen in Sections 2.3, 2.4, and 3.6 that it is impossible to generate a parallel bundle of light and transmit it over arbitrary distances without change of its cross section. A truly parallel beam of light is possible only in the limit of vanishingly small wavelength or with an infinitely large aperture. Both of these conditions can of course never be realized. However, because of the short wavelength of light, it is relatively easy to use large apertures (Large in terms of wavelength) and thus produce a fairly well-collimated light beam. If we have a receiver at a moderate distance from the light transmitter, we may well be able to intercept practically all the light that was transmitted. But if the distances that we want to bridge become very large, we will no longer be able to intercept all of the transmitted light. For example, a coherent light beam that emerges from a circular aperture of 30 cm diameter on the earth illuminates a circle of 1 km radius on the moon (disregarding the effect of the earth atmosphere).

The problem of light transmission for earth-based communications is limited not so much by the tendency of light to spread by diffraction but rather by the influence of the earth's atmosphere. A laser beam that has traveled several miles through the atmosphere arrives at its destination badly distorted. If this beam is intercepted and viewed on a screen, it gives the impression of a lively burning flame. The reason for the unsteady behavior of the received light is random disturbances in the atmosphere that deflect and diffract the light beam so that the diffraction limit of light collimation cannot be reached even approximately. If we consider in addition that there are plenty of obstacles that must be circumvented along an earthbound transmission path, we appreciate the need for a light guidance system. There are of course many different ways of guiding light. We shall discuss a few of them in this book. The present chapter is devoted to a particularly promising approach, among others suggested by Goubau—the lens waveguide[43] (Goubau called it the beam waveguide). As the name suggests, this light waveguide uses lenses to accomplish the guidance effect. We will treat the lens waveguide by means of ray optics as well as wave optics. In addition to describing the principle of the perfectly straight waveguide, we shall also consider bent lens waveguides and waveguides with statistical irregularities.

5.2 RAY OPTICS OF THE PERFECT LENS WAVEGUIDE

The geometry of the lens waveguide is shown in Figure 5.2.1. The lens spacing is shown as being twice the focal length. This is a special configuration; many others are possible, as we shall see during the course of our discussion. The arrangement shown in Figure 5.2.1 is called a confocal lens waveguide, since the focal points of the various lenses coincide. The confocal geometry gives rise to particularly simple ray trajectories. Two (of infinitely many) rays traveling through the waveguide are shown in the figure.

The analysis of the lens waveguide will be restricted to the paraxial approximation. Only in this approximation is it possible to solve the ray equation analytically.

The action of the lens waveguide can be understood in terms of the quantum theory of light rays. According to the uncertainty principle (3.6-56), we lose all knowledge of the ray position when we force the light rays to be perfectly parallel. A perfectly collimated beam allows us to determine the slope (in terms of the quantum theory of light rays, the ray momentum) with certainty. The uncertainty principle requires $\Delta x = \infty$ if $\Delta p_x = 0$. In order to have a beam of finite extension (that is, finite values of its position uncertainty Δx), we must allow for some uncertainty of the ray's "momentum" Δp_x. The lens waveguide forces the light beam to diverge and to converge

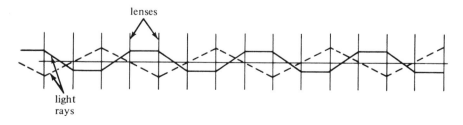

Figure 5.2.1   Two typical ray trajectories in the confocal lens waveguide.

as it travels from lens to lens. The ray slopes are never identical for all the rays of a bundle, so that there exists a finite uncertainty Δp_x of the ray "momentum." This gives us the possibility of confining the ray position to within a given region Δx without violating the uncertainty principle for light rays.

In order to be able to determine the ray trajectories in the lens waveguide, we derive a difference equation for the ray positions at each lens.[44] Consider three of the lenses of the waveguide as shown in Figure 5.2.2. The distance of the ray from the optical axis taken at the position of the lens is r_n. The angle between the ray and the direction of the guide axis immediately to the right of each lens is indicated by α_n. All lenses have the same focal length f, and are spaced a distance L from each other. The paraxial lens equation (4.2-4) allows us to write.

$$\alpha_n - \alpha_{n-1} = -\frac{r_n}{f} \tag{5.2-1}$$

The angles are counted positive, as shown in the figure; this explains the difference in sign between (4.2-4) and (5.2-1). The position of the ray at the nth lens can be expressed in terms of its position at the n-1th lens and the ray angle by

$$r_n = r_{n-1} + \alpha_{n-1} L \tag{5.2-2}$$

This equation holds also only in the paraxial approximation. In order to obtain a difference equation for only the r_n's, we must eliminate the angles from (5.2-1) and (5.2-2). This is accomplished by using (5.2-2) with $n+1$ instead of n

$$r_{n+1} = r_n + \alpha_n L \tag{5.2-3}$$

Taking the difference of (5.2-3) and (5.2-2) and using (5.2-1) yields the desired equation

$$r_{n+1} + \left(\frac{L}{f} - 2\right)r_n + r_{n-1} = 0 \tag{5.2-4}$$

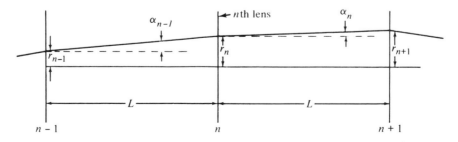

Figure 5.2.2   This diagram indicates the definition of the ray angles α_n and the ray positions r_n.

What we need of course is a solution of this difference equation that allows us to determine the position of the ray at any arbitrary lens if we know its position at the first two lenses.

The solution of (5.2-4) can be found with the help of a trial solution of the form

$$r_n = Ae^{in\Theta} \tag{5.2-5}$$

Substitution of the trial solution into (5.2-4) yields an equation for the determination of Θ

$$\cos \Theta = 1 - \frac{L}{2f} \tag{5.2-6}$$

Equation (5.2-5) with (5.2-6) is indeed a solution of the difference equation (5.2-4). Another solution is obviously obtained by simply changing the sign of Θ, since (5.2-6) holds for positive as well as negative signs of Θ. The most general solution can therefore be written in the form

$$r_n = Ae^{in\Theta} + Be^{-in\Theta} \tag{5.2-7}$$

The constants A and B can be expressed in terms of the ray position at the first two lenses

$$r_1 = Ae^{i\Theta} + Be^{-i\Theta} \tag{5.2-8}$$

$$r_2 = Ae^{2i\Theta} + Be^{-2i\Theta} \tag{5.2-9}$$

The solution of these two equations is

$$A = -\frac{e^{-i\Theta}}{2i} \frac{r_1 e^{-i\Theta} - r_2}{\sin \Theta} \tag{5.2-10}$$

$$B = \frac{e^{i\Theta}}{2i} \frac{r_1 e^{i\Theta} - r_2}{\sin \Theta} \tag{5.2-11}$$

Equation (5.2-7) can now be written in its final form

$$r_n = -r_1 \frac{\sin(n-2)\Theta}{\sin \Theta} + r_2 \frac{\sin(n-1)\Theta}{\sin \Theta} \tag{5.2-12}$$

Equation (5.2-12) allows us to determine the trajectory of every ray in the lens waveguide in the paraxial approximation. In particular, we can verify the two ray trajectories shown in Figure 5.2.1. This figure was drawn for the special case $L = 2f$. In this confocal geometry, we obtain from (5.2-6) $\cos \Theta = 0$ or $\sin \Theta = 1$. This allows us to simplify (5.2-12) for the confocal lens waveguide

$$r_n = r_1 \sin n\frac{\pi}{2} - r_2 \cos n\frac{\pi}{2} \tag{5.2-13}$$

The ray trajectory shown by the solid line in Figure 5.2.1 has the initial conditions

$$r_1 = r_o \qquad r_2 = -r_o \qquad (5.2\text{-}14)$$

so that we can write

$$r_n = r_o \left(\sin n \frac{\pi}{2} + \cos n \frac{\pi}{2} \right) \qquad (5.2\text{-}15)$$

It can easily be seen that (5.2-15) describes the ray trajectory indicated by the solid line in Figure 5.2.1.

The ray trajectory that is shown by a dotted line in that same figure has the initial conditions

$$r_1 = -r_o \qquad r_2 = 0 \qquad (5.2\text{-}16)$$

so that its trajectory in the confocal waveguide becomes

$$r_n = -r_o \sin n \frac{\pi}{2} \qquad (5.2\text{-}17)$$

Again it is immediately apparent that (5.2-17) results in the ray trajectory shown by the broken line in Figure 5.2.1.

The most important result of our analysis is Equation (5.2-6) because it allows us to determine the range of L/f ratios that are admissible for lens waveguides. It is apparent from the solution (5.2-12) that guided rays can exist only for real values of the parameter Θ. If Θ becomes complex, the sine functions would become superpositions of hyperbolic and ordinary sine and cosine functions, and the rays would have amplitudes that grow indefinitely. A guided ray could thus not exist in the lens waveguide. Since the cosine function has values only between -1 and 1, we obtain for the admissible range of the L/f ratio

$$0 \leq \frac{L}{f} \leq 4 \qquad (5.2\text{-}18)$$

For L/f values outside the range (5.2-18), guided rays do not exist. Spacing the lenses only an infinitesimal distance further apart than $L = 4f$ results in a waveguide whose rays grow in amplitude and eventually must leave the area of the lenses and disappear from the lens waveguide.

Let us study this case in more detail. For $L > 4f$, (5.2-6) becomes negative but larger than unity in its absolute value.
A solution is possible only in the form

$$\Theta = i\theta' + \pi \qquad (5.2\text{-}19)$$

with real values of θ'. For $L > 4f$, we can write (5.2-6) in the following form

$$\cosh \theta' = \frac{L}{2f} - 1 \qquad \text{for } \frac{L}{f} > 4 \qquad (5.2\text{-}20)$$

(5.2-20) indicates that there are indeed solutions outside the region where guided waves are possible. The corresponding ray trajectories are obtained from (5.2-12) and (5.2-19)

$$r_n = (-1)^n \left\{ r_1 \frac{\sinh(n-2)\theta'}{\sinh \theta'} + r_2 \frac{\sinh(n-1)\theta'}{\sinh \theta'} \right\} \qquad (5.2\text{-}21)$$

Since the hyperbolic sine function does not have an oscillatory behavior, but grows monotonically with increasing argument, we see that (5.2-21) describes rays that oscillate around the optical axis on account of the factor $(-1)^n$ but whose amplitudes increase without limit as the rays travel along the lens waveguide.

To compare the behavior of the waveguide just below the region where ray guidance is still possible and just above the region where it ceases to exist, we write (5.2-12) for $\Theta = \pi$

$$r_n = (-1)^n [(n-2)r_1 + (n-1)r_2] \qquad (5.2\text{-}22)$$

The same result is obviously also obtained from the equation for unguided rays (5.2-21). The transition between the two regions is thus continuous. In fact, the critical border region with $\Theta = \pi$ or $\theta' = 0$ is interesting because guided and unguided rays exist simultaneously. Whether a ray is guided or unguided depends only on the initial conditions. For general values of r_1 and r_2, we see that the ray amplitudes grow linearly, so that the ray wanders off the lenses as it follows its oscillatory path. However, for the special case $r_1 = -r_2 = r_o$, we still obtain the guided ray

$$r_n = -(-1)^n r_o \qquad (5.2\text{-}23)$$

The ray optics treatment of the lens waveguide provides us with a great deal of information about the possible regions of guided and unguided rays or waves.

As another interesting example of a lens waveguide, let us study a succession of lenses similar to those in Figure 5.2.1 but with the difference that the focal length of the lenses alternates from lens to lens. Let the even-numbered lenses have focal length f_1, while the odd-numbered lenses have focal length f_2. We can even allow for the case that either f_1 or f_2 (but not both) become negative. The lenses can alternate between converging and diverging lenses. We shall see that guided rays can exist even in this case. This latter case is called alternating gradient focusing.[45,46]

The ray equation follows from Figure 5.2.3.

Using the paraxial lens equations similar to (5.2-1), we obtain

$$\alpha_{2n} - \alpha_{2n-1} = -\frac{r_n}{f_1} \qquad (5.2\text{-}24)$$

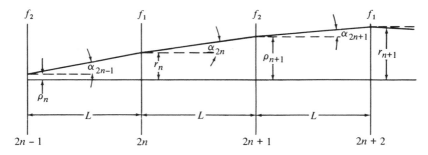

Figure 5.2.3   Definition of ray angles and ray positions for a lens waveguide with alternating lenses of two different focal lengths.

and

$$\alpha_{2n+1} - \alpha_{2n} = -\frac{\rho_{n+1}}{f_2} \tag{5.2-25}$$

The rays at adjacent lenses are again connected by Equation (5.2-2)

$$r_n = \rho_n + \alpha_{2n-1} L \tag{5.2-26}$$

and

$$\rho_{n+1} = r_n + \alpha_{2n} L \tag{5.2-27}$$

Taking the difference of (5.2-27) and (5.2-26), we obtain with the help of (5.2-24)

$$\rho_{n+1} + \rho_n = \left(2 - \frac{L}{f_1}\right) r_n \tag{5.2-28}$$

Replacing n with $n + 1$ in (5.2-26) and subtracting (5.2-27) from the resulting equation yields with the help of (5.2-25)

$$r_{n+1} + r_n = \left(2 - \frac{L}{f_2}\right) \rho_{n+1} \tag{5.2-29}$$

We write (5.2-29) again with n replaced by $n - 1$ and add the resulting equation to (5.2-29). Using (5.2-28) to eliminate p results in an equation containing only r

$$r_{n+1} + \left[2 - \left(2 - \frac{L}{f_1}\right)\left(2 - \frac{L}{f_2}\right)\right] r_n + r_{n-1} = 0 \tag{5.2-30}$$

Similarly, we replace n with $n + 1$ in (5.2-28) and add the resulting equation to the original one. Using (5.2-29) to eliminate r results in an equation containing only ρ

$$\rho_{n+1} + \left[2 - \left(2 - \frac{L}{f_1}\right)\left(2 - \frac{L}{f_2}\right)\right]\rho_n + \rho_{n-1} = 0 \qquad (5.2\text{-}31)$$

The two paraxial ray equations (5.2-30) and (5.2-31) are identical in form to the ray equation (5.2-4) of the simpler lens waveguide. We obtain their solutions directly from (5.2-5) and (5.2-6)

$$r_n = Ae^{in\Delta} + Be^{-in\Delta} \qquad (5.2\text{-}32)$$

and

$$\rho_n = Ce^{in\Delta} + De^{-in\Delta} \qquad (5.2\text{-}33)$$

with the definition of the parameter Δ

$$\cos \Delta = -\frac{1}{2}\left[2 - \left(2 - \frac{L}{f_1}\right)\left(2 - \frac{L}{f_2}\right)\right] \qquad (5.2\text{-}34)$$

Before discussing this solution, let us check if we obtain the original theory in the limit $f_1 = f_2$

$$\cos \Delta = 2\left(\frac{L}{2f} - 1\right)^2 - 1 = 2\cos^2 \Theta - 1$$

or

$$\cos \Delta = \cos 2\Theta \qquad (5.2\text{-}35)$$

This allows us to write in this special case

$$\left.\begin{array}{l} r_n = Ae^{i2n\Theta} + Be^{-i2n\Theta} \\ \rho_n = Ce^{i\Theta}e^{i(2n-1)\Theta} + De^{-i\Theta}e^{-i(2n-1)\Theta} \end{array}\right\} \text{ for } f_1 = f_2 \qquad (5.2\text{-}36)$$

We must remember that the labeling of the variables r_n and ρ_n is different from the labeling used to describe the simple lens waveguide. It is apparent from Figure 5.2.3 that our present designation of r_n corresponds in the earlier notation to r_{2n} and that ρ_n corresponds to r_{2n-1}. Keeping this in mind, we see from a comparison of (5.2-7) with (5.2-36) that the special case $f_1 = f_2$ reduces, indeed, to the earlier case of the simple lens waveguide with $Ce^{i\Theta} = A$ and $De^{-i\Theta} = B$.

Equation (5.2-32) is formally exactly identical to (5.2-7) and (5.2-12), so that we can write immediately

$$r_n = -r_1\frac{\sin(n-2)\Delta}{\sin \Delta} + r_2\frac{\sin(n-1)\Delta}{\sin \Delta} \qquad (5.2\text{-}37)$$

Similarly, we obtain for ρ_n

$$\rho_n = -\rho_1\frac{\sin(n-2)\Delta}{\sin \Delta} + \rho_2\frac{\sin(n-1)\Delta}{\sin \Delta} \qquad (5.2\text{-}38)$$

The lens waveguide with alternating lenses of different focal length is thus described in a peculiar way. Equation (5.2-37) describes the ray position at all even-numbered lenses provided the position at the second and fourth lenses (r_1 and r_2) is known. Equation (5.2-38) describes the ray position at the odd-numbered lenses provided the ray position at the first and third lenses is known. Of course the ray trajectories on the even- and odd-numbered lenses are not independent of each other. The connection between the ray positions at even- and odd-numbered lenses is given by (5.2-28) and (5.2-29). We are thus able to predict the ray position at any lens in this waveguide.

It remains to study the condition at which guided rays are possible.[46] Using the abbreviations

$$x = 1 - \frac{L}{2f_1} \qquad (5.2\text{-}39)$$

and

$$y = 1 - \frac{L}{2f_2} \qquad (5.2\text{-}40)$$

we can write (5.2-34) in the form

$$\cos \Delta = 2xy - 1 \qquad (5.2\text{-}41)$$

The cosine function cannot be larger than unity in its absolute value. In order to determine the boundaries of the range of values that x and y are allowed to assume, we set $\cos \Delta = 1$ and obtain

$$xy = 1 \qquad (5.2\text{-}42)$$

At the other extreme, $\cos \Delta = -1$, we obtain from (5.2-41)

$$xy = 0 \qquad (5.2\text{-}43)$$

The boundaries of the allowed range of x and y values are shown in Figure 5.2.4. The hyperbolas are defined by (5.2.42). Equation (5.2-43) states that the x-axis and y-axis are the remaining boundary lines. The allowed values of x and y lie in the shaded region of Figure 5.2.4.

Several points of this diagram are of particular interest.

The point $x = 1$, $y = 1$ is the point where the lens waveguide disappears. Both types of lenses assume infinite focal length. The point $x = -1$, $y = -1$ is the limit $f = L/4$ of the simple lens waveguide with identical lenses. The focal length of these lenses assumes its shortest possible value compatible with guided rays. Along the x-axis the regions of focal length for the even-numbered lenses (f_1) is indicated in Figure 5.2.4. The point $x = 1$ corresponds to infinite focal length. To the right of this point extends the region of negative focal length. This is the region of alternating gradient focusing, $f_1 < 0$ and $f_2 > 0$. Similar regions exist of course along the y-axis. The entire negative x-axis belongs to the region from $f_1 = 0$ ($x = -\infty$) to $f_1 = L/2$ ($x = 0$).

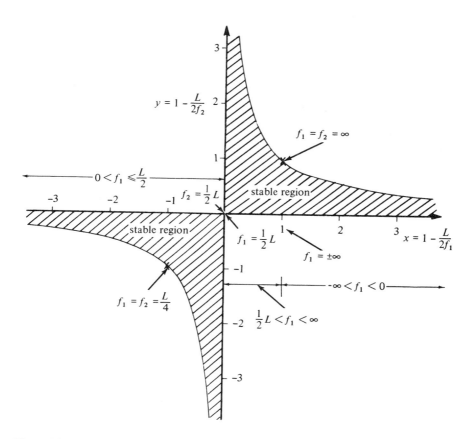

Figure 5.2.4   Stability diagram of a lens waveguide with alternating lenses of two different focal lengths or of a laser cavity with unequal mirrors. The hatched area indicates the region with stable (low loss) operation. Outside the hatched area the waveguide or cavity suffers from high loss. (From G. D. Boyd and H. Kogelnik,Generalized Confocal Resonator Theory, *B.S.T.J.*, Vol. 41, No. 4, July 1962, pp. 1347–1369. Copyright 1962, The American Telephone and Telegraph Co., reprinted by permission.)

The point $x = y = 0$ deserves special attention. As indicated in Figure 5.2.4, it corresponds to the confocal simple lens waveguide with $f_1 = f_2 = L/2$. When we studied the properties of the simple lens waveguide, the confocal geometry was peculiar only because of its simple ray trajectories. The study of the lens waveguides composed of two types of alternating lenses presents this point in an entirely new light. We see that, from this point of view, the confocal lens waveguide lies at the border between the region of stable ray trajectories and that of unstable (growing) ray trajectories. This fact is not very important for actual lens waveguides. The chances are very slim that the lenses will accidentally alternate in focal length in just the right way to make our present theory apply. However, our discussion of the lens waveguide has very important applications to the theory of laser resonators. The confocal laser resonator is also described by the point $x = 0$, $y = 0$, and the instability indicated by its precarious position at the edge of the stable region has a direct significance for the laser cavity.

5.3 LASER RESONATORS[47,48,49]

Laser resonators typically have the appearance of Figure 5.3.1. Two mirrors of high reflectivity oppose each other. Usually these mirrors are curved. Both mirrors may have the same or different curvature. We can describe the action of such a laser resonator in terms of lens waveguides by the following argument. The laser cavity (resonator) can be transformed as indicated by Figure 5.3.2.

The curved mirror is optically equivalent to the combination of one plane

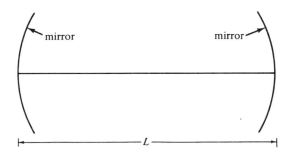

Figure 5.3.1   Diagram of a laser cavity.

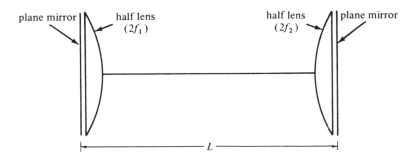

Figure 5.3.2   Equivalent laser cavity. The curved mirrors are replaced with a combination of a plane mirror and a lens.

mirror and a lens placed immediately in front of it. The resonators shown in Figures 5.3.1 and 5.3.2 are optically completely equivalent.

We shall see in our discussion of the wave optics of lens waveguides that these structures have modes. The modes are field distributions that repeat themselves, except perhaps for their phase, at each lens. A mode traveling in a lens waveguide is fully equivalent to the corresponding mode of a laser resonator provided we can neglect the losses of either device. Consider a resonant mode of the cavity of Figure 5.3.1 or Figure 5.3.2. Let us start in the middle of the cavity. The standing wave field of the resonant mode can be decomposed into two traveling waves moving in opposite directions. Following one of these traveling waves components to the right, we see it propagate toward the mirror, reflect back from the mirror, and proceed in the opposite direction toward the other mirror. As far as this field distribution is concerned, the cavity looks exactly like a lens waveguide. In fact, the structure of Figure 5.3.2 transforms into a lens waveguide if we ignore the fact that the wave reverses its direction at the plane mirror but if we let it travel on unreflected but provide another identical lens immediately behind the first one. The actual reflected wave traverses each lens of Figure 5.3.2 twice in quick succession. As it approaches the lens-mirror combination, it passes through the lens, reflects from the plane mirror, and passes the lens once more. This wave propagates exactly as if it found itself inside the structure shown in Figure 5.3.3. The lenses in Figure 5.3.2 have the focal lengths $2f_1$ and $2f_2$. However, since they are traversed by the wave twice in immediate succession, they act like two closely spaced lenses which, combined, have the focal length f_1 or f_2. The resonator in its unfolded form shown in Figure 5.3.3 is indeed identical to the lens waveguide of Figure 5.2.3.

So far we have concentrated on one of the traveling wave components of the standing wave pattern in the resonator. The other component suffers precisely the same fate with the exception that it travels in the opposite direction. Both traveling wave components superimpose each other at each point in space, forming the standing wave pattern of the resonator. It is of course necessary that the waves retrace each other exactly as they travel back

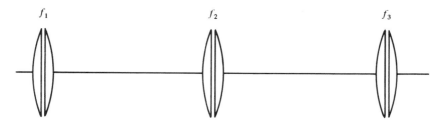

Figure 5.3.3   Diagram of an unfolded laser cavity showing the equivalence between laser cavities and lens waveguides.

and forth inside the resonator, and that they superimpose themselves on their own patterns after each round trip in such a way that constructive interference results. This requirement forces the resonator to operate only at a set of discrete frequencies dependent on its length. This is the only difference between the resonator field and a mode traveling in the corresponding lens waveguide. The lens waveguide can operate at a continuum of frequencies, while the resonator can operate only at certain discrete frequencies. However, the lens waveguide is capable of supporting modes just as the resonator is. Their field patterns are identical to the field pattern of the corresponding resonator if we consider it as an unfolded structure. The relationship between the lens waveguide and the laser cavity will become even more striking when we formulate the wave optics solution of these problems in Sections 5.6 and 6.6. The laser resonator can be described by an integral eigenvalue equation. The iterative solution of this equation is exactly equivalent to a wave traveling inside a lens waveguide.

Having established the equivalence between laser resonators and lens waveguides, we can proceed to apply the findings of our lens waveguide discussion to the resonator case.

We have seen in Figure 5.2.4 that the confocal lens waveguide corresponds to the origin $x = y = 0$ of that figure. This special point is precariously located between stable and unstable regions of the structure. We also remarked that this instability would not be very serious for an actual lens waveguide, since its lenses would have to alternate precisely between two focal length values f_1 and f_2 if there should be any danger of the lens waveguide slipping accidentally into the forbidden region of Figure 5.2.4. However, consider a laser resonator that was constructed to operate as a confocal cavity. In that case, the resonator would correspond to a confocal lens waveguide. Since the many lenses of the waveguide are represented by the same two mirrors of the resonator, any accidental deviation from equal focal length can either cause the device to operate inside the shaded region close to the origin of Figure 5.2.4 or, equally likely, the resonator might accidentally slip into the unshaded region. In that case, its loss would increase very rapidly, and its performance would therefore be seriously degraded. For this reason, laser cavities are usually not built with the confocal arrangement, but are intentionally designed to operate at a point in the shaded region of Figure 5.2.4 to avoid their operation inside the high loss region caused by an unwanted departure from exactly identical mirrors. For some special applications in high-gain lasers, operation in the unstable region of the resonator has been suggested as a means of obtaining preferential mode selection.

We shall return to the discussion of laser resonators in connection with the wave theory of lens waveguides. However, it should have become apparent what valuable insight into the performance of lens waveguides and laser resonators can already be obtained from a discussion based on geometrical optics.

5.4 LENS WAVEGUIDE WITH CURVED AXIS

In the previous sections, we considered lens waveguides whose axes were perfectly straight. This is of course an idealization that is not realizable in practice. In fact, one of the desirable features of lens waveguides is that they are capable of guiding light beams around bends. In order to be able to study the ability of lens waveguides to lead light around curved paths and also to be able to study lens waveguides with random deviations from perfect straightness, we formulate the ray theory of lens waveguides with arbitrarily curved axes.[50,51]

The geometry of a lens waveguide section with curved axis is shown in Figure 5.4.1. We assume in this section that the lenses of the guide are all identical. The ray position at each lens measured from its center is again indicated by r_n. The lens centers are connected by straight lines that form the distorted axis of the structure. The angles γ_n indicate the change in direction of the guide axis at the nth lens. The direction of the light ray is again indicated by α_n. This angle is defined by the direction of the ray immediately to the right of the nth lens and the straight line connecting the lens centers. The ray equation can be obtained from the following paraxial equations

$$r_{n+1} = r_n + \alpha_n L \tag{5.4-1}$$

and

$$\alpha_n - \alpha_{n-1} - \gamma_n = -\frac{r_n}{f} \tag{5.4-2}$$

Equation (5.4-1) is identical with (5.2-2). Equation (5.4-2) is the same as (5.2-1) with the exception that we must subtract the angle γ_n that describes the change in the direction of the waveguide axis from the angle difference $\alpha_n - \alpha_{n-1}$, since these angles are measured against the guide axis, while the

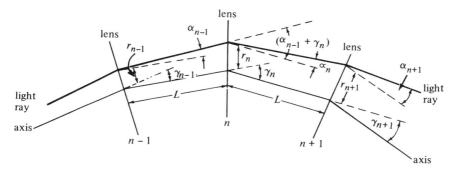

Figure 5.4.1 Diagram of a lens waveguide with curved axis.

change of the ray direction is determined only by the power of the lens and the ray position on the lens, independent of the changing direction of the waveguide axis. We eliminate the angles again by taking the difference of (5.4-1) with a similar equation that is obtained by replacing n with $n - 1$ and by using (5.4-2)

$$r_{n+1} - 2\left(1 - \frac{L}{2f}\right)r_n + r_{n-1} = \gamma_n L \tag{5.4-3}$$

The lens waveguide with curved axis is described by an inhomogeneous difference equation. The solutions of difference equations can be obtained very similar to the solutions of differential equations. Inhomogeneous differential equations can be solved by a method known as variation of constants. Adopting this approach to the solution of our difference equation, we attempt a trial solution of the following form

$$r_n = a_n e^{in\Theta} \tag{5.4-4}$$

with Θ given by (5.2-6). For constant values of a_n (that is, a_n independent of n) the trial solution (5.4-4) would be a solution of the homogeneous difference equation ($\gamma_n = 0$). Using the notation of (5.2-6), we can write the equation system (5.4-3) in the following form for the special case $r_1 = r_2 = 0$ and $\gamma_1 = 0$

$$a_3 e^{i\Theta} \qquad\qquad\qquad\qquad = L\gamma_2 e^{-2i\Theta}$$

$$a_4 e^{i\Theta} - 2a_3 \cos\Theta \qquad\qquad = L\gamma_3 e^{-3i\Theta}$$

$$a_5 e^{i\Theta} - 2a_4 \cos\Theta + a_3 e^{-i\Theta} = L\gamma_4 e^{-4i\Theta}$$

$$a_6 e^{i\Theta} - 2a_5 \cos\Theta + a_4 e^{-i\Theta} = L\gamma_5 e^{-5i\Theta}$$

$$\cdot\quad\cdot\quad\cdot\quad\cdot\quad\cdot\quad\cdot\quad\cdot\quad\cdot\quad\cdot\quad\cdot\quad\cdot$$

$$a_n e^{i\Theta} - 2a_{n-1} \cos\Theta + a_{n-2} e^{-i\Theta} = L\gamma_{n-1} e^{-(n-1)i\Theta}$$

Adding all the equations of this system leads to the expression

$$a_n e^{i\Theta} - a_{n-1} e^{-i\Theta} = L \sum_{v=2}^{n-1} \gamma_v e^{-iv\Theta} \tag{5.4-5}$$

The trial solution (5.4-4) has reduced the inhomogeneous difference equation of second order (5.4-3) to an inhomogeneous difference equation of first order. This reduction of the order of the equation is completely analogous to the corresponding method of variation of constants as applied to differential equations. Once more we write down the equation system (5.4-5)

$$a_3 e^{i\Theta} \qquad\qquad = L \sum_{\nu=2}^{2} \gamma_\nu e^{-i\nu\Theta} \Bigg|$$

$$a_4 e^{i\Theta} - a_3 e^{-i\Theta} = L \sum_{\nu=2}^{3} \gamma_\nu e^{-i\nu\Theta} \Bigg| e^{2i\Theta}$$

$$a_5 e^{i\Theta} - a_4 e^{-i\Theta} = L \sum_{\nu=2}^{4} \gamma_\nu e^{-i\nu\Theta} \Bigg| e^{4i\Theta}$$

$$\cdot \quad \cdot \quad \cdot \quad \cdot \quad \cdot \quad \cdot \quad \cdot \quad \cdot \quad \cdot \quad \cdot$$

$$a_n e^{i\Theta} - a_{n-1} e^{-i\Theta} = L \sum_{\nu=2}^{n-1} \gamma_\nu e^{-i\nu\Theta} \Bigg| e^{i2(n-3)\Theta}$$

It is apparent that all but the term a_n on the left-hand side of these equations cancel each other when we multiply each equation by the exponential factor indicated to the right of the vertical bar and add all equations. This procedure results in

$$a_n e^{i(2n-5)\Theta} = L \sum_{\nu=2}^{n-1} \gamma_\nu e^{-i\nu\Theta} \sum_{\mu=\nu+1}^{n} e^{i2(\mu-3)\Theta} \qquad (5.4\text{-}6)$$

The double summation on the right-hand side of (5.4-6) is obtained by considering a particular term of each sum occurring on the right-hand side of the equation system and adding all these corresponding terms after multiplication with the exponential factors.

After carrying out the summation and substituting the result into (5.4-4), we obtain as a particular solution of the inhomogeneous second order difference equation

$$r_n = \frac{L}{\sin\Theta} \sum_{\nu=2}^{n-1} \gamma_\nu \sin(n-\nu)\Theta \qquad \text{for } n \geq 3 \qquad (5.4\text{-}7)$$

The complete solution of the inhomogeneous difference equation (5.4-3) is the sum of the solution (5.2-12) of the homogeneous equation and the particular solution (5.4-7) of the inhomogeneous equation

$$r_n = \frac{1}{\sin\Theta} \left\{ -r_1 \sin(n-2)\Theta + r_2 \sin(n-1)\Theta \right.$$
$$\left. + L \sum_{\nu=2}^{n-1} \gamma_\nu \sin(n-\nu)\Theta \right\} \text{ for } n \geq 3 \quad (5.4\text{-}8)$$

This solution of the ray trajectory through a lens waveguide with curved axis is a paraxial approximation. However, it holds for waveguides whose axes follow an arbitrary path as long as the change of direction of the waveguide axis γ_n remains small.

The angle γ_n can be used to express the radius of curvature of the lens waveguide. Figure 5.4.2 shows how the angle γ_n is related to the radius of

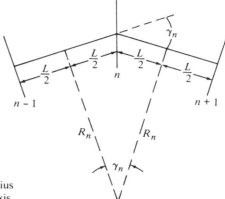

Figure 5.4.2 Definition of the radius of curvature of the lens waveguide axis.

curvature of the waveguide axis. This relation can be expressed in the form

$$\gamma_n = \frac{L}{R_n} \tag{5.4-9}$$

Equation (5.4-9) holds of course also only in the paraxial approximation.

Another useful way of describing the lens waveguide with curved axis is by referring the ray position as well as the location of the lens centers to a certain straight line. This representation is particularly useful for the description of lens waveguides whose axes deviate only very slightly from a perfectly straight line. The geometry of this representation is sketched in Figure 5.4.3. The displacement of the lenses is greatly exaggerated in this figure. The ray positions are now no longer referred to the center of each lens but to the straight reference line. The distance from the reference line to the lens center is designated by S_n, so that the ray position ρ_n is given by the relation

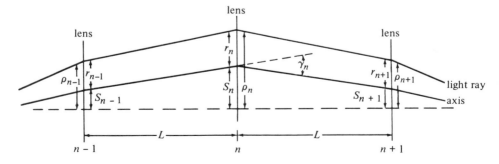

Figure 5.4.3 Alternate description of a lens waveguide with curved axis. The position of the lens centers and the ray position is referred to the dotted line.

$$\rho_n = r_n + S_n \tag{5.4-10}$$

A relation between the distance S_n and the angle γ_n can be found in the following way. Consider the triangle that results from the straight line sections connecting the lens centers of the lenses $n - 1$, n, and $n + 1$ and a third straight line (not shown in Figure 5.4.3) connecting the center of lens $n - 1$ with the center of lens $n + 1$. The angles inside this triangle are approximately $(S_n - S_{n-1})/L$ at lens $n - 1$, $\pi - \gamma_n$ at lens n, and $(S_n - S_{n+1})/L$ at lens $n + 1$. The requirement that the sum of these angles must equal π leads to the condition

$$\gamma_n = \frac{1}{L}(2S_n - S_{n-1} - S_{n+1}) \tag{5.4-11}$$

Substitution of (5.4-10) and (5.4-11) into (5.4-7) allows us to write

$$\rho_n = S_n + \frac{1}{\sin \Theta} \sum_{v=2}^{n-1} (2S_v - S_{v-1} - S_{v+1})\sin(n - v)\Theta \tag{5.4-12}$$

Regrouping of the terms in the sum leads to the expression

$$\rho_n = \frac{1}{\sin \Theta}\left\{ -S_1 \sin(n - 2)\Theta + S_2 \sin(n - 1)\Theta + \sum_{v=2}^{n-1} S_v[2 \sin(n - v)\Theta \right.$$
$$\left. - \sin(n - v - 1)\Theta - \sin(n - v + 1)\Theta] \right\} \tag{5.4-13}$$

It is our purpose to construct a particular solution of the inhomogeneous difference equation that corresponds to the problem in this modified reference system. The first two terms outside the summation sign of (5.4-13) are apparently solutions of the homogeneous difference equation, and may be incorporated into the homogeneous solution that must be added to obtain the complete solution of the inhomogeneous difference equation. Omitting the first two terms from (5.4-13) and using the addition theorems for the sine functions allows us to write a particular solution of the inhomogeneous problem as follows

$$\rho_n = \frac{2(1 - \cos \Theta)}{\sin \Theta} \sum_{v=2}^{n-1} S_v \sin(n - v)\Theta \qquad n \geq 3 \tag{5.4-14}$$

Adding the solution of the homogeneous equation, in analogy to the procedure leading to (5.4-8), finally yields the complete solution of our modified problem

$$\rho_n = \frac{1}{\sin \Theta}\left\{ -\rho_1 \sin(n - 2)\Theta + \rho_2 \sin(n - 1)\Theta + \frac{L}{f} \sum_{v=2}^{n-1} S_v \sin(n - v)\Theta \right\}$$
$$\tag{5.4-15}$$

Equation (5.2-6) was used to express cos Θ in terms of L/f. This second formulation of the problem of the lens waveguide with curved axis has the disadvantage that it is not suitable to describe waveguides whose axis departs appreciably from its original direction. In those cases, (5.4-8) must be used. However, (5.4-15) is useful for certain statistical studies of the behavior of light rays in lens waveguides with random lens displacements.

We are now capable of predicting the trajectory of any ray in a lens waveguide with arbitrarily deformed axis as long as the radius of curvature of the waveguide axis remains large enough to justify the approximations that were required to obtain (5.4-8).

There are several simple cases that can easily be described by our theory. Consider a lens waveguide whose axis is bent into a circle. We may drop the index n from the radius of curvature and write instead of (5.4-9)

$$\gamma_n = \frac{L}{R} \tag{5.4-16}$$

The summation in (5.4-8) can now easily be performed with the result

$$r_n = \frac{1}{\sin \Theta} \left\{ -r_1 \sin(n-2)\Theta + r_2 \sin(n-1)\Theta \right.$$

$$\left. - \frac{L^2}{2R} \left[\frac{\cos \frac{\Theta}{2}}{\sin \frac{\Theta}{2}} \{\cos[(n-1)\Theta] - 1\} + \sin(n-1)\Theta \right] \right\} \tag{5.4-17}$$

The only term not dependent on n is

$$r_c = \frac{L^2}{2R} \frac{\cos \frac{\Theta}{2}}{\sin \Theta \sin \frac{\Theta}{2}} \tag{5.4-18}$$

With the help of (5.2-6), we can rewrite it in the simple form

$$r_c = \frac{Lf}{R} \tag{5.4-19}$$

It can easily be shown that we obtain $r_n = r_c$ if we set $r_1 = r_2 = r_c$ in (5.4-17). However, instead of showing this explicitly, we will prove more generally that the light ray travels in the lens waveguide with circular axis in a way very similar to the trajectory of a light ray in a perfectly straight waveguide with the only difference that the ray does not oscillate around the axis of the guide but around a line that is parallel to the axis but displaced from it by an amount r_c.

We prove this assertion by introducing the new variable

$$w_n = r_n - r_c \tag{5.4-20}$$

The new variable is the ray position as measured from the displaced reference line a distance r_c from the waveguide axis. Introducing (5.4-20) in (5.4-17) leads to the equation

$$w_n = \frac{1}{\sin\Theta} \left\{ -w_1 \sin(n-2)\Theta + w_2 \sin(n-1)\Theta \right.$$
$$\left. + \frac{L^2}{2R} \left[\frac{\cos\dfrac{\Theta}{2}}{\sin\Theta \sin\dfrac{\Theta}{2}} (1 - \cos\Theta) - 1 \right] \sin(n-1)\Theta \right\} \tag{5.4-21}$$

This expression was obtained by expressing $\sin(n-2)\Theta$ in terms of $\sin(n-1)\Theta$ and $\cos(n-1)\Theta$. The term with $\cos(n-1)\Theta$ cancels out. With the help of the addition theorems of the circular functions, it is easy to show that the expression inside the bracket in (5.4-21) vanishes. We see therefore that the transformation (5.4-20) transforms (5.4-17) into the simpler expression

$$w_n = \frac{1}{\sin\Theta} \{ -w_1 \sin(n-2)\Theta + w_2 \sin(n-1)\Theta \} \tag{5.4-22}$$

This expression for the ray trajectory in a lens waveguide that is bent into a circle is formally identical to the ray trajectory (5.2-12). The ray inside the circular lens waveguide travels just as a ray does inside the straight lens waveguide if the ray trajectory is referred not to the guide axis but to a line displaced from the guide axis by the amount r_c.

This is a very interesting result. It proves that the lens waveguide is capable of guiding light beams along curved trajectories. It shows, furthermore, that a light ray can travel in a circularly deformed lens waveguide without oscillations ($w_1 = w_2 = 0$) if it follows the trajectory $w_n = 0$ or $r_n = r_c$ with r_c of (5.4-19). The constant ray trajectory is displaced more for smaller values of the radius of curvature R—that is, for more sharply bent guides. We also see that spacing the lenses closely (small L) allows us to negotiate sharper bends. The ability of a lens waveguide to guide light rays through circular paths is limited only by the size of the lenses. The radius of the nonoscillatory trajectory can of course be no larger than the radius a of the lenses. In fact, for a practical system, we must allow a to be considerably larger than r_c, since the ray must be given room to follow an oscillatory trajectory.

The trajectories of rays in lens waveguides with axis of more complicated shape can be obtained from (5.4-8), but the sum appearing in this equation can usually not be evaluated in closed form. For computer solutions of the ray trajectory, we can either use (5.4-8) or it may be just as easy to let the computer solve the difference equation (5.4-3).

Another deviation from perfect straightness that can easily be handled by our analytical solution is a simple tilt of an otherwise straight lens waveguide. In this case we have

$$\gamma_v = \gamma \delta_{v\mu} \tag{5.4-23}$$

which means that all but one of the angles γ_n are equal to zero. If the ray started out on axis ($r_1 = r_2 = 0$), we obtain the ray trajectory at points behind the tilt from (5.4-8)

$$r_n = \frac{\gamma L}{\sin \Theta} \sin(n - \mu)\Theta \tag{5.4-24}$$

The tilt with tilt angle γ causes a ray that approaches the tilt on axis to oscillate around the axis of the guide with an amplitude $\gamma L/\sin \Theta$.

As a final example, we consider the case of a straight lens waveguide with one displaced lens. In this case, we have

$$S_v = S \delta_{v\mu} \tag{5.4-25}$$

and obtain for an initially on-axis ray ($\rho_1 = \rho_2 = 0$) from (5.4-15)

$$\rho_n = \frac{LS}{f \sin \Theta} \sin(n - \mu)\Theta \tag{5.4-26}$$

A single displaced lens, just like a tilt, causes an on-axis beam to oscillate with an amplitude $SL/(f \sin \Theta)$.

The discussion in this section has shown that lens waveguides are able to guide light beams along curved paths with a radius of curvature that is determined by the focal length and the radius of the lenses. However, we have also seen that any imperfection from either perfect straightness or perfect circular shape causes the light beam to oscillate. These oscillations, once they have started, are not likely to decrease at later points of the waveguide. It is of course possible that another accidental tilt of another displaced lens may reduce the amplitude of an oscillating ray. However, on the average, the ray oscillations tend to grow, as we shall see in the next section.

The property of tilts and displaced lenses to change the trajectory of a light ray can be used to advantage to guide a light beam back on-axis if it has started to oscillate as it propagated through the lens waveguide. However, this suppression of ray oscillations is possible only for particular rays whose trajectory is known. It is not possible on a statistical basis, as will be shown in a later section. The suppression of ray oscillations requires sensors that detect the positon of the light ray on at least two lenses. The output of these sensors can be used to steer subsequent lenses in a controlled way designed to bring the beam back on an on-axis path.

5.5 LENS WAVEGUIDES WITH RANDOM LENS DISPLACEMENTS

We have seen in the previous section that lens waveguides, even though they have the ability to guide rays along curved paths, do this by allowing the ray to follow an oscillatory trajectory. Since the lenses have finite radii, the amplitude of the lens oscillations must not grow to values in excess of the radii of the lenses, since the ray will then be lost. We study the statistical behavior of the ray trajectories of light beams in lens waveguides to obtain information about the danger of random lens displacements.

We begin our discussion of lens waveguide statistics by considering a guide with randomly displaced lenses with no correlation between the displacements of adjacent lenses.[50,52] The assumption of random lens displacements is most realistic, since it is impossible to construct a lens waveguide with perfectly positioned lenses. Assuming no correlation between the lens displacements may be an idealization, but it represents a first approximation to any real situation.

We study a lens waveguide that is nominally straight. The ray is injected into this guide on axis, so that we have

$$\rho_1 = \rho_2 = 0 \tag{5.5-1}$$

Next, we introduce the variance σ_n^2 of the ray position

$$\sigma_n^2 = <\rho_n^2> \tag{5.5-2}$$

The symbol $< >$ indicates an ensemble average. From (5.4-15) we obtain

$$\sigma_n^2 = \frac{L^2}{f^2 \sin^2 \Theta} \sum_{v=2}^{n-1} \sum_{\mu=2}^{n-1} <S_v S_\mu> \sin(n-v)\Theta \sin(n-\mu)\Theta \tag{5.5-3}$$

The assumption of random, uncorrelated lens displacements can be mathematically expressed in the form

$$< S_v S_\mu > = s^2 \delta_{v\mu} \tag{5.5-4}$$

The quantity s is the rms value of the lens displacements. The condition (5.5-4) allows us to obtain from (5.5-3)

$$\sigma_n^2 = \frac{s^2 L^2}{f^2 \sin^2 \Theta} \sum_{v=2}^{n-1} \sin^2(n-v)\Theta \tag{5.5-5}$$

The sum over $\sin^2(n-v)\Theta$ can be carried out with the result

$$\sigma_n^2 = \frac{s^2 L^2}{2f^2 \sin^2\Theta} \left\{ n - 2 - \frac{\cos(n-1)\Theta \sin(n-2)\Theta}{\sin \Theta} \right\} \tag{5.5-6}$$

Of real interest is only the case of very large lens number n. If n is large enough to allow us to neglect the terms appearing with it inside the bracket, we obtain for the rms deviation (square root of the variance) of the lens displacement

$$\sigma_n = \frac{sL}{\sqrt{2f}\sin\Theta}\sqrt{n} \qquad \text{for } n \gg 1 \qquad (5.5\text{-}7)$$

Using (5.2-6), we can write this expression also as

$$\sigma_n = \frac{s\sqrt{\dfrac{L}{f}}}{\sqrt{2 - \dfrac{L}{2f}}}\sqrt{n} \qquad (5.5\text{-}8)$$

In the important case of a confocal lens waveguide ($L = 2f$), the rms deviation of the beam is simply

$$\sigma_n = s\sqrt{2n} \qquad (5.5\text{-}9)$$

We have seen that a lens waveguide with $L = 4f$ represents a limiting case in which only certain special rays can still be guided by the lens waveguide. This result is again obtained from (5.5-8), which becomes infinite for $L = 4f$. The beam is unstable in this case, and is lost if the lenses are misaligned even slightly.

To obtain a feeling for the seriousness of the random lens displacement, let us consider an example. Let us assume that we want to guide a light beam by means of a lens waveguide over a path of 10 km. Using low-loss gas lenses, it is not unreasonable to space these lenses 1 m apart. The total number of lenses is then $n = 10{,}000$. If we can allow the beam to deviate approximately $\sigma_n = 3$ mm from the optical axis (a reasonable assumption for gas lenses), we must keep

$$s < 2.12 \cdot 10^{-2} \text{ mm} = 21.2\mu \qquad (5.5\text{-}10)$$

This means that, on the average, the lenses must not depart more than approximately 20 microns from their nominal position if we want to be sure that the light beam can travel over the 10 km distance in the lens waveguide. We remember that one of the attractive features of gas lenses is that it is possible to space them closely without introducing an undue amount of loss into the transmission path. Our discussion of curved lens waveguides showed us that we could negotiate sharper bends with lenses that were spaced closely, (5.4-19). These considerations made it appear desirable to use very many closely spaced gas lenses to form a lens waveguide. Our present result, however, warns us not to use too many lenses in order to keep the tolerance requirements within reasonable bounds. A tolerance requirement of 20μ is surely too severe to be practical. Let us assume that we can tolerate an rms lens deviation of $s = 0.1$ mm. This means that we can allow the light beam to pass through 450 lenses or, at 1 m lens spacing, through a distance of 450 m.

Equation (5.4-15) with random S_v is an ideal application for the central limit theorem of probability theory.[54] The central limit theorem states that a random variable that is the sum of a large number of uncorrelated random variables is distributed according to the Gaussian probability distribution regardless of the probability distribution of the random variables in the sum. We can assume, therefore, that the probability of finding a ray at any given position in the lens waveguide with random lens displacements is governed by the Gaussian probability distribution. This result holds of course only after the light ray has passed through a large number of lenses. However, it holds for arbitrary values of n if the probability distribution of the lens displacements is also Gaussian. The Gaussian probability distribution is completely determined by its variance. We thus have the result that the probability of finding a ray at position r in the interval dr is given by[53]

$$p_n(r)dr = \frac{1}{\sqrt{2\pi}\,\sigma_n}\, e^{-(r^2/2\sigma_n^2)}dr \tag{5.5-11}$$

This probability distribution of finding a ray at a certain radius at the position of the nth lens holds of course only for infinitely large lenses. The probability distribution (5.5-11) is still approximately correct for lenses with a finite radius a provided that $\sigma_n \ll a$.

If we want to discuss the realistic case of lens waveguides with lenses whose radii are not necessarily much larger than the rms deviation of the beam, we must look for probability distributions that hold at least approximately for that case. In order to understand the limitations of the probability distribution (5.5-11), it is important to realize that a ray that is being observed at a given radius r may get there in a number of ways. It may be that this ray happens to be near its maximum departure on its undulating trajectory when we encounter it at r at the nth lens. More likely, however, this ray is intercepted at lens n during any other portion of its cycle. In other words, we observe rays at any given point that most likely have reached much larger values of r during the last half cycle of their oscillatory trajectory. If the ray amplitude A (that is, the amplitude of the oscillatory trajectory) has a value $A > a$, the ray would not reach the nth lens in a realistic lens waveguide with lenses of finite radius a. The probability distribution (5.5-11) is particularly vulnerable to the presence of finite lens apertures, since it provides information about rays at a given radius $r < a$, some of which are sure to have ray amplitudes A that are larger than the lens radius a.

The preceding discussion shows that it would be much more reasonable to consider the probability distribution for the ray amplitudes rather than the probability for the ray positions. The probability for the amplitude of an oscillatory process whose instantaneous value is known to be Gaussian is well known in the theory of thermal noise of electrical circuits. It is the Rayleigh distribution[55]

$$P_n(A)dA = \frac{A}{\sigma_n^2} e^{-A^2/2\sigma_n^2} dA \qquad (5.5\text{-}12)$$

It is not hard (and may be instructive) to prove the validity of our assertion. The ray trajectory can be written as

$$r_n = A \sin(n\Theta + \phi) \qquad (5.5\text{-}13)$$

In a lens waveguide with randomly displaced lenses, A and ϕ are not constant but are random variables. For any given ray, they are slowly varying functions of n. The probability distribution of the phase must be constant. Limiting ourselves for the moment to the single-valued portion of the ray trajectory that extends from $\sin(n\Theta + \phi) = -1$ to $\sin(n\Theta + \phi) = +1$, we can safely assume that the probability distribution of ϕ is

$$\Phi(\phi) = \frac{1}{\pi} \qquad (5.5\text{-}14)$$

Letting A be constant and allowing ϕ to vary causes r_n to run through its range of values from $-A$ to A. The probability of finding r_n in the interval dr is $G_A(r_n)dr$. G_A is the conditional probability density of finding r_n provided it is known that A has a certain definite value. The relation between G_A and Φ is given by the requirement that if ϕ is found at a certain value in the interval $d\phi$ the radius r must be found at its corresponding point in the interval dr. We express this requirement by the equation

$$G_A(r)dr = \Phi(\phi)d\phi \qquad (5.5\text{-}15)$$

or by

$$G_A(r) = \frac{1}{\dfrac{dr}{d\phi}} \Phi(\phi) \qquad (5.5\text{-}16)$$

Using (5.5-13) and (5.5-14) and eliminating ϕ allows us to write

$$G_A(r) = \begin{cases} \dfrac{1}{\pi\sqrt{A^2 - r^2}} & \text{for } |r| < A \\[2ex] 0 & \text{for } |r| > A \end{cases} \qquad (5.5\text{-}17)$$

The probability density for the ray position r follows simply from the rules of probability theory[54,55]

$$p_n(r) = \int_0^\infty G_A(r)P_n(A)dA \qquad (5.5\text{-}18)$$

$P_n(A)$ is of course the probability density for the ray amplitude A. Using (5.5-17), we can express this relation as

$$p_n(r) = \frac{1}{\pi} \int_r^\infty \frac{P_n(A)}{\sqrt{A^2 - r^2}} \, dA \qquad (5.5\text{-}19)$$

The reader can convince himself that the pair of probability distributions (5.5-11) and (5.5-12) satisfies equation (5.5-19).

The relation (5.5-19) holds not only for the case of infinitely large lenses but also in the realistic case of lenses with finite radius. However, in this case, $p_n(r)$ and $P_n(A)$ are no longer given by (5.5-11) and (5.5-12).

For practical applications, it is more important to know the probability of finding either the ray or the ray amplitude below certain values. The cumulative probability of finding the ray with a radius less than r_n is given by

$$W_n(r_n) = 2 \int_0^{r_n} p_n(r) \, dr \qquad (5.5\text{-}20)$$

Because p_n is an even function, it is possible to limit the integration range to positive values. This explains the factor 2 in (5.5-20). The cumulative probability of finding the amplitude with a value less than A is likewise

$$V_n(A) = \int_0^A P_n(x) \, dx \qquad (5.5\text{-}21)$$

The factor 2 is absent in this case, since the probability P_n is defined only for positive values of A.

Since it is very hard to formulate analytical expression for the various probabilities in case of lenses with finite radius, the author conducted computer-simulated experiments to find the probability of ray positions in a realistic lens waveguide with finite lens apertures.[53] A confocal lens waveguide with 100 lenses was simulated on the computer. The lenses were randomly displaced with the help of random numbers. In each experiment, the ray was started on axis and traced through the lenses. If the ray assumed a value larger than the lens radius a, it was dropped. For each ray tracing the random distribution of lenses was changed. It was thus possible to compute the probability of finding a ray at lens 100 with a radius less than r. This probability distribution is shown in Figure 5.5.1. There are several curves shown in this figure. Each curve belongs to a different value of a/σ_n. The value of σ_n, according to (5.5-9), indicates the degree of lens misalignment. The horizontal axis of the plot is also normalized with respect to σ_n. The dotted line in the figure corresponds also to $a/\sigma_n = 1.77$, but it was obtained by tracing the ray through 1000 instead of 100 lenses. It is surprising how nearly alike the two curves for the same a/σ_n value are. This test shows that the probability curves, if represented with the normalization shown, are not strongly dependent on the number of lenses.

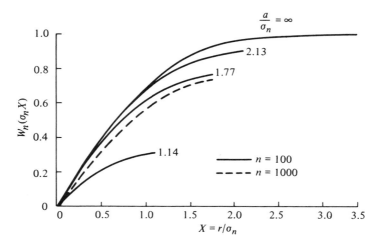

Figure 5.5.1   Cumulative probability of beam position, W_n, for various values of the ratio a/σ_n; a = lens aperture half width; σ_n = rms beam deviation. (From Probability of Ray Position in Beam Waveguides, *IEEE Trans. on MTT*, Vol. 15, No. 3, pp. 167–171, March 1967. Reproduced with permission.)

Figure 5.5.1 shows that the probability W_n is strongly dependent on the ratio of lens radius to rms deviation. This was to be expected. It would be useful to know whether the probability distribution V_n for A is less sensitive to the presence of the lens apertures. With the help of (5.5-19), it is possible to obtain a relation between V_n and W_n. We remove the variable r from the lower limit of the integral in (5.5-19) by introducing the new integration variable $u = A/r$

$$p_n(r) = \frac{1}{\pi} \int_1^\infty \frac{P_n(ru)}{\sqrt{u^2 - 1}}\, du \qquad (5.5\text{-}22)$$

Substitution of (5.5-22) into (5.5-20) leads to

$$W_n(r) = \frac{2}{\pi} \int_1^\infty \frac{V_n(ru)}{u\sqrt{u^2 - 1}}\, du \qquad (5.5\text{-}23)$$

Equation (5.5-21) was used to express one of the two integrals—that appear in this procedure—by V_n. Returning to the integration variable A allows us to write

$$W_n(r) = \frac{2r}{\pi} \int_r^\infty \frac{V_n(A)}{A\sqrt{A^2 - r^2}}\, dA \qquad (5.5\text{-}24)$$

The cumulative probability V_n has the property

$$V_n(A) = V_n(a) \qquad \text{for } A \geq a \tag{5.5-25}$$

Since all rays must be found between $-a \leq r \leq a$, we have the additional relationship

$$V_n(a) = W_n(a) \qquad \text{for} \begin{cases} r > a \\ A > a \end{cases} \tag{5.5-26}$$

It is thus possible to perform the integration over A in two steps. First, we integrate the integral in (5.5-24) from r to a. To this integral, we add the integral from a to ∞. In this latter integral, $V_n(A)$ assumes the constant value $W_n(a)$, so that the integration can be carried out. We thus obtain the following integral equation for $V_n(A)$

$$W_n(r) - \frac{2}{\pi} W_n(a)\arcsin\frac{r}{a} = \frac{2r}{\pi} \int_r^a \frac{V_n(A)}{A\sqrt{A^2 - r^2}}\, dA \tag{5.5-27}$$

This integral equation can be used to determine $V_n(A)$ once $W_n(r)$ is known. The result of the numerical solution of this integral equation is shown in Figure 5.5.2. The function $W_n(r)$ of Figure 5.5.1 was used to obtain the curves of this figure.[53]

It is apparent that $V_n(A)$ is much less dependent on the ratio of a/σ_n than is W_n. In fact, for probabilities $V_n(A)$ higher than 0.8 (or 80 percent), the function is very little different from the corresponding function for infinitely large lenses. It is therefore possible to use the function that results from substitution of (5.5-12) into (5.5-21)

$$V_n(A) = 1 - e^{-A^2/2\sigma_n^2} \text{ for } a/\sigma_n \to \infty \tag{5.5-28}$$

The assumption that the cumulative probability $V_n(A)$ of the ray amplitude

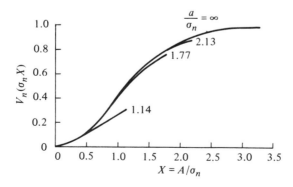

Figure 5.5.2 Cumulative probability for beam amplitudes, V_n, for various values of the ratio a/σ_n; a = lens aperture half width; σ_n = rms beam deviation. (From Probability of Ray Position in Beam Waveguides, *IEEE Trans. on MTT*, Vol. 15, No. 3, pp. 167–171, March 1967. Reproduced with permission.)

would be less sensitive to the presence of lens apertures than the cumulative probability $W_n(r)$ of the ray position has been proved correct.

The design of a lens waveguide for communications purposes must of course require that the light beam reaches the end of the waveguide with high probability. In terms of wave optics, it is not realistic to assume that the entire beam would be lost when the ray optics beam hits the apertures. The field distribution in the lens waveguide has a finite width, so that some of the field remains inside the guide even if the corresponding light ray of geometrical optics has hit the lens aperture. However, it is apparent that we must expect high losses for the light transmission if the ray optics picture predicts that the light ray will not reach the end of the lens waveguide.

Let us consider the tolerance requirements for a lens waveguide that is designed to pass a light ray with 99 percent probability. We consider again a confocal lens waveguide ($L = 2f$). The requirement of finding the ray at the last lens of the guide with 99 percent probability with an amplitude A less than a leads to the condition $A = a = 3\sigma_n$. (The value $A/\sigma_n = 3$ makes $V_n(A) = 0.989$ according to [5.5-28].) The value of σ_n is related to the rms lens displacement s via (5.5-9). The ratio of s/a is thus uniquely determined by the requirement of 99 percent transmission probability. For our confocal lens waveguide, we obtain the following values for the ratio of the rms lens displacement (that can be tolerated) to the radius of the lenses

$$\frac{s}{a} = \frac{1}{3\sqrt{2n}} = \begin{cases} 2.36 \ 10^{-2} & \text{for } n = 100 \\ 6.47 \ 10^{-3} & \text{for } n = 1000 \\ 2.36 \ 10^{-3} & \text{for } n = 10,000 \end{cases} \tag{5.5-29}$$

Up to this point, we have assumed that the lens displacements were uncorrelated among each other. This assumption is too stringent for many practical situations. Consider, for example, a lens waveguide whose lenses are mounted in a tube. The tube axis cannot be perfectly straight, and must be assumed to depart from perfect straightness in a random way. Because of the stiffness of the tube, the random deviations from perfect straightness cannot be arbitrarily abrupt. This causes the displacement of neighboring lenses to be correlated. If one lens deviates in a certain direction from its nominal position, its next neighbors most likely will be displaced in the same direction because of the stiffness of the tube. Superimposed on this correlated departure from perfect alignment is a truly random, uncorrelated deviation from the nominal position that is caused by the mounting of each individual lens inside of the tube. Such more or less correlated departures from perfect alignment can statistically be described by a correlation factor.

Let us consider the ensemble average of the product of two lens displacements

$$R_{\nu\mu} = \langle S_\nu S_\mu \rangle \tag{5.5-30}$$

In the absence of any correlation, $R_{\nu\mu}$ would vanish for $\nu \neq \mu$. If, however, there is a relation between the departures of lens ν and lens μ from their nominal positions, that means that if a certain value of S_ν will cause S_μ to depart in a certain way, then $R_{\nu\mu}$ will no longer vanish even if ν and μ are different. However, it is unreasonable to expect that the correlation factor $R_{\nu\mu}$ will assume considerable values even when $\nu - \mu$ is very large. If the correlation factor does not vanish for $\nu - \mu \to \infty$, we must conclude that there are systematic deviations of the lens positions. For truly random processes, we expect $R_{\nu\mu} \to 0$ as $\nu - \mu \to \infty$. Finally, it is often permissible to assume that the correlation factor depends only on $\rho = \nu - \mu$. Processes for which this assumption is true are called stationary random processes.[54,55] For a stationary random process, we are allowed to write

$$R_{\nu\mu} = R_{\nu - \mu} \qquad (5.5\text{-}31)$$

This notation is intended to indicate that $R_{\nu\mu}$ depends only on the difference between, but not on the particular values of, ν and μ.

After this digression into the nature of the correlation factor, we are ready to calculate the variance of the ray position for the case that certain correlations exist between the lens displacements.[52] Let us return to equation (5.5-3), and express it with the help of the correlation factor in the form

$$\sigma_n^2 = \frac{L^2}{2f^2 \sin^2\Theta} \sum_{\nu=2}^{n-1} \sum_{\mu=2}^{n-1} R_{\nu-\mu}\{\cos(\nu - \mu)\Theta - \cos[2n - (\nu + \mu)]\Theta\} \qquad (5.5.32)$$

The product of the sine functions was converted to the sum of cosine functions appearing in (5.5-32). We introduce new summation variables $\sigma = \nu - \mu$ and $\rho = \nu$. The range of the summation indices ν and μ is the square that is indicated in Figure 5.5.3 by dotted lines. It must be remembered that all the summation indices can only assume integer values. This fact is not indicated in the figure. The solid lines labeled σ and ρ indicate the regions in ν, μ space where σ or ρ assumes constant values. Of the large number of parallel lines on which σ and ρ can assume integer values, only two of each kind are shown. The figure allows us to find the range of variation of the new summation indices. The transformation to these new summation labels results in the following expression

$$\sigma_n^2 = \frac{L^2}{2f^2 \sin^2\Theta} \left\{ \sum_{\sigma=-(n-3)}^{0} R_\sigma \sum_{\rho=2}^{\sigma+n-1} [\cos \sigma\Theta - \cos(2n - 2\rho + \sigma)\Theta] \right.$$
$$\left. + \sum_{\sigma=1}^{n-3} R_\sigma \sum_{\rho=2+\sigma}^{n-1} [\cos \sigma\Theta - \cos(2n - 2\rho + \sigma)\Theta] \right\} \qquad (5.5.33)$$

The summation limits are determined as follows. When the σ line forms the diagonal of the square in Figure 5.5-3, σ assumes the value zero. As the σ

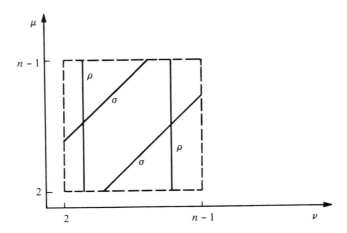

Figure 5.5.3 This diagram shows the transformation from the old summation indices ν and μ to new summation indices σ and ρ.

line moves upward in the square, the value of σ decreases, and reaches $\sigma = -(n-3)$ as the σ line reaches the upper left corner of the square with the co-ordinates $\nu = 2$ and $\mu = n - 1$. These limits are indicated on the first sum over σ in (5.5-33). To determine the limits of the sum over ρ, we must consider a fixed value of σ. The values of ρ on the σ line are found by first considering the point of intersection of the σ line with the left side of the square. Here we find $\nu = 2$, so that we have $\rho = 2$. This is the lower limit of the first sum over ρ. Its upper limit is found by considering the intersection of the σ line with the upper side of the square. Here we have $\mu = n - 1$ and $\sigma = \nu - (n - 1)$, so that we find $\rho = \sigma + n - 1$. This is the upper limit of the first sum over ρ. The limits of the second double summation are found by similar considerations in the lower right part of the square. The two double summations can be combined into one by introducing the new summation variable $\rho' = \rho - \sigma$ into the second sum

$$\sigma_n^2 = \frac{L^2}{2f^2\sin^2\Theta} \sum_{\sigma=-(n-3)}^{n-3} R_\sigma \sum_{\rho=2}^{n-1-|\sigma|} [\cos\sigma\Theta - \cos(2n - 2\rho - |\sigma|)\Theta] \quad (5.5\text{-}34)$$

The summation over ρ can now be carried out, with the result

$$\sigma_n^2 = \frac{L^2}{2f^2\sin^2\Theta} \sum_{\sigma=-(n-3)}^{n-3} R_\sigma \left[(n - |\sigma| - 2)\cos\sigma\Theta \right.$$
$$\left. - \frac{\cos(n-1)\Theta\,\sin(n-2-|\sigma|)\Theta}{\sin\Theta}\right] \quad (5.5\text{-}35)$$

In the important case that R_σ decreases rapidly with increasing values of σ, and particularly if n is so large that the σ values, for which R_σ gives an appreciable contribution, can be neglected with respect to n, we obtain the approximation

$$\sigma_n^2 = \frac{nL^2}{2f^2 \sin^2 \Theta} \sum_{\sigma = -\infty}^{\infty} R_\sigma \cos \sigma\Theta \qquad (5.5\text{-}36)$$

The limits on the summation sign are arbitrary as long as they are large, because R_σ vanishes for large values of σ.

In the absence of any correlation, we have

$$R_\sigma = s^2 \delta_{o\sigma} \qquad (5.5\text{-}37)$$

In this limiting case, (5.5-35) reduces to (5.5-6), while (5.5-36) reduces to (5.5-7).

The proper form of the correlation factor depends on the construction details of each particular lens waveguide. We must be content to use mathematical models if we want to study the consequences of our theory. A particularly simple correlation factor is

$$R_\sigma = s^2 e^{-|\sigma|/B} \qquad (5.5\text{-}38)$$

The factor s determines the rms value of the lens displacement. The factor B occurring in the exponent of the exponential function has the physical interpretation of a correlation distance. B determines the exponential decrease of R_σ with increasing values of $|\sigma|$. If the correlation factor decays rapidly with increasing values of σ, the correlation length between the lenses is short. If the decay of the function takes place slowly, we speak of a long correlation length. In terms of actual distance, we can define

$$D = BL \qquad (5.5\text{-}39)$$

as the correlation length. The shape of the correlation factor and the values of s and B describe the details of the statistics of the lens misalignment. Substitution of (5.5-38) into (5.5-36) yields the following value of the rms ray displacement

$$\sigma_n = \frac{L\sqrt{n}\,s}{\sqrt{2}f \sin \Theta} \left\{ \frac{\sinh \dfrac{1}{B}}{\cosh \dfrac{1}{B} - \cos \Theta} \right\}^{1/2} \qquad (5.5\text{-}40)$$

For confocal lens spacing ($\cos \Theta = 0$, $\sin \Theta = 1$, $L = 2f$), (5.5-40) simplifies to

$$\sigma_n = \sqrt{2n}\, s \left\{ \tanh \frac{1}{B} \right\}^{1/2} \qquad (5.5\text{-}41)$$

For vanishing correlation length, $B = 0$, (5.5-41) reduces to (5.5-9). It is thus apparent that an increase of the correlation length improves the performance of the lens waveguide by reducing the rms ray deviation. In the limit of $B \to \infty$, we have an ideal lens waveguide, with $\sigma_n = 0$. The degree of improvement over completely uncorrelated lens displacements is shown in Figure 5.5.4. For large values of B, the curve decreases as $1/\sqrt{B}$. A confocal lens waveguide, whose statistics are described by the correlation factor (5.5-38), causes only 1/5 of the beam deflection of a confocal guide with completely uncorrelated lenses (for identical values of s) if the correlation between the lens displacements reaches over approximately 25 lens spacings. This discussion describes some of the general features of correlated lens displacements. The actual details of the ray deflection caused by randomly displaced lenses depends on the circumstances of each individual case. To obtain the appropriate correlation factor for a practical lens waveguide represents a formidable problem.

An important feature of randomly displaced lenses is the increase of the beam deflection proportional to the square root of the number of lenses. If the lenses happen to be displaced systematically according to a sinusoidal function, the resulting beam displacement can be far more severe, and can increase in direct proportion to the number of lenses.[52] To see this, let us assume that the lens displacements follow the law

$$S_v = A \sin v\psi \tag{5.5-42}$$

Substitution of this formula into (5.4-15), with $r_1 = r_2 = 0$, leads to the equation

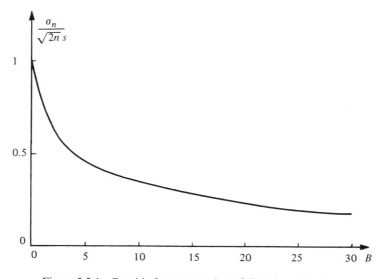

Figure 5.5.4 Graphical representation of Equation (5.5-41).

$$\rho_n = \frac{AL}{2f\sin\Theta}\left\{\frac{\cos\left(n\frac{\psi-\Theta}{2}+\frac{\psi+\Theta}{2}\right)\sin(n-2)\frac{\psi+\Theta}{2}}{\sin\frac{\psi+\Theta}{2}}\right.$$

$$\left.-\frac{\cos\left(n\frac{\psi+\Theta}{2}+\frac{\psi-\Theta}{2}\right)\sin(n-2)\frac{\psi-\Theta}{2}}{\sin\frac{\psi-\Theta}{2}}\right\} \qquad (5.5\text{-}43)$$

As long as ψ is significantly different from Θ, the beam deflection ρ_n does not become very serious. However, if $\psi = \Theta$, the situation is radically different. In this case, Equation (5.5-43) becomes

$$\rho_n = -\frac{(n-2)AL}{2f\sin\Theta}\left\{\cos n\Theta - \frac{\cos\Theta\sin(n-2)\Theta}{(n-2)\sin\Theta}\right\} \qquad (5.5\text{-}44)$$

The amplitude of the ray oscillation grows linearly with the lens number n. This shows that a sinusoidal lens displacement is much more effective in forcing the beam off-axis than a random displacement of the lenses. As an example, let us again consider the confocal case. We obtain

$$\rho_n = -(n-2)A\cos n\frac{\pi}{2} \qquad (5.5\text{-}45)$$

Using the same numbers that were used to discuss the effect of random lens displacement earlier in this chapter, we find with $n = 10,000$ and, if we are limited to a ray displacement of no more than 3 mm, that we can tolerate only a lens displacement amplitude of

$$A \le 3 \times 10^{-4}\text{mm} = 0.3\mu \qquad (5.5\text{-}46)$$

This is a very small displacement amplitude, indeed, for it is equal to roughly half the wavelength of red light.

It may appear as if a sinusoidal displacement of the lenses of just this period ($\psi = \Theta$) is not very likely to happen. We must remember, however, that an arbitrary lens displacement can always be expressed in the form of a Fourier series, and that a Fourier component with the period $\psi = \Theta$ must always exist. The restriction expressed in (5.5-46) can thus be interpreted as a restriction on the Fourier amplitude of the critical Fourier component.

This last remark may cause the reader to wonder if there is not a contradiction in this discussion. We found earlier that randomly displaced lenses cause the ray amplitude to grow proportionally to the square root of n. We now seem to claim that the ray grows proportionally to n driven by a sinusoidal Fourier component of the lens displacement. The apparent discrepancy is removed if one studies the dependence of the amplitude of the Fourier components of random functions on their length. It can be shown that the amplitudes of Fourier components of random functions decrease inversely proportional to the square root of the length of the function (lens number), so that the dependence of ρ_n (or its rms value) on the square root of n follows also from the discussion in terms of Fourier analysis.

The general features of random processes that we have discovered in this section hold quite generally. We shall meet similar behavior again when we discuss the scattering losses of dielectric light waveguides in a later chapter.

We have seen that light rays move inside the lens waveguide on oscillatory trajectories. Any wanted or unwanted deviation of the guide axis from perfect straightness " excites " the light ray to oscillations. The situation is reminiscent of a harmonic oscillator that is driven by external forces. The oscillatory behavior of the light ray causes little difficulty as long as its amplitude remains well below the radius of the lens apertures. However, when the ray oscillations grow in amplitude, the ray may hit the lens apertures and is lost. It would be very desirable if a lens waveguide could be invented that causes the ray oscillations to decay with distance of travel along the guide. A ray in such a lens waveguide would correspond to a damped harmonic oscillator.

It is not hard to design a (alas, incorrect but plausible) mathematical model of a lens waveguide that seems to force the rays to undergo damped oscillations.[24] Using thin lenses shaped like those shown in Figure 4.5.11 and application of the thin lens formula (4.2-4) results in oscillatory ray trajectories with decaying amplitudes. Our discussion of Liouville's theorem in Section 3.7 has already prepared us to know that the thin lens formula (4.2-4) can be applied only if the light rays penetrate the thin lens at or near normal incidence. Instead of (4.2-4), we must use (4.2-3). The rays treated with the help of the correct thin lens formula no longer follow oscillatory trajectories with decaying amplitudes but behave exactly like the rays discussed in Section 5.3.

The reader may wonder whether the failure to construct a lens waveguide with damped ray oscillations is caused by lack of ingenuity. Liouville's theorem provides the answer that any attempts to invent a passive lossless ray oscillation suppressor are futile. Let us assume that we have a ray oscillation suppressor. We start at one point with a bundle of rays whose representation points in phase space fill a certain volume. Remember that the coordinates of phase space represent the positions and the slopes of the rays. If our ray oscillation suppressor works, it would have the tendency to reduce the amplitudes of all rays. This would cause the phase space volume to shrink in contradiction to Liouville's theorem. A ray oscillation suppressor that acts as an attenuator for the ray oscillations is thus not possible. Liouville's theorem does allow us to build devices that trade ray slope for ray amplitudes. Such a device can be realized for example by gradually increasing the lens power along the guide and decreasing the lens spacing so that their focal points remain coincident. However, decreasing the ray amplitudes at the expense of the ray slopes is of little value, since it is impossible to increase the lens powers indefinitely. At some point, it is necessary to return to lenses with lower power. As soon as the ray leaves the guide with increasingly powerful lenses, it would immediately increase its amplitude at the expense of the ray slopes and nothing would be gained. The equivalent of a damped harmonic oscillation for light rays does not exist. It is possible to reduce the ray amplitudes if we allow the light beam to lose power. However, ray oscillation suppression at the expense of light power is not likely to be of much benefit.

The only practical solution to the ray oscillation problem appears to be an active ray redirector.[56] Such a device would sense the position of a particular ray and direct lenses or prisms to change in such a way as to force the light beam to move back on axis. Such a device does not violate Liouville's theorem, since it does not cause a bundle of rays to reduce their volume in phase space but, instead, acts on one individual ray whose phase space volume is zero.

5.6 NORMAL MODES OF THE LENS WAVEGUIDE

We saw in the preceding sections that lens waveguides are capable of guiding light rays. However, we know that ray optics is only an approximate description of light, so that we must now confirm that our ray optics results are, indeed, correct. The wave description of light inside the lens waveguide will confirm our previous results. However, in addition to the information about ray trajectories that we are about to rederive, we also gain information concerning the spread of the field distribution and the losses suffered by the light beam. The wave optics treatment of the lens waveguide will be limited to guides with perfectly aligned lenses. The study of random lens displacements in terms of wave optics is far more complicated. Our ray optics results have provided us with a good deal of information to estimate the importance of lens alignment.

The wave optics treatment of the lens waveguide is based on the scalar wave theory. This approach is sufficient to describe most of the phenomena of interest to us. We are about to demonstrate that the lens waveguide is capable of supporting normal modes. Field distributions of more general form can then be expressed as superpositions of normal modes.

In order to obtain the normal modes of the lens waveguide, we look for field distributions that repeat their shape on each lens of the guide except for a phase factor. This definition of a normal mode differs from the modes of cylindrical metallic waveguides.[2,58] Those modes are field distributions that maintain their shape at every cross section of the guide with the exception of a phase factor. The difference in the two cases can be described as follows. The cylindrical metallic waveguide is a structure that remains invariant with respect to arbitrary displacements in the direction of its axis. We can shift the metallic waveguide by an arbitrary amount in z direction without being able to detect any difference in the structure. The system is invariant with respect to arbitrary displacements in z direction. The lens waveguide is a structure of a very different type. An arbitrary displacement along its axis changes its appearance for a stationary observer. Only if we displace the guide by multiples of the lens spacing, do we maintain its appearance for a stationary observer. The lens waveguide is a period structure, and only displacements that are multiples of the period length leave the structure invariant. Group theory tells us that any function $\exp(i\beta z)$ is a representation of the group of continuous displacements.[59] However, the representations of periodic structures must be periodic functions whose period equals the period of the structure.

For readers who are familiar with wave mechanics, the following analogy might be helpful. An electron in free space can be described by plane waves. These can be regarded as normal modes that have the same appearance at each cross section perpendicular to the direction of propagation. The only change along the direction of propagation is the phase variation. In a periodic crystal, the electron wave function is drastically different. It now assumes the form of a Bloch function.[60] That is, it is represented by the product of a function that is periodic with the period of the crystal lattice and a z-dependent phase factor. The normal modes of the lens waveguide are analogous to the Bloch function of electrons in crystal lattices.

The problem of finding mode solutions for systems of apertures led Goubau to the invention of the lens waveguide. He tells the story of how, on a vacation trip, he wrestled with the problem of finding a field solution for a system of periodic apertures. He was looking for fields that would repeat periodically at every aperture of the system. He had formulated the problem but was unable to solve the resulting integral equation. It occurred to him that he could easily solve his integral equation if he were allowed to modify it in a certain way. When he examined the physical meaning of this simplifying modification, he discovered that it meant inserting lenses into the apertures. He had thus not only simplified the solution of a mathematical problem but had, at the same time, invented a very useful and important, low-loss light transmission system.

We find the normal mode solutions of the lens waveguide with the help of the Kirchhoff-Huygens diffraction theory.[47,48] We concentrate on two successive lenses of the infinitely long lens waveguide as shown in Figure 5.6.1. There are many different ways to describe the lens waveguide. We could, for example, choose a reference plane located midway between the lenses, and follow the field from one of these midway planes to the next. Another alternative would be to choose the reference plane directly to the right or to the left of each lens. We choose a description that has the advantage of being easily adaptable to the description of laser resonators. We split each lens in

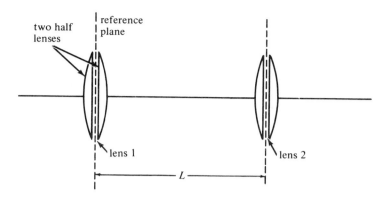

Figure 5.6.1 Lens waveguide. Each lens is replaced by two half lenses. The field distribution is computed at the reference planes between the half lenses.

half, and place our reference plane in the middle between the two half lenses. The transmission of the field through the half lens is described by the transformation (4.3-1) and (4.3-8)

$$\bar{\psi}_1 = e^{-i(k/4f)(a^2 - x^2 - y^2)}\psi_1(x, y) \qquad (5.6\text{-}1)$$

The field at the reference plane 1 is designated by ψ_1, while the field immediately to the right of the first lens is $\bar{\psi}_1$. The lens radius is now called a (a replaces ρ in [4.3-8]). The factor 4, instead of the 2 appearing in (4.3-8), is required, since the focal length of the half lens is twice the value of the lens combination; f is the focal length of the two half lenses that make up each lens. The field $\bar{\psi}_2$ immediately to the left of the second lens is obtained by transforming $\bar{\psi}_1$ with the help of the Kirchhoff-Huygens diffraction integral (2.2-31) in its Fresnel approximation. Choosing the left reference plane as the coordinate origin and approximating $\cos \alpha = \cos \beta = 1$, we obtain

$$\bar{\psi}_2 = \frac{i}{\lambda L} e^{-ikL} \int_{A_1} \bar{\psi}_1 \exp\left\{-i\frac{k}{2L}[(x' - x)^2 + (y' - y)^2]\right\} dx\, dy \quad (5.6\text{-}2)$$

The field ψ_2 at the second reference plane is finally obtained by a transformation similar to (5.6-1)

$$\psi_2(x', y') = e^{-i(k/4f)(a^2 - x'^2 - y'^2)}\bar{\psi}_2 \qquad (5.6\text{-}3)$$

The object of this calculation is to find a mode solution. We thus require that the field ψ_2 be equal to ψ_1 except for a factor γ. For lenses with very large apertures ($a \to \infty$), the field does not lose power in going from the first to the second reference plane. In this case, γ is simply a phase factor. If the lenses have sufficiently small apertures, part of the field will arrive outside the lens aperture, so that less power passes reference plane 2 inside the lens than was present at reference plane 1. In that case, γ is a complex number with magnitude less than unity. The condition for a mode field can be written as

$$\psi_2 = \gamma\psi_1 \tag{5.6-4}$$

Equation (5.6-4) formulates an eigenvalue problem. Not every field satisfies this condition. In fact, general fields do not obey the eigenvalue relation (5.6-4). This relation has an infinite number of solutions, and every arbitrary field that can exist in the lens waveguide can be described as a superposition of infinitely many modes (solutions of [5.6-4]) of the structure. With the help of (5.6-1) through (5.6-3), we can write the eigenvalue equation (5.6-4) as follows

$$\gamma\psi(x', y') = \frac{i}{\lambda L} e^{-ik(L + a^2/2f)} \cdot \int_{A_1} \psi(x, y)\exp\left\{i\,\frac{k}{2L}\left[(x^2 + y^2 + x'^2\right.\right.$$
$$\left.\left. + y'^2)\left(\frac{L}{2f} - 1\right) + 2xx' + 2yy'\right]\right\} dx\, dy \tag{5.6-5}$$

The subscript 1 of the ψ function has been dropped. The general eigenvalue equation (5.6-5) is hard to solve. However, if we specialize this equation to the case of the confocal lens waveguide

$$L = 2f \tag{5.6-6}$$

the equation simplifies to a form whose analytical solutions are easy to obtain. Absorbing the phase factor appearing in front of the integral into the eigenvalue γ, and calling this new eigenvalue κ^2, allows us to obtain the eigenvalue equation for the modes of the confocal lens waveguide in the simple form $(k = 2\pi/\lambda)$

$$\kappa^2\psi(x', y') = \frac{1}{2\lambda f}\int_{A_1} \psi(x, y)e^{i(\pi/\lambda f)(xx' + yy')}dx\, dy \tag{5.6-7}$$

The problem is simplified further if we ask for solutions of the form[47]

$$\psi(x, y) = f(x)g(y) \tag{5.6-8}$$

Equation (5.6-7) can now be decomposed into two equations

$$\kappa f(x') = \frac{1}{\sqrt{2\lambda f}}\int_{-a}^{a} f(x)e^{i(\pi/\lambda f)xx'}dx \tag{5.6-9}$$

and

$$\kappa g(y') = \frac{1}{\sqrt{2\lambda f}}\int_{-a}^{a} g(y)e^{i(\pi/\lambda f)yy'}dy \tag{5.6-10}$$

In the process of decomposing the eigenvalue equation (5.6-7) into the two eigenvalue equations (5.6-9) and (5.6-10), we had to restrict the integration interval to a square. This gives the lens apertures a somewhat unusual shape. However, for apertures that are much larger than the resulting field distributions, the shape of the aperture is of no importance, since we can then increase the integration range from $-\infty$ to $+\infty$ without changing the value of the integral. For lenses whose apertures are small enough so that the field intensity at the location of the aperture is not negligible, the shape of the aperture influences the solution. In this case, our results hold for lenses with a square mask. Results for lenses with the usual round apertures have been obtained by numerical solution of the integral equation (5.6-7).[48] We discuss the results of these numerical solutions later in this section.

The two eigenvalue equations (5.6-9) and (5.6-10) are both of the same form (for $a \to \infty$, they both are Fourier transformations). It is therefore sufficient to concentrate on the solution of one of them. The decomposition of the three-dimensional eigenvalue equation (5.6-7) reduces the problem to two dimensions. We could have obtained (5.6-9) by considering the two-dimensional problem with cylindrical lenses from the start.

It turns out that the mode solutions result in field distributions that are tightly concentrated around the axis of the lens waveguide. The features of the mode solutions cause them to be independent of the lens apertures provided these are large enough. It is therefore meaningful to increase the limit of integration from $-a$ to a to the new limits $-\infty$ to $+\infty$ without changing the result. For lenses with large apertures, we therefore consider the solutions of the equation

$$\kappa f(x') = \frac{1}{\sqrt{2\lambda f}} \int_{-\infty}^{\infty} f(x)e^{i(\pi/\lambda f)xx'}dx \qquad (5.6\text{-}11)$$

The Fresnel approximation that was used to obtain (5.6-11) holds in spite of the infinite limits on the integral, because of the tightly concentrated field distribution. If the solution of (5.6-11) were to extend to large distances from the optical axis, we would have to disregard it as being meaningless, since it would be in contradiction to the assumptions that led to the Fresnel approximation.

The solution of our integral equation (5.6-11) can be obtained by inspection of the integration tables of Gradshteyn and Ryzhik,[61] p. 838. We find there the integral relation

$$(i^n)e^{-(y^2/2)}H_n(y) = \frac{1}{\sqrt{2\pi}} \int_{-\infty}^{\infty} e^{ixy}e^{-(x^2/2)}H_n(x)dx \qquad (5.6\text{-}12)$$

The functions $H_n(x)$ are called Hermite polynomials. They are defined by the relation

$$H_n(x) = (-1)^n e^{x^2} \frac{d^n}{dx^n} e^{-x^2} \qquad (5.6\text{-}13)$$

The first five Hermite polynomials are

$$H_0(x) = 1$$
$$H_1(x) = 2x$$
$$H_2(x) = 4x^2 - 2 \tag{5.6-14}$$
$$H_3(x) = 8x^3 - 12x$$
$$H_4(x) = 16x^4 - 48x^2 + 12$$

A comparison of (5.6-11) and (5.6-12) shows that the eigenvalue equation has the solution

$$f_n(x) = A_n H_n\left(\sqrt{\frac{\pi}{\lambda f}}\, x\right) e^{-(\pi x^2/2\lambda f)} \qquad \text{for } n = 0, 1, 2 \ldots \infty \tag{5.6-15}$$

with the eigenvalue

$$\kappa_n = i^n \tag{5.6-16}$$

The functions (5.6-15) are called Hermite-Gaussian polynomials. They form a complete set of functions. It is convenient to normalize them, so that

$$\int_{-\infty}^{\infty} f_n^2(x)dx = 1 \tag{5.6-17}$$

Using again the tables of Gradshteyn and Ryzhik,[61] we find that this normalization is accomplished if we choose

$$A_n = \frac{1}{(2^n n!)^{1/2}(\lambda f)^{1/4}} \tag{5.6-18}$$

The lowest order mode, $n = 0$, is of particular importance. It has the simple form

$$f_0(x) = \frac{e^{-(\pi x^2/2\lambda f)}}{(\lambda f)^{1/4}} \tag{5.6-19}$$

The normal modes are orthogonal among each other and, with the normalization (5.6-18), form an orthonormal set

$$\int_{-\infty}^{\infty} f_n(x)f_m(x)dx = \delta_{nm} \tag{5.6-20}$$

All the normal modes decay rapidly with increasing values of x. Their decay is governed by the exponential function. The half width of this function is defined by the parameter

$$w' = \sqrt{\frac{2\lambda f}{\pi}} \tag{5.6-21}$$

At $x = w'$, the field distribution (5.6-19) has decayed to e^{-1} of its maximum value at $x = 0$. The beam width parameter w' does not define the narrowest portion of the beam. It determines the beam width at the lenses. Halfway between the lenses, the field distribution is most tightly concentrated around the axis. We return to a discussion of the field shape between the lenses in the chapter on Gaussian beams.

The first three functions f_n of (5.6-15) are plotted in Figure 5.6.2. The modes of the lens waveguide with very large apertures follow from (5.6-8), (5.6-15), and (5.6-18)

$$\psi_{nm} = \frac{1}{(2^{n+m}n!m!\lambda f)^{1/2}} H_n\left(\sqrt{\frac{\pi}{\lambda f}}\, x\right) H_m\left(\sqrt{\frac{\pi}{\lambda f}}\, y\right) e^{-(\pi/2\lambda f)(x^2+y^2)} \quad (5.6\text{-}22)$$

The eigenvalue follows from (5.6-16)

$$\kappa_{nm}^2 = \kappa_n \kappa_m = i^{n+m} \quad (5.6\text{-}23)$$

It is important to keep in mind that this solution of the mode problem holds for confocal lens waveguides. The general solution is presented in Section 6.3, Equation (6.3-20).

Having found the solution of the mode problem of confocal lens waveguides with very large apertures, we return to the problem of waveguides with small but rectangular apertures. The integral equation (5.6-9) can be solved with the help of prolate spheroidal wave functions that have been studied in great

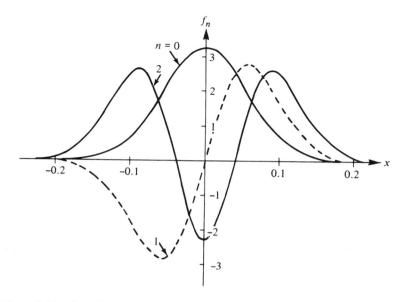

Figure 5.6.2   Graphical representation of the first three Hermite-Gaussian functions.

detail by Slepian and Pollak,[62] Landau and Pollak,[63] and Slepian and Sonnenblick.[41] (See also Section 4.6.) These authors investigated functions that are solutions of the integral equation (5.6-9). (In order to see the equivalence of the integral equation of reference 62, equation 29, p. 58, we have to use the following relations between their parameters and ours: $\psi_n(c, t) = f_n(x')$, $\omega T/(2\Omega) = x$, $t = x'$, $T = 2a$, $c = \pi a^2/(\lambda f)$.) The eigenvalue κ_n is related to the parameter λ_n used in reference 62 by the relation

$$\kappa_n = i^n \sqrt{\lambda_n} \qquad (5.6\text{-}24)$$

The subscripted parameter λ_n must not be confused with the wavelength λ. The parameter λ_n is extensively tabulated in reference 41. The normalization (5.6-17) is maintained. However, in addition to the orthogonality relation (5.6-20), there exists another orthogonality relation of the mode functions f_n in the domain $-a$ to $+a$ (compare (4.6-12) of section 4.6)

$$\int_{-a}^{a} f_n(x)f_m(x)dx = \lambda_n \delta_{nm} \qquad (5.6\text{-}25)$$

This orthogonality over the infinite as well as over the finite domain is a most remarkable property of the modes of the confocal lens waveguide with square apertures.

The power carried through the lens aperture is proportional to

$$P = \int_{-a}^{a} dx \int_{-a}^{a} dy\, |\psi_{nm}(x, y)|^2 \qquad (5.6\text{-}26)$$

The field at the reference plane 1 of Figure 5.6.1 is transformed into the field

$$\psi_{nm} \rightarrow \kappa_{nm}^2 \psi_{nm} \qquad (5.6\text{-}27)$$

This relation follows from the integral equation (5.6-7). The power P_2 crossing through reference plane 2 is thus related to the power P_1 crossing through reference plane 1 by the equation

$$P_2 = |\kappa_{nm}^2|^2 P_1 \qquad (5.6\text{-}28)$$

In terms of the parameter λ_n, we obtain from (5.6-23) and (5.6-24)

$$P_2 = \lambda_n \lambda_m P_1 \qquad (5.6\text{-}29)$$

The parameter λ_n describes the loss of the modes of the lens waveguide. The properties of λ_n were already briefly discussed in Section 4.6. We saw there that the parameter λ_n for given values of (compare [4.6-15])

$$c = \pi \frac{a^2}{\lambda f} \qquad (5.6\text{-}30)$$

is very nearly unity for $n < 2c/\pi$, but that it decreases very rapidly for increasing values of n with $n > 2c/\pi$. It is thus evident that the losses of the normal modes are small for small mode numbers, but that they increase very rapidly for modes whose mode numbers exceed $2c/\pi$. The dependence of λ_n on n is plotted in Figure 5.6.3 from the tables of reference 41, for $c = 10$. The dependence of the loss λ_0 of the lowest order mode on the parameter c is plotted in Figure 5.6.4. The mode losses are caused by diffraction. In traveling from one lens to the next, the field spreads apart. Some of it is intercepted by the aperture of the following lens and is thus kept from contributing to the flow of electromagnetic power along the lens waveguide. If the lens apertures are kept large enough, the diffraction losses can be extremely low. Figure 5.6.4 shows that the diffraction losses of the lowest order mode become small for $c > 4$.

The relation between the integral equation (5.6-9) and Equation (4.6-13) can be established as follows. We use Equation (5.6-9) to express $f(x)$, and substitute it into the integrand of this same equation. We obtain

$$\kappa_n^2 f_n(x') = \frac{1}{2\lambda f} \int_{-a}^{a} f_n(x'') \int_{-a}^{a} e^{i(\pi/\lambda f)x(x'+x'')}dx\, dx'' \qquad (5.6\text{-}31)$$

The integral over x can be carried out, with the result

$$\kappa_n^2 f_n(x') = \frac{1}{\pi} \int_{-a}^{a} \frac{\sin\dfrac{\pi a}{\lambda f}(x'+x'')}{x'+x''} f_n(x'')dx'' \qquad (5.6\text{-}32)$$

The functions $f_n(x)$ have the property

$$f_n(-x) = (-1)^n f_n(x) \qquad (5.6\text{-}33)$$

That means that they are even or odd functions, depending on whether n is even or odd. That this property is correct can be seen when we consider that for $a \to \infty$ the present wave function for finite values of a must change continuously to the functions (5.6-15). The property of being either an even or odd function cannot change during this continuous transition. The functions (5.6-15) have the property (5.6-33), as can be seen from (5.6-13) or (5.6-14). Using (5.6-33), we can write (5.6-32) as follows

$$(-1)^n \kappa_n^2 f_n(x') = \int_{-a}^{a} \frac{\sin\dfrac{\pi a}{\lambda f}(x'-x'')}{\pi(x'-x'')} f_n(x'')dx'' \qquad (5.6\text{-}34)$$

Since κ_n and λ_n are related by (5.6-24), it is apparent that our f_n functions do indeed satisfy (4.6-13), establishing their identity with the ψ_n functions of Section 4.6.

We have thus found a complete description of the modes of a confocal lens waveguide with square apertures. The modes of the confocal lens waveguide with circular apertures can be expressed in terms of hyperspheroidal functions.[120] The losses of a lens waveguide with circular aperture have been

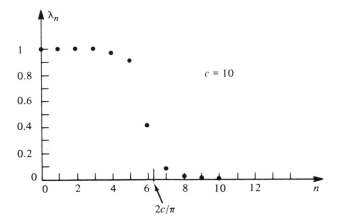

Figure 5.6.3   Graphical representation of the eigenvalue λ_n as a function of the mode number n for $c = 10$. The point $2c/\pi$ is indicated on the n-axis.

calculated numerically by Fox and Li.[48] Actually, their calculation was performed with a laser resonator in mind. However, we have seen in Section 5.3 that there exists a very close relationship between laser resonators and lens waveguides. The diffraction losses of both structures are identical. We shall return to this comparison between lens waveguides and laser resonators in Section 6.6. Figure 5.6.5 shows the loss curve for the first three modes of the confocal lens waveguide with circular apertures. Also shown in Figure 5.6.5 is the power loss of the lowest order mode of the lens waveguide, with square apertures as the dotted line. In comparing the two losses, it is

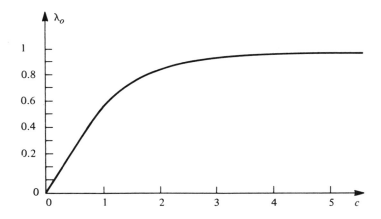

Figure 5.6.4   Eigenvalue of the lowest order mode as a function of c.

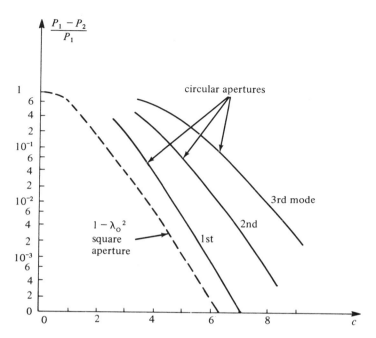

Figure 5.6.5 Comparison of the losses of the first three modes of a laser cavity with circular apertures with the lowest order mode of a cavity with a square aperture. (From A. G. Fox and T. Li, Resonant Modes in a Maser Interferometer, *B.S.T.J.*, Vol. 40, No. 2, March 1961 pp. 453–488. Copyright 1961, The American Telephone and Telegraph Co., reprinted by permission.)

important to keep in mind that a entering the parameter c is the radius of the spherical aperture for the case of the lens waveguide with round apertures, while it is the half width (Equation [5.6-30]) of the square apertures in the other case. The area of the lenses with square apertures is thus larger for the same values of c. The diffraction losses of the lenses with square apertures are consequently smaller.

5.7 WAVE TRAJECTORY IN A CONFOCAL LENS WAVEGUIDE

We have found the normal modes of the lens waveguide in the preceding section. However, we have not yet shown that the results of our ray optical treatment can be obtained by means of wave optics. Once it is established that the ray trajectories of geometrical optics correspond to the paths of light waves in the lens waveguide, we can use the ray optics method with more confidence. It will become apparent that the ray optics of lens waveguides is far simpler than its wave optical treatment. It is thus desirable to use ray optics whenever possible.

We study the path of a light beam in the lens waveguide with infinite apertures by letting an off-axis field distribution travel through the lenses. We begin by assuming that the field at the first reference plane of Figure 5.6.1 is given by the expression

$$\psi_1 = F_1(x)g_1(y) \tag{5.7-1}$$

with

$$F_1(x) = \frac{1}{(\lambda f)^{1/4}} e^{-(\pi/2\lambda f)(x+\xi)^2} \tag{5.7-2}$$

and

$$g_1(y) = \frac{1}{(\lambda f)^{1/4}} e^{-(\pi/2\lambda f)y^2} \tag{5.7-3}$$

The field at the next reference plane is given by (5.6-7), except that we do not now require that the field repeats itself, but use the integral on the right-hand side of this equation to calculate the field ψ_2 at the second reference plane. Because the field distribution is given as a product of an x-dependent function times a y-dependent function, the integral decomposes into a product of an integral over x times another integral over y. The g function (5.7-3) represents the lowest order mode of the integral (5.6-10). It satisfies this equation with $a \to \infty$ and propagates through successive lenses without change of its shape. The y dependence of the wave field is therefore always given by (5.7-3), except for phase factors. It suffices, therefore, to study the x dependence of the function (5.7-1) separately.

The F function on the next lens is given by

$$F_2 = DF_1 \tag{5.7-4}$$

where D indicates symbolically the integral operator occurring on the right-hand side of (5.6-9). For a normal mode, we obtain

$$\kappa_n f_n = Df_n \tag{5.7-5}$$

The field distribution at the mth lens is given by

$$F_m = D^{m-1}F_1 \tag{5.7-6}$$

The notation D^m indicates the m-fold repeated application of the operator D. For a normal mode, we have simply

$$\kappa_n^{m-1}f_n = D^{m-1}f_n \tag{5.7-7}$$

The difficulty with our problem is that (5.7-2) is not a normal mode of the lens waveguide. It would be a normal mode for $\xi = 0$. For nonvanishing values of ξ, (5.7-2) represents a field distribution that resembles a normal mode that is displaced by an amount $-\xi$ from the axis of the guide.

In order to find the field distribution (5.7-6) at the mth lens, we express the initial field F_1 as a superposition of normal modes. We assume that the lens apertures of the lens waveguide are sufficiently large so that we can approximate all the normal modes, which contribute appreciably to our expansion, by the expression (5.6-15).

We obtain quite generally

$$F_1 = \sum_{v=0}^{\infty} a_v f_v \tag{5.7-8}$$

With the help of (5.7-7), we obtain the field at the mth lens from (5.7-6) and (5.7-8)

$$F_m = \sum_{v=0}^{\infty} a_v \kappa_v^{m-1} f_v \tag{5.7-9}$$

Our problem is thus solved in principle. It remains to find the expansion coefficients a_v and to sum the series (5.7-9).

The expansion coefficients are obtained with the help of the orthogonality relations (5.6-20). Multiplying (5.7-8) with f_μ and integrating results in

$$a_\mu = \frac{1}{\sqrt{\lambda f}} \frac{1}{\sqrt{2^\mu \mu!}} \int_{-\infty}^{\infty} H_\mu \left(\sqrt{\frac{\pi}{\lambda f}} x \right) e^{-(\pi/2\lambda f)[x^2+(x+\xi)^2]} dx \tag{5.7-10}$$

The integral can be solved with the aid of the integration tables by Gradshteyn and Ryzhik,[61] p. 837. We obtain

$$a_\mu = \frac{1}{\sqrt{2^\mu \mu!}} \left(\frac{\pi}{\lambda f} \right)^{\mu/2} (-\xi)^\mu e^{-(\pi/4\lambda f)\xi^2} \tag{5.7-11}$$

Using (5.6-15) and (5.6-16), we obtain from (5.7-9)

$$F_m = \frac{1}{(\lambda f)^{1/4}} \sum_{\mu=0}^{\infty} \frac{i^{\mu(m-1)}}{2^\mu \mu!} \left(\frac{\pi}{\lambda f} \right)^{\mu/2} (-\xi)^\mu H_\mu \left(\sqrt{\frac{\pi}{\lambda f}} x \right) e^{-(\pi/2\lambda f)(x^2+1/2\xi^2)} \tag{5.7-12}$$

The sum in (5.7-12) can be evaluated with the help of the generating function of the Hermite polynomials (reference 61, p. 1034)

$$e^{-t^2+2tz} = \sum_{\mu=0}^{\infty} \frac{t^\mu}{\mu!} H_\mu(z) \tag{5.7-13}$$

Using this expression, we obtain

$$F_m = \frac{1}{(\lambda f)^{1/4}} \exp\left\{ -\frac{\pi}{2\lambda f} \left[x^2 + \frac{1}{2}(i^{2(m-1)}+1)\xi^2 + 2i^{m-1}x\xi \right] \right\} \tag{5.7-14}$$

The expression in brackets appearing in the exponent of the exponential function can be rewritten in the following form

$$x^2 + \frac{1}{2}(i^{2(m-1)} + 1)\xi^2 + 2i^{m-1}x\xi$$

$$= x^2 + 2x\xi\frac{e^{im(\pi/2)} - e^{-im(\pi/2)}}{2i} + \xi^2\frac{2 - e^{2im(\pi/2)} - e^{-2im(\pi/2)}}{4}$$

$$+ 2x\xi\frac{e^{im(\pi/2)} + e^{-im(\pi/2)}}{2i}$$

$$= \left(x + \xi\sin m\frac{\pi}{2}\right)^2 - 2ix\xi\cos m\frac{\pi}{2} \qquad (5.7\text{-}15)$$

The field distribution at the mth lens can thus be written as

$$F_m = \frac{1}{(\lambda f)^{1/4}}\exp\left\{-\frac{\pi}{2\lambda f}\left[\left(x + \xi\sin m\frac{\pi}{2}\right)^2 - 2ix\xi\cos m\frac{\pi}{2}\right]\right\} \qquad (5.7\text{-}16)$$

The imaginary part of the exponent indicates an x-dependent phase shift at even-numbered lenses. This means physically that the phase fronts at even-numbered lenses are slanted. At odd-numbered lenses, the phase fronts are parallel to the reference planes. These results are of course valid only for this special confocal lens waveguide with a field launched at the first lens with a phase front parallel to the reference plane. The maximum of the magnitude of the field distribution occurs at the point where the real part of the exponent vanishes. The peak of the field distribution occurs thus at

$$x = -\xi\sin m\frac{\pi}{2} \qquad (5.7\text{-}17)$$

Comparison with the ray trajectory (5.2-17) in the confocal lens waveguides proves the important fact that the peak of the field distribution moves through the lens waveguide like a ray of the geometrical optics solution.

We have thus proved for the special case of a confocal lens waveguide and for a special ray trajectory that the center of gravity of the field distribution travels in the lens waveguide like a ray. This important result tells us that the ray optics solution of our problem is, indeed, a valid approximation to the actual wave problem, since it tells us how the center of gravity of the light field moves through the lens waveguide. Since the geometrical optics approach is substantially simpler than the wave optics approach, it is preferable to study the motion of wave, fields inside the lens waveguide by means of geometrical optics. We have also proved that the wave field maintains its original shape as it travels through the lens waveguide. This is true because we have chosen the input field in such a way that it corresponds to the displaced zero order mode of the lens waveguide. We shall show in the chapter on Gaussian beams that a more general field distribution, even though its center of gravity still moves like a ray, changes its beam width periodically as it moves through the lens waveguide. The frequency of the beam width changes is twice the frequency of the ray undulations.

It is also important to realize that the wave field maintained its shape only because it propagates in a lens waveguide with ideal lenses. We shall show in the next section that a wave field in a lens waveguide with imperfect lenses does not maintain its shape but breaks up and deforms itself as it travels through the lens waveguide. This behavior is caused by lenses whose focal length is a function of the radius. Gas lenses have this property, as we saw in Section 4.5.

We proved in Section 3.6 that rays in a dielectric medium whose index of refraction contains only terms up to second order in x and y travel like the center of gravity of wave fields. This fact was established with the help of Ehrenfest's theorem of the quantum theory of light rays. This theorem can now be extended to lens waveguides. It states that only in lens waveguides with perfect lenses—that is, lenses described by (4.2-4) and (4.3-8)—does the center of gravity of a light field move like a ray of geometrical optics. In general, lens waveguides with arbitrary distorting lenses—gas lenses, for example—there is no simple relationship between the trajectory of the wave field and the ray solutions of geometrical optics. For short distances, we can still expect to find that the rays of geometrical optics describe the trajectory of the light field approximately. When we pass through hundreds or thousands of distorting lenses, we must not expect to find a simple relationship between the wave field and the ray solutions of geometrical optics. The results of geometrical optics calculations can no longer be interpreted in such cases. However, an experiment with 100 simple commercial glass lenses of 2.5 cm focal length proved that the observed light ray did not break up after passing through the entire lens waveguide and that it moved through it as predicted by geometrical optics. On the other hand, it was found, by passing wave fields and rays through several hundred simulated gas lenses in a computer-simulated experiment, that there was no obvious interpretation of the wave field in terms of the ray tracing results. In this case of badly distorting lenses, the geometrical optics solution proved fairly useless.

5.8 BEAM BREAKUP IN IMPERFECT LENS WAVEGUIDES

The lens waveguide constructed with perfect, thin lenses allows a light beam to propagate essentially without change. The beam may follow an oscillatory trajectory. However, if its initial shape conforms to one of the normal modes of the lens waveguide, it does not change even if the beam does not follow a straight path, Computer-simulated experiments as well as real experiments have shown that light beams in lens waveguides consisting of imperfect lenses do not keep their shape but may distort substantially. Gas lenses are sufficiently imperfect to cause light beams to deform on traveling through relatively few lenses.[28]

Deformation of light beams inside the lens waveguide is undesirable for two reasons. First of all, one would like to detect the light beam at the end of the waveguide. This task is simpler if the beam arrives in a well-defined mode. It is then possible to act on it with beam transforming devices to match it to a local oscillator beam for most efficient heterodyne detection. Or it may be desirable to focus the received light beam to as small a spot as possible for detection with a small area photo diode. Focusing to small spots also requires essentially single mode light beams. Finally, it may be necessary to use beam redirectors in order to prevent the light beam from growing to such large amplitudes that it can no longer travel in the lens waveguide.[56,57] Beam redirectors need to work on a well-defined beam. A beam that has broken up into several parts can no longer be redirected on axis.

The distortion problem is most severe for off-axis beams. A light beam that travels perfectly on-axis does not suffer serious distortions as it travels in the lens waveguide. Particularly, it is always possible to find a normal mode for any perfectly periodic structure even if the lenses are of poor image-forming quality and may thus be designated as imperfect. However, we have seen in Section 5.5 that on-axis beams cannot remain in this preferred position for long because of unavoidable lens misalignment. It is not surprising that an off-axis beam in an imperfect lens waveguide suffers distortions. We call a lens imperfect when its focal length varies with the distance of the light ray from the axis of the guide. If the focal length of the lens is the same for all rays, a mode of the structure remains undeformed even if it travels along an oscillatory, off-axis trajectory. The mode "fits" into the lens waveguide provided that the lenses have the same focal lengths at any radius. This, however, is not the case for imperfect lenses. Their focal lengths depend on the radius at which the beam intersects the lens. A beam that is a mode of the guide on-axis no longer "fits" into the guide if it becomes displaced from the axis. Field distributions that do not "fit" the waveguide (that are not its modes) suffer distortions even in a perfect lens waveguide. This explains why field distributions cannot travel off-axis in an imperfect lens waveguide without suffering distortions.

We have seen in Section 4.5 that gas lenses can be represented by equivalent thin lenses whose shape is shown in Figure 4.5.11. The computer simulation of gas lens waveguides used the equivalent thin lenses.[64] The process of passing wave fields through warped lenses of the type of Figure 4.5-11 is quite complicated. The complication stems from the form (2.2-41) of the two-dimensional Kirchhoff-Huygens diffraction integral. The two-dimensional problem was used since the full three-dimensional problem requires double integrations that are not only exceedingly time-consuming to perform on a computer but also tax its storage capabilities beyond the capacity of an IBM 7094 computer. The complication with the diffraction integral arises because both the function and its derivatives need to be known at the surface of integration. The simplification (2.2-30) (or its two-dimensional equivalent) was possible only since the derivative of the function normal to its phase fronts can approximately be replaced by multiplication with $-ik$. The derivatives required by the diffraction integral are in the direction normal to the surface of integration—that is, in our case, normal to the shape of the warped lens. This direction is not normal to the phase fronts of the field, so that the derivative of the function is not easily obtainable. Taking the derivative numerically would require computation of the field at two closely neighboring surfaces. This procedure would not only double the amount of time required for the computation but would also be of doubtful accuracy. The additional time is more wisely spent by evaluating not only the function itself but also its derivative (obtained directly by differentiating the expression (2.2-41)) at each lens surface. Tracing the field from one lens to the next thus involves the numerical integration of two integrals of the type (2.2-41). The evaluation of the phase transformation, which the lens imposes on the field, was based on Equation (3.7-37) and is described in detail in reference 64. A peculiar phenomenon arises in passing the field distribution through the lens. The required phase shift can be obtained unambiguously from ray optics. However, if one investigates the power of the wave that has just passed the infinitesimal distance through the lens, it is found that conservation of power has been violated. The change in power is not very substantial. In the computer simulation, it has been taken care of by reassigning a new amplitude to each field element after passing through the lens in such a way that the power flow density at each point of the lens is preserved. This example shows that one must be very careful in arbitrarily prescribing certain properties to a thin optical transformer. The pitfall of inventing ray oscillation suppressors was avoided by using Liouville's theorem as a guide to find the proper form of the transforming equations.[24] (See Section 3.7 and the end of Section 5.5.) However, even this precaution does not guarantee that the principle of conservation of energy is not being violated.

The field distribution whose progress in the lens waveguide was investigated was chosen to be the lowest order mode of the lens waveguide if located on-axis. The properties of the lenses were chosen to coincide with those of actual gas lenses. The results of this computer simulation are shown in the following figures.[64]

Figure 5.8.1 shows the focal length as measured from the deformed principal surface as a function of radius. Figure 5.8.2 shows the shape of the principal surface of the lenses. The field distribution is started off-axis at the first lens of the guide. Its position at the first three lenses is shown in Figure 5.8.3. No distortion is yet discernible. This shows that the individual lenses do not cause significant field distortion. The field distortion that is caused by the cumulative effect of many lenses is shown in Figure 5.8.4. This figure again shows the power density of the field at lens 100 and two successive lenses. The main portion of the field is still clearly visible, but it is also apparent

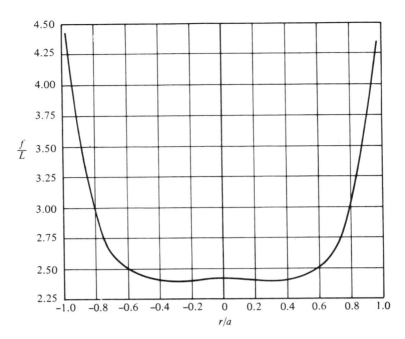

Figure 5.8.1   Focal length dependence on ray position of a gas lens. (From D. Marcuse, Deformation of Fields Propagating Through Gas Lenses, *B.S.T.J.*, Vol. 45, No. 8, October 1966, pp. 1345–68. Copyright 1966; The American Telephone and Telegraph Co., reprinted by permission.)

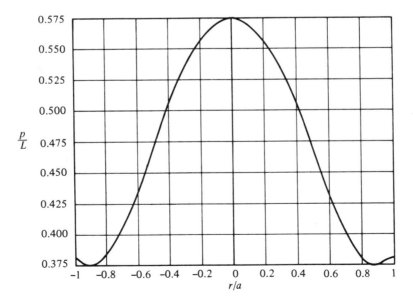

Figure 5.8.2   Shape of the principal plane in the gas lens. (From D. Marcuse, Deformation of Fields Propagating Through Gas Lenses, *B.S.T.J.*, Vol. 45, No. 8, October 1966, pp. 1345–68. Copyright 1966; The American Telephone and Telegraph Co., reprinted by permission.)

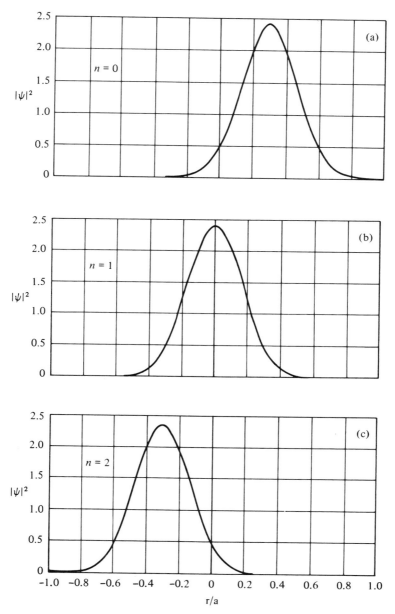

Figure 5.8.3 Field distribution on the first three lenses of a lens waveguide composed of gas lenses. (From D. Marcuse, Deformation of Fields Propagating Through Gas Lenses, *B.S.T.J.*, Vol. 45, No. 8, October 1966, pp. 1345–68. Copyright 1966; The American Telephone and Telegraph Co., reprinted by permission.)

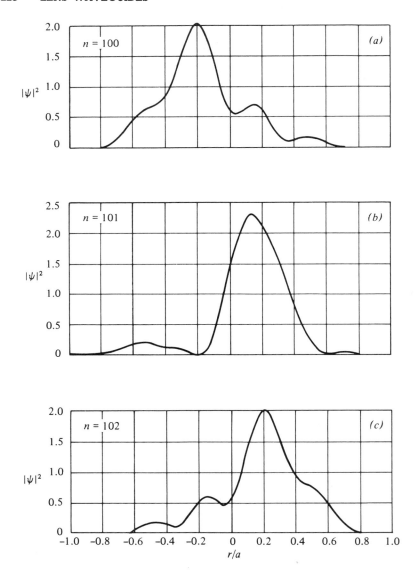

Figure 5.8.4   Field distribution on three successive lenses near the 100th lens of a gas lens waveguide. The field distribution begins to show signs of distortion. (From D. Marcuse, Deformation of Fields Propagating Through Gas Lenses, *B.S.T.J.*, Vol. 45, No. 8, October 1966, pp. 1345–68. Copyright 1966; The American Telephone and Telegraph Co., reprinted by permission.)

that some of the power has moved over to the opposite side of the lens. At lens 149, Figure 5.8.5, the field has broken up into two nearly equally intense light beams. Surprising is the shape at the next lens, also shown in Figure 5.8.5. The field has been refocused, and looks much less distorted. These two shapes alternate on the following lenses. The imperfect gas lenses cause the light beam to break up into two beams after traversing only 150 lenses. The severity of the beam distortion is not a function of the degree of lens imperfection. Every imperfection, however slight, causes similar beam deformation. The significant difference in the behavior of lens waveguides with differing degrees of lens distortion is the number of lenses that must be traversed before the beam breakup becomes noticeable. Lenses whose principal surface is plane but which have a certain amount of focal length distortion also cause the beam to deform in a very similar way. E. A. J. Marcatili[65] has found analytically, with the help of a continuously focusing medium with variable index of refraction, the same beam deformations that we have shown here. The deformation of light beams traveling through gas lenses has experimentally been observed by P. Kaiser.[28]

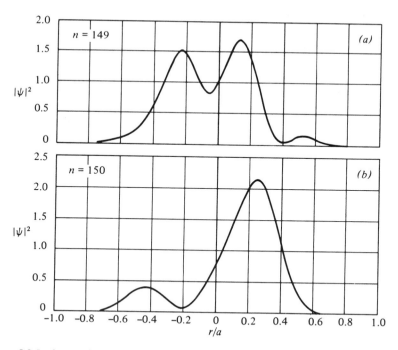

Figure 5.8.5 Severe field distortion at the 149th lens of the gas lens waveguide. At the next lens (No. 150), the field distortion is far less severe. (From D. Marcuse, Deformation of Fields Propagating Through Gas Lenses, B.S.T.J., Vol. 45, No. 8, October 1966, pp. 1345–68. Copyright 1966; The American Telephone and Telegraph Co., reprinted by permission.)

CHAPTER

6

GAUSSIAN BEAMS

6.1 INTRODUCTION

Our discussion of wave propagation in the lens waveguide showed that the normal modes of lens waveguides with very large apertures are products of Hermite polynomials with Gaussian functions. A Gaussian function is of the form $\exp(-ax^2)$. If this were the only place where Hermite-Gaussian field distributions occurred, it would hardly warrant an entire chapter of its own. However, we have repeatedly pointed out that laser resonators are very closely related to lens waveguides. As a consequence of this close relationship, laser modes also have the shape of Hermite-Gaussian functions. Since lasers are an important source of coherent light, it is not surprising that Gaussian beams assumed considerable importance with the advent of the laser. The output of a good, single mode laser is a pure Gaussian beam. The lowest order mode of lasers and lens waveguides is purely Gaussian, since the lowest order Hermite polynomial is equal to unity. (See [5.6-14].)

In communications applications with lens waveguides, it is important to insert a laser mode into a lens waveguide in the most efficient way. This means that it is desirable to be able to inject the lowest order laser mode into the waveguide in such a way that only the lowest order waveguide mode is excited. We thus face the problem of matching Gaussian beams to each other. The transformation of Gaussian beams is thus discussed in this chapter. Before we can consider matching of Gaussian beams, it is important to study their properties in more detail. In particular, we need to know how these beams propagate through free space.

The treatment of Gaussian beam propagation will be handled by two methods. We begin by using the Kirchhoff-Huygens diffraction integral. It is of course also possible to study the propagation of Gaussian beams directly with the help of the scalar wave equation. We do this in the paraxial approximation.

We shall see that all Hermite-Gaussian modes propagate essentially in the same way. It is thus sufficient to concentrate on the properties of the lowest order Gaussian mode.

There is a remarkable transformation property of the parameters of Gaussian beams. We shall show that the transformation of the radius of curvature and the beam width obeys formally the same law as the transformation of light rays.

6.2 PROPAGATION OF GAUSSIAN BEAMS IN FREE SPACE

Let us assume that we have somehow generated a light beam with a Gaussian intensity distribution that happens to have a plane phase front at the plane $z = 0$. This light beam can be described by the equation

$$\psi(x, y, 0) = Ae^{-(x^2+y^2)/w_o^2} \tag{6.2-1}$$

The parameter w_o is the beam half width at $z = 0$. It is the distance from the peak of the field distribution at which the function decays to $1/e$ of its maximum value.

In order to find out how this field distribution looks as it propagates along the z-axis, we use the diffraction integral in the form (5.6-2) and obtain

$$\psi(x', y', z) = \frac{iA}{\lambda z} e^{-ikz} \int\!\!\int_{-\infty}^{\infty} e^{-(x^2+y^2)/w_o^2} \exp\{-i\frac{k}{2z}[(x'-x)^2 + (y'-y)^2]\} dx\,dy \tag{6.2-2}$$

The integral can be written as the product of an integral over x times an integral over y. Both integrals are of exactly the same form. It is thus sufficient to concentrate on one of them. The x integral has the form

$$I_x = \int_{-\infty}^{\infty} e^{-(x^2/w_o^2)} e^{-i(k/2z)(x'-x)^2} dx$$

$$= \sqrt{\pi} \frac{w_o\sqrt{2z}}{\sqrt{2z + ikw_o^2}} \exp\left[-i\frac{2kzx'^2}{4z^2 + (kw_o^2)^2}\right] \exp\left[-\frac{(kw_o x')^2}{4z^2 + (kw_o^2)^2}\right] \tag{6.2-3}$$

The integration of the integral is easily accomplished with the help of integration tables by writing the exponent in the form $-(ax' + b)^2 + c$ and introducing the new integration variable $u = ax' + b$.

The field at any given distance z can now be written as

$$\psi(x', y', z) = \frac{2i\pi w_o^2}{\lambda(2z + ikw_o^2)} A e^{-ikz} \exp\left[-i\frac{2kzr'^2}{4z^2 + (kw_o^2)^2}\right]$$

$$\exp\left[-\frac{(kw_o r')^2}{4z^2 + (kw_o^2)^2}\right] \quad (6.2\text{-}4)$$

with

$$r'^2 = x'^2 + y'^2 \quad (6.2\text{-}5)$$

The first exponential factor describes the phase of a plane wave. The second exponential factor is responsible for phase front curvature. The last exponential factor determines the field intensity in transverse direction.

We introduce the beam width parameter again by the definition that it is equal to the distance (in transverse direction) at which the field amplitude decays to $1/e$ of its maximum value. The square of the beam half width is thus given by

$$w^2(z) = w_o^2\left[1 + \left(\frac{2z}{kw_o^2}\right)^2\right] \quad (6.2\text{-}6)$$

Using $k = 2\pi/\lambda$, we can write this equation also in the following form

$$w^2(z) = w_o^2\left[1 + \left(\frac{\lambda z}{\pi w_o^2}\right)^2\right] \quad (6.2\text{-}7)$$

Remember that the beam at $z = 0$ had a plane phase front and half width w_o. From that point, it propagates toward increasing values of z, with an increase in its width. We could have used the diffraction integral to find the field to the left of the point $z = 0$, and would have found that its behavior is exactly the same. In fact, (6.2-7) holds for positive as well as negative values of z, and describes the beam width throughout all space. It is thus apparent that the point $z = 0$, at which the Gaussian beam has a plane phase front, is also the point at which it assumes its narrowest half width w_o. The minimum beam half width w_o is apparently arbitrary. However, the spread of the field past its narrowest point depends on the minimum beam width. We can define a half angle of the beam far from $z = 0$ by the equation

$$\theta = \lim_{z\to\infty} \frac{w(z)}{z}. \quad (6.2\text{-}8)$$

This beam half angle is thus given by the expression

$$\theta = \frac{\lambda}{\pi w_o}. \quad (6.2\text{-}9)$$

This formula for the half width of a Gaussian light beam is very similar to the corresponding expression (2.4-11) for the spread of a beam that originated at a uniformly illuminated circular aperture. However, the Gaussian beam spreads somewhat less rapidly than a beam originating at a uniformly illuminated aperture.

A Gaussian beam at a given point in space is not completely determined by its beam width alone. To describe the beam completely, we need to know one more parameter—for example, the radius of curvature of its phase front. We obtain it from the phase front curvature term of (6.2-4). The radius of curvature of the field can be obtained with the help of Figure 6.2.1

The phase of the wave is constant on the spherical surface. The phase shift from the curved surface to the plane that touches this surface at $r' = 0$ is given by

$$kd = \frac{2kzr'^2}{4z^2 + (kw_o^2)^2} \tag{6.2-10}$$

The distance d is defined in Figure 6.2.1, and the right-hand side of this equation is obtained from (6.2-4). In the paraxial approximation, we obtain from Figure 6.2.1 the relationship

$$r'^2 + R^2 = (R + d)^2 \tag{6.2-11}$$

or to a similar approximation

$$R = \frac{r'^2}{2d} \tag{6.2-12}$$

Using (6.2-10), we obtain for the radius of curvature of the Gaussian beam

$$R(z) = z \left[1 + \left(\frac{\pi w_o^2}{\lambda z} \right)^2 \right]. \tag{6.2-13}$$

Beam width and radius of curvature determine the Gaussian beam completely at a given point along the axis. Equation (6.2-13) holds also for

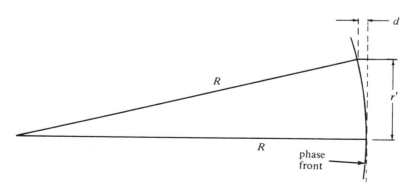

Figure 6.2.1   A spherical wave front is compared with a plane wave. The phase shift for the distance d between the two phase fronts is kd.

positive as well as negative values of z. The change of sign of $R(z)$ that occurs for a change of sign of z indicates the reversal of the curvature of the phase front as we pass the beam waist.

Even though Equations (6.2-7) and (6.2-13) are both continuous in z, we have no reason to expect that they describe the Gaussian beam accurately in the vicinity of $z = 0$, since the Fresnel approximation of the diffraction integral breaks down in the near zone of the field—that is, close to $z = 0$. However, amazingly, both formulas assume the correct value at $z = 0$. The beam width becomes $w(0) = w_o$, and the radius of curvature becomes $R(0) = \infty$, so that they seem to hold for all values of z in spite of the limitations of the diffraction integral from which they were derived. We show in the next section that (6.2-4) is an approximate solution of the wave equation. There is no limitation similar to the near field problem of the Fresnel approximation of the diffraction integral involved in a general solution of the wave equation, so that we can be confident that the expressions (6.2-4), (6.2-7), and (6.2-13) hold for all values of z.

Our solution (6.2-4) can be expressed in a simpler form with the help of the expression for the beam width and the radius of curvature of the phase fronts. We express the complex factor in front of the exponential functions by

$$\frac{2i\pi w_o^2}{\lambda(2z + ikw_o^2)} = \frac{1}{1 - i\dfrac{2z}{kw_o^2}} = \frac{w_o}{w(z)} e^{i\,\arctan(\lambda z/\pi w_o^2)} \qquad (6.2\text{-}14)$$

With the help of (6.2-6), (6.2-13), and (6.2-14), we can write (6.2-4) in the following form

$$\psi(x, y, z) = \frac{w_o}{w(z)} A \exp\left\{-i\left[kz + \frac{\pi r^2}{\lambda R(z)} - \arctan\left(\frac{\lambda z}{\pi w_o^2}\right)\right]\right\} e^{-(r/w(z))^2} \qquad (6.2\text{-}15)$$

This expression is not only simpler in appearance but it also shows the phase shift of the field as it travels along the z-axis. Close to the beam waist near $z = 0$, the phase of the field is equal to that of a plane wave. With increasing values of z, the phase of the wave departs somewhat from that of a plane· wave as $1/R(z)$ and the arctangent function show increasing values. However, in the far field at $z \to \infty$, the field front curvature $1/R(z)$ vanishes, while the arctangent function approaches the value $\pi/2$, and the phase of the field again resembles that of a plane wave.

Following Kogelnik,[67] we can introduce the complex beam parameters

$$\frac{1}{q} = \frac{1}{R} - i\frac{\lambda}{\pi w^2(z)} \qquad (6.2\text{-}16)$$

and

$$P = -\arctan\left(\frac{\lambda z}{\pi w_o^2}\right) + i\,ln\left(\frac{w_o}{w(z)}\right) \qquad (6.2\text{-}17)$$

With these abbreviations, we can write the wave field in the simple form

$$\psi(x, y, z) = A \exp\left[-i\left(P + \frac{\pi}{\lambda q} r^2\right)\right] e^{-ikz} \qquad (6.2\text{-}18)$$

In order to obtain the parameter q in a different form that will be of importance in later discussions, we first divide (6.2-7) by (6.2-13) and obtain the relation

$$\frac{w^2(z)}{R(z)} = \left(\frac{\lambda}{\pi w_o}\right)^2 z \qquad (6.2\text{-}19)$$

We obtain from (6.2-16)

$$q = \frac{R\left(1 + i\,\dfrac{\lambda R}{\pi w^2(z)}\right)}{1 + \left(\dfrac{\lambda R}{\pi w^2(z)}\right)^2} \qquad (6.2\text{-}20)$$

With the help of (6.2-13) and (6.2-19), we obtain finally

$$q = z + i\,\frac{\pi w_o^2}{\lambda} \qquad (6.2\text{-}21)$$

The origin, $z = 0$, is located at the narrowest point (waist) of the beam. We have thus found a very simple expression for the inversion of Equation (6.2-16).

6.3 ALTERNATE DERIVATION OF THE GAUSSIAN BEAMS

We are about to rederive the law of propagation of a Gaussian beam in free space with the help of the wave equation. Our starting point is the reduced wave equation (2.2-2).

$$\nabla^2 \psi + k^2 \psi = 0 \qquad (6.3\text{-}1)$$

We introduce the trial solution

$$\psi = u(x, y, z) e^{-ikz} \qquad (6.3\text{-}2)$$

We are interested in solutions that are not too different from plane waves. The rapid z variation of the function is expressed by the exponential factor in (6.3-2). The z dependence of the u function can be assumed to be slow compared to the rapid variation of the exponential factor. Application of the ∇^2 operator to (6.3-2) results in

$$\nabla^2 \psi = \left(\frac{\partial^2 u}{\partial x^2} + \frac{\partial^2 u}{\partial y^2} + \frac{\partial^2 u}{\partial z^2} - 2ik\frac{\partial u}{\partial z} - k^2 u\right) e^{-ikz} \qquad (6.3\text{-}3)$$

Our assumption regarding the z dependence of the u function allows us to neglect the second derivative of u with respect to z compared to its first derivative and the large factor $k^2 u$. We thus obtain the following partial differential equation for u from the reduced wave equation[68]

$$\frac{\partial^2 u}{\partial x^2} + \frac{\partial^2 u}{\partial y^2} - 2ik\frac{\partial u}{\partial z} = 0 \qquad (6.3\text{-}4)$$

The same equation is also obtained by substituting (6.3-2) into the paraxial wave equation (3.6-9). However, with the choice of sign of (3.5-13), the Hamiltonian (3.5-17) describes waves traveling in negative z direction, so that we must use k with the opposite sign in (6.3-2) and (6.3-4) to obtain agreement.

It is easy to prove that the Gaussian beam, which we obtained in the previous section, is a solution of this partial differential equation and thus solves (to the approximation inherent in the step from [6.3-1] to [6.3-4]) the scalar wave equation. To prove this, we write (6.2-18) in the form (6.3-2), so that we obtain

$$u = A\exp[-i(P + \frac{\pi}{\lambda q}r^2)] \qquad (6.3\text{-}5)$$

Since P and q are only functions of z, it is easy to take the derivatives that enter (6.3-4). We obtain by substitution of (6.3-5) into (6.3-4) $(k = 2\pi/\lambda)$

$$2k\left(P' + \frac{i}{q}\right) + \left(\frac{kr}{q}\right)^2(1 - q') = 0 \qquad (6.3\text{-}6)$$

The prime indicates the derivative with respect to z. Since each of the two terms must be independently equal to zero if the equation is to vanish for all values of z and r, we obtain the two equations

$$q' = 1 \qquad (6.3\text{-}7)$$

and

$$P' = -\frac{i}{q} \qquad (6.3\text{-}8)$$

It is apparent from (6.2-21) that (6.3-7) is satisfied. That (6.3-8) is also satisfied can be seen with the help of (6.2-7), (6.2-13), (6.2-16), (6.2-17), and (6.2-19).

We have thus proved that the field (6.2-4), which was obtained with the help of the Fresnel approximation of the Kirchhoff-Huygens diffraction integral, is, indeed, an approximate solution of the reduced wave equation. However, the approximation required to derive (6.3-4) from (6.3-1) appears quite plausible, so that we expect our Gaussian beam (6.2-15) to be a reasonable approximation for all values of z.

The Gaussian field distribution that we have studied so far resembles the lowest order mode of the lens waveguide that we encountered in Section 5.

We also learned in that section that the higher order modes of the lens waveguides are products of the Gaussian function with Hermite polynomials. In order to study the propagation of these higher order modes of the lens waveguide in free space, we substitute the trial solution

$$u(x, y, z) = f\left(\frac{x}{w}\right)g\left(\frac{y}{w}\right)\exp\{-i[P + \frac{\pi}{\lambda q}r^2 + \Phi(z)]\} \qquad (6.3\text{-}9)$$

into the differential equation (6.3-4). The parameters w, P, and q are functions of z, and are given by (6.2-7), (6.2-16), and (6.2-17). Equation (6.3-4) is thus transformed into the following form

$$\frac{f''}{f} - 2ikx\frac{w}{q}\frac{f'}{f} + 2ikx\frac{\partial w}{\partial z}\frac{f'}{f}$$

$$+ \frac{\ddot{g}}{g} - 2iky\frac{w}{q}\frac{\dot{g}}{g} + 2iky\frac{\partial w}{\partial z}\frac{\dot{g}}{g} - 2kw^2\frac{\partial \Phi}{\partial z} = 0 \qquad (6.3\text{-}10)$$

Equation (6.3-6) has been used to simplify this equation. The primes and dots indicate derivatives with respect to the argument of the function. The argument of f is x/w, while the argument of g is y/w.

With the help of (6.2-7), (6.2-16), and (6.2-21), we can eliminate q and $\partial w/\partial z$ from (6.3-10) and obtain

$$\frac{f''}{f} - 4\xi\frac{f'}{f} + \frac{\ddot{g}}{g} - 4\eta\frac{\dot{g}}{g} - 2kw^2\frac{\partial \Phi}{\partial z} = 0 \qquad (6.3\text{-}11)$$

with

$$\xi = \frac{x}{w} \text{ and } \eta = \frac{y}{w} \qquad (6.3\text{-}12)$$

In order not to lose sight of our goal, let us remember at this point that we are trying to obtain a solution of the approximate wave equation that describes Hermite Gaussian beams propagating in free space. The f and g functions must thus be identifiable with Hermite polynomials. The differential equation of an Hermite polynomial of order n is of the form[61]

$$\frac{d^2H_n}{dt^2} - 2t\frac{dH_n}{dt} + 2nH_n = 0 \qquad (6.3\text{-}13)$$

with n being a positive integer.

The differential equation (6.3-11) does contain terms that resemble the differential equation (6.3-13). To achieve an even closer resemblance, we introduce the variables

$$t = \sqrt{2}\,\xi \text{ and } \tau = \sqrt{2}\,\eta \qquad (6.3\text{-}14)$$

The differential equation (6.3-11) is thus transformed into

$$\frac{1}{f}\left(\frac{d^2f}{dt^2} - 2t\frac{df}{dt}\right) + \frac{1}{g}\left(\frac{d^2g}{d\tau^2} - 2\tau\frac{dg}{d\tau}\right) - kw^2\frac{d\Phi}{dz} = 0 \qquad (6.3\text{-}15)$$

It is apparent that we can identify f and g with the following Hermite polynomials

$$f = H_n\left(\sqrt{2}\,\frac{x}{w}\right) \tag{6.3-16}$$

and

$$g = H_m\left(\sqrt{2}\,\frac{y}{w}\right) \tag{6.3-17}$$

if we require

$$kw^2\frac{d\Phi}{dz} = -2(m+n) \tag{6.3-18}$$

The requirement that f and g are Hermite polynomials leads to a determination of the phase factor Φ that, so far, had remained arbitrary. We require that $\Phi(0) = 0$ and obtain with the help of (6.2-7) as the solution of the differential equation (6.3-18)

$$\Phi = -(m+n)\arctan\left(\frac{\lambda z}{\pi w_o^2}\right) \tag{6.3-19}$$

Collecting our results, we have found the following approximate solution of the reduced wave equation (6.3-1)

$$\psi(x,y,z) = A\frac{w_o}{w}H_n\left(\sqrt{2}\,\frac{x}{w}\right)H_m\left(\sqrt{2}\,\frac{y}{w}\right)$$

$$\exp\left\{-i\left[kz - (m+n+1)\arctan\left(\frac{\lambda z}{\pi w_o^2}\right) + \frac{\pi r^2}{\lambda R}\right]\right\}e^{-(r/w)^2} \tag{6.3-20}$$

This solution proves the important result that the beam width parameter w, (6.2-7), and the phase front curvature R, (6.2-13), are the same for every Hermite-Gaussian beam mode. The phase of the wave does depend on the mode number n and m. The transformation of Hermite-Gaussian modes can be done independently of the mode number, since all modes have the same beam width parameter and the same phase front curvature. The discussion of the transformation of Gaussian beams can thus be limited to the lowest order Gaussian beam. The form of the general solution (6.3-20) will be important for our discussion of laser resonators.

The Hermite-Gaussian beam modes form a complete system of orthogonal functions. It is thus possible to express every wave field as a series expansion in terms of these modes. However, we must also remember that the field expressed by (6.3-20) is only an approximate solution of the wave equation. The approximation becomes poorer with increasing values of $m + n$, and begins to break down as the contribution of the $m + n + 1$ term in this equation begins to be comparable to the kz term. A wave field expressed as a series expansion of the modes (6.3-20) is therefore not automatically (even an approximate) solution of the wave equation. If terms of very high order make an appreciable contribution to the series expansion, it would be necessary to adjust the expansion coefficients in such a way that the series expansion as a whole becomes a solution of the wave equation, since this is not true for the individual terms of very high order.

6.4 TRANSFORMATION OF GAUSSIAN BEAMS

It is often desirable to inject a Gaussian light beam into an optical system in such a way that only one pure mode of the system is excited. Examples of such systems are lens waveguides and optical resonators. This mode matching problem requires the transformation of a given Gaussian beam—for example, the output of a laser—to be transformed into the Gaussian beam that forms the mode of the structure into which we want to inject the beam. We have seen in the last section that the Gaussian beam parameters, beam width and phase front curvature, are the same for all the modes. The beam transformation to be described in this section thus holds for all Hermite-Gaussian beam modes simultaneously.

We begin the discussion by transforming Gaussian beams from one point to another along the optical axis of the system. The transformation of the beam as it travels in free space is given by (6.2-7) and (6.2-13). Both beam parameters have already been combined into the complex parameter q of (6.2-16) or (6.2-21). For any given complex value of q, the beam half width w and phase front curvature R are uniquely determined. Calling the complex beam parameter at a point z_1 on the optical axis q_1 and designating by q_2 its value at a later point allows us to accomplish the beam transformation in free space by the simple relation $(d_1 = z_2 - z_1)$

$$q_2 = q_1 + d_1 \qquad (6.4-1)$$

Next, we need to determine how the complex beam parameter is transformed as the Gaussian beam passes through an ideal thin lens. The parameter immediately to the left of the lens is q_2, and immediately to its right q_3. The lens simply changes the phase of the beam but does not influence its width. The beam width parameter w is thus unchanged, while the phase front curvature changes. If (6.2-15) represents the field immediately to the left of the lens, we obtain the field immediately to its right by multiplication with the phase shift term (4.3-8)

$$e^{i\gamma} = e^{i(\pi/\lambda f)r^2} \qquad (6.4-2)$$

An unimportant constant phase shift was omitted. The product of (6.2-15) with (6.4-2) must again result in a Gaussian beam with a different radius of curvature R_3 (R_2 is the phase front radius to the left of the lens). Thus we obtain immediately

$$\frac{1}{R_3} = \frac{1}{R_2} - \frac{1}{f} \tag{6.4-3}$$

Equation (6.4-3) together with $w_3 = w_2$ represent the transformation law of a Gaussian beam that is passing through a thin ideal lens. It was assumed that the axis of the Gaussian beam coincides with the axis of the lens and that the beam is incident perpendicular to the lens surface.

In terms of the complex beam parameter (6.2-16), we can express the transformation law of a lens as follows

$$\frac{1}{q_3} = \frac{1}{q_2} - \frac{1}{f} \tag{6.4-4}$$

Or, in inverted form

$$q_3 = \frac{q_2}{-\dfrac{1}{f} q_2 + 1} \tag{6.4-5}$$

The two transformation laws can be used to transform the beam from a distance d_1 to the left in front of the lens to a point at distance d_2 to the right behind the lens. The beam parameter at distance d_2 behind the lens is designated by q_4. We have

$$q_4 = q_3 + d_2 \tag{6.4-6}$$

Combination of (6.4-1), (6.4-5), and (6.4-6) results in

$$q_4 = \frac{\left(1 - \dfrac{d_2}{f}\right) q_1 + d_1 + d_2 - \dfrac{d_1 d_2}{f}}{-\dfrac{1}{f} q_1 + 1 - \dfrac{d_1}{f}} \tag{6.4-7}$$

The transformation law of Gaussian beams has an interesting matrix property that becomes clear if we write (6.4-7) in the form[67]

$$q_4 = \frac{A_4 q_1 + B_4}{C_4 q_1 + D_4} \tag{6.4-8}$$

with

$$\begin{pmatrix} A_4 & B_4 \\ \\ C_4 & D_4 \end{pmatrix} = \begin{pmatrix} 1 - \dfrac{d_2}{f} & d_1 + d_2 - \dfrac{d_1 d_2}{f} \\ \\ -\dfrac{1}{f} & 1 - \dfrac{d_1}{f} \end{pmatrix} \tag{6.4-9}$$

We express the transformations that led up to the form (6.4-7) similarly. This can be done by preserving the form (6.4-8) for all transformations and assigning the matrix

$$\begin{pmatrix} A_1 & B_1 \\ C_1 & D_1 \end{pmatrix} = \begin{pmatrix} 1 & d_1 \\ 0 & 1 \end{pmatrix} \tag{6.4-10}$$

to (6.4-1), the matrix

$$\begin{pmatrix} A_2 & B_2 \\ C_2 & D_2 \end{pmatrix} = \begin{pmatrix} 1 & 0 \\ -\dfrac{1}{f} & 1 \end{pmatrix} \tag{6.4-11}$$

to the lens transformation (6.4-5), and, finally, the matrix

$$\begin{pmatrix} A_3 & B_3 \\ C_3 & D_3 \end{pmatrix} = \begin{pmatrix} 1 & d_2 \\ 0 & 1 \end{pmatrix} \tag{6.4-12}$$

to the transformation (6.4-6). We verify, by using the laws of matrix multiplication, that the matrix (6.4-9) describing the combined transformation (6.4-7) and (6.4-8) is obtained as the matrix product of the three individual matrices

$$\begin{pmatrix} A_4 & B_4 \\ C_4 & D_4 \end{pmatrix} = \begin{pmatrix} A_3 & B_3 \\ C_3 & D_3 \end{pmatrix}\begin{pmatrix} A_2 & B_2 \\ C_2 & D_2 \end{pmatrix}\begin{pmatrix} A_1 & B_1 \\ C_1 & D_1 \end{pmatrix} \tag{6.4-13}$$

More generally, if we have two transformations

$$q_k = \frac{A_j q_j + B_j}{C_j q_j + D_j} \tag{6.4-14}$$

and

$$q_j = \frac{A_i q_i + B_i}{C_i q_i + D_i} \tag{6.4-15}$$

we obtain the matrix that belongs to the combined transformation

$$q_k = \frac{A_k q_i + B_k}{C_k q_i + D} \tag{6.4-16a}$$

by forming the matrix products of the matrices of the component transformations

$$\begin{pmatrix} A_k & B_k \\ C_k & D_k \end{pmatrix} = \begin{pmatrix} A_j & B_j \\ C_j & D_j \end{pmatrix}\begin{pmatrix} A_i & B_i \\ C_i & D_i \end{pmatrix} \tag{6.4-16b}$$

This relation is easily verified by substitution of (6.4-15) into (6.4-14). The transformation laws of the q parameters of Gaussian beams are Kogelnik's

ABCD law [67,49] This is a very handy formalism for describing the transmission of Gaussian beams through complicated combinations of lenses.

We have shown that the ABCD law of Gaussian beam transformation holds for the propagation of Gaussian beams through free space and through thin lenses. We have also shown that the transmission of Gaussian beams through a succession of free space elements and lenses is described by the successive multiplication of the corresponding matrices. We can thus immediately state that the ABCD law holds for any optical system that can be described by a succession of thin lenses and free space elements. This covers a wide range of applications. Most optical instruments are combinations of compound lenses. Since the ABCD law holds for the passage of the Gaussian beam through any individual element of the system, and since the matrix of beam transmission through combinations of these elements is obtained by multiplication of the individual matrices, we see immediately that the ABCD law holds for the entire optical system. Lens waveguides are combinations of thin lenses and free space sections, so that the ABCD law holds for them. Even square law media can be considered as an infinite succession of infinitesimally closely spaced lenses. We can thus apply the ABCD law for this lens combination by first assuming that the number of lenses is finite and then allowing it to increase without limit. We thus see that the ABCD law holds for square law media. However, it is also important to keep in mind that the ABCD law does not hold for transmission of Gaussian beams through media with arbitrary non-square law index distributions or for beam transmission through lenses with distortions. Its applicability is limited to systems that can be approximated by combinations of ideal thin lenses. Other more general systems distort the shape of the Gaussian beam, as we have seen in Section 5.8, so that the beam leaving such a system is no longer Gaussian and consequently cannot be described by only two parameters—beam half width, and phase front curvature.

There is an interesting relation between ray optics and Gaussian beam optics. Light rays are also characterized by two parameters—the beam position r, and its slope r'. The transformation of a light ray from one point in space to another or its passage through a lens or any other optical device can be described by the following matrix relation

$$\begin{pmatrix} r_2 \\ r_2' \end{pmatrix} = \begin{pmatrix} A & B \\ C & D \end{pmatrix} \begin{pmatrix} r_1 \\ r_1' \end{pmatrix} \qquad (6.4\text{-}17)$$

We are considering rays that stay in a plane containing the optical axis of the system. The two parameters—ray deflection from the optical axis r, and ray slope r'—are thus sufficient to describe the ray completely. Ray trajectories remaining wholly inside a plane are always possible in rotationally symmetric optical systems of the type we are interested in here. For a beam in free space, the relation between beam slope and position at two different points spaced a distance d along the optical axis can be expressed in the paraxial approximation by

$$r_2 = r_1 + r_1' d_1 \tag{6.4-18}$$

and

$$r_2' = r_1' \tag{6.4-19}$$

The matrix corresponding to this transformation is obviously the same as for the transformation of a Gaussian beam between two points in space— that is, Equation (6.4-10).

The passage of a light ray through an ideal thin lens is described by the transformation

$$r_2 = r_1 \tag{6.4-20}$$

and by (4.2-4), which, again in the paraxial approximation, can also be written as

$$r_2' = r_1' - \frac{r_1}{f} \tag{6.4-21}$$

(Note that in the notation of Section 4.2 $r' = -\alpha$). In matrix notation, this transformation is also identical to the transformation of a Gaussian beam by a lens (6.4-11). The repeated application of transformations of light rays is obviously also described by the multiplication of the transformation matrices. We have thus the important result that the transformation of Gaussian beams through optical systems, which can be synthesized by a combination of thin lenses, is known if the corresponding ray optics problem has been solved.

Our result allows us immediately to solve the problem of the passage of (on-axis) Gaussian beams through lens waveguides. The ray matrix describing the ideal lens waveguide follows from (5.2-7). Instead of expressing the ray position at the nth lens by its position at the first two lenses, we express it now by the position and slope (taken at the left of the lens) at the first lens

$$r_1 = A e^{i\theta} + B e^{-i\theta} \tag{6.4-22}$$

The expression for the ray slope r_1' follows from (6.4-21), with $r_2' = (r_2 - r_1)/L$

$$
\begin{aligned}
r_1' &= \frac{r_2 - r_1}{L} + \frac{r_1}{f} \\
&= \frac{1}{L}\left\{ A\left[\left(\frac{L}{f} - 1\right)e^{i\theta} + e^{2i\theta}\right] + B\left[\left(\frac{L}{f} - 1\right)e^{-i\theta} + e^{-2i\theta}\right]\right\} \\
&= \frac{1}{L}\left\{ A(e^{i\theta} - 1) + B(e^{-i\theta} - 1)\right\}
\end{aligned}
\tag{6.4-23}
$$

The last line of (6.4-23) was obtained with the help of (5.2-6). These two equations, (6.4-22) and (6.4-23), can be used to solve for A and B

$$A = \frac{(1 - e^{-i\theta})r_1 + Lr_1'e^{-i\theta}}{2i \sin \theta} \tag{6.4-24}$$

and

$$B = \frac{(e^{i\theta} - 1)r_1 - Lr_1'e^{i\theta}}{2i \sin \theta} \tag{6.4-25}$$

The ray position at the nth lens, expressed in terms of position and slope at the first lens, follows from (5.2-7)

$$r_n = \frac{[\sin n\theta - \sin (n - 1) \theta]r_1 + Lr_1' \sin (n - 1)\theta}{\sin \theta} \tag{6.4-26}$$

$$r_n' = \frac{r_n - r_{n-1}}{L}$$

$$= \frac{[\sin n\theta - 2 \sin(n - 1)\theta + \sin(n - 2)\theta]r_1 + Lr_1'[\sin(n - 1) \theta - \sin(n - 2)\theta]}{L \sin \theta} \tag{6.4-27}$$

Rewriting the sum and difference of the sine functions, we obtain the matrix of the ray transformation in the form

$$\begin{pmatrix} A & B \\ C & D \end{pmatrix} = \begin{pmatrix} \dfrac{\cos\left(n\theta - \dfrac{\theta}{2}\right)}{\cos \dfrac{\theta}{2}} & L\dfrac{\sin(n - 1)\theta}{\sin \theta} \\[4mm] \dfrac{2 \sin[(n - 1)\theta](-1 + \cos \theta)}{L \sin \theta} & \dfrac{\cos\left[(n - 1) \theta - \dfrac{\theta}{2}\right]}{\cos \dfrac{\theta}{2}} \end{pmatrix} \tag{6.4-28}$$

The same result can also be obtained by constructing the matrix of one element of the lens waveguide that describes the progress of the ray from the left of one lens to the position immediately to the left of the next lens. The matrix of the ray at the nth lens is then obtained as the $(n - 1)$th power of the matrix of one of the elements of the lens waveguide. This calculation is left as an exercise for the reader.

Knowing the ABCD matrix of the lens waveguide enables us to calculate the behavior of the beam width parameter as the Gaussian beam travels through the guide. A pure mode has the same beam width at every lens. A Gaussian beam that is not a mode of the structure can be expected to change its width from lens to lens.

The beam width parameter can be obtained from the imaginary part of $1/q$ with the help of (6.2-16). In the form (6.2-21), q is expressed as a function of the distance z from the narrowest point on the beam. Let us assume that we are launching a Gaussian beam into the lens waveguide, as shown in Figure 6.4.1. The waist of the beam entering the lens waveguide is located at distance d to the left of the first lens of the guide. Before we can calculate the behavior of the beam width parameter at each lens of the guide, we must know its relation to the complex beam parameter q_1 at the first lens. This is accomplished by replacing $z = d$ in (6.2-21) with the help of (6.2-7). We write

$$q_1 = d + ib \tag{6.4-29}$$

with

$$b = \frac{\pi w_o^2}{\lambda} \tag{6.4-30}$$

From (6.2-7), we obtain

$$d^2 = \left(\frac{\pi w_o}{\lambda}\right)^2 (w_1^2 - w_0^2) = b^2 \left(\frac{w_1^2}{w_0^2} - 1\right) \tag{6.4-31}$$

We are now ready to calculate $1/q_n$ at the nth lens with the help of the matrix (6.4-28). From (6.4-14), we have

$$\frac{1}{q_n} = \frac{Cq_1 + D}{Aq_1 + B} = \frac{Cd + D + ibC}{Ad + B + ibA} \tag{6.4-32}$$

The imaginary part of $1/q_n$ is thus

$$Im \frac{1}{q_n} = \frac{BC - AD}{(dA + B)^2 + b^2 A^2} b \tag{6.4-33}$$

The expression in the numerator is the negative determinant of the ABCD matrix. It can be checked, by direct calculation, that its value is -1. (See the discussion at the end of this section). Using (6.2-16), we obtain the desired expression for the beam width parameter at the nth lens

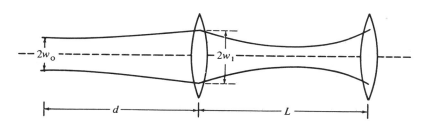

Figure 6.4.1   Launching of a Gaussian beam into a lens waveguide.

$$\frac{\pi}{\lambda} w_n^2 = b\left[\left(\frac{d}{b} A + \frac{B}{b}\right)^2 + A^2\right] \tag{6.4-34}$$

It remains for us to try to cast this expression into a useful form which allows us to determine the behavior of the beam width. Using (6.4-30) and (6.4-31) allows us to express (6.4-34) as follows

$$w_n^2 = w_1^2 A^2 + \frac{2\lambda}{\pi} \sqrt{\frac{w_1^2}{w_0^2} - 1} \; AB + \left(\frac{\lambda}{\pi w_0}\right)^2 B^2 \tag{6.4-35}$$

The matrix elements A and B are obtained from (6.4-28). By expressing the squares and products of sine and cosine functions by circular functions of twice the angle, we obtain

$$
w_n^2 = \frac{1}{2\cos^2\dfrac{\theta}{2}} \left\{ w_1^2 \left[\cos\{(2n-1)\theta\} + 1\right] \right.
$$
$$
+ \frac{\lambda L}{\pi} \sqrt{\frac{w_1^2}{w_0^2} - 1} \; \frac{\sin\left(2n - \dfrac{3}{2}\right)\theta - \sin\dfrac{\theta}{2}}{\sin\dfrac{\theta}{2}}
$$
$$
\left. + \left(\frac{\lambda L}{\pi w_0}\right)^2 \frac{1 - \cos(2n-2)\theta}{4\sin^2\dfrac{\theta}{2}} \right\} \tag{6.4-36}
$$

Let us introduce the length coordinate z_n along the lens waveguide by the definition

$$z_n = nL \tag{6.4-37}$$

and the length of the oscillation period Λ by

$$\frac{\theta}{L} = \frac{2\pi}{\Lambda} \tag{6.4-38}$$

Equation (6.4-26) shows that the light ray oscillates in the lens waveguide with a period Λ. However, we see from (6.4-36) that the width of the Gaussian beam inside the lens waveguide oscillates with the period length $\Lambda/2$. The beam width oscillations are thus twice as rapid as the oscillations of the ray.

Equation (6.4-36) can be written in a form that displays its dependence on the oscillatory terms more clearly

$$w_n^2 = U^2 \cos(2n-1)\theta + V^2 \sin(2n-1)\theta + W^2 \tag{6.4-39}$$

with

$$U^2 = \frac{1}{2\cos^2\frac{\theta}{2}} \left[w_1^2 - \frac{\lambda L}{\pi} \sqrt{\frac{w_1^2}{w_0^2} - 1} - \left(\frac{\lambda L}{\pi w_0}\right)^2 \frac{\cos\theta}{4\sin^2\frac{\theta}{2}} \right] \qquad (6.4\text{-}40)$$

$$V^2 = \frac{\cos\frac{\theta}{2}}{2\cos^2\frac{\theta}{2}\sin\frac{\theta}{2}} \frac{\lambda L}{\pi} \left[\sqrt{\frac{w_1^2}{w_0^2} - 1} - \frac{1}{2}\frac{\lambda L}{\pi w_0^2} \right] \qquad (6.4\text{-}41)$$

and

$$W^2 = \frac{1}{2\cos^2\frac{\theta}{2}} \left[w_1^2 - \frac{\lambda L}{\pi} \sqrt{\frac{w_1^2}{w_0^2} - 1} + \left(\frac{\lambda L}{\pi w_0}\right)^2 \frac{1}{4\sin^2\frac{\theta}{2}} \right] \qquad (6.4\text{-}42)$$

The Gaussian beam is a mode of the lens waveguide if its width remains the same at every lens. The Gaussian beam that is injected into the lens waveguide is characterized by its width w_o at the beam waist and by its width w_1 at the first lens. These two parameters can be varied at will. A normal mode is obtained if $U^2 = V^2 = 0$ and at the same time $w_n = w_1 = W$. The requirement $V^2 = 0$ leads to the condition

$$\sqrt{\frac{w_1^2}{w_0^2} - 1} = \frac{\lambda L}{2\pi w_0^2} \qquad (6.4\text{-}43)$$

The requirement $W = w_1$ together with (6.4-43) leads to the condition

$$w_1 = \frac{1}{2\sin\frac{\theta}{2}} \frac{\lambda L}{\pi w_o} \qquad (6.4\text{-}44)$$

It is easy to show that substitution of (6.4-43) and (6.4-44) into (6.4-40) results in $U^2 = 0$. It is thus possible to solve the three simultaneous equations with only two variables. Substitution of (6.4-44) into (6.4-43) allows us to determine w_1

$$w_1^2 = \frac{1}{\sin\theta} \frac{\lambda L}{\pi} \qquad (6.4\text{-}45)$$

and w_o

$$w_0^2 = \frac{\lambda L}{2\pi} \cot\frac{\theta}{2} \qquad (6.4\text{-}46)$$

We have thus solved the mode problem of the general lens waveguide with large apertures. In Section 5.6, we were able to solve the integral equation only for the special case of the confocal lens waveguide. Kogelnik's transformation theory[67] of Gaussian modes enables us to solve the general problem in a simple and elegant way. The modes of the general lens waveguide are given by (6.3-20), with w_o of (6.4-46), w of (6.2-7), and R determined by (6.2-13). In applying these formulas, we must remember that the length coordinate z has its origin at the beam waist. Because of the symmetry of the structure, the beam waist appears midway between the lenses.

In order to express the beam width parameter in terms of the waveguide dimensions and the focal length of the lenses, we eliminate θ from (6.4-45) and (6.4-46) with the help of (5.2-6). We obtain

$$w_1 = \left[\frac{2\lambda f}{\pi \sqrt{\frac{4f}{L} - 1}} \right]^{1/2} \tag{6.4-47}$$

as the half width of the general modes at each lens. The beam waist midway between two lenses has the half width

$$w_o = \left[\frac{\lambda L}{2\pi} \sqrt{\frac{4f}{L} - 1} \right]^{1/2} \tag{6.4-48}$$

In the special case of the confocal lens waveguide, $L = 2f$, we obtain

$$w_o = \sqrt{\frac{\lambda f}{\pi}} \tag{6.4-49}$$

and

$$w_1 = \sqrt{2}\, w_o \tag{6.4-50}$$

Equation (6.4-50) agrees with (5.6-21).

Because of the close relationship between the lens waveguide and laser resonators, our present solution of the general mode problem of the lens waveguide applies also to the modes of laser resonators with arbitrarily curved mirrors. It would of course also be possible to solve the problem of the modes in a lens waveguide with alternating lenses (compare Section 5.2). However, we shall not discuss this problem. The relation between the modes of laser resonators and lens waveguides will be discussed in more detail in Section 6.6.

The mode problem can also be solved in a different way. Considering Figure 6.4.1, we can determine the matrix (of either the rays or the beam parameter) of one element of the periodic structure. We start at a plane immediately to the left of the first lens and proceed to a plane immediately to the left of the next lens. The matrix for this waveguide element is given as the product of (6.4-10) and (6.4-11)

$$\begin{pmatrix} A & B \\ C & D \end{pmatrix} = \begin{pmatrix} 1 & L \\ 0 & 1 \end{pmatrix} \begin{pmatrix} 1 & 0 \\ -\dfrac{1}{f} & 1 \end{pmatrix} = \begin{pmatrix} 1 - \dfrac{L}{f} & L \\ -\dfrac{1}{f} & 1 \end{pmatrix} \qquad (6.4\text{-}51)$$

The condition for obtaining a mode is expressed by the requirement that the transformed complex beam parameter at the second lens be identical to the corresponding parameter at the first lens. The condition for a mode can thus be expressed by the equation

$$q = \frac{Aq + B}{Cq + D} \qquad (6.4\text{-}52)$$

The solution of this problem leads to the same results that we obtained above. The actual calculation is suggested as an exercise for the reader.

We conclude this section with a proof that the determinant of the ABCD matrix always has the value unity. The proof is based on Liouville's theorem.

Since we know that the ABCD matrix of Gaussian beam optics also governs the behavior of light rays, we can use our knowledge of ray optics to study the properties of the matrix. We write (6.4-17) in the form

$$r_2 = Ar_1 + Br_1' \qquad (6.4\text{-}53)$$

$$r_2' = Cr_1 + Dr_1' \qquad (6.4\text{-}54)$$

In the two-dimensional case of interest here, we can write (3.7-30) in the following form that holds in the paraxial approximation, with $n = 1$

$$\begin{vmatrix} \dfrac{\partial r_2}{\partial r_1} & \dfrac{\partial r_2'}{\partial r_1} \\[2ex] \dfrac{\partial r_2}{\partial r_1'} & \dfrac{\partial r_2'}{\partial r_1'} \end{vmatrix} = 1 \qquad (6.4\text{-}55)$$

In (6.4-55) the prime indicates the ray slope r', contrary to its meaning in (3.7-30). Using (6.4-53) and (6.4-54) to form the derivatives appearing in (6.4-55), we obtain

$$\begin{vmatrix} \dfrac{\partial r_2}{\partial r_1} & \dfrac{\partial r_2'}{\partial r_1} \\[2ex] \dfrac{\partial r_2}{\partial r_1'} & \dfrac{\partial r_2'}{\partial r_1'} \end{vmatrix} = \begin{vmatrix} A & C \\ B & D \end{vmatrix} = \begin{vmatrix} A & B \\ C & D \end{vmatrix} = 1$$

We have thus established the fact that it was no accident that the determinants of the matrices encountered in this section all had the value 1.

6.5 MODE MATCHING

We have seen in the last section that it is possible to inject a Gaussian beam into a lens waveguide. If the beam is not a mode of the guide, it changes its cross section periodically as it travels in the waveguide. The period of the beam width oscillations was found to be half the period of the oscillation of an off-axis ray. If the Gaussian beam is properly chosen, so that its half width at its waist is given by (6.4-48), while its half width at the first lens is given by (6.4-47), it propagates through the lens waveguide without change in its appearance. Its half width at all the other lenses assumes the same value that it had on the first lens. A beam of this kind is properly matched to the lens waveguide—it is a mode.

In most practical situations the experimenter has at his disposal a Gaussian beam that is the output of a laser. The mode of the lens waveguide or resonator that he wishes to excite with the laser beam is not likely to be of the same form as the beam produced by the laser. The laser and the lens waveguide (or other device) to be excited are mismatched. It is possible to transform the Gaussian beam of any laser by means of a lens into the mode of the structure. We have seen that every Gaussian beam mode is characterized by two parameters. For our present purposes, we choose the minimum beam width and its distance from a given reference point to characterize the beam uniquely. The geometry of the problem is sketched in Figure 6.5.1. As the reference point from which the distance to the beam waists is being measured, we choose the position of the mode matching lens. The minimum beam waists w_{o1} and w_{o2} are determined by the nature of the problem. We assume that we have a certain lens with focal length f at our disposal. The mode-matching problem consists in determining the required distances d_1 and d_2 at which the lens must be placed from the beam waist of the desired beam

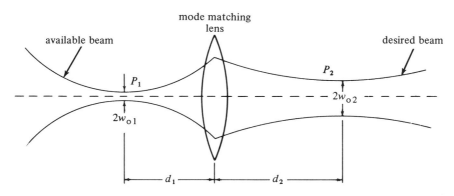

Figure 6.5.1   Mode matching of Gaussian beams by means of a lens.

and at which the waist of the available beam must be positioned from the lens.

The complex beam parameter q assumes the value (see [6.2-21])

$$q_{o1} = i\frac{\pi}{\lambda}w_{o1}^2 = ib_1 \qquad (6.5\text{-}1)$$

at point P_1 and

$$q_{o2} = i\frac{\pi}{\lambda}w_{o2}^2 = ib_2 \qquad (6.5\text{-}2)$$

at point P_2 of Figure 6.5.1. The object is to find a transformation such that the following relation holds

$$q_{o2} = \frac{Aq_{o1} + B}{Cq_{o1} + D} \qquad (6.5\text{-}3)$$

The required ABCD matrix has already been evaluated in Section 6.4, Equation (6.4-9). The matrix elements are real, while the parameters q_{o1} and q_{o2} are purely imaginary. Using the notation of (6.5-1) and (6.5-2), we can separate (6.5-3) into its real and imaginary parts

$$B + b_1 b_2 C = 0 \qquad (6.5\text{-}4)$$

and

$$b_1 A - b_2 D = 0 \qquad (6.5\text{-}5)$$

Substitution of the matrix elements (6.4-9) and elimination of d_1 and d_2 from the two equations leads to

$$d_1 = f \pm \frac{w_{o1}}{w_{o2}}\sqrt{f^2 - f_o^2} \qquad (6.5\text{-}6)$$

and

$$d_2 = f \pm \frac{w_{o2}}{w_{o1}}\sqrt{f^2 - f_o^2} \qquad (6.5\text{-}7)$$

with

$$f_o = \frac{\pi}{\lambda}w_{o1}w_{o2} \qquad (6.5\text{-}8)$$

The signs on the right-hand side must either both be positive or both be negative. It is thus possible to satisfy the mode-matching condition in two different ways. The focal length of the matching lens is arbitrary. It is required only that

$$f \geq f_o \qquad (6.5\text{-}9)$$

The focal length cannot be arbitrarily small.

Mode matching is of particular importance for the application of a scanning Fabry-Perot interferometer,[70] which is quite similar to a laser cavity. It is often used to display the frequency content of a multimode laser. For this purpose, the interferometer cavity changes its length periodically in time. If properly matched, light can pass the interferometer only at its resonance frequencies. By displaying the output of a detector, which is positioned behind the scanning Fabry-Perot interferometer, on an oscilloscope that is swept in syncronism with the scanning cavity, it is possible to display the spectrum of the laser output on the face of the cathode-ray tube. The resolution of such an instrument is very high.

The point of importance in connection with our present discussion is the multimode aspect of the interferometer cavity. If improperly used, the entering laser beam excites a great number of modes. Each mode has a different resonance frequency, so that each frequency component of the laser output would result in many responses on the oscilloscope screen. Operated under multimode conditions, it is impossible to use the scanning Fabry-Perot interferometer as a spectrum analyzer. It is apparent that one must match the modes of the laser carefully to the modes of the interferometer cavity to avoid multiple frequency responses. The matching of modes is accomplished with a suitable lens that satisfies condition (6.5-9). The proper position of the lens in relation to the laser beam and the interferometer is accomplished with the help of (6.5-6) and (6.5-7). Unless other optical components are placed in the path of the laser beam, its beam waist is located halfway between its two mirrors (provided the laser cavity uses two identical mirrors). The same is true for the beam waist of the Fabry-Perot cavity. The distances d_1 and d_2 must thus be measured from the center of the laser cavity and from the center of the Fabry-Perot cavity. The width of the beam waists is determined by the radius of curvature and spacing of the mirrors of the laser and Fabry-Perot resonator cavities. The geometry of this mode matching problem is shown in Figure 6.5.2. The beam distortion that results from the passage of the beams through the plane faces on the outside of the cavity mirrors may have to be taken into account. It is neglected in the figure.

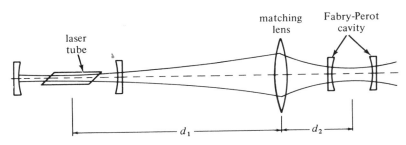

Figure 6.5.2   A laser beam is matched to a Fabry-Perot interferometer by means of a lens.

6.6 LASER CAVITIES

A brief discussion of the stability condition of laser resonators in Section 5.3 was based on the results of the transmission characteristics of lens waveguides with unequal lenses.

The discussion of the modes of laser resonators in this section will also draw heavily on the analogy between laser resonators and lens waveguides. However, before utilizing the results obtained from our discussion of lens waveguides, let us consider several aspects of laser resonators.

E. A. J. Marcatili* pointed out that a laser resonator can be regarded as part of an elliptical cavity. If one considers the modes of the elliptical cavity shown in Figure 6.6.1, one would find solutions whose field distribution can be indicated by the dotted lines in the figure.

The crosshatched area indicated in Figure 6.6.1 contains most of the field energy. There are of course other modes that would fill other portions of the elliptical cavity. The modes whose field distribution is shown in the figure do not utilize the portion of the reflecting enclosure outside the crosshatched area. Cutting these portions away has no noticable influence on their field distribution. This consideration provides an insight into the relation between closed resonators and the open resonator structures used for lasers.

In Section 5.6, we derived an integral equation for the modes of the lens waveguides. In order to see the close relationship between these two structures, we formulate the eigenvalue problem of the laser resonators modes. It will become apparent that the laser mode problem is mathematically identical to the lens waveguide mode problem.

Consider the laser cavity of Figure 6.6.2. We postulate the existence of a field over the surface of the first reflector. This field propagates to the second reflector. Its propagation from the reference plane at mirror 1 to the reference

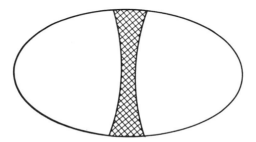

Figure 6.6.1   An elliptical metallic cavity can be used to explain the modes in an open laser cavity.

---

* Private communication.

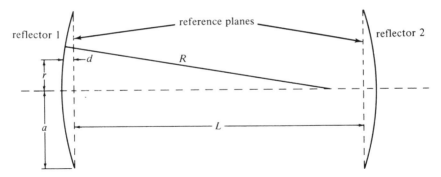

Figure 6.6.2   Laser resonator showing the reference planes used for computing the modes.

plane at mirror 2 is described by the diffraction integral (5.6-2). The progress of the field from the curved surface of the first mirror to the first reference plane can approximately be described simply by the phase shift suffered by a plane wave as it travels from the surface of the mirror to the reference plane. This phase shift is equal to

$$\phi = -ikd \tag{6.6-1}$$

In the paraxial approximation, d can be expressed in terms of the radius a of the first mirror, its radius of curvature R_1, and the distance r of the point under consideration from the axis of the structure

$$d = \frac{a^2 - r^2}{2R} \tag{6.6-2}$$

The radius of curvature of the mirror is related to the focal length of the curved reflector by the equation

$$f = \frac{1}{2}R \tag{6.6-3}$$

The field ψ_1 at the curved surface of mirror 1 is thus transformed into the field $\bar{\psi}_1$ at the reference plane by the equation

$$\bar{\psi}_1 = e^{-i(k/4f)(a^2 - r^2)}\psi_1 \tag{6.6-4}$$

An analogous expression relates the field at the reference plane at the second mirror to the field at the curved surface of that mirror. The transformation (6.6-4) is exactly the same as the corresponding transformation (5.6-1) that was caused by the passage of the field through the half lens of the lens waveguide.

In order to obtain a mode of the resonator, we require that the field at the first mirror be transmitted unchanged to the second mirror except for a phase shift. This condition for obtaining a resonator mode leads again to equation

(5.6-5). This proves the mathematical equivalence of the resonator problem with the lens waveguide. All the modes of one structure are automatically also modes of the other structure.

The formulation of the mode problem of the laser resonator used the paraxial approximation. The same approximation was also used to obtain the modes of the lens waveguide. In this approximation, the spherical reflectors of the cavity are indistinguishable from parabolic reflectors. This approximation holds only if the curvature of the mirrors is slight or, in a mathematical formulation, if

$$\frac{a^2}{2R} \ll a \qquad (6.6\text{-}5)$$

and

$$a \ll L \qquad (6.6\text{-}6)$$

The lens radius a appearing in these equations need not actually be the physical radius of the lens, but is the radius at which the field distribution has dropped to negligible values compared to the field strength on-axis.

The eigenvalue problem (5.6-5) provides the solution of the mode problem for the lens waveguide as well as for the laser resonator. The approximate solutions of this eigenvalue equation for large mirror or lens apertures is given by (6.3-20). For finite mirror or lens apertures, several approximate methods have been considered for solving the eigenvalue problem (5.6-5). (See references 115 and 116.) In the confocal case, $L = 2f$, a closed form solution exists, as discussed in Section 5.6. The analytical solutions for the confocal case with square apertures are expressed in terms of spheroidal wave functions.[47] The solutions of the confocal lens waveguide with circular apertures can be expressed by hyperspheroidal functions.[120]

Numerical solutions of the general eigenvalue problem (5.6-5) can be obtained by the method of successive approximations.[48] This iterative method works in the following way. We substitute an arbitrary function for ψ into the right-hand side of (5.6-5) and calculate (numerically) the value of the integral, thus obtaining a new function $\gamma\psi$. It is of course not known what value γ has. Since a constant multiplier does not influence the shape of the function, the fact that γ is not known does not concern us. We simply use this first approximation of ψ to obtain a second approximation by again evaluating the integral, using the result of the first integration. This process, if repeated many times, in general results in the lowest order solution of the integral equation. The problem whether or not the iteration procedure converges need not worry us too much. In those practical cases that have been tried, a solution by successive approximation was always obtained. There are moreover, physical reasons why convergence of the iteration process can be expected.

The mathematical process of successive approximations can be justified on physical grounds. Let us assume either one of two cases. We might imagine that we feed a certain field distribution into the lens waveguide and follow its progress through the guide. Alternately, we can also assume that we allow a field distribution that happens to exist inside a laser resonator to travel back and forth between its two mirrors. In either case, the field distribution will alter its shape from lens to lens or, at each alternate bounce, from mirror to mirror. The mathematical process of describing the fate of our arbitrary input field corresponds exactly to the mathematical procedure of successive approximations, because we must apply the diffraction integral (5.6-5) (the right-hand side of this equation) to obtain the field distribution at the next lens or at the next mirror of the resonator. In systems with finite apertures, we lose power after each transit through the resonator or after passing from one mirror to the next. However, if we go far enough through the lens waveguide or follow the field distribution long enough as it bounces back and forth between the two laser mirrors, we find that the field distribution settles down to the lowest order mode of the structure. The fact that laser oscillators can settle down in modes other than the lowest order mode is caused by the gain mechanism. The mode that possesses the highest overall gain will win out in the competition between the different modes. Often there is enough gain available to allow more than one cavity mode to oscillate simultaneously. However, in a lossy structure, only the lowest order mode—that is, the one with the lowest loss—is obtained by the method of successive approximations.

This mode selection process can be understood more easily if we imagine that the arbitrary input field is expressed as a linear superposition of infinitely many modes of the structure. As the field travels down the lens waveguide, each mode of the superposition travels without change. But after passing from one lens to the next, each mode gets multiplied with its eigenvalue. After passing through n lenses, the νth mode of the series expansion finds itself multiplied by γ_ν^n. These powers of γ_ν become vanishingly small for increasing values of n. Only the lowest order γ_ν^n makes an appreciable contribution if n is a large number. (Compare [5.7-9].)

This last argument, based on a decomposition of the input field into modes of the structure, is somewhat circular, since we invoke the concept of modes to explain why a mode is obtained by successive approximation. However, it helps to see why the lowest order (that is, the lowest loss) mode survives after passing through many lenses and why every arbitrary field distribution tends to settle down into the shape of the lowest order mode. Only if the coefficient of the lowest order mode in the original expansion happens to vanish, do we obtain the mode with the next lowest loss. Experimental experience has shown that arbitrary field distributions traveling in lossy waveguides always settle down to the lowest loss guided mode. It is also an experimental fact that modes always exist in practical periodic structures. (Even ordinary waveguides can be considered to be periodic structures.) It is this practical experience that makes us confident that modes must exist even though this may be hard to prove mathematically in any given case.

We return to the discussion of the iteration process. We have now convinced ourselves that we expect to obtain the field distribution of the lowest order, lowest loss mode of the lens waveguide or laser resonator by the method of successive approximations. Having repeated the iteration a sufficient number of times, so that no further change of the field shape can be observed, allows us to find not only the shape of the lowest order solution of the eigenvalue problem but also the eigenvalue itself. We need only observe the decrease of the field amplitude and the change of the phase of the field after one more iteration to obtain the value of γ. Iterative solutions of the eigenvalue problem (5.6-5) have been obtained by Fox and Li.[48] Some of their results are shown in Figure 5.6-5 for confocal lens waveguides or laser resonators with circular apertures. The higher order modes were obtained by the same iterative method by taking care that the mode content of the lowest order mode of the initial field distribution is vanishingly small. For large values of c of (5.6-30), the iteration converges slowly. Many iteration steps are required, and the numerical analysis becomes quite costly. Other approximate methods for obtaining the eigenvalues and eigenfunctions consist in converting the integral equation (5.6-5) to a system of simultaneous algebraic equations, which is then solved by standard matrix techniques.[116]

The reader may have wondered why the discussion of resonator modes so far has never involved any mention of the resonance condition. The iterative solution of the eigenvalue equation does not involve the resonance condition for cavity modes. However, in the physical resonator, the situation is slightly different from the mathematical assumptions that led to (5.6-5). Not only is there a field right at the mirrors of the resonator, but the field extends throughout the space between the two reflectors. In fact, we must assume that a long field train approaches the reflector, gets reflected, and moves back on itself until it is reflected from the second mirror and now begins to interfere with itself. This self-interference must be constructive, since otherwise there is no chance of any buildup of field energy between the two reflectors. This more complete picture of the physics of the resonator brings in the requirement of a resonance condition. We must require that, after a complete round trip, the field adds up in phase with the existing field, or, in mathematical terms

$$2\Delta\phi = 2N\pi \tag{6.6-7}$$

The phase shift $\Delta\phi$ applies to the transit from one mirror to the next. Twice this amount of phase shift, or the phase shift suffered in one round trip, must be equal to an integer multiple of 2π.

The buildup of field energy in an actual laser cavity can be visualized in the same way that we described the method of successive approximations. Many of the atoms that make up the gain medium of the laser emit energy spontaneously. A small amount of this energy travels in the right direction to strike one or the other of the two mirrors and gets reflected back into the gain region. It traverses this gain region located between the mirrors and reaches the opposite mirror a little stronger. The wave form that achieves the highest round trip gain manages to obtain the most power, and grows more rapidly than any other wave form in the cavity. Its ultimate victory over competing wave trains traveling inside the same resonator is assured by the process of saturation. The gain medium can provide linear gain only for relatively weak electromagnetic fields. As soon as the field has grown strong enough to saturate the gain medium, the gain is decreased. A field form inside the resonator can exist for any length of time only when it gains more power during one round trip than it loses to various loss mechanisms. Only certain wave forms manage to experience enough gain to offset their losses after the gain medium has saturated. The vast majority of wave forms that started out initially end up having more loss than the saturated gain medium can provide. These wave forms lose out and do not develop. A relatively small number of (the infinitely many) modes is thus capable only of oscillation in any given laser. By introducing apertures into the laser beam inside the resonator, it is possible to increase the losses of higher order modes so much that only the lowest order mode can oscillate. However, the lowest order solution of the eigenvalue equation (5.6-5) can still exist at several frequencies, each satisfying the resonance condition (6.6-7). There are therefore usually several so-called longitudinal cavity modes possible that are all solutions of the eigenvalue equation (5.6-5), having the same field shape in the transverse plane (each such field shape is often referred to as a transverse mode) but slightly different frequencies. How many longitudinal modes are simultaneously capable of oscillation depends on the width (in frequency) of the gain curve. The frequency spacing of the longitudinal cavity modes is determined by the spacing of the cavity mirrors. Far-spaced mirrors result in closely spaced resonator modes. If single mode operation is desired, it is sometimes possible to design a laser that is so short that only one mode exists on the gain curve of the laser medium. This method of laser operation reduces the available power. If higher power output is desirable and single frequency operation is required, other methods of mode selection must be employed.[71]

Most lasers operate with sufficiently large apertures so that their output is described to a good approximation by the Hermite-Gaussian beam modes discussed in the last section.

We saw in Section 6.5 that Hermite-Gaussian beams have the tendency to reach a narrowest possible beam waist and spread from that point in both directions. The phase front of the beam is plane at the waist but is approximately spherical at all other points along the beam. A typical Gaussian beam is shown in Figure 6.6.3. The solid lines indicate the beam width, while the dotted lines perpendicular to the axis show the phase fronts at various points along the beam. We know that a Hermite-Gaussian beam is an approximate solution of the wave equation. It is thus apparent that we obtain an admissible solution of the resonator problem (with large apertures) by placing reflectors anywhere into the beam. We must only take care that the reflectors are shaped like the phase fronts. If this is the case, the field is reflected back on itself, and a possible resonator solution is obtained. This consideration shows that a wide variety of resonators are possible for any given Hermite-Gaussian beam. Placing resonators with the proper curvature anywhere in the path of the beam results in an admissible laser resonator. It is not necessary that both mirrors have the same curvature as long as their curvatures conform to the phase front curvature of the Hermite-Gaussian beam at the appropriate point in space. In actuality, it is of course the existing mirrors that produce the proper Hermite-Gaussian beam modes. But in order to visualize the procedure, it is convenient to reverse the argument and pretend that the beam exists before mirrors are provided to reflect it back on itself.

A typical laser cavity is shown in Figure 6.6.4. The mirrors could of course also be chosen to be of equal curvature, or they could be chosen so that one is concave while the other one is convex. In that case, they would both be located on the same side of the beam waist.

The on-axis phase shift of the Hermite-Gaussian beam for one transit through the cavity is, according to (6.3-20), $(r = 0)$

$$\Delta\phi = \phi(d_2) - \phi(d_1) = kL - (m + n + 1)\left[\arctan\frac{\lambda d_2}{\pi w_o^2} + \arctan\frac{\lambda d_1}{\pi w_o^2}\right] \quad (6.6\text{-}8)$$

The mirrors are spaced at distance d_1 and d_2 from the beam waist, $L = d_1 + d_2$. The resonance condition follows from (6.6-7) $(k = 2\pi/\lambda)$

$$2\pi\frac{L}{\lambda} - (m + n + 1)\left[\arctan\frac{\lambda d_2}{\pi w_o^2} + \arctan\frac{\lambda d_1}{\pi w_o^2}\right] = N\pi \quad (6.6\text{-}9)$$

The wavelength λ appearing in these equations is not the wavelength of the Gaussian beam mode but the wavelength that a corresponding plane wave would have if it traveled in the same medium with the same frequency as the

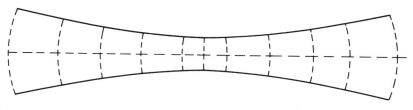

Figure 6.6.3   Gaussian beam with lines indicating the beam width (solid lines) and the surfaces of constant phase (dotted lines).

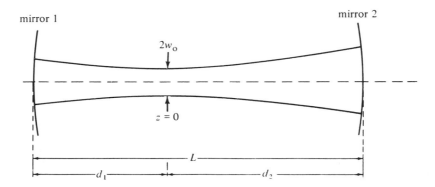

Figure 6.6.4   Laser resonator with unequal mirrors.

actual Gaussian beam mode. The wavelength of the Gaussian beam could be defined as the on-axis distance that corresponds to a phase shift of 2π. This wavelength is not constant, however, but depends slightly on the position along the Hermite-Gaussian mode.

It is possible to express the resonance condition (6.6-9) in terms of the radius of curvature of the laser mirrors instead of the distances d_1 and d_2 of the beam waist from the mirrors. To achieve this transformation, we start out by expressing the distances d_1, d_2, and the beam half width w_1 at the left mirror, w_2 at the right mirror, and w_o at the beam waist in terms of the radius of curvature R_1 of the left mirror, R_2 of the right mirror, and the distance L between the two mirrors. The relation between the radius of curvature of the phase fronts, which is equal to the curvature of the mirrors, and the distances of the mirrors from the beam waist is given by (6.2-13)

$$R_1 = d_1 + \left(\frac{\pi w_o^2}{\lambda}\right)^2 \frac{1}{d_1} \qquad (6.6\text{-}10)$$

and

$$R_2 = d_2 + \left(\frac{\pi w_o^2}{\lambda}\right)^2 \frac{1}{d_2} \qquad (6.6\text{-}11)$$

Elimination of w_o from these two equations results in

$$\frac{R_1 - d_1}{R_2 - d_2} = \frac{d_2}{d_1} \qquad (6.6\text{-}12)$$

Using

$$d_1 + d_2 = L \qquad (6.6\text{-}13)$$

immediately allows us to obtain

$$d_1 = \frac{L(R_2 - L)}{R_1 + R_2 - 2L} \tag{6.6-14}$$

and

$$d_2 = \frac{L(R_1 - L)}{R_1 + R_2 - 2L} \tag{6.6-15}$$

Substitution of (6.6-14) and (6.6-15) into either (6.6-10) or (6.6-11) allows us to obtain the half width of the beam at its waist

$$w_0^4 = \left(\frac{\lambda}{\pi}\right)^2 \frac{L(R_1 - L)(R_2 - L)(R_1 + R_2 - L)}{(R_1 + R_2 - 2L)^2} \tag{6.6-16}$$

In the special case $R_1 = R_2 = 2f$, (6.6-16) becomes identical with (6.4-48). Substitution of (6.6-14) and (6.6-16) into (6.2-19) (with $z = -d_1$) results in

$$w_1^4 = \left(\frac{\lambda R_1}{\pi}\right)^2 \frac{L(R_2 - L)}{(R_1 - L)(R_1 + R_2 - L)} \tag{6.6-17}$$

Using similarly (6.6-15) or simply interchanging the indices 1 and 2 results in

$$w_2^4 = \left(\frac{\lambda R_2}{\pi}\right)^2 \frac{L(R_1 - L)}{(R_2 - L)(R_1 + R_2 - L)} \tag{6.6-18}$$

Equations (6.6-14) through (6.6-18) express the Gaussian beam parameters in terms of the dimensions of the resonant cavity.

We use these expressions to transform the resonance condition (6.6-9) into a form that depends more directly on the cavity parameters. We use the addition theorem for the arc tangent function

$$\arctan \alpha + \arctan \beta = \arccos\left(\frac{1 - \alpha\beta}{\sqrt{1 + \alpha^2 + \beta^2 + \alpha^2\beta^2}}\right) \tag{6.6-19}$$

with

$$\alpha = \frac{\lambda}{\pi} \frac{d_1}{w_o^2} = \sqrt{\frac{L(R_2 - L)}{(R_1 - L)(R_1 + R_2 - L)}} \tag{6.6-20}$$

and

$$\beta = \frac{\lambda}{\pi} \frac{d_2}{w_o^2} = \sqrt{\frac{L(R_1 - L)}{(R_2 - L)(R_1 + R_2 - L)}} \tag{6.6-21}$$

A little algebra allows us to write (6.6-19) in the following form

$$\arctan\left(\frac{\lambda}{\pi} \frac{d_1}{w_o^2}\right) + \arctan\left(\frac{\lambda}{\pi} \frac{d_2}{w_o^2}\right) = \arccos\sqrt{\left(1 - \frac{L}{R_1}\right)\left(1 - \frac{L}{R_2}\right)} \tag{6.6-22}$$

The resonance condition (6.6-9) now assumes the form

$$\frac{2L}{\lambda} - \frac{m+n+1}{\pi} \arccos \sqrt{\left(1 - \frac{L}{R_1}\right)\left(1 - \frac{L}{R_2}\right)} = N \qquad (6.6\text{-}23)$$

The resonant frequency, $f = c/\lambda$, is therefore [49,72]

$$f = \frac{c}{2L}\left\{N + \frac{m+n+1}{\pi} \arccos \sqrt{\left(1 - \frac{L}{R_1}\right)\left(1 - \frac{L}{R_2}\right)}\right\} \qquad (6.6\text{-}24)$$

For a confocal cavity with

$$R_1 = R_2 = L \qquad (6.6\text{-}25)$$

(6.6-24) specializes to

$$f = \frac{c}{2L}\left[N + \frac{1}{2}(m + n + 1)\right] \qquad (6.6\text{-}26)$$

The integer N determines the number of half waves that exist between the mirrors. The number N is thus the mode number of the longitudinal mode of the cavity. The integer numbers n and m determine the transverse mode pattern of the Hermite-Gaussian field distribution. For $n = m = 0$, the lowest order Gaussian mode is obtained.

It is important to remember that the Hermite-Gaussian modes apply only to laser resonators whose mirrors are so large that the diffraction losses are negligible. If the mirrors are small enough so that diffraction losses must be considered, the approximation that leads to the Hermite-Gaussian resonator modes breaks down. The mode solution must then be obtained by means of numerical solutions of the eigenvalue problem (5.6-5). (See reference 48.)

The stability problem of the laser resonator modes was discussed in Section 5.2. The shaded regions of Figure 5.2.4 indicate regions of stability for the laser resonator modes. Outside the stable region, the mode losses become very much higher. However, for special applications, operation in the unstable region of the resonator has been suggested.[117]

7

LIGHT PROPAGATION IN SQUARE LAW MEDIA

7.1 INTRODUCTION

In Chapter 5, we studied the wave and ray guiding properties of the lens waveguide. There are of course many other ways of guiding light beams. If we allow the spacing between the ideal thin lenses of the lens waveguide to approach zero, we obtain a continuous guiding medium. This continuous medium can be described by an index of refraction that assumes a maximum value on-axis and decreases monotonically with increasing distance from the optical axis. A medium with an index of refraction of the form

$$n = n_o - \frac{1}{2} n_1 (x^2 + y^2) \tag{7.1-1}$$

is called a square law medium. The guidance of light waves and light rays in square law media is very closely analogous to the light guidance in the lens waveguide.[69,76,77,78]

Optical waveguides using square law media may become of practical importance as methods of producing such media become perfected. Optical glass fibers are presently available that have a graded index of refraction closely resembling a square law medium.[23,112] Our purpose in devoting an entire chapter to light guidance in square law media is to acquaint the reader with light guidance in continuous dielectric distributions.

Square law media have the desirable property that the center of intensity of paraxial light beams move according to the laws of ray optics (see Section 3.6). The ray optics treatment of these structures is thus physically meaningful, providing easy and valuable insight into the propagation properties of light beams.

The square law medium of (7.1-1) suffers from one serious shortcoming. For sufficiently large values of x and y, the index not only becomes less than unity but even reaches negative values. Even though an index of refraction of less than unity can occur in ionized gases, most optical media have refractive indices that are larger than unity. We must thus be aware of the fact that the index distribution (7.1-1) reaches into unphysical regions. As long as the light rays do not reach into the region of $n < 1$ and as long as the field distribution inside these regions has negligible values, we are reasonably safe and can use the index distribution (7.1-1) as a good first approximation. We shall limit ourselves to the paraxial approximation. Paraxial beams travel with ray slopes that are always very nearly parallel to the optical axis. These rays are less likely to depart far from the optical axis than are rays with large slopes.

The square law medium (7.1-1) is always a first approximation to a more complicated symmetrical index distribution. If we expand an arbitrary symmetrical index distribution in a Taylor series, we may obtain terms higher than the second order. But in many applications, the Taylor series expansion converges rapidly, so that the first two terms are a reasonable first approximation. It would not be any harder to allow the coefficients of x^2 and y^2 to be different from each other. However, since not much is gained by this slight generalization, we shall limit ourselves to index distributions of the form (7.1-1). It has been shown that a medium with a $1/\cosh x$ distribution is an ideal focusing structure even for planar nonparaxial rays.[75,118,119] However, we are dealing here only with the square law medium in paraxial approximation.

7.2 RAY OPTICS OF THE SQUARE LAW MEDIUM

We derive the ray trajectory in the square law medium from the ray trajectory of the lens waveguide with infinitely closely spaced lenses. In (5.2-4), we let L go toward zero and set

$$\frac{r_{n+1} - r_n}{L} = \frac{dr}{dz} \tag{7.2-1}$$

and

$$\frac{(r_{n+1} - r_n)/L - (r_n - r_{n-1})/L}{L} = \frac{d^2r}{dz^2} \tag{7.2-2}$$

The difference equation (5.2-4) can thus be written

$$\frac{d^2r}{dz^2} = -\frac{1}{Lf}r \tag{7.2-3}$$

We now let $f \to \infty$ such that

$$\frac{1}{Lf} = \frac{n_1}{n_o} \tag{7.2-4}$$

This limiting process transforms the difference equations for the ray trajectory of the lens waveguide into the paraxial differential equation

$$\frac{d^2r}{dz^2} = -\frac{n_1}{n_o}r \tag{7.2-5}$$

of the square law medium. The differential equation (7.2-5) is apparently identical to the paraxial differential equation (3.5-36) with (7.1-1) inserted on the right-hand side. We have thus proved that the lens waveguide becomes a square law medium in the limit of infinitely closely spaced lenses. The solution of the ray trajectory of the lens waveguide can likewise be transformed to the limit of the square law medium. From (5.2-6), we obtain $\cos \theta = 1$ in the limit $L \to 0$ and $f \to \infty$, but we have

$$\sin \theta = \theta = \sqrt{\frac{L}{f}} = \sqrt{\frac{n_1}{n_o}}\, L \tag{7.2-6}$$

Using

$$z = nL \tag{7.2-7}$$

we obtain from (6.4-26)

$$r = r_o \cos \sqrt{\frac{n_1}{n_o}}\, z + \sqrt{\frac{n_o}{n_1}}\, r_o' \sin \sqrt{\frac{n_1}{n_o}}\, z \tag{7.2-8}$$

It is apparent that (7.2-8) is indeed a solution of the paraxial ray equation (7.2-5). The ray position and slope r_1 and r_1' have now been designated by r_o and r_o' to indicate that they are the initial values of the ray position and slope at $z = 0$. Our derivation of the ray trajectory in the square law medium stressed the close relationship between the square law medium and the lens waveguide. It is clear, however, that (7.2-5) and its solution (7.2-8) could have been obtained from the paraxial ray equation (3.5-36) and the index distribution (7.1-1) without ever mentioning the lens waveguide. The square law medium is a light waveguide in its own right.

Because of the close relationship, we can obtain the ray trajectory of the ray in a square law medium with curved axis from the results of the lens waveguide treatment. Using (5.4-8) and (5.4-9), we obtain immediately, with the help of (7.2-6) and (7.2-7) ($dz = L$ with $L \to 0$)

$$r(z) = r_o \cos \sqrt{\frac{n_1}{n_o}} z + \sqrt{\frac{n_o}{n_1}} r_o' \sin \sqrt{\frac{n_1}{n_o}} z$$

$$+ \sqrt{\frac{n_o}{n_1}} \int_o^z \frac{1}{R(u)} \sin \sqrt{\frac{n_1}{n_o}} (z - u) du \quad (7.2\text{-}9)$$

The solution of the homogeneous equation (the first two terms of [7.2-9]) was written in the form (7.2-8). The function $R(u)$ is the radius of curvature of the axis of the square law medium at $z = u$.

Using the form (5.4-15) of the lens waveguide with curved axis, we obtain for the square law medium with curved axis

$$\rho(z) = \rho_o \cos \sqrt{\frac{n_1}{n_o}} z + \sqrt{\frac{n_o}{n_1}} \rho_o' \sin \sqrt{\frac{n_1}{n_o}} z$$

$$+ \sqrt{\frac{n_1}{n_o}} \int_o^z S(u) \sin \sqrt{\frac{n_1}{n_o}} (z - u) du \quad (7.2\text{-}10)$$

The function $S(z)$ describes the departure of the axis of the square law medium from a straight line.

The equivalence of the two solutions (7.2-9) and (7.2-10) of the ray trajectory in the square law medium with curved axis can easily be demonstrated. The radius of curvature of the axis of the square law medium is related to the displacement of the axis from perfect straightness $S(z)$ by the following approximate relation

$$\frac{1}{R(z)} = -\frac{d^2 S}{dz^2} \quad (7.2\text{-}11)$$

This approximation is valid if the first derivative dS/dz is always much less than unity. The negative sign in (7.2-11) takes care of the definition of the positive radius of curvature used in Figure 5.4.2. Substitution of (7.2-11) into (7.2-9) leads, after two partial integrations, to the result

$$r(z) = r_o \cos \sqrt{\frac{n_1}{n_o}} z + \sqrt{\frac{n_o}{n_1}} r_o' \sin \sqrt{\frac{n_1}{n_o}} z - S(z) + S(o) \cos \sqrt{\frac{n_1}{n_o}} z$$

$$+ \sqrt{\frac{n_o}{n_1}} \left(\frac{dS}{dz}\right)_{z=o} \sin \sqrt{\frac{n_1}{n_o}} z + \sqrt{\frac{n_1}{n_o}} \int_o^z S(u) \sin \sqrt{\frac{n_1}{n_o}} (z - u) du \quad (7.2\text{-}12)$$

According to (5.4-10), the relation

$$\rho(z) = r(z) + S(z) \quad (7.2\text{-}13)$$

holds. It is apparent from (7.2-13) that (7.2-12) is identical to (7.2-10), showing the equivalence of the two solutions.

The rms beam deviation σ caused by random displacements of the axis of the square law medium is obtained from (5.5-36). This equation is already an approximation of the more general expression (5.5-35). However, in most practical cases the approximation is well justified, so that it is more convenient to use the simpler equation. In the limit of vanishing lens spacings, (5.5-36) becomes

$$\sigma^2(z) = \frac{1}{2} \frac{n_1}{n_o} z \int_{-\infty}^{\infty} R_c(u) \cos \sqrt{\frac{n_1}{n_o}} u \, du \qquad (7.2\text{-}14)$$

The correlation function of the displacement $S(z)$ of the optical axis is defined as the ensemble average of the product of $S(z)$ and $S(z - u)$

$$R_c(u) = \langle S(z)S(z - u) \rangle \qquad (7.2\text{-}15)$$

Equation (7.2-14) shows that the variance of the beam deviation is proportional to the distance along the guide at which the beam is observed. It is, furthermore, proportional to the Fourier transform of the correlation function. The Fourier transform of the correlation function contains only the cosine terms, since $R_c(u)$ is an even function.

The behavior of the square law medium is in all respects very similar to the behavior of the lens waveguide. This similarity makes it unnecessary to discuss the ray trajectory in the square law medium in detail. All the results for the square law medium can immediately be obtained from the corresponding results for the lens waveguide.

7.3 MODES OF THE SQUARE LAW MEDIUM

We continue our discussion of the wave-guiding properties of the square law medium by considering its wave optical properties. Strictly speaking, it would be necessary to find solution of Maxwell's equation in a medium with the index distribution (7.1-1). However, we have seen in Section 1.3 that the scalar wave equation (1.3-6) must be satisfied by every component of the electromagnetic field. For optical media with varying index of refraction, the wave equation is not an exact equivalent of Maxwell's equations. The approximation that is involved in using the wave equation instead of Maxwell's equations is good if the index of refraction (or the dielectric constant) varies only very slightly over the distance of one optical wavelength. This requirement means that R of (1.3-28) must be much less than unity. This requirement is of a similar nature as the paraxial approximation that was used to solve the ray propagation problem in the square law medium. If we were content with ray solutions that were valid in the paraxial approximation, we may also use the wave equation instead of Maxwell's equations to determine the wave propagation properties of the square law medium. In fact, we shall discuss only the propagation of scalar waves in the square law medium.[68,69] This scalar wave problem is a good approximation to the electromagnetic field problem as long as polarization effects are not being considered. We know that each component of the electromagnetic field must approximately be a solution of the scalar wave equation. The scalar wave problem in the square law medium also has applications to the problem of sound propagation in the ocean. There are density changes in certain layers in the ocean that give rise to guidance of sound waves. In a first approximation, we may assume that the ocean at certain depth behaves like a square law medium.

Assuming, as always, that our waves have a harmonic time dependence, we use the reduced wave equation in the form

$$\nabla^2\psi + n^2 k_o^2 \psi = 0 \tag{7.3-1}$$

with the free space propagation constant

$$k_o = \omega\sqrt{\varepsilon_o\mu_o} = \frac{2\pi}{\lambda} \tag{7.3-2}$$

and the relative dielectric constant (square of the refractive index)

$$n^2 = n_o^2 - n_o n_1 (x^2 + y^2) \tag{7.3-3}$$

With (7.1-1) as our starting point, (7.3-3) is only approximately correct. However, we may also consider (7.3-3) as the correct expression of the relative dielectric constant and consider (7.1-1) as the approximation. This latter approach is advantageous for the purpose of this section, since n^2, rather than n, is the primary quantity in this discussion.

We try to solve the reduced wave equation with the trial solution

$$\psi = f(x)g(y)e^{-i\beta z} \tag{7.3-4}$$

Not only are the functions f and g undetermined at this point but we also do not know the propagation constant β. A solution of the form (7.3-4) is a mode of the square law medium. Substitution of (7.3-3) and (7.3-4) into (7.3-1) yields

$$\left[\frac{1}{f}\frac{\partial^2 f}{\partial x^2} + n_o^2 k_o^2 - \beta^2 - k_o^2 n_o n_1 x^2\right] + \left[\frac{1}{g}\frac{\partial^2 g}{\partial y^2} - k_o^2 n_o n_1 y^2\right] = 0 \tag{7.3-5}$$

This equation must hold for all values of x and y. This is possible only if the x-dependent and the y-dependent parts are each equal to a constant. The sum of these two constants must vanish. We thus obtain the two equations

$$\frac{d^2 f}{dx^2} + \left(n_o^2 k_o^2 - \kappa^2 - \beta^2 - k_o^2 n_o n_1 x^2\right) f = 0 \tag{7.3-6}$$

and

$$\frac{d^2 g}{dy^2} + \left(\kappa^2 - k_o^2 n_o n_1 y^2\right) g = 0 \tag{7.3-7}$$

In order to transform these two equations into a well-known standard form, we introduce the following new variables and parameters

$$\xi = \sqrt{k_o}\,(n_o n_1)^{1/4} x \tag{7.3-8}$$

$$\eta = \sqrt{k_o}\,(n_o n_1)^{1/4} y \tag{7.3-9}$$

$$\sigma = \frac{n_o^2 k_o^2 - \kappa^2 - \beta^2}{k_o \sqrt{n_o n_1}} \tag{7.3-10}$$

and

$$\rho = \frac{\kappa^2}{k_o \sqrt{n_o n_1}} \tag{7.3-11}$$

The differential equations assume the form

$$\frac{d^2 f}{d\xi^2} + (\sigma - \xi^2) f = 0 \tag{7.3-12}$$

and

$$\frac{d^2 g}{d\eta^2} + (\rho - \eta^2) g = 0 \tag{7.3-13}$$

The differential equations (7.3-12) and (7.3-13) are well known in the theory of the quantum mechanical harmonic oscillator.[22] There they determine the energy eigenfunction and eigenvalues that the harmonic oscillator can assume. In our theory as well as in the theory of the quantum mechanical harmonic oscillator, we must require boundary conditions. The modes of the square law medium must be guided near the axis of the structure. That means we must require that the solutions of the f and g functions must vanish for $x \rightarrow \infty$ and $y \rightarrow \infty$. The same requirements must be imposed on the wave function of the harmonic oscillator, since they are supposed to represent bound states. It is demonstrated in most books on quantum mechanics[22] that bound solutions of this type can exist only if the following conditions are met

$$\sigma = 2p + 1 \quad \text{with } p = 0, 1, 2, 3 \ldots \tag{7.3-14}$$

$$\rho = 2q + 1 \quad \text{with } q = 0, 1, 2, 3 \ldots \tag{7.3-15}$$

These conditions together with (7.3-10) and (7.3-11) determine the possible values of the propagation constant β. They are therefore the eigenvalue equations of the modes of the square law medium. We shall return to this important point. The solutions of the differential equations (7.3-12) and (7.3-13) are the well-known Hermite-Gaussian functions that we have already encountered repeatedly. To show this, we set

$$f = H_p(\xi)e^{-(1/2)\xi^2} \tag{7.3-16}$$

and

$$g = H_q(\eta)e^{-(1/2)\eta^2} \tag{7.3-17}$$

Substitution into (7.3-12) and (7.3-13) yields

$$\frac{d^2 H_p}{d\xi^2} - 2\xi \frac{dH_p}{d\xi} + 2pH_p(\xi) = 0 \tag{7.3-18}$$

and

$$\frac{d^2 H_q}{d\eta^2} - 2\eta \frac{dH_q}{d\eta} + 2qH_q(\eta) = 0 \tag{7.3-19}$$

These equations are identical with the differential equation (6.3-13) of the Hermite polynomials. We have thus demonstrated that the mode solutions of the square law medium are again Hermite-Gaussian functions. This result is hardly surprising, since we have seen that the square law medium is very closely related to the lens waveguide, whose modes are Hermite-Gaussian functions. The general mode solution of the square law medium is thus

$$\psi_{pq}(x, y, z) = \frac{\sqrt{2/\pi}}{\sqrt{2^{p+q} p! q!}\, w} H_p\left(\sqrt{2}\frac{x}{w}\right)$$
$$H_q\left(\sqrt{2}\frac{y}{w}\right)e^{-(x^2+y^2)/w^2}e^{-i\beta_{pq}z} \tag{7.3-20}$$

with the beam half width given by

$$w^2 = \frac{2}{k_o\sqrt{n_o n_1}} \tag{7.3-21}$$

The mode function (7.3-20) is already properly normalized.

The value of the propagation constant β_{pq} follows from (7.3-10), (7.3-11), (7.3-14), and (7.3-15)

$$\beta_{pq} = \left\{n_o^2 k_o^2 - k_o\sqrt{n_o n_1}\,[(2p + 1) + (2q + 1)]\right\}^{1/2} \tag{7.3-22}$$

For a comparison with the modes of the lens waveguide, we let $L \to 0$ and $f \to \infty$ in (6.4-47) and (6.4-48). Using (7.3-4), we obtain, indeed, the half width of the modes of the square law medium (7.3-21). For this comparison, we must keep in mind that the index of refraction of the square law medium on axis is n_o. This means that we must use $\lambda = 2\pi/(n_o k_o)$ in (6.4-47) and (6.4-48). A comparison of (7.3-20) with (6.3-20) shows that the modes of the square law medium are, indeed, identical to the modes of the lens waveguide with $L \to 0$. To make the comparison complete, we need only to show that the phase factor in (6.3-20) reduces to (7.3-22) in the limit of vanishing lens separation. We must of course replace k with $n_o k$ in (6.3-20), for the above-mentioned reason. We also set $m = p$ and $n = q$ and obtain from (6.3-20) in the limit $L \to 0$ ($z < L$ in [6.3-20]) using (7.3-21).

$$\beta_L = n_o k_o - \frac{1}{2}\sqrt{\frac{n_1}{n_o}} (2p + 1 + 2q + 1) \qquad (7.3\text{-}23)$$

According to (6.6-10), the phase front radius becomes infinite for $d_1 = L = 0$. The two propagation constants (7.3-22) and (7.3-23) apparently agree only to a first order approximation. We obtain (7.3-23) if we expand the square root of (7.3-22) and keep only the first two terms of the expansion. It is apparent from (7.3-23) that to this approximation and neglecting dispersion in the medium the modes of the square law medium have the remarkable property that their group velocity $\left(\dfrac{\partial \beta}{\partial \omega}\right)^{-1}$ is independent of the mode number.

We have thus obtained a very interesting result. The modes of the square law medium agree with the modes of the lens waveguide in the limit of vanishing lens spacings except for the phase factor. The phase factors agree only to first order of approximation. This result is revealing for the following reason. The modes of the square law medium (7.3-20) through (7.3-22) are exact solutions of the wave equation (7.3-1) with the index distribution (7.3-3). However, we were careful to point out that the Hermite-Gaussian mode fields (6.3-20) are only an approximate solution of the wave equation. The approximation entered the derivation when we neglected the second derivative of u in (6.3-3). It is thus not surprising that the solutions are not in exact agreement when we go to the limit of vanishing lens spacings. However, for most practical applications

$$Q = \frac{\sqrt{n_o n_1}}{n_o^2 k_o} \qquad (7.3\text{-}24)$$

is usually very small, so that (7.3-22) and (7.3-23) are in good agreement at least for small mode numbers. We did already point out in Section 6.3 that the Hermite-Gaussian fields are poor solutions of the wave equation for very high mode numbers. Incidentally, our present theory becomes questionable when Q of (7.3-24) approaches unity. We can use the scalar wave equation only when the refractive index varies very little over the distance of one wavelength. This requirement also leads to $Q \ll 1$. To see this, we form, in accordance with (1.3-28), using (7.3-3)

$$R = \frac{n_o^2 - n^2}{2\pi n_o^2} = \frac{1}{2\pi} \frac{n_o n_1 x^2}{n_o^2} \qquad (7.3\text{-}25)$$

We use the distance of one wavelength, $x = \lambda/n_o$, for the point at which the comparison of n^2 with n_o^2 is being made. For simplicity, $y = 0$ was assumed.

We thus have

$$R = \frac{n_o n_1}{2\pi_o n_o^4} \lambda^2 = 2\pi \frac{n_o n_1}{n_o^4 k_o^2} \tag{7.3-26}$$

Comparison with (7.3-24) leads to

$$R = 2\pi Q^2 \tag{7.3-27}$$

This little calculation shows that Q^2 is of the same order of magnitude as R. Since R must be much less than unity for the wave equation to be meaningful, we see that $Q \ll 1$ has to be required. The agreement between (7.3-22) and (7.3-23) is thus good for low order modes as long as the scalar wave approach is applicable. However, it is a curious fact that the Hermite-Gaussian mode solution (7.3-20) with the phase constant (7.3-22) is an exact solution of the scalar wave equation. The approximate nature of our theory lies hidden in the question of the applicability of the scalar wave equation.

7.4 OFF-AXIS BEAMS IN THE SQUARE LAW MEDIUM

We have proved quite generally in Section 3.6 that the center of intensity of the light field moves according to the laws of ray optics in a square law medium. This proof was based on the paraxial approximation. In this section, we prove by a direct calculation that a Gaussian beam can move off-axis in the square law medium. The peak of the field distribution moves like a light ray. The width of the field distribution varies periodically with a period length that is half the length of the ray oscillation period.

We assume that at $z = 0$ the following field distribution exists in the square law medium

$$F(x, y, o) = Ae^{-[(x-\xi)/W]^2 - (y/w)^2} \tag{7.4-1}$$

The y-dependent part of the field distribution conforms to the lowest order mode, but the x dependence shows that the field distribution is shifted by an amount ξ in x direction. (This ξ is not the same as in [7.3-8]). In addition, we have given the function an arbitrary half width W in x direction. The initial field distribution at $z = 0$ can be expanded in an infinite series with the help of the modes (7.3-20) of the square law medium

$$F(x, y, z) = \sum_{p=0}^{\infty} C_p \psi_{po}(x, y, z) \tag{7.4-2}$$

The expansion (7.4-2) gives the field at all points along the z axis once the expansion coefficients have been determined at $z = 0$. Using the orthogonality of the modes, we obtain

$$C_p = \int_{-\infty}^{\infty} dx \int_{-\infty}^{\infty} dy \, F(x, y, o)\psi^*_{po}(x, y, o) \qquad (7.4\text{-}3)$$

In case that n_o and n_1 of (7.3-3) are complex quantities, ψ^* indicates not the complex conjugate field but only a change of sign in the exponent $\exp[i(\omega t - \beta z)]$. (Compare the discussion below [10.2-13]). The determination of C_p is based on the orthogonality of the Hermite-Gaussian functions, which is a purely mathematical relation that holds even for complex parameters provided $\exp\left(-\dfrac{x^2}{w^2}\right) \to 0$ as $x \to \infty$. The integrals can be solved with the help of Gradshteyn and Ryzhik's integral tables,[61] p. 837, with the result

$$C_p = \frac{\sqrt{\pi}\,A}{2^{p/2}\sqrt{p!}}\frac{wW}{\sqrt{w^2 + W^2}}\left(\frac{w^2 - W^2}{w^2 + W^2}\right)^{p/2} H_p\left(\frac{\sqrt{2}\,w\xi}{\sqrt{w^4 - W^4}}\right)e^{-\xi^2/(w^2 + W^2)} \qquad (7.4\text{-}4)$$

The field distribution at any arbitrary point in the square law medium can now be expressed by the equation

$$F(x, y, z) = \frac{\sqrt{2}\,WA}{\sqrt{w^2 + W^2}}e^{-\xi^2/(w^2 + W^2)}e^{-(x^2 + y^2)/w^2}e^{-i(n_o k_o - \sqrt{n_1/n_o})z}$$
$$\times \sum_{p=o}^{\infty}\left\{\frac{1}{p!}\left[\frac{1}{2}\sqrt{\frac{w^2 - W^2}{w^2 + W^2}}\,e^{i\sqrt{n_1/n_o}\,z}\right]^p H_p\left(\sqrt{2}\,\frac{x}{w}\right)H_p\left(\frac{\sqrt{2}\,w\xi}{\sqrt{w^4 - W^4}}\right)\right\} \qquad (7.4\text{-}5)$$

The propagation constant β_{po} was used in the form (7.3-23). The infinite series can be evaluated with the help of one of the generating functions of the Hermite polynomials.[73] We thus obtain

$$F(x, y, z) = \frac{WA}{\sqrt{W^2 \cos \gamma z - iw^2 \sin \gamma z}}e^{-\xi^2/(w^2 + W^2)}e^{-(x^2 + y^2)/w^2}e^{-i(n_o k_o - \gamma/2)z}$$
$$\cdot \exp\left\{\frac{2x\xi - \left[\left(1 - \dfrac{W^2}{w^2}\right)x^2 + \dfrac{w^2}{w^2 + W^2}\xi^2\right]e^{i\gamma z}}{W^2 \cos \gamma z - iw^2 \sin \gamma z}\right\} \qquad (7.4\text{-}6)$$

with

$$\gamma = \sqrt{\frac{n_1}{n_o}} \qquad (7.4\text{-}7)$$

Equation (7.4-6) is an approximate solution of the reduced wave equation. Substitution of (7.4-6) satisfies (7.3-1) (to the paraxial approximation) even if n_o and n_1 are complex. We simplify the discussion of the form of the wave field by breaking it into two parts. We begin by assuming that the half width of the displaced initial field distribution is equal to that of the modes of the structure

$$W = w \tag{7.4-8}$$

With this specialization, we can express (7.4-6) as follows

$$F(x, y, z) = A \exp\left[-\frac{(x - \xi \cos \gamma z)^2 + y^2}{w^2}\right] e^{+i\phi(z)} e^{-i(n_o k_o - \gamma)z} \tag{7.4-9}$$

with

$$\phi(z) = \frac{1}{w^2}\left(2x\xi \sin \gamma z - \frac{1}{2}\xi^2 \sin 2\gamma z\right) \tag{7.4-10}$$

Equation (7.4-9) proves that an off-axis field distribution, which has the shape of a mode of the structure, travels in the square law medium essentially without distortion. The center of the field distribution follows the ray trajectory

$$x = \xi \cos \gamma z \tag{7.4-11}$$

in accordance with (7.2-8). The tilted phase fronts of the off-axis field are described by the phase function (7.4-10).

Next, we assume an on-axis field whose width does not conform to the modes of the structure. Setting

$$\xi = 0 \tag{7.4-12}$$

we obtain from (7.4-6)

$$F(x, y, z) = A\sqrt{\frac{W}{\overline{W}(z)}}\, e^{-y^2/w^2} e^{-(x/\overline{W}(z))^2} e^{i\psi(z)} e^{-i(n_o k_o - \gamma)} \tag{7.4-13}$$

The square of the beam half width in x direction is given by

$$\overline{W}^2(z) = \frac{1}{2W^2}[(W^4 + w^4) + (W^4 - w^4)\cos 2\gamma z] \tag{7.4-14}$$

and the phase factor is defined as

$$\psi(z) = \frac{1}{2}\left[\arctan\left[\left(\frac{w}{W}\right)^2 \tan \gamma z\right] - \gamma z + \frac{W^4 - w^4}{w^2 W^2 \overline{W}^2(z)} x^2 \sin 2\gamma z\right] \tag{7.4-15}$$

Equation (7.4-14) shows that the field distribution changes its width periodically with a period whose length is one-half the ray oscillation period.

The result (7.4-14) can also be obtained from the treatment of Gaussian beams in the lens waveguide. In Section 6.4, we studied the width of a Gaussian beam traveling in the lens waveguide. In order to be able to compare (6.4-36) with (7.4-14), we must adjust the beam so that it enters the lens waveguide with infinite phase front curvature. This is necessary because the beam in the square law medium was assumed to have a plane phase front at $z = 0$. Setting $w_1 = w_0 = W$, letting $L \to 0$, and $f \to \infty$, and using (7.2-4), (7.2-6), and (7.3-21) converts (6.4-36) to the form (7.4-14).

Our solution (7.4-6) is an approximation. The modes of the square law medium are exact solutions of the wave equation. (The wave equation itself is only an approximate description of the electromagnetic field.) But in order to be able to sum the series in (7.4-5), we had to use the approximation (7.3-23) of the propagation constant. These approximations are equivalent to the paraxial approximation of ray optics, since we found that the solution (7.4-9) describes a field distribution whose center of intensity moves like a paraxial ray.

We have found that, to the paraxial approximation, a Gaussian beam moves through a square law medium without field distortion. The beam describes an oscillatory trajectory undulating around the optical axis. But the beam remains well collimated. Its width may change periodically if it is not identical to the width of the modes of the structure, but no field breakup occurs. E. A. J. Marcatili[65] has shown that this stability of Gaussian beams can exist only in media whose index is given exactly by (7.3-3). Any higher order terms in an expansion of the index in powers of x and y cause the beam to break up in a manner similar to that discussed in Section 5.8. The breakup of the beam into several distinct maxima may require a considerable length of waveguide. Its occurrence, however, is not dependent on the degree of aberrations of the medium. That beam breakup should occur in dielectric media whose dielectric constants do not obey the strict parabolic law is not surprising in view of our study of beam breakup in lens waveguide with imperfect lenses. The similarity between continuous dielectric media and lens waveguides is not limited only to square law media. The fact that Gaussian beams do not maintain their shapes in nonideal media means in practice that it is impossible to send Gaussian beams undistorted over arbitrarily long distances. No medium can ever be made to be absolutely perfect. Our discussion of propagation in perfect lens waveguides and perfect square law media is therefore of necessity an idealization that, strictly speaking, cannot occur in practical waveguides. However, knowledge of the ideal case is still very useful. Good transmission media approximate the ideal case in some respect, so that the behavior of Gaussian beams can be expected to follow our prediction at least over some finite distance. For extremely long distances, the beam can be expected to behave quite differently from the ideal case. What happens in a realistic situation was shown in Section 5.8.

7.5 SQUARE LAW MEDIA WITH LOSS OR GAIN

In the previous sections, we investigated the modes of a square law medium that exhibited neither loss nor gain. This treatment is clearly an idealization. Aside from the fact that it is unlikely that any guiding medium will ever have a precise square law dependence of its dielectric constant, it is even less likely that the dielectric material is completely lossless. Most dielectrics have relatively high losses in the optical region of the spectrum. Losses in the order of 100 to 1000 db/km are not at all unusual. We have seen in Section 2.6, Equations (2.6-18) and (2.6-19), that lossy dielectric media can be described by complex dielectric constants.

However, it is important to consider dielectric media not only with losses but also with gain. The active material in lasers can be described as dielectric media exhibiting gain rather than loss. Gain can also be described by a complex dielectric constant. Whether a medium exhibits gain or loss is determined only by the sign of the imaginary part of its refractive index.

In the present section, we investigate the modes of square law media with complex dielectric constant. We have seen in our discussions of the square law medium with real dielectric constant that mode guidance was produced by the fact that the real refractive index had its highest value on-axis and decreasing values at increasing distances from the optical axis. This guidance mechanism is so well known that people are surprised when they learn that it is possible to obtain guided modes in media with constant real part of the refractive index. Even in the case that the real part of the refractive index has its lowest value on-axis and increasing values at increasing distance from the axis, is it possible to obtain stable guided modes. All that is required is that the loss of the medium has its lowest value on-axis and increasing loss values at increasing distance from the axis. In the case of a medium with gain, stable modes are obtained if the gain has its highest value on-axis. The dependence of the real part of the index is arbitrary to quite an extent. In particular, it is not at all necessary for the real part of the refractive index to decrease with increasing distance from the optical axis. It thus appears that, for the existence of stable guided modes, the imaginary part of the refractive index is more important than its real part.[74]

We have mentioned repeatedly the existence of stable guided modes without explaining what is meant by this term. We call a mode stable if it continues to travel along the axis of the structure even if it is displaced from it by a small amount. An unstable mode, on the other hand, may exist as a mathematical solution of the mode problem, but it will not travel along the optical axis indefinitely if it is displaced from it even by an infinitesimal amount. Unstable modes occur whenever the losses of the medium decrease with increasing distance from the optical axis or, in case of a gain medium, if the gain is lowest on-axis. The real part of the refractive index plays no part in this question of mode stability. However, it is important to mention in this connection that the question of mode stability can be somewhat academic in certain cases. Imagine, for example, a dielectric medium whose real part of the refractive index is well behaved as far as mode guidance is concerned. The real part of the refractive index of our medium decreases with increasing distance from the axis, while its loss decreases with increasing distance from the optical axis. By our criterion of mode stability, such a mode is unstable. However, it is clear that this instability cannot have a drastic effect if the change in loss across the guiding medium is only very slight. Even though, in principle, modes of this structure may be unstable if they are allowed to travel extremely long distances along the structure, this instability may not even be observable if the power of the field has decayed to insignificant values before the tendency of the mode to depart from the axis has become apparent. The question of stability or instability of the modes can be one of principle rather than of practical importance. However, on the other hand, in media with gain rather than loss, the mode instability is bound to have far more drastic consequences, since the field grows rather than decays, so that the mode instability must become apparent if the mode is allowed to travel a sufficiently long distance along the structure.

A mode in a medium with inverted real part of the refractive index ("inverted," in this context, means that the refractive index increases away from the axis) behaves quite differently from a mode that is guided by a normal index distribution. The mode guidance in a normal medium of the type discussed in the previous sections is caused by total internal reflection. In terms of ray optics, we can think of the mechanism of mode guidance as being caused by the deflection of the ray toward the axis of the structure. No power outflow away from the axis takes place. In an inverted medium, mode guidance is possible with the help of the imaginary part of the refractive index. This guidance, however, is of quite a different type. Imagine a medium that exhibits some gain on-axis. The gain decreases with increasing distance from the axis, and turns to loss far enough from the axis of the structure. A structure of this type possesses true modes in the mathematical sense. The mode problem has stable mathematical solutions. However, the power of the mode is not contained near the axis of the guiding medium. Power outflow radially away from the axis does actually occur. The fact that a stable mode is possible is caused not by total reflection of the field energy back toward the guide axis but rather by creation of field energy in the gain medium at a rate high enough to maintain the mode distribution with higher field energy on-axis and decaying values of the field farther from the axis. The same argument still holds if the medium exhibits loss rather than gain. As long as the loss increases away from the axis, there are stable mode solutions. The field shape is now maintained by the fact that more power is dissipated away from the axis than right on-axis resulting in a field shape that peaks on-axis. Some reflection of power toward the axis does, however, take place even in a medium with constant real part of the refractive index. A change in the imaginary part of the index also tends to reflect power at a dielectric interface.

After these introductory remarks, we turn to a discussion of the mode problem in a dielectric medium with complex index of refraction. Fortunately, we do not need to solve the wave equation all over again. The solution obtained in Section 7.3 is valid for all possible values of n_o and n_1, and holds also when these constants assume complex values. We need only to examine the implications of complex parameters for our solutions. Complex values for

n_o and n_1 cause ξ and η of (7.3-8) and (7.3-9) to become complex. The Hermite polynomials (7.3-16) and (7.3-17) are analytic functions that remain well defined by their polynomial expressions (5.6-14) even for complex values of their argument.

We assume that n_o has the complex value

$$n_o = n_{or} - i n_{oi} \tag{7.5-1}$$

Positive values of n_{oi} mean that the medium described by n_o exhibits loss, while negative values of n_{oi} mean that the medium has gain. In all reasonable physical media, we expect to find that

$$|n_{oi}| \ll n_{or} \tag{7.5-2}$$

so that we can write approximately

$$n_o^2 = n_{ro}^2 - 2 i n_{or} n_{oi} \tag{7.5-3}$$

Similarly, we introduce the complex value

$$\sqrt{n_o n_1} = a + ib \tag{7.5-4}$$

The expression for w^{-2} (7.3-21) now assumes a complex value. Its meaning as the square of the inverse beam half width is still applicable to its real part. Keeping in mind that w is complex, we can write the mode solution (7.3-20) of the square law medium with complex dielectric constant as follows

$$\Psi(x, y, z) = \frac{\sqrt{2}}{\sqrt{\pi 2^{p+q} p! q!}\, w} H_p\left(\sqrt{2}\, \frac{x}{w}\right)$$

$$\times H_q\left(\sqrt{2}\, \frac{y}{w}\right) e^{-(k_o/2)ar^2} e^{-i(k_o/2)br^2} e^{-i\beta_{pq}z} \tag{7.5-5}$$

We require of a guided mode that its field distribution decays with increasing values of

$$r = (x^2 + y^2)^{1/2} \tag{7.5-6}$$

This definition of a guided mode shows us that the only requirement that we must make of the square law medium is the condition

$$a > o \tag{7.5-7}$$

If (7.5-7) is satisfied, we obtain a guided mode regardless of all other values of the dielectric constant.

We see immediately from our mode solution (7.5-5) that the phase fronts in a medium with complex refractive index are no longer plane. Their shape is rather complicated, since they are determined not only by the exponents of the exponential functions but also by the complex values of the Hermite polynomials. For the lowest order mode, $p = 0$ and $q = 0$, we see immediately that the phase front for positive values of b is concave if we view it looking in the direction of the positive z-axis. This means that power is flowing radially away from the axis for positive values of b. In order to see what this means in terms of the refractive index, we consider Equation (7.1-1). Using (7.5-1) and (7.5-4), we can write for its real part

$$Re\; n = n_{or} - \frac{1}{2n_{or}}(a^2 - b^2)r^2 \qquad (7.5\text{-}8)$$

and for its imaginary part

$$Im\; n = -n_{oi} - \frac{1}{n_{or}}abr^2 \qquad (7.5\text{-}9)$$

Products of n_{oi} with a^2, b^2, and ab were neglected on the assumption that the neglected terms are much smaller than the terms appearing in (7.5-8) and (7.5-9).

Since a must be positive, we see that a positive value of b implies that the medium increases in loss for increasing values of r. Let us assume that n_{oi} is negative. The medium then exhibits gain on-axis at $r = 0$. For a positive value of b, the gain decreases away from the axis, and it is understandable that power flows away from the axis, where power is being created by the gain mechanism, toward the outside, where it is being absorbed.

The behavior of the imaginary part of the refractive index (in other words, whether it increases or decreases with increasing values of r) depends entirely on the sign of b, because a must be positive in order for a guided mode to exist. The behavior of the real part of the refractive index depends on whether a^2 is larger or smaller than b^2. The medium has a normal guidance behavior—$Re\; n$ decreases with increasing r— if $a^2 > b^2$; and it has an anomalous guidance behavior—$Re\; n$ increases with increasing r—if $a^2 < b^2$. It is apparent that a guided mode exists in all cases, that is, for normal as well as anomalous behavior of the real part of the refractive index as well as for media whose imaginary part either increases or decreases with increasing values of r. This surprising result does not tell us, however, whether the modes are stable or unstable. It says only that mathematical mode solutions do exist in all four cases.

In order to check the question of mode stability, we study the behavior of the lowest order mode ($p = q = 0$) if it is displaced from the optical axis. This calculation has also already been performed in Section 7.4. We need only to consider the implication of Equation (7.4-9) for complex values of n_o and n_1.

From (7.3-21), and (7.5-4), we obtain

$$\frac{1}{w^2} = \frac{k_o}{2}(a + ib) \qquad (7.5\text{-}10)$$

and from (7.4-7) and (7.5-4), we obtain approximately

$$\gamma = \frac{a + ib}{n_{or}} \tag{7.5-11}$$

Products of the small quantity n_{oi} with the small quantities a and b have again been neglected. The sine and cosine functions assume the following complex values

$$\cos \gamma z = \cos \frac{a}{n_{or}} z \cosh \frac{b}{n_{or}} z - i \sin \frac{a}{n_{or}} z \sinh \frac{b}{n_{or}} z \tag{7.5-12}$$

and

$$\sin \gamma z = \sin \frac{a}{n_{or}} z \cosh \frac{b}{n_{or}} z + i \cos \frac{a}{n_{or}} z \sinh \frac{b}{n_{or}} z \tag{7.5-13}$$

We form the absolute value of $F(x, y, z)$ of (7.4-9) and obtain

$$|F(x, y, z)| = A e^{-(ak_o/2)(x^2 + y^2)} e^{-(n_{oi}k_o + (b/n_{or}))z} \exp\left[x\xi k_o e^{-(b/n_{or})z} \left(a \cos \frac{a}{n_{or}} z \right. \right.$$
$$\left. - b \sin \frac{a}{n_{or}} z \right) \right] \cdot \exp\left\{ -\frac{k_o}{4} \xi^2 \left[a \right. \right.$$
$$\left. \left. + e^{-2(b/n_{or})z} \left(a \cos 2 \frac{a}{n_{or}} z - b \sin 2 \frac{a}{n_{or}} z \right) \right] \right\} \tag{7.5-14}$$

The stability or instability of the displaced mode is determined by the factor $\exp(-bz/n_{or}^2)$ that appears in the exponents of the exponential functions. For positive values of b, the exponential factor decays with increasing values of z, so that the field distribution approaches the stable value

$$|F(x, y, z)| = A e^{-(ak/4)\xi^2} e^{-(ak_o/2)(x^2 + y^2)} e^{-(n_{oi}k_o + (b/n_{or}))z} \tag{7.5-15}$$

Equation (7.5-15) shows that for positive values of b the off-axis field distribution moves back on-axis and grows or decays depending on whether $-(n_{oi}k_o + b/n_{or})$ is positive or negative. Positive values of b and therefore stable mode guidance are obtained in media whose gain decreases with increasing values of r or whose loss increases with increasing values of r.

In the opposite case, b is negative, and the exponential factor $\exp(-bz/n_{or}^2)$ grows with increasing values of z. The field distribution becomes very complicated in this unstable case. However, it is obvious that the field distribution spreads further and further away from the axis. This is apparent from the third exponential term in (7.5-14), which tends to counteract the first term that, with its dependence on $-r^2$, causes the mode to remain concentrated closely to the axis of the structure. Even though a guided mode is apparently possible if ξ vanishes completely, it is only necessary to displace the field distribution by an infinitesimal amount to assure its complete disintegration. As mentioned in the introduction to this section, this disintegration may take place relatively slowly depending on the values of the constants and in particular on b. An almost stable mode may exist that is still useful for practical mode guidance purposes. In principle, however, modes with negative values of b are unstable.

This behavior of stable or unstable mode guidance is not limited to the square law medium with complex index of refraction. It can be shown that the same principle behavior exists for a layered medium of the type to be discussed in the next chapter. I have conducted a numerical analysis of the complex eigenvalue equation of the layered dielectric slab waveguide, and found that, here too, formal mode solutions exist in all four cases of index distributions. The existence of stable and unstable guided modes of dielectric waveguides with complex refractive index is thus not an isolated phenomenon that is limited to the square law medium. It has been demonstrated for the stratified dielectric waveguide, and probably occurs for many other types of dielectric waveguides.

Returning to the modes of the square law medium, we consider the loss or gain of the modes in case of complex refractive index. The propagation constant (7.3-23) can be decomposed into its real and imaginary parts

$$Re\,\beta_{pq} = n_{or}k_o - \frac{a}{n_{or}}(p + q + 1) \tag{7.5-16}$$

and

$$Im\,\beta_{pq} = -n_{oi}k_o - \frac{b}{n_{or}}(p + q + 1) \tag{7.5-17}$$

Products of n_{oi} with a and b were again neglected. The loss or gain is determined by the imaginary part of β_{pq}. The mode experiences gain if

$$Im\,\beta_{pq} > 0 \tag{7.5-18}$$

and it suffers loss if

$$Im\,\beta_{pq} < 0 \tag{7.5-19}$$

The question of whether a given mode experiences loss or gain depends on three factors: the sign of n_{oi}, the sign of b, and the mode number $p + q + 1$.

Let us begin by considering stable modes, $b > 0$. We distinguish two cases beginning with $n_{oi} > 0$. In this case, the medium is lossy on-axis with increasing loss for increasing values of r. $Im\,\beta_{pq}$ is negative with increasing absolute values for increasing mode numbers. This means that the stable modes in a lossy square law medium suffer increasingly more loss with increasing mode order. This behavior is typical for normal guiding media. It is responsible for the stability of the modes. If an arbitrary off-axis field distribution is expanded in terms of normal modes of the structure, the higher terms of the expansion decay more rapidly with increasing values of z, so that the lowest order mode remains if we proceed far enough along the guide. Any arbitrary field distribution has the tendency to change into the lowest order mode that appears in it series expansion.

Next, we consider the case $b > 0$ but $n_{oi} < 0$. A medium of this kind has gain on-axis, but the gain decreases with increasing values of r turning to loss if r becomes sufficiently large. Whether the mode experiences gain or loss depends on the relative values of $n_{o}k_{o}$ and $b(p + q + 1)$. The mode experiences gain if

$$|n_{oi}k_{o}| > \frac{b}{n_{or}} (p + q + 1) \tag{7.5-20}$$

and it suffers loss if

$$|n_{oi}k_{o}| < \frac{b}{n_{or}} (p + q + 1) \tag{7.5-21}$$

It may be that the on-axis gain is insufficient for any of the modes to experience gain. If the on-axis gain is large enough, the lower order modes experience gain. However, with increasing mode order, the gain decreases, and turns to loss for sufficiently large values of $p + q + 1$. It is again apparent that any arbitrary field distribution ends up in the shape of the lowest order mode appearing in its series expansion.

We now consider the unstable case $b < 0$. First, we assume that the medium has loss on-axis, $n_{oi} > 0$. If b is sufficiently small, the lower order modes will suffer loss whose values decrease with increasing mode number. For sufficiently high mode numbers, the modes experience gain. If $|b|$ is large enough so that all modes experience gain, we find again that the amount of gain increases with increasing mode numbers. This anomalous situation explains why the field distributions in a guide with $b < 0$ are unstable. Consider again a series expansion of an arbitrary input field. As the field progresses along the guide, the higher order terms in its series expansion grow more and more and become increasingly dominant in determining the value of the field. The entire convergence behavior of the series is altered, so that it is not surprising to find complete field disintegration of a mode that initially was displaced only by an infinitesimal amount from the axis of the structure.

It may be well to point out that a square law medium whose gain increases indefinitely with increasing values of r is unphysical. For any physically realizable medium, there may be an initial increase in gain for increasing values of r. However, this trend cannot persist indefinitely. For this reason, we must expect that the gain of the higher order modes cannot grow indefinitely with increasing mode number. Eventually, modes of high order must show decreasing gain and even loss. The series expansions of arbitrary fields do not, in actual cases, diverge. However, even though the field expansions do converge in physically realizable media, considerable field distortion can still occur if the gain is increasing with increasing mode numbers at least for the modes of lower order. Our discussion of stable and unstable guiding media tells us the trends that are to be expected in practical cases.

7.6 LENS PROPERTIES OF SQUARE LAW MEDIA

We conclude this chapter on the square law medium with a discussion of the lens properties of a lossless square law medium.[76] To simplify the discussion, we shall assume that a light ray can enter and leave the square law medium without suffering additional deflection. This simplifying assumption can be justified by the following argument. The index of refraction changes only very slightly in practical square law media. If we immerse the square law medium in a homogeneous medium whose index matches that of the square law medium on-axis, the index match can be expected to be quite good at all points of the interface between the two media. Refraction on passing the interface between the homogeneous medium and the square law medium is thus sufficiently weak to be negligible. If we do not employ an index-matching technique there will be refraction of light rays entering and leaving the square law medium. These refraction effects can approximately be considered to be similar to refraction at the interface between two homogeneous media. It is well known that rays entering or leaving a homogeneous medium from a point source propagate in the second medium as if they came from a point source that appears to be either closer or farther than its actual distance. For simplicity of discussion, we ignore these trivial refraction effects.

It is easy to see that square law media must have image-forming properties. Consider the series expansion (7.4-2) of an arbitrary field at $z = 0$ in terms of the modes (7.3-20). As the field progresses in the medium, its shape is changing because of the change in the relative phases of the different terms of the expansion. The phases of the terms of the series expansions are given by (7.3-23)

$$\beta_{pq}z = n_o k_o z - (p + q + 1)\gamma z \tag{7.6-1}$$

with γ of (7.4-7). For arbitrary values of z, the phases of each mode are different, so that the superposition of the modes in the series expansion leads to ever-changing field shapes. However, there are points along the axis of the structure where the relation

$$\gamma z = 2N\pi \text{ with } N = 1, 2, 3 \ldots \tag{7.6-2}$$

holds. At these points, all phases are integral multiples of 2π with the exception of a common phase factor $n_o k_o z$. At the points given by (7.6-2), all terms of the series expansion add again in exactly the same way as they did at the initial plane $z = 0$. The original field distribution thus restores itself periodically at the planes.

$$z = \frac{2N\pi}{\gamma} \tag{7-6-3}$$

At all planes whose z value is given by (7.6-3), we find an exact replica of the original field distribution. The square law medium is thus capable of forming images of objects placed at its input. In the cases just discussed, the two-dimensional image is undistorted and unmagnified. We see from (7.4-11) that the image planes appear at multiples of the ray oscillation period.

A section of square law medium of finite length acts as a lens. To show this, consider the geometry of Figure 7.6.1. A light ray entering the square law medium at $z = 0$ with a horizontal tangent, $r'(0) = 0$ and at a distance r_o from the axis leaves the medium at $z = L$ at a distance (see [7.2-8] or [7.4-11])

$$r = r_o \cos \gamma L \qquad (7.6\text{-}4)$$

The slope of the ray at $z = L$ is

$$r' = -\gamma r_o \sin \gamma L \qquad (7.6\text{-}5)$$

Neglecting additional refraction at the interface, we find that the ray crosses the optical axis at a distance f from the back portion of the square law medium that is given by

$$f = -\frac{r}{r'} = \frac{\cot \gamma L}{\gamma} \qquad (7.6\text{-}6)$$

It is remarkable that the distance f is independent of the distance r_o, at which the ray entered the square law medium. All parallel input rays cross the optical axis at the same point. The square law medium acts as a true lens. The focal distance f can be positive or negative. Positive values of f indicate that the medium acts as a positive lens capable of forming real images. Negative values of f mean that the medium acts as a negative or dispersive lens capable of forming virtual images. Whether the square law medium acts as a positive or negative lens depends on its length and on its focusing power described by γ of (7.4-7). The positive lens shown in Figure 7.6.1 could be reproduced with the help of two ordinary thin lenses.

There are very thin optical fibers on the market that have a radial index

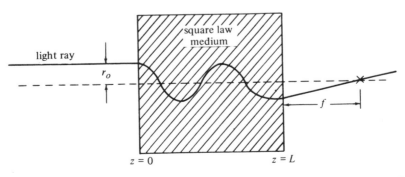

Figure 7.6.1   The square law medium of finite length acts like a lens. All horizontal input rays pass through the focal point.

gradient approximating a square law medium. These fibers act as lenses.[23,112] It is possible to see images of real objects placed in front of the fiber. Such fibers could be very convenient to look around corners into unaccessible places. They may also become important in medical applications for looking into body cavities. They are of course also useful for transmitting light over some distances. At present, they are still too lossy to be useful as light waveguides for long-distance optical communications. There is hope, however, that their losses may be decreased to the point where long-distance light transmission through such square law media becomes feasible. The problem of beam deviation caused by random displacements of the optical axis of such fibers requires extreme care in positioning these optical waveguides.

8

OPTICAL FIBERS AND DIELECTRIC WAVEGUIDES

8.1 INTRODUCTION

The term "optical fiber" describes a certain type of dielectric waveguide for guiding light waves.[79] These waveguides are called fibers because of their filamentary appearance. Two classes of optical fibers are usually distinguished. We have already met the optical fiber whose index of refraction obeys the square law in radial direction. The more common type of optical fiber functions also on the principle of radial variation of the refractive index. The index variation is not continuous, however, but occurs abruptly, defining regions of space with different but homogeneous refractive indices. The present chapter is devoted to a discussion of the properties of optical fibers of this latter type. Whenever the term optical fiber is used in this and the following chapters, a fiber whose cross section is shown in Figure 8.1.1 is being considered.

The type of optical waveguide that is depicted in the figure is called a cladded fiber. A dielectric cylinder of index n_1 is surrounded by a concentric dielectric cylinder of refractive index n_2. The two refractive indices obey the relation

$$n_1 > n_2 \qquad (8.1\text{-}1)$$

The second dielectric region, with index n_2, is not really essential for the principle of wave guidance in this type of dielectric waveguide. The optical fiber still functions even when $n_2 = 1$. There are two reasons why it is desirable to use a cladded optical fiber rather than a bare dielectric cylinder. The fields of a dielectric waveguide are not fully contained inside the dielectric region

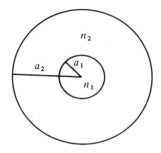

Figure 8.1.1 Cross section through a cladded optical fiber.

with index n_1, but they reach out into the outside region, where they decay exponentially. Since the fiber must somehow be supported in space, it is advantageous to surround the inner core region with an outer cladding to avoid scattering and field distortion by the supporting mechanism coming in touch with the guided field. Since the field decays exponentially inside region 2, with index n_2, practically no field exists outside the cladding.

The second reason for using cladded rather than naked fibers has to do with the mechanism of field guidance in the fiber. At a given frequency, an optical fiber is capable of supporting a finite number of modes.[80,81] If the diameter a_1 of the core is much larger than the wavelength λ of the guided radiation, a very large number of guided modes is possible. For purposes of light transmission, it is often desirable to limit the number of guided modes and keep it as small as possible. Single guided mode operation is possible by properly dimensioning the guide. The dimensions of the inner core that permit single mode operation depend critically on the ratio of n_1/n_2. The larger this ratio, the smaller a_1 must be to ensure that only one guided mode can propagate. It is often inconvenient to produce optical fibers with a core diameter of 0.5μ. The inner core can be much larger and still allow single mode operation when the index ratio n_1/n_2 becomes very nearly unity. It is then possible to allow $2a_1$ to be several microns wide and still obtain a single mode waveguide.

Optical fibers have some features in common with hollow metallic waveguides. Both can support a finite number of guided modes at any given frequency. Both structures suffer from mode conversion problems if the guide departs in any way from perfect geometry.[82] However, whereas metallic waveguides can support only guided modes, and mode conversion is limited to interchange of power between the finite number of guided modes, the mode spectrum of dielectric waveguides and optical fibers, in addition to having a finite number of guided modes, also possesses a continuum of unguided radiation modes. These radiation modes are also legitimate solutions of Maxwell's equations that satisfy the boundary conditions imposed on the fields by the presence of the dielectric interfaces. However, whereas the guided modes can have only discrete values of their propagation constants, there are infinitely many radiation modes with propagation constants that form a continuous spectrum. Deviations from perfect geometry not only convert power among the guided modes of the dielectric waveguide but also scatter power into the continuous spectrum of radiation modes. This scattering of power into the continuous spectrum appears as radiation on the outside of the dielectric optical waveguide.

We shall discuss the modes of the dielectric optical waveguides on the basis of Maxwell's equations. Polarization effects are important for these structures, and can no longer be ignored. Consequently, it is no longer admissible to consider the problems of mode guidance in these layered dielectric waveguides on the basis of the scalar wave equation. In addition to discussing the properties of the perfect structures, we shall concern ourselves in Chapter 9 with mode conversion and radiation effects caused by imperfections of the waveguide boundaries.

The mathematical treatment of radiation losses in optical fibers of cylindrical geometry is quite complicated. There are related structures whose geometry and mathematical treatment are far simpler than those of the cylindrical optical fibers. Such a structure is the slab waveguide. The geometry of the slab waveguide is shown in Figure 8.1.2. The waves are supposed to travel in z direction. The slab is infinitely extended in z as well as in y direction (perpendicular to the plane of the figure). The principle of mode guidance and the presence of guided as well as radiation modes are the same in the slab waveguide and in the optical fibers. Most radiation and mode conversion phenomena can be studied more simply in the slab waveguide than in the optical fiber. The results obtained from treatments using the slab waveguide are, however, directly applicable to optical fibers. The amount of radiation losses caused by surface ripples is very nearly the same for the slab waveguide and for the cylindrical fiber. For this reason, we shall discuss mode guidance in the slab waveguide to quite an extent in order to learn about the properties of dielectric optical fibers. This approach saves the reader the trouble of having to wade through a swamp of complicated mathematical derivations, but allows us to demonstrate

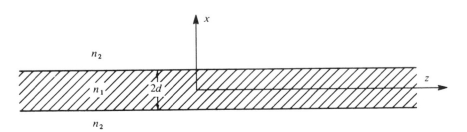

Figure 8.1.2   Cross sectional view of a slab waveguide.

all the relevant features of fiber optics and provides us with numerical results that are directly applicable to the more complicated cylindrical fibers.

8.2 GUIDED MODES OF ROUND OPTICAL FIBERS

In this section, we derive the expressions for the guided modes of a clad optical fiber. The cladding works properly only if its radius is so large that practically no field intensity exists at the interface between the cladding and the surrounding air. It is thus sufficient to assume that the radius of the cladding is infinite, $a_2 \to \infty$. The difference between the modes in an actual fiber and one with infinite cladding should be negligible for any well-designed optical fiber.

The field equations for optical systems of cylindrical geometry were derived in Section 1.4. For the purpose of the present section, we need these equations in cylindrical coordinates r, ϕ, z. The transformation from Cartesian to cylindrical coordinates is accomplished by a coordinate transformation. The z coordinate in the direction of the optical axis of the system is common to both the Cartesian and the cylindrical polar coordinates, and need not be altered by the transformation from one system to the other. The relation between the two coordinate systems is shown in Figure 8.2.1. It is apparent from this figure that the following relations hold between the two coordinate systems

$$x = r \cos \phi \tag{8.2-1}$$

$$y = r \sin \phi \tag{8.2-2}$$

The transformation of the vector components F_x, F_y to the vector components F_r, F_ϕ is shown in Figure 8.2.2.

$$F_r = F_x \cos \phi + F_y \sin \phi \tag{8.2-3}$$

$$F_\phi = -F_x \sin \phi + F_y \cos \phi \tag{8.2-4}$$

The derivatives of the field components with respect to r and ϕ are obtained as follows

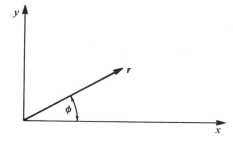

Figure 8.2.1 Cartesian and polar coordinate system.

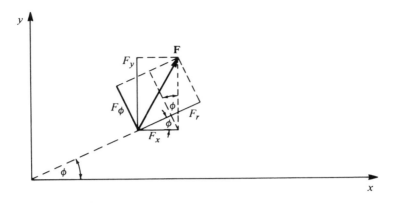

Figure 8.2.2   The vector **F** is decomposed into components parallel to the Cartesian and polar coordinate directions.

$$\frac{\partial f}{\partial r} = \frac{\partial f}{\partial x}\frac{\partial x}{\partial r} + \frac{\partial f}{\partial y}\frac{\partial y}{\partial r} = \frac{\partial f}{\partial x}\cos\phi + \frac{\partial f}{\partial y}\sin\phi \qquad (8.2\text{-}5)$$

$$\frac{\partial f}{\partial \phi} = \frac{\partial f}{\partial x}\frac{\partial x}{\partial \phi} + \frac{\partial f}{\partial y}\frac{\partial y}{\partial \phi} = -r\frac{\partial f}{\partial x}\sin\phi + r\frac{\partial f}{\partial y}\cos\phi \qquad (8.2\text{-}6)$$

The function f stands for either E_z or H_z. With the aid of the transformations (8.2-3) through (8.2-6), we can transform the expressions for the x and y coordinates of **E** and **H**, Equations (1.4-16) through (1.4-19), to cylindrical polar coordinates

$$E_r = -\frac{i}{\kappa^2}\left(\beta\frac{\partial E_z}{\partial r} + \omega\mu\frac{1}{r}\frac{\partial H_z}{\partial \phi}\right) \qquad (8.2\text{-}7)$$

$$E_\phi = -\frac{i}{\kappa^2}\left(\beta\frac{1}{r}\frac{\partial E_z}{\partial \phi} - \omega\mu\frac{\partial H_z}{\partial r}\right) \qquad (8.2\text{-}8)$$

$$H_r = -\frac{i}{\kappa^2}\left(\beta\frac{\partial H_z}{\partial r} - \omega\varepsilon\frac{1}{r}\frac{\partial E_z}{\partial \phi}\right) \qquad (8.2\text{-}9)$$

$$H_\phi = -\frac{i}{\kappa^2}\left(\beta\frac{1}{r}\frac{\partial H_z}{\partial \phi} + \omega\varepsilon\frac{\partial E_z}{\partial r}\right) \qquad (8.2\text{-}10)$$

with

$$\kappa^2 = k^2 - \beta^2 \qquad (8.2\text{-}11)$$

and

$$k^2 = \omega^2\,\varepsilon\mu \qquad (8.2\text{-}12)$$

The parameter β is the propagation constant in z direction. Finally, we must transform the wave equations (1.4-22) and (1.4-23) to cylindrical polar coordinates. We invert the transformation (8.2-1) and (8.2-2)

$$r = \sqrt{x^2 + y^2} \qquad (8.2\text{-}13)$$

$$\phi = \arctan \frac{y}{x} \qquad (8.2\text{-}14)$$

and, with their help, form the first derivatives of E_z

$$\frac{\partial E_z}{\partial x} = \frac{\partial E_z}{\partial r}\frac{\partial r}{\partial x} + \frac{\partial E_z}{\partial \phi}\frac{\partial \phi}{\partial x} = \frac{x}{r}\frac{\partial E_z}{\partial r} - \frac{y}{r^2}\frac{\partial E_z}{\partial \phi} \qquad (8.2\text{-}15)$$

$$\frac{\partial E_z}{\partial y} = \frac{\partial E_z}{\partial r}\frac{\partial r}{\partial y} + \frac{\partial E_z}{\partial \phi}\frac{\partial \phi}{\partial y} = \frac{y}{r}\frac{\partial E_z}{\partial r} + \frac{x}{r^2}\frac{\partial E_z}{\partial \phi} \qquad (8.2\text{-}16)$$

and also the second derivatives

$$\frac{\partial^2 E_z}{\partial x^2} = \left(\frac{1}{r} - \frac{x^2}{r^3}\right)\frac{\partial E_z}{\partial r} + \frac{x}{r}\left(\frac{x}{r}\frac{\partial^2 E_z}{\partial r^2} - \frac{y}{r^2}\frac{\partial^2 E_z}{\partial r \partial \phi}\right) \qquad (8.2\text{-}17)$$

$$+ \frac{2xy}{r^4}\frac{\partial E_z}{\partial \phi} - \frac{y}{r^2}\left(\frac{x}{r}\frac{\partial^2 E_z}{\partial r \partial \phi} - \frac{y}{r^2}\frac{\partial^2 E_z}{\partial \phi^2}\right)$$

$$\frac{\partial^2 E_z}{\partial y^2} = \left(\frac{1}{r} - \frac{y^2}{r^3}\right)\frac{\partial E_z}{\partial r} + \frac{y}{r}\left(\frac{y}{r}\frac{\partial^2 E_z}{\partial r^2} + \frac{x}{r^2}\frac{\partial^2 E_z}{\partial r \partial \phi}\right)$$

$$- \frac{2xy}{r^4}\frac{\partial E_z}{\partial \phi} + \frac{x}{r^2}\left(\frac{y}{r}\frac{\partial^2 E_z}{\partial \phi \partial r} + \frac{x}{r^2}\frac{\partial^2 E_z}{\partial \phi^2}\right) \qquad (8.2\text{-}18)$$

Substitution of these equations into (1.4-22) leads to the wave equation in cylindrical polar coordinates

$$\frac{\partial^2 E_z}{\partial r^2} + \frac{1}{r}\frac{\partial E_z}{\partial r} + \frac{1}{r^2}\frac{\partial^2 E_z}{\partial \phi^2} + \kappa^2 E_z = 0 \qquad (8.2\text{-}19)$$

A similar equation for H_z follows from (1.4-23)

$$\frac{\partial^2 H_z}{\partial r^2} + \frac{1}{r}\frac{\partial H_z}{\partial r} + \frac{1}{r^2}\frac{\partial^2 H_z}{\partial \phi^2} + \kappa^2 H_z = 0 \qquad (8.2\text{-}20)$$

In the present application, these wave equations for E_z and H_z are precisely correct, since we apply them only in regions with homogeneous refractive index. We now have collected all the necessary equations to calculate the problem of the guided modes in the cladded optical fiber. For the z component of the electric field, we attempt the trial solution

$$E_z = AF(r)e^{i\nu\phi} \qquad (8.2\text{-}21)$$

The time- and z-dependent factor

$$e^{i(\omega t - \beta z)} \qquad (8.2\text{-}22)$$

that multiplies all the field components is again omitted from the equations. The constant v in (8.2-21) can be either positive or negative, but it must be an integer in order to ensure that the fields are periodic in ϕ with the period 2π.

Substitution of (8.2-21) into (8.2-19) leads to a differential equation for $F(r)$

$$\frac{d^2F}{dr^2} + \frac{1}{r}\frac{dF}{dr} + \left(\kappa^2 - \frac{v^2}{r^2}\right)F = 0 \qquad (8.2\text{-}23)$$

Equation (8.2-23) is the well-known differential equation for Bessel functions.[11] Since it is a differential equation of second order, there must be two independent solutions of this equation. Several choices exist for the two independent solutions of (8.2-23). We may use the Bessel function $J_v(\kappa r)$ and the Neumann function $N_v(\kappa r)$. These two functions behave like standing waves for very large real values of their argument. The Bessel function J_v remains finite at the coordinate origin, while the Neumann function N_v has a singularity at $r = 0$. Another set of independent solutions is the Hankel functions of the first and second kind, $H_v^{(1)}(\kappa r)$ and $H_v^{(2)}(\kappa r)$. Both of these functions have singularities at $r = 0$. For large real values of the argument, $H_v^{(1)}$ represents a traveling wave that, with our choice of the time dependence (8.2-22), travels in the direction of negative r. The Hankel function of the second kind $H_v^{(2)}$ represents a wave that travels in the direction of positive r—that is, away from the axis of the structure. The Hankel functions are also important for imaginary values of κ

$$\kappa = i\gamma \qquad (8.2\text{-}24)$$

In this case, $H_v^{(1)}(i\gamma r)$ becomes proportional to $e^{-\gamma r}$ for large values of the argument, while $H_v^{(2)}(i\gamma r)$ becomes proportional to $e^{\gamma r}$. It is apparent that only $H_v^{(1)}(i\gamma r)$, with its exponentially decaying behavior, is suitable to describe guided modes outside the core of the fiber. $H_v^{(2)}(i\gamma r)$ must be rejected, because it grows exponentially for increasing values of r and thus does not describe a field distribution that is tightly bound to the core of the fiber.

Collectively, all four functions, or any combination of them, are known as cylinder functions. If we do not want to single out a particular member of the family, we write $Z_v(\kappa r)$ for the solutions of (8.2-23). The notation used for the Neumann functions is that of Jahnke-Emde[11] and Gradshteyn and Ryzhik.[61] Other authors sometimes use the symbol Y_v for the Neumann function. The Hankel functions with imaginary arguments are often denoted by a new symbol. They are proportional to the so-called modified Hankel function K_v. We refrain from using this notation, and follow Jahnke-Emde[11] in writing $H_v^{(1)}(i\gamma r)$. There is already a sufficient abundance of different cylinder functions, so that it seems prudent to use the same symbol for the same functions, indicating their arguments explicitly instead of changing the symbol used for the function. Each of the four functions can be expressed as a linear combination of two of the other functions. The enormous number of relations as well as approximations of cylinder functions are listed in several excellent books. Important references are Jahnke-Emde,[11] Gradshteyn and Ryzhik,[61] and the *Handbook of Functions*, published by the National Bureau of Standards.[8]

We need different solutions of (8.2-23) for the regions inside and outside the core of the fiber. The solutions inside the core must remain finite at $r = 0$, while the solutions on the outside must decay for $r \to \infty$ if we want to find guided mode solutions.

Having collected all the pertinent facts, we are now in a position to write down the expressions for the electromagnetic field of the optical fiber. The r and ϕ components follow immediately from the z components via the relations (8.2-7) through (8.2-10).

We have for $r < a$ (since the radius of the cladding is assumed to be infinite, we denote the radius of the core simply by a)

$$E_z = AJ_\nu(\kappa r)e^{i\nu\phi} \tag{8.2-25}$$

$$H_z = BJ_\nu(\kappa r)e^{i\nu\phi} \tag{8.2-26}$$

$$E_r = -\frac{i}{\kappa^2}\left[\beta\kappa AJ_\nu'(\kappa r) + i\omega\mu_0\frac{\nu}{r}BJ_\nu(\kappa r)\right]e^{i\nu\phi} \tag{8.2-27}$$

$$E_\phi = -\frac{i}{\kappa^2}\left[i\beta\frac{\nu}{r}AJ_\nu(\kappa r) - \kappa\omega\mu_0 BJ_\nu'(\kappa r)\right]e^{i\nu\phi} \tag{8.2-28}$$

$$H_r = -\frac{i}{\kappa^2}\left[-i\omega\varepsilon_1\frac{\nu}{r}AJ_\nu(\kappa r) + \kappa\beta BJ_\nu'(\kappa r)\right]e^{i\nu\varphi} \tag{8.2-29}$$

$$H_\phi = -\frac{i}{\kappa^2}\left[\kappa\omega\varepsilon_1 AJ_\nu'(\kappa r) + i\beta\frac{\nu}{r}BJ_\nu(\kappa r)\right]e^{i\nu\phi} \tag{8.2-30}$$

The prime indicates differentiation with respect to the argument κr (not r) of the Bessel function. The relation between κ, β, and k_1 is given by

$$\kappa^2 = k_1^2 - \beta^2 \tag{8.2-31a}$$

$$k_1^2 = \omega^2\varepsilon_1\mu_0 \tag{8.2-31b}$$

The dielectric constant of the core is related to the index of refraction by the equation

$$n_1^2 = \frac{\varepsilon_1}{\varepsilon_0} \tag{8.2-32}$$

The field on the outside, $r > a$, is given by*

---

*Note that the factor (8.2-22) is omitted from the equations.

$$E_z = CH_\nu^{(1)}(i\gamma r)e^{i\nu\phi} \tag{8.2-33}$$

$$H_z = DH_\nu^{(1)}(i\gamma r)e^{i\nu\phi} \tag{8.2-34}$$

$$E_r = -\frac{1}{\gamma^2}\left[\beta\gamma CH_\nu^{(1)\prime}(i\gamma r) + \omega\mu_0\frac{\nu}{r}DH_\nu^{(1)}(i\gamma r)\right]e^{i\nu\phi} \tag{8.2-35}$$

$$E_\phi = -\frac{1}{\gamma^2}\left[\beta\frac{\nu}{r}CH_\nu^{(1)}(i\gamma r) - \gamma\omega\mu_0 DH_\nu^{(1)\prime}(i\gamma r)\right]e^{i\nu\phi} \tag{8.2-36}$$

$$H_r = -\frac{1}{\gamma^2}\left[-\omega\varepsilon_2\frac{\nu}{r}CH_\nu^{(1)}(i\gamma r) + \gamma\beta DH_\nu^{(1)\prime}(i\gamma r)\right]e^{i\nu\phi} \tag{8.2-37}$$

$$H_\phi = -\frac{1}{\gamma^2}\left[\gamma\omega\varepsilon_2 CH_\nu^{(1)\prime}(i\gamma r) + \beta\frac{\nu}{r}DH_\nu^{(1)}(i\gamma r)\right]e^{i\nu\phi} \tag{8.2-38}$$

The prime indicates again differentiation with respect to the argument, which, in this case, means differentiation with respect to $i\gamma r$. The relation between γ, β, and k_2 is given by

$$\gamma^2 = \beta^2 - k_2^2 \tag{8.2-39a}$$

$$k_2^2 = \omega^2\varepsilon_2\mu_0 \tag{8.2-39b}$$

The constants A, B, C, and D are not determined by Maxwell's equations. The equations for the field components are solutions of Maxwell's equations but, in order to be modes of the optical fiber,[80] they must also satisfy the boundary conditions (1.5-3) and (1.5-4). Since there are two tangential components for the E field and two for the H field, the boundary conditions provide us with four equations. Since we have four undetermined constants in our field expressions, the number of equations matches the number of unknowns. However, in addition to the amplitude coefficients, there is also the propagation constant β that needs to be determined. Its determination presents no problem. The boundary conditions result in four homogeneous equations. Homogeneous equations have solutions only if the determinant of the equation system vanishes. This additional condition is just what is needed to determine the propagation constant β. The condition for the determinant to be zero is called an eigenvalue equation, since it determines the eigenvalue β of the eigenvalue problem. The prefix "eigen" is borrowed from the German language, and means "proper." The eigenvalue problem thus determines the proper values for the propagation constants of the guided modes.

The boundary conditions require us to set the expressions for E_z, E_ϕ, H_z, and H_ϕ on the inside and outside of the fiber core equal to each other at $r = a$. The following equation system results

$$AJ_\nu(\kappa a) \qquad\qquad -CH_\nu^{(1)}(i\gamma a) \qquad\qquad = 0 \tag{8.2-40}$$

$$\frac{\beta}{\kappa^2}\frac{v}{a}AJ_v(\kappa a) + i\frac{\omega\mu_0}{\kappa}BJ_v'(\kappa a) + \frac{\beta}{\gamma^2}\frac{v}{a}CH_v^{(1)}(i\gamma a) - \frac{\omega\mu_0}{\gamma}DH_v^{(1)\prime}(i\gamma a) = 0$$

$$(8.2\text{-}41)$$

$$BJ_v(\kappa a) \qquad\qquad\qquad - DH_v^{(1)}(i\gamma a) = 0$$

$$(8.2\text{-}42)$$

$$-i\frac{\omega\varepsilon_1}{\kappa}AJ_v'(\kappa a) + \frac{\beta}{\kappa^2}\frac{v}{a}BJ_v(\kappa a) + \frac{\omega\varepsilon_2}{\gamma}CH_v^{(1)\prime}(i\gamma a) + \frac{\beta}{\gamma^2}\frac{v}{a}DH_v^{(1)}(i\gamma a) = 0$$

$$(8.2\text{-}43)$$

Equations (8.2-40) and (8.2-42) connect the coefficients A,C and B,D

$$C = \frac{J_v(\kappa a)}{H_v^{(1)}(i\gamma a)}\,A \qquad\qquad (8.2\text{-}44)$$

and

$$D = \frac{J_v(\kappa a)}{H_v^{(1)}(i\gamma a)}\,B \qquad\qquad (8.2\text{-}45)$$

The coefficients A and B are connected by (8.2-43), (8.2-44), and (8.2-45)

$$B = \frac{i}{v}\frac{a\kappa\gamma[\varepsilon_1\gamma J_v'(\kappa a)H_v^{(1)}(i\gamma a) + i\varepsilon_2\kappa J_v(\kappa a)H_v^{(1)\prime}(i\gamma a)]}{\omega(\varepsilon_1 - \varepsilon_2)\mu_0\beta J_v(\kappa a)H_v^{(1)}(i\gamma a)}\,A \qquad (8.2\text{-}46)$$

The relation

$$\kappa^2 + \gamma^2 = k_1^2 - k_2^2 = \omega^2(\varepsilon_1 - \varepsilon_2)\mu_0 \qquad (8.2\text{-}47)$$

was used to simplify this expression. Equation (8.2-41) could have been used instead of (8.2-43). The resulting expression for B/A could be transformed into the form of (8.2-46) with the help of the eigenvalue equation. (See [8.2-53].)

The eigenvalue equation is obtained from the requirement that the determinant of the equation system (8.2-40) through (8.2-43) must vanish

$$\begin{vmatrix} J_v(\kappa a) & 0 & -H_v^{(1)}(i\gamma a) & 0 \\ \dfrac{v}{a}\dfrac{\beta}{\kappa^2}J_v(\kappa a) & i\dfrac{\omega\mu_0}{\kappa}J_v'(\kappa a) & \dfrac{v}{a}\dfrac{\beta}{\gamma^2}H_v^{(1)}(i\gamma a) & -\dfrac{\omega\mu_0}{\gamma}H_v^{(1)\prime}(i\gamma a) \\ 0 & J_v(\kappa a) & 0 & -H_v^{(1)}(i\gamma a) \\ -i\dfrac{\omega\varepsilon_1}{\kappa}J_v'(\kappa a) & \dfrac{v}{a}\dfrac{\beta}{\kappa^2}J_v(\kappa a) & \dfrac{\omega\varepsilon_2}{\gamma}H_v^{(1)\prime}(i\gamma a) & \dfrac{v}{a}\dfrac{\beta}{\gamma^2}H_v^{(1)}(i\gamma a) \end{vmatrix} = 0$$

$$(8.2\text{-}48)$$

Evaluation of the determinant results in the eigenvalue equation

$$\left(\frac{\varepsilon_1}{\varepsilon_2}\frac{a\gamma^2}{\kappa}\frac{J'_\nu(\kappa a)}{J_\nu(\kappa a)} + i\gamma a\frac{H_\nu^{(1)'}(i\gamma a)}{H_\nu^{(1)}(i\gamma a)}\right)\left(\frac{a\gamma^2}{\kappa}\frac{J'_\nu(\kappa a)}{J_\nu(\kappa a)} + i\gamma a\frac{H_\nu^{(1)'}(i\gamma a)}{H_\nu^{(1)}(i\gamma a)}\right)$$

$$= \left[\nu\left(\frac{\varepsilon_1}{\varepsilon_2} - 1\right)\frac{\beta k_2}{\kappa^2}\right]^2 \quad (8.2\text{-}49)$$

Equation (8.2-47) has again been used to simplify this equation. The collection of equations presented in this section solves the problem of the guided modes of the cladded optical fiber. Approximate solutions of the eigenvalue equation are given in Section 8.6. We could have used $e^{-i\nu\phi}$, instead of $e^{i\nu\phi}$, in (8.2-21). This would have resulted in a change of the sign of ν in all the equations except for the index ν of the cylinder functions, which would remain the same because ν appears in (8.2-23) only in the form ν^2. Since ν appears in the eigenvalue equation (8.2-49) also only as ν^2 (aside from the index of the cylinder function) this change would not affect the eigenvalue β. By adding the new modes with the changed sign of ν to the old ones, we would obtain field expressions that would contain $\cos \nu\phi$ and $\sin \nu\phi$ instead of the exponential function. Subtraction of the new modes from the old ones results in another set of modes, with sine and cosine functions of $\nu\phi$ interchanged. The modes are often written in terms of sine and cosine functions of $\nu\phi$ instead of the exponential function used here. This form of the field expression can be found in (8.6-59) through (8.6-64).

It is apparent that the modes of the dielectric waveguide have six field components. It is not possible, in general, to separate them into transverse electric modes and transverse magnetic modes. The modes of dielectric waveguides are thus more complicated than the modes of hollow metallic waveguides. There is, however, one exception to the rule that the modes of the optical fiber are hybrids. In the special case $\nu = 0$, the right-hand side of (8.2-49) vanishes. In this case, we obtain two different eigenvalue equations

$$\text{TM modes:} \frac{\varepsilon_1}{\varepsilon_2}\frac{\gamma}{\kappa}\frac{J_1(\kappa a)}{J_0(\kappa a)} + i\frac{H_1^{(1)}(i\gamma a)}{H_0^{(1)}(i\gamma a)} = 0 \quad (8.2\text{-}50)$$

and

$$\text{TE modes:} \frac{\gamma}{\kappa}\frac{J_1(\kappa a)}{J_0(\kappa a)} + i\frac{H_1^{(1)}(i\gamma a)}{H_0^{(1)}(i\gamma a)} = 0 \quad (8.2\text{-}51)$$

The relation

$$Z'_0 = -Z_1 \quad (8.2\text{-}52)$$

was used to obtain these relations. For modes that satisfy the eigenvalue equation (8.2-51), we find from (8.2-46) that $B = \infty$ if $\nu = 0$. In order to keep B finite, we must set $A = 0$. This means that in this case the longitudinal E component, E_z, vanishes. The modes become transverse electric or TE modes. In case the other eigenvalue equation, (8.2-50), is satisfied, we must first use (8.2-49) to remove ν from the denominator of (8.2-46). We obtain

$$B = iv \frac{\omega(\varepsilon_1 - \varepsilon_2)\beta J_\nu(\kappa a) H_\nu^{(1)}(i\gamma a)}{\kappa\gamma a[\gamma J_\nu'(\kappa a) H_\nu^{(1)}(i\gamma a) + i\kappa J_\nu(\kappa a) H_\nu^{(1)'}(i\gamma a)]} A \qquad (8.2\text{-}53)$$

The same expression would have been obtained had we used (8.2-41) instead of (8.2-43) to derive the expression for B. We are now allowed to let $\nu \to 0$ and find $B = 0$ provided that the denominator of (8.2-53) does not also vanish. However, if the mode satisfies the eigenvalue equation (8.2-50), the denominator of (8.2-53) does not vanish. $B = 0$ means that the longitudinal magnetic field component H_z vanishes. Modes satisfying the eigenvalue equation (8.2-50) are thus transverse magnetic or TM modes.

An important parameter for each mode is its cutoff frequency. A mode is called cut off when its field no longer decays on the outside of the core. The rate of decay of the field with increasing r is determined by the value of the constant γ. We mentioned earlier that the function $H_\nu^{(1)}(i\gamma r)$ decays like an exponential function for large values of its argument. The asymptotic approximation for large argument is[11,61]

$$H_\nu^{(1)}(i\gamma r) = \sqrt{\frac{2}{\pi i \gamma r}} e^{-i(\pi\nu/2 + \pi/4)} e^{-\gamma r} \qquad \text{for} \qquad \gamma r \gg 1 \qquad (8.2\text{-}54)$$

For large values of γ, the field is tightly concentrated inside and close to the core. With decreasing values of γ, the field reaches further out into the space outside the core. Finally, for $\gamma = 0$, the field detaches itself from the guide. The frequency at which this happens is called cutoff frequency. The cutoff condition is thus

$$\gamma = \sqrt{\beta^2 - k_2^2} = 0 \qquad (8.2\text{-}55)$$

The solutions of the eigenvalue equation right at cutoff can be obtained by a procedure demonstrated by Schlesinger, Diament, and Vigants.[83] Following the example of these authors, we transform the eigenvalue equation to a different form. We begin by introducing the abbreviations

$$J^+ = \frac{1}{\kappa a} \frac{J_{\nu+1}(\kappa a)}{J_\nu(\kappa a)} \qquad (8.2\text{-}56)$$

$$J^- = \frac{1}{\kappa a} \frac{J_{\nu-1}(\kappa a)}{J_\nu(\kappa a)} \qquad (8.2\text{-}57)$$

$$H^+ = \frac{1}{i\gamma a} \frac{H_{\nu+1}^{(1)}(i\gamma a)}{H_\nu^{(1)}(i\gamma a)} \qquad (8.2\text{-}58)$$

$$H^- = \frac{1}{i\gamma a} \frac{H_{\nu-1}^{(1)}(i\gamma a)}{H_\nu^{(1)}(i\gamma a)} \qquad (8.2\text{-}59)$$

Using the following functional relations of the cylinder functions[11,61]

$$Z'_v = \frac{1}{2}(Z_{v-1} - Z_{v+1}) \tag{8.2-60}$$

and the abbreviation

$$\varepsilon = \frac{\varepsilon_1}{\varepsilon_2} \tag{8.2-61}$$

we write the eigenvalue equation (8.2-49) after division by $a^4\gamma^4$ in the following form

$$[\varepsilon(J^- - J^+) - (H^- - H^+)][(J^- - J^+) - (H^- - H^+)]$$
$$= \left[\frac{2v(\varepsilon - 1)\beta k_2}{a^2\gamma^2\kappa^2}\right]^2 \tag{8.2-62}$$

Regrouping of terms leads to

$$-(\varepsilon J^- - H^-)(J^+ - H^+) - (\varepsilon J^+ - H^+)(J^- - H^-)$$
$$+(\varepsilon J^+ - H^+)(J^+ - H^+) + (\varepsilon J^- - H^-)(J^- - H^-) = \left[\frac{2v(\varepsilon - 1)\beta k_2}{a^2\gamma^2\kappa^2}\right]^2 \tag{8.2-63}$$

Using another functional relation of the cylinder functions[11,61]

$$Z_{v+1}(z) + Z_{v-1}(z) = \frac{2v}{z} Z_v(z) \tag{8.2-64}$$

we obtain

$$J^+ + J^- = \frac{2v}{(\kappa a)^2} \tag{8.2-65}$$

and

$$H^+ + H^- = -\frac{2v}{(\gamma a)^2} \tag{8.2-66}$$

With the help of these relations, (8.2-63) assumes the following form

$$-2(\varepsilon J^- - H^-)(J^+ - H^+) - 2(\varepsilon J^+ - H^+)(J^- - H^-)$$
$$+\frac{2v}{a^2}\left[\frac{\varepsilon}{\kappa^2} + \frac{1}{\gamma^2}\right][(J^+ - H^+) + (J^- - H^-)] = \left[\frac{2v(\varepsilon - 1)\beta k_2}{a^2\gamma^2\kappa^2}\right]^2 \tag{8.2-67}$$

With the help of (8.2-65) and (8.2-66), we obtain

$$(J^+ - H^+) + (J^- - H^-) = \frac{2v}{a^2}\left[\frac{1}{\kappa^2} + \frac{1}{\gamma^2}\right] \tag{8.2-68}$$

Using the relations between the constants κ, γ, β, k_1, and k_2, we derive the equation

$$\left(\frac{\varepsilon}{\kappa^2} + \frac{1}{\gamma^2}\right)\left(\frac{1}{\kappa^2} + \frac{1}{\gamma^2}\right) = \left[\frac{(\varepsilon - 1)\beta k_2}{\kappa^2 \gamma^2}\right]^2 \tag{8.2-69}$$

The eigenvalue equation (8.2-67) now assumes the simple form

$$(\varepsilon J^- - H^-)(J^+ - H^+) + (\varepsilon J^+ - H^+)(J^- - H^-) = 0 \tag{8.2-70}$$

The two forms of the eigenvalue equation, (8.2-49) and (8.2-70), are of course fully equivalent. The form (8.2-70) is better suited to study its cutoff solutions.

We know that cutoff requires $\gamma = 0$. Since the argument of the Hankel functions vanishes in this case, we need their approximation for small arguments[11,61]

$$\lim_{\gamma \to 0} H_0^{(1)}(i\gamma a) = 1 + \frac{2i}{\pi} \ln \frac{i\Gamma \gamma a}{2} \qquad \text{with} \qquad \Gamma = 1.781672$$

$$= \frac{2i}{\pi} \ln \frac{\Gamma \gamma a}{2} \tag{8.2-71}$$

and

$$\lim_{\gamma \to 0} H_\nu^{(1)}(i\gamma a) = -\frac{i(\nu - 1)!}{\pi}\left(\frac{2}{i\gamma a}\right)^\nu \qquad \text{for} \qquad \nu = 1, 2, 3 \ldots \tag{8.2-72}$$

From these equations follow the relations

$$\lim_{\gamma \to 0} H^+ = -\frac{2\nu}{(a\gamma)^2} \qquad \text{for} \qquad \nu = 1, 2, 3 \ldots \tag{8.2-73}$$

$$\lim_{\gamma \to 0} H^- = -\ln \frac{\Gamma \gamma a}{2} \qquad \text{for} \qquad \nu = 1 \tag{8.2-74}$$

$$\lim_{\gamma \to 0} H^- = \frac{1}{2(\nu - 1)} \qquad \text{for} \qquad \nu = 2, 3, 4 \ldots \tag{8.2-75}$$

With the help of (8.2-73), we obtain from (8.2-70) for small values of γ and for $\nu \neq 0$

$$(\varepsilon J_{\nu-1} - \kappa a H^- J_\nu)(a^2 \gamma^2 J_{\nu+1} + 2\nu \kappa a J_\nu)$$

$$+ (\varepsilon a^2 \gamma^2 J_{\nu+1} + 2\nu \kappa a J_\nu)(J_{\nu-1} - \kappa a H^- J_\nu) = 0 \tag{8.2-76}$$

As $\gamma \to 0$, we must distinguish between the cases $\nu = 1$ and $\nu > 1$. Starting with $\nu = 1$, we obtain from (8.2-76) for $\gamma \to 0$

$$[2(\kappa a)J_1(\kappa a)]^2 \ln \frac{\Gamma \gamma a}{2} = 0 \tag{8.2-77}$$

The solution of this equation is

$$J_1(\kappa a) = 0 \tag{8.2-78}$$

The other possible solution, $\kappa a = 0$, is already included in (8.2-78).

For $\nu > 1$, we obtain from (8.2-76) in the limit $\gamma \to 0$

$$J_\nu(\kappa a)\left[(\varepsilon + 1)J_{\nu-1}(\kappa a) - \frac{\kappa a}{\nu - 1}\,J_\nu(\kappa a)\right] = 0 \qquad (8.2\text{-}79)$$

This equation admits two solutions

$$J_\nu(\kappa a) = 0 \qquad \text{for} \qquad \kappa a \neq 0 \qquad \text{and} \qquad \nu = 2, 3, 4 \dots \quad (8.2\text{-}80)$$

and

$$(\varepsilon + 1)J_{\nu-1}(\kappa a) = \frac{\kappa a}{\nu - 1}\,J_\nu(\kappa a) \qquad \text{for} \qquad \nu = 2, 3, 4 \dots \quad (8.2\text{-}81)$$

We have indicated in (8.2-80) that the "solution" $\kappa a = 0$ must be excluded. This very important fact follows from (8.2-70). To see this, we need the approximation for the Bessel functions of small argument[11,61]

$$J_0(\kappa a) = 1 \qquad (8.2\text{-}82)$$

and

$$J_\nu(\kappa a) = \frac{1}{\nu!}\left[\frac{\kappa a}{2}\right]^\nu \qquad \text{for} \qquad \nu = 1, 2, 3 \dots \quad (8.2\text{-}83)$$

With these approximations, we obtain

$$\lim_{\kappa a \to 0} J^+ = \frac{1}{2(\nu + 1)} \qquad \text{for} \qquad \nu = 1, 2, 3 \dots \quad (8.2\text{-}84)$$

and

$$\lim_{\kappa a \to 0} J^- = \frac{2\nu}{(a\kappa)^2} \qquad \text{for} \qquad \nu = 1, 2, 3 \dots \quad (8.2\text{-}85)$$

If both γ and κ become simultaneously vanishingly small, (8.2-70) becomes for $\nu > 1$

$$\left(\frac{2\varepsilon\nu}{(a\kappa)^2} - \frac{1}{2(\nu - 1)}\right)\left(\frac{1}{2(\nu + 1)} + \frac{2\nu}{(a\gamma)^2}\right)$$

$$+ \left(\frac{\varepsilon}{2(\nu + 1)} + \frac{2\nu}{(a\gamma)^2}\right)\left(\frac{2\nu}{(a\kappa)^2} - \frac{1}{2(\nu - 1)}\right) = 0 \qquad (8.2\text{-}86)$$

As κ and γ approach zero, (8.2-86) reduces to

$$\frac{4\nu^2(\varepsilon + 1)}{(a^2\kappa\gamma)^2} = 0 \qquad \text{for} \qquad \nu = 2, 3, 4 \dots \quad (8.2\text{-}87)$$

This equation cannot be satisfied, proving that $\kappa a = 0$ is not a solution of the cutoff equation (8.2-80). For $\nu = 1$, on the other hand, we obtain from (8.2-70)

$$\left(\frac{2\varepsilon v}{(a\kappa)^2} - \ln\frac{2}{a\gamma\Gamma}\right)\left(\frac{1}{2(v+1)} + \frac{2v}{(a\gamma)^2}\right)$$

$$+ \left(\frac{\varepsilon}{2(v+1)} + \frac{2v}{(a\gamma)^2}\right)\left(\frac{2v}{(a\kappa)^2} - \ln\frac{2}{a\gamma\Gamma}\right) = 0 \qquad (8.2\text{-}88)$$

As κ and γ approach zero, we obtain

$$\frac{4v}{(a\gamma)^2}\left(\frac{v(\varepsilon+1)}{(a\kappa)^2} - \ln\frac{2}{a\gamma\Gamma}\right) = 0 \qquad (8.2\text{-}89)$$

This equation can be satisfied because the logarithm goes to infinity as γ goes to zero. The solution is

$$\kappa a = 0 \qquad \text{for} \qquad v = 1 \qquad (8.2\text{-}90)$$

Finally, we need to investigate the case $v = 0$. For small values of γ, (8.2-50) becomes

$$\frac{1}{\varepsilon}\frac{\kappa}{\gamma}\frac{J_0(\kappa a)}{J_1(\kappa a)} = -a\gamma \ln\frac{2}{a\gamma\Gamma} \qquad (8.2\text{-}91)$$

Since γa vanishes more rapidly than the logarithm grows to infinity, the product of the right-hand side vanishes as $\gamma \to 0$. The solution of this equation is thus

$$\left.\begin{array}{c}\text{TE}\\\text{TM}\end{array}\right\}\text{modes}: J_0(\kappa_c a) = 0 \qquad (8.2\text{-}92)$$

Vanishing κ does not lead to a solution of (8.2-91). Since (8.2-51) differs from (8.2-50) only by a factor ε, the cutoff condition for this equation is again (8.2-92).

We have thus found the complete cutoff behavior of all the modes. Let us summarize the cutoff solutions of the eigenvalue equation of the guided modes (with $v \neq 0$) of the optical fiber. We have the following conditions

$$HE_{11}: \kappa_c a = 0 \qquad \text{for} \qquad v = 1 \qquad (8.2\text{-}93)$$

$$\left.\begin{array}{c}EH_{v\mu}:\\(\text{for } v=1)\ HE_{1\mu}:\end{array}\right\}\kappa_c a = w_{v\mu} \qquad \text{for} \qquad v = 1, 2, 3\ldots \qquad (8.2\text{-}94)$$

and the implicit condition ($\varepsilon = \varepsilon_1/\varepsilon_2$)

$$HE_{v\mu}: (\varepsilon+1)J_{v-1}(\kappa_c a) = \frac{a\kappa_c}{v-1}J_v(\kappa_c a) \qquad \text{for} \qquad v = 2, 3, 4\ldots \qquad (8.2\text{-}95)$$

The parameter $w_{v\mu}$ is the μth root of the equation

$$J_v(w_{v\mu}) = 0 \qquad w_{v\mu} \neq 0 \qquad (8.2\text{-}96)$$

The most important result of our investigation of the cutoff behavior of the guided modes is the fact that there is one mode whose cutoff frequency is zero. The cutoff frequency f_c follows from $\gamma = 0$, with the help of (8.2-31), (8.2-39), and (8.2-55)

$$f_c = \frac{\kappa_c}{2\pi\sqrt{(\varepsilon_1 - \varepsilon_2)\mu_0}} \qquad (8.2\text{-}97)$$

Only the lowest order mode of symmetry $v = 1$ has a vanishing value of κ_c and consequently a cutoff frequency $f_c = 0$. This mode can thus exist at any frequency and at any rod diameter. All other modes are cut off below their cutoff frequencies. It is thus possible to operate the optical fiber with a single guided mode. This requires a fiber that is so thin that all other guided modes are below their cutoff frequencies.

It is apparent from (8.2-94) and (8.2-95) that there are two types of modes for each integral value of $v > 1$. The modes whose cutoff frequencies are determined by the condition (8.2-94) are designated as $EH_{v\mu}$ modes, while the condition (8.2-95) determines the cutoff frequencies of the $HE_{v\mu}$ modes.[80] The only exception are the $HE_{1\mu}$ modes, whose cutoffs are also determined by (8.2-94), and the HE_{11} mode, whose cutoff is determined by (8.2-93). Both sets of modes have six field components. The existence of two sets of modes corresponds to the TE and TM modes of metallic waveguides.

For $v = 1$, there is only one cutoff condition. There are, however, two sets of modes also in this case. Both the $EH_{1\mu}$ and $HE_{1\mu}$ modes have the same cutoff frequency. At frequencies other than cutoff, the two sets of modes have different propagation constants. They are not degenerate. For $v = 0$, we have the nondegenerate TE and TM modes, whose identical cutoff condition is given by (8.2-92).

The values of $\kappa_c a$ for a few v and μ values are shown in Table 8.2.1. For the HE modes of (8.2-95), the values in the table hold for $\varepsilon_1/\varepsilon_2 = 1.1$. It is apparent from this table that with increasing frequency the following modes appear. The first mode to propagate without a cutoff frequency is HE_{11}.

TABLE 8.2.1 The cutoff value $\kappa_c a$ is listed for a number of modes. The values for the HE modes (except $HE_{1\mu}$ which does not depend on it) are calculated for $\varepsilon_1/\varepsilon_2 = 1.1$.

v \ μ	1	2	3	
0	2.405	5.52	8.654	TE, TM
1	0.000	3.832	7.016	HE
1	3.832	7.016	10.173	EH
2	2.445	5.538	8.665	HE
2	5.136	8.417	11.620	EH

$$\varepsilon_1/\varepsilon_2 = 1.1$$

As the frequency is increased, the TE_{01} and TM_{01} modes can exist. The HE_{21} mode appears at only a slightly higher frequency than the TE_{01} and TM_{01} modes. The frequency range of single guided mode operation is given by

$$0 < f < \frac{2.405}{2\pi a \sqrt{(\varepsilon_1 - \varepsilon_2)\mu_0}} \tag{8.2-98}$$

Figure 8.2.3 shows the ratio β/k_2 as a function of $k_2 a$ for the HE_{11}, TE_{01}, and TM_{01} modes. (Approximation solutions of the eigenvalue equation can be found in Section 8.6). This figure was drawn for $n_2 = 1$ and $n_1 = 1.01$. The propagation constants of the modes TE_{01} and TM_{01} are so nearly the same for this slight index difference that they have been drawn as one line. There are other modes possible in the range of $k_2 a$ that are not shown in the figure. However, the other modes have cutoff values that are larger than those of the TE_{01} mode. The HE_{11} mode is not cut off. It appears from the figure as though the β/k_2 line for the HE_{11} modes ends at $k_2 a = 5$. This is a misleading impression that is caused by the fact that β/k_2 becomes so nearly unity for $k_2 a < 5$ that it appears to have a horizontal tangent. The curve does continue to $k_2 a = 0$. All modes start out at the line $\beta/k_2 = 1$ and approach $\beta/k_2 = n_1/n_2$ asymptotically. To show the asymptotic behavior of the HE_{11} mode more clearly, we have drawn a qualitative plot of ω as a function of β in Figure 8.2.4. A plot of this type is called an $\omega - \beta$ diagram. Its advantage consists in the fact that the slope

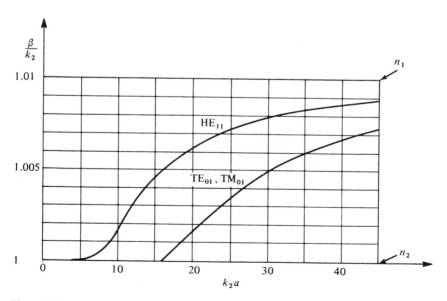

Figure 8.2.3 Ratio of propagation constant β, of guided modes, to plane wave propagation constant k_2 in the medium of the cladding plotted as a function of $k_2 a$.

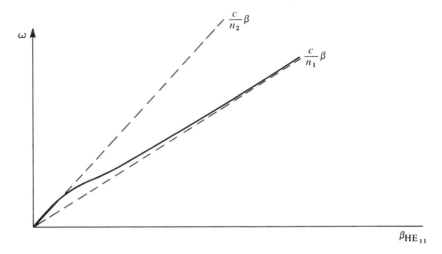

Figure 8.2.4 $\omega - \beta$ diagram for the HE_{11} mode.

$$v_g = \frac{\partial \omega}{\partial \beta} \qquad (8.2\text{-}99)$$

represents the group velocity of the mode in the optical fiber. It is apparent that the group velocity is a function of frequency. However, the curve has a point of inflection where the second derivative

$$\frac{\partial^2 \omega}{\partial \beta^2} = \frac{\partial v_g}{\partial \beta} = \frac{\partial v_g}{\partial \omega} v_g \qquad (8.2\text{-}100)$$

vanishes. In the vicinity of this point, the group velocity is approximately independent of frequency. This means that pulses propagating in this mode keep their shape approximately undistorted. The point of inflection appears at a frequency where more than one guided mode can propagate in the fiber. The possible number of modes is still low, however. It may thus be possible to operate a fiber at this frequency and transmit pulses with a minimum of distortion. The dispersion of the dielectric media of the core and cladding has been ignored in this discussion. Taking it into consideration would shift the point of inflection.[94]

Let us summarize our results. The cladded (or uncladded) optical fiber is capable of supporting guided modes. The number of possible guided modes depends on the value of $V = a \sqrt{k_1^2 - k_2^2}$ (see [8.2-97]). For large values of V, many modes can propagate. However, it is possible to design optical fibers so that at any given frequency only one guided mode, the HE_{11} mode, can propagate. The electromagnetic energy of the guided modes is transported partially inside and partially outside the fiber core. The farther a mode is from its cutoff frequency, the more tightly its energy is concentrated inside the core. The modes of the fiber show some dispersion. The group velocity as well as the phase velocity depend on the operating frequency. However, the HE_{11} mode makes it possible to choose an operating frequency at which the group velocity is approximately constant over a narrow range of frequencies. The term "narrow" is used here with respect to the ratio of frequency range to operating frequency. The frequency range over which the HE_{11} mode possesses a constant group velocity may be extremely large in terms of conventional microwave frequencies.

Optical fibers show great promise as light waveguides for communications purposes. At the time of this writing, fibers with a loss of only 20 dB/km have been reported.[95] Theoretical considerations make it seem possible to lower the loss of glasses to approximately 5 to 10 dB/km at visible frequencies. Losses in that range would make optical fibers very attractive for light transmission over long distances. An important field of application for light fibers is the transmission of light and images over short distances. It is possible to produce bundles containing large numbers of optical fibers that are capable of transmitting images of good quality.[79] Each fiber carries the light corresponding to one image point. This mode of image transmission is quite different from the image transmission through optical materials with graded index of refraction that was discussed in Chapter 7. Transmission of incoherent light through bundles of optical fibers without the intent of image formation finds applications in monitoring light sources. The operation of the rear lights in an automobile can be checked by means of an optical fiber that transmits a small portion of the output of the rear lights to the instrument panel.

8.3 GUIDED MODES OF THE SLAB WAVEGUIDE

We have studied the guided modes of round optical fibers in the previous section. We found that the mathematical description of these modes is rather complicated. The situation gets even worse when we try to analyze the radiation losses of round optical fibers. However, in order to learn about the properties of light transmission in dielectric waveguides, it is not necessary to study the most complicated structure. There are simpler dielectric waveguides, whose physical properties are very nearly the same as those of the round optical fiber, that are much easier to analyze. For this reason, we now turn to a discussion of the modes of the dielectric slab waveguide. This structure is particularly simple, and allows us to study the radiation and mode conversion properties of dielectric waveguides. The results obtained with the help of the slab waveguide model are usually directly applicable to the round optical fiber.

The slab waveguide is sketched in Figure 8.1.2. Our treatment of the slab waveguide as a model for optical fibers is analogous to our practice of treating diffraction problems in two dimensions whenever possible. The slab waveguide resembles a longitudinal cross section through the cladded optical fiber, and can be regarded as its two-dimensional analog.

We simplify the discussion further by assuming that there is no variation of either the waveguide geometry or the field distributions in y direction. We express this limitation on guide symmetry by the symbolic relation

$$\frac{\partial}{\partial y} = 0 \tag{8.3-1}$$

The modes of the optical fiber are hybrids in the general case, but they reduce to TE and TM modes in the special case $v = 0$. The restriction (8.3-1) also allows us to decompose the field of the slab waveguide into TE and TM modes.

We begin with the derivation of the properties of the TE modes. We have by definition $E_z = 0$. Using (8.3-1), we see from (1.4-16) through (1.4-19) that only the following field components are nonzero: H_z, H_x, and E_y. We express H_z and H_x in terms of E_y, using Maxwell's equations (1.4-13) and (1.4-15)

$$H_x = -\frac{i}{\omega\mu}\frac{\partial E_y}{\partial z} \tag{8.3-2}$$

and

$$H_z = \frac{i}{\omega\mu}\frac{\partial E_y}{\partial x} \tag{8.3-3}$$

The E_y component is obtained as a solution of the reduced wave equation*

$$\frac{\partial^2 E_y}{\partial x^2} + \frac{\partial^2 E_y}{\partial z^2} + n^2 k_0^2 E_y = 0 \tag{8.3-4}$$

with

$$n^2 = \frac{\varepsilon}{\varepsilon_0} \tag{8.3-5}$$

and

$$k_0 = \omega\sqrt{\varepsilon_0\mu_0} = \frac{2\pi}{\lambda_0} \tag{8.3-6}$$

With the time- and z-dependence

---

* Note that (8.3-4) holds even at the interface $x = \pm d$. The term $\mathbf{E}\cdot\nabla\epsilon$ appearing in (1.3-4) vanishes because of the restriction imposed by (8.3-1) and the fact that E_y is the only nonvanishing component of $\mathbf{E}$.

$$e^{i(\omega t - \beta z)} \tag{8.3-7}$$

we obtain from (8.3-4)

$$\frac{\partial^2 E_y}{\partial x^2} + (n^2 k_0^2 - \beta^2)E_y = 0 \tag{8.3-8}$$

The solution of this equation inside the slab is different from the solution in the surrounding medium (cladding). We simplify the treatment by separating the modes from the start into even and odd modes. It is of course possible to start with general field expressions and obtain the even and odd modes from the resulting eigenvalue problem. The treatment is simplified, however, by considering even and odd modes from the start. (Compare the derivation of the TE and TM modes in Section 8.2.)

Even Guided TE Modes

The mode solution for even modes inside the slab, $|x| < d$, is*

$$E_y = A_e \cos \kappa x \tag{8.3-9}$$

and

$$H_z = -\frac{i\kappa}{\omega\mu_0} A_e \sin \kappa x \tag{8.3-10}$$

with

$$\kappa^2 = n_1^2 k_0^2 - \beta^2 \tag{8.3-11}$$

The H_x component will not be required. It can be obtained from (8.3-2). The field outside the slab at $|x| > d$ is

$$E_y = A_e \cos \kappa d e^{-\gamma(|x|-d)} \tag{8.3-12}$$

and

$$H_z = -\frac{x}{|x|} \frac{i\gamma}{\omega\mu_0} A_e \cos \kappa d \, e^{-\gamma(|x|-d)} \tag{8.3-13}$$

with

$$\gamma^2 = \beta^2 - n_2^2 k_0^2 \tag{8.3-14}$$

Both κ^2 and γ^2 can be positive quantities since $n_1 > n_2$. For positive values of γ^2, the field on the outside of the slab decays with increasing values of $|x|$. The condition for a guided mode is thus

$$\gamma^2 > 0 \tag{8.3-15}$$

---

* The factor (8.3-7) is omitted from all the field components.

The amplitude constant of E_y in (8.3-12) was chosen to make the E_y component continuous at $x = \pm d$. We must also require that the H_z component cross the interface between the two media continuously. We thus obtain from (8.3-10) and (8.3-13) the eigenvalue equation

$$\tan \kappa d = \frac{\gamma}{\kappa} \qquad (8.3\text{-}16)$$

The amplitude coefficient can be expressed in terms of the power P carried by the mode We obtain from (1.2-12) with the help of (8.3-2) and (8.3-7)

$$P = \frac{1}{2} \int_{-\infty}^{\infty} (\mathbf{E} \times \mathbf{H}^*)_z dx = -\frac{1}{2} \int_{-\infty}^{\infty} E_y H_x^* \, dx = \frac{\beta}{\omega\mu_0} \int_0^{\infty} |E_y|^2 dx \qquad (8.3\text{-}17)$$

The quantity P expresses the power flowing in z direction per unit length of the y direction. Substitution of (8.3-9) and (8.3-12) into (8.3-17) leads to the relation

$$A_e = \left(\frac{2\omega\mu_0}{\beta d + \dfrac{\beta}{\gamma}} P \right)^{1/2} \qquad (8.3\text{-}18)$$

The eigenvalue equation (8.3-16) was used to obtain A_e in this simple form.

Odd Guided TE Modes

The odd modes are obtained very similarly. The field inside the slab for $|x| < d$ becomes now

$$E_y = A_0 \sin \kappa x \qquad (8.3\text{-}19)$$

and

$$H_z = \frac{i\kappa}{\omega\mu_0} A_0 \cos \kappa x \qquad (8.3\text{-}20)$$

The field outside the slab at $|x| > d$ is given by

$$E_y = \frac{x}{|x|} A_0 \sin \kappa d \, e^{-\gamma(|x|-d)} \qquad (8.3\text{-}21)$$

and

$$H_z = \frac{-i\gamma}{\omega\mu_0} A_0 \sin \kappa d \, e^{-\gamma(|x|-d)} \qquad (8.3\text{-}22)$$

The constants κ and γ are given by (8.3-11) and (8.3-14). The E_y component is again continuous because of our choice of the amplitude coefficients. The continuity requirement for the H_z component leads to the eigenvalue equation

$$\tan \kappa d = -\frac{\kappa}{\gamma} \tag{8.3-23}$$

The amplitude coefficient can be expressed in terms of the power flow with the help of (8.3-17), (8.3-19), (8.3-21), and (8.3-23), with the result

$$A_0 = \left(\frac{2\omega\mu_0}{\beta d + \dfrac{\beta}{\gamma}} P \right)^{1/2} \tag{8.3-24}$$

Equations (8.3-18) and (8.3-24) are formally identical. The constants β and γ entering these equations are solutions of (8.3-16) for the even modes and of (8.3-23) for the odd TE modes.

Eigenvalues of TE Modes

The eigenvalue equations for the even and odd TE modes of the dielectric slab waveguide are very much simpler than those of the optical fiber. We can get a good feeling for their solutions by plotting the left-hand side and the right-hand side of (8.3-16) and (8.3-23). Figure 8.3.1 shows this plot. Solutions are obtained at the intersections between the curves. It is immediately apparent that, for a given frequency, there are a finite number of guided mode solutions. The function γ/κ is obtained from (8.3-11) and (8.3-14)

$$\frac{\gamma}{\kappa} = \frac{\sqrt{(n_1^2 - n_2^2)k_0^2 - \kappa^2}}{\kappa} \tag{8.3-25}$$

Keeping d constant but varying the frequency causes the point at which γ/κ ends on the κd axis, and the identical point at which $-\kappa/\gamma$ becomes infinite, to move. For increasing frequencies, this point moves to larger values of κd. Both curves γ/κ as well as $-\kappa/\gamma$ cross more lines of $\tan \kappa d$, indicating that more guided modes can propagate in the slab waveguide. Decreasing the frequency causes a decrease in guided modes. The figure also shows us that the lowest order even TE mode can propagate at arbitrarily small frequencies. It is the only mode that is never cut off. This behavior contrasts with that of the optical fiber. The TE and TM modes of the fiber all had cutoff frequencies larger than zero. Only the HE_{11} mode did not suffer a cutoff. For purposes of comparison, we must assume that the lowest order even TE mode of the slab waveguide corresponds to the HE_{11} mode of the round optical fiber.

The cutoff condition for the even TE modes is obtained from the requirement that the end point of γ/κ on the κd axis coincides with the zero crossing

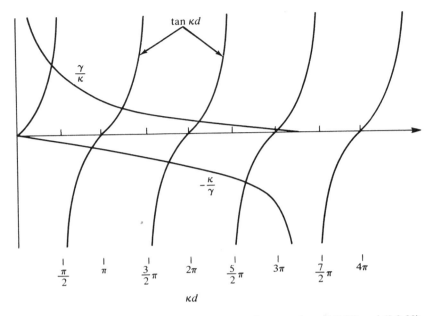

Figure 8.3.1   Graphical solution of the eigenvalue equations (8.3-16) and (8.3-23).

of the tangent function. We thus must require that $\kappa_c d = v\pi$ (with integer v) at the point where $\gamma_c d = 0$. We thus find the cutoff condition from (8.3-25)

$$\sqrt{n_1^2 - n_2^2}\, k_0 d = v\pi = \kappa_c d \qquad (8.3\text{-}26)$$

At cutoff, we have $\beta = n_2 k_0$. The ratio of β over κ at the cutoff point $\gamma = 0$ is thus

$$\frac{\beta_c}{\kappa_c} = \frac{n_2}{\sqrt{n_1^2 - n_2^2}} \qquad (8.3\text{-}27)$$

Equation (8.3-27) holds for the even TE as well as the even TM modes. (See [8.3-38].)

The mechanism of mode guidance in the slab waveguide can be explained by a simple physical argument. Consider the even TE mode (8.3-9). Reinserting the omitted factor (8.3-7), we can write this expression in the following form

$$E_y = \frac{1}{2} A_e \left[e^{i(\omega t + \kappa x - \beta z)} + e^{i(\omega t - \kappa x - \beta z)} \right] \qquad (8.3\text{-}28)$$

In this form, it becomes apparent that the E_y component inside the dielectric slab can be expressed as the superposition of two plane waves. The direction of propagation of these plane waves is given by

$$\tan \alpha = \pm \frac{\beta}{\kappa} \tag{8.3-29}$$

The angle α is formed by the direction of plane wave propagation and the normal to the interface between the slab and the outside medium. Our study of total internal reflection tells us that the plane waves cannot leave the medium if their angle of incidence with the interface is larger than the total internal reflection angle. In the limit given by (1.6-23), we obtain

$$\tan \alpha_i = \frac{\sin \alpha_i}{\sqrt{1 - \sin^2 \alpha_i}} = \frac{n_2}{\sqrt{n_1^2 - n_2^2}} \tag{8.3-30}$$

A comparison of (8.3-27) and (8.3-30) identifies the cutoff condition as being identical with the point at which the plane waves traveling inside the dielectric slab assume the limiting angle α_i for total internal reflection.

The physical explanation for the existence of a guided mode can now be given as follows. Inside the dielectric slab, a plane wave travels at an angle to the interface at the slab boundary. The plane wave is totally reflected at the slab boundary, and bounces back and forth between the two dielectric interfaces. The evanescent field that exists outside the slab was encountered in Section 1.6 as belonging to a wave that is totally reflected at a dielectric interface. This explanation of the mechanism of mode guidance in the dielectric slab also holds for other dielectric waveguides and in particular for the optical fiber. Because of the complicated geometry, the mechanism of total internal reflection is not as readily apparent for the round dielectric rod as it is for the slab waveguide. This explanation fails, however, to provide any inside into the existence of a cutoff frequency for some modes and the lack of a cutoff for other modes. The angle at which the plane waves propagate inside of the slab is obtained by a complete solution of the guided wave problem.

Even TM Modes

In addition to the TE modes studied so far, there are also TM modes in the dielectric slab waveguide. The TM modes are obtained by setting $H_z = 0$. The only nonvanishing field components can again be found from (1.4-16) through (1.4-19) and (8.3-1). They are E_z, E_x, and H_y. The two electric components can be expressed in terms of the H_y component with the help of (1.4-10) and (1.4-12)

$$E_x = \frac{i}{n^2 \omega \varepsilon_0} \frac{\partial H_y}{\partial z} \tag{8.3-31}$$

and

$$E_z = -\frac{i}{n^2 \omega \varepsilon_0} \frac{\partial H_y}{\partial x} \tag{8.3-32}$$

The H_y component is obtained as a solution of the reduced wave equation*

$$\frac{\partial^2 H_y}{\partial x^2} + (n^2 k_0^2 - \beta^2)H_y = 0 \tag{8.3-33}$$

The even TM modes inside the slab, $|x| < d$, have the following field solutions

$$H_y = B_e \cos \kappa x \tag{8.3-34}$$

and

$$E_z = \frac{i\kappa}{n_1^2 \omega \varepsilon_0} B_e \sin \kappa x \tag{8.3-35}$$

with κ given by (8.3-11). The field components outside the slab at $|x| > d$ are

$$H_y = B_e \cos \kappa d \, e^{-\gamma(|x|-d)} \tag{8.3-36}$$

and

$$E_z = \frac{x}{|x|} \frac{i\gamma}{n_2^2 \omega \varepsilon_0} B_e \cos \kappa d \, e^{-\gamma(|x|-d)} \tag{8.3-37}$$

The eigenvalue equation is obtained from the requirement that E_z remain continuous at $x = \pm d$

$$\tan \kappa d = \frac{n_1^2}{n_2^2} \frac{\gamma}{\kappa} \tag{8.3-38}$$

The power flow in z direction per unit length of the y direction is obtained from (1.2-12)

$$P = \frac{1}{2} \int_{-\infty}^{\infty} (\mathbf{E} \times \mathbf{H}^*)_z \, dx = \frac{1}{2} \int_{-\infty}^{\infty} E_x H_y^* \, dx = \frac{\beta}{\omega \varepsilon_0} \int_0^{\infty} \frac{1}{n^2} |H_y|^2 dx \tag{8.3-39}$$

The amplitude coefficient can thus be expressed as

$$B_e = \left\{ \frac{2\omega \varepsilon_0 n_1^2 P}{\beta \left[d + \frac{(n_1 n_2)^2}{\gamma} \frac{\kappa^2 + \gamma^2}{n_2^4 \kappa^2 + n_1^4 \gamma^2} \right]} \right\}^{1/2} \tag{8.3-40}$$

Odd TM Modes

The field of the odd TM modes inside the slab is

$$H_y = B_0 \sin \kappa x \tag{8.3-41}$$

and

$$E_z = -\frac{i\kappa}{n_1^2 \omega \varepsilon_0} B_0 \cos \kappa x \tag{8.3-42}$$

---

* Note that (8.3-33) does not hold at the interfaces $x = \pm d$ (Compare [9.5-2].) The region $x = \pm d$ is taken care of by the boundary conditions.

The field for $|x| > d$ is given by

$$H_y = \frac{x}{|x|} B_0 \sin \kappa d \; e^{-\gamma(|x|-d)} \qquad (8.3\text{-}43)$$

and

$$E_z = \frac{i\gamma}{n_2^2 \omega \varepsilon_0} B_0 \sin \kappa d \; e^{-\gamma(|x|-d)} \qquad (8.3\text{-}44)$$

The eigenvalue equation follows from the boundary condition in the usual way

$$\tan \kappa d = -\frac{n_2^2}{n_1^2} \frac{\kappa}{\gamma} \qquad (8.3\text{-}45)$$

The coefficient B_0 can be expressed as

$$B_0 = \left\{ \frac{2\omega\varepsilon_0 n_1^2 P}{\beta \left[d + \dfrac{(n_1 n_2)^2}{\gamma} \dfrac{\kappa^2 + \gamma^2}{n_2^4 \kappa^2 + n_1^4 \gamma^2} \right]} \right\}^{1/2} \qquad (8.3\text{-}46)$$

We find again that the expressions for the amplitudes of the even TM modes are equal in form to the corresponding expression for the amplitude of the odd TM modes.

Within the limitation set by the condition (8.3-1), the TE and TM modes provide a complete set of modes of the slab waveguide. However, the condition (8.3-1) is an arbitrary restriction on the problem. By omitting it and allowing the fields to vary also in y direction, we would obtain many more modes. These additional modes are hybrid modes having all six field components. Since we want to give a simplified description of radiation loss and mode conversion phenomena, we shall not derive the more complicated hybrid modes of the slab waveguide. By imposing the condition (8.3-1), we limit the problem to such an extent that the TE and TM modes describe the waveguide completely.

8.4   RADIATION MODES OF THE SLAB WAVEGUIDE

The set of guided modes that we found in the last section is sufficient to describe any guided field distribution in the slab waveguide provided the condition (8.3-1) is imposed. It is not sufficient, however, to describe radiation phenomena. The complete set of modes of a dielectric waveguide includes a finite number of guided modes and an infinite number of radiation modes. This situation is analogous to the problem of the hydrogen atom. There is an infinite number of bound states of the hydrogen atom, corresponding to the guided modes of the dielectric waveguide. The hydrogen energy spectrum also contains a continuum of unbound states that are necessary to describe scattering phenomena. These unbound states of the hydrogen atom correspond to the radiation modes of dielectric waveguides. A good description of radiation modes is given in V. V. Shevchenko's book.[113]

The fact that unguided modes must be solutions of the field problem of the dielectric waveguide can be seen by the following argument. Consider a plane wave incident from the outside onto the slab waveguide. The plane wave will suffer refraction and reflection, so that a standing wave exists inside as well as outside the waveguide. This radiation field does not decay on the outside of the structure. It is not bound to the dielectric slab, and persists undiminished throughout all space. It is, however, a legitimate solution of the electromagnetic field problem, since it satisfies Maxwell's equations and the boundary conditions on the surface of the slab. By using two sources on either side of the slab, we can excite symmetric and antisymmetric radiation modes. The fact that there are standing waves reaching out to infinity should not be disturbing, since the radiation modes must not be considered as being excited by sources inside the slab but by an infinitely extended source on its outside at $z = -\infty$. The sources that excite the guided modes must also be considered to exist at $z = -\infty$. However, sources of guided modes provide only a finite amount of energy, while the energy of each individual radiation mode is infinite. The normalization of the radiation modes does not involve a finite amount (or the unit) of power but requires a delta function.

With these introductory remarks in mind, let us formulate the radiation modes mathematically.

Even TE Radiation Modes

The number of field components of the TE radiation modes is the same as those of the guided modes. The E_y component must satisfy (8.3-8), while the magnetic field components follow from (8.3-2) and (8.3-3). The only difference between the modes of Section 8.3 and our present radiation modes is that we no longer require the fields to decay exponentially on the outside of the slab. It is not sufficient to allow γ of (8.3-14) to become imaginary. Imaginary values of γ would result in traveling waves on the outside of the slab. It is impossible to satisfy the boundary conditions with traveling waves. Standing waves are needed to satisfy the boundary value problem. The mathematical reason for the need for standing waves can be seen as follows. If we did nothing else but require imaginary values of γ without otherwise changing the formulation of the mode problem presented in the previous section, we would again have to solve the homogenous equation that results from requiring continuity of H_z at $x = \pm d$. We would again obtain the eigenvalue equation (8.3-16). However, it turns out that there are no solutions of (8.3-16) that result in real values of β but imaginary γ. In order to obtain mode solutions different from the ones already found in the previous section, we need to satisfy the boundary condition without obtaining a restriction on the possible β values. This means that we must try to obtain an inhomogeneous equation for the amplitude coefficients. This can be achieved only by adding one more wave to the field on the outside of the slab. Letting γ be imaginary in (8.3-12) results in a traveling wave at $|x| > d$. Adding to it another traveling wave moving in the opposite direction provides us with an additional amplitude coefficient, thus causing the boundary value problem to result in an inhomogeneous equation.

We again use the fields*

$$E_y = C_e \cos \sigma x \qquad (8.4\text{-}1)$$

and

$$H_z = -\frac{i\sigma}{\omega\mu_0} C_e \sin \sigma x \qquad (8.4\text{-}2)$$

with

$$\sigma^2 = n_1^2 k_0^2 - \beta^2 \qquad (8.4\text{-}3)$$

on the inside of the slab. On the outside, $|x| > d$, we set

$$E_y = D_e e^{-i\rho|x|} + F_e e^{i\rho|x|} \qquad (8.4\text{-}4)$$

and

$$H_z = \frac{x}{|x|}\frac{\rho}{\omega\mu_0}\left(D_e e^{-i\rho|x|} - F_e e^{i\rho|x|}\right) \qquad (8.4\text{-}5)$$

with

$$\rho^2 = n_2^2 k_0^2 - \beta^2 \qquad (8.4\text{-}6)$$

Compared to (8.3-9) through (8.3-14), we now have an additional undetermined amplitude coefficient in our field expressions. The boundary conditions requiring continuity of E_y and H_z at $x = \pm d$ lead to the following system of equations (we need to consider only $x = d$ since the conditions at $x = -d$ are identical)

$$D_e e^{-i\rho d} + F_e e^{i\rho d} = C_e \cos \sigma d \qquad (8.4\text{-}7)$$

$$D_e e^{-i\rho d} - F_e e^{i\rho d} = -i\frac{\sigma}{\rho} C_e \sin \sigma d \qquad (8.4\text{-}8)$$

Since we have only two equations for the determination of three unknowns, we must consider one of them, C_e, as given, and solve the resulting inhomogeneous equation system. An inhomogeneous system of equations

---

* The factor (8.3-7) has been omitted.

has solutions only when the system determinant does not vanish. We do not obtain an eigenvalue equation in this case, and the propagation constant β remains arbitrary. The coefficients D_e and F_e can be obtained from the above equations

$$D_e = \frac{C_e}{2} e^{i\rho d} \left(\cos \sigma d - i \frac{\sigma}{\rho} \sin \sigma d\right) = \bar{D}_e C_e \qquad (8.4\text{-}9)$$

$$F_e = D_e^* \qquad (8.4\text{-}10)$$

The amplitude coefficient C_e must be related to the power carried by the radiation mode. We mentioned earlier that the power in each radiation mode is infinite. We thus require, instead of (8.3-17)

$$P\delta(\rho - \rho') = \frac{\beta}{\omega\mu_0} \int_0^\infty E_y(\rho) E_y^*(\rho') dx \qquad (8.4\text{-}11)$$

Equation (8.4-11) indicates an orthogonality condition for the two different modes labeled with the parameter ρ and ρ'. For different values of these parameters the integral must vanish, while for $\rho = \rho'$ it assumes the value infinity. The evaluation of the integral requires some care, since it is necessary to identify the delta function. For this reason, it seems advisable to carry out the calculation in some detail. From (8.4-1), (8.4-4), (8.4-9), and (8.4-10) we obtain

$$I = \int_0^\infty E_y(\rho) E_y^*(\rho') dx = C_e C_e^{*'} \left\{ \int_0^d \cos \sigma x \cos \sigma' x \, dx \right.$$

$$\left. + \int_d^\infty (\bar{D}_e e^{-i\rho x} + \bar{D}_e^* e^{i\rho x})(\bar{D}_e^{*'} e^{i\rho' x} + \bar{D}_e' e^{-i\rho' x}) \, dx \right\} \qquad (8.4\text{-}12)$$

The primed quantities are obtained by replacing ρ with ρ'. The integrations can easily be performed. The symbol ∞ is here used as a very large number that tends to infinity

$$I = C_e C_e^{*'} \left\{ \frac{\sigma \sin \sigma d \cos \sigma' d - \sigma' \cos \sigma d \sin \sigma' d}{\sigma^2 - \sigma'^2} \right.$$

$$- \frac{\bar{D}_e \bar{D}_e^{*'}}{i(\rho - \rho')} \left(e^{-i(\rho - \rho')\infty} - e^{-i(\rho - \rho')d}\right)$$

$$- \frac{\bar{D}_e \bar{D}_e'}{i(\rho + \rho')} \left(e^{-i(\rho + \rho')\infty} - e^{-i(\rho + \rho')d}\right)$$

$$+ \frac{\bar{D}_e^* \bar{D}_e^{*'}}{i(\rho + \rho')} \left(e^{i(\rho + \rho')\infty} - e^{i(\rho + \rho')d}\right)$$

$$\left. + \frac{\bar{D}_e^* \bar{D}_e'}{i(\rho - \rho')} \left(e^{i(\rho - \rho')\infty} - e^{i(\rho - \rho')d}\right) \right\} \qquad (8.4\text{-}13)$$

Using (8.4-9) and the relation

$$\sigma^2 - \sigma'^2 = \rho^2 - \rho'^2 = \beta'^2 - \beta^2 \tag{8.4-14}$$

we obtain

$$
\begin{aligned}
I = \frac{1}{2} C_e C_e^{*\prime} \Bigg[& \left(\cos \sigma d \cos \sigma' d + \frac{\sigma \sigma'}{\rho \rho'} \sin \sigma d \sin \sigma' d \right) \frac{\sin (\rho - \rho')(\infty - d)}{\rho - \rho'} \\
+ & \left(\frac{\sigma}{\rho} \sin \sigma d \cos \sigma' d - \frac{\sigma'}{\rho'} \cos \sigma d \sin \sigma' d \right) \frac{\cos (\rho - \rho')(\infty - d)}{\rho - \rho'} \\
+ & \left(\cos \sigma d \cos \sigma' d - \frac{\sigma \sigma'}{\rho \rho'} \sin \sigma d \sin \sigma' d \right) \frac{\sin(\rho + \rho')(\infty - d)}{\rho + \rho'} \\
+ & \left(\frac{\sigma}{\rho} \sin \sigma d \cos \sigma' d + \frac{\sigma'}{\rho'} \cos \sigma d \sin \sigma' d \right) \frac{\cos (\rho + \rho')(\infty - d)}{\rho + \rho'} \Bigg]
\end{aligned}
\tag{8.4-15}
$$

Before we evaluate this expression further, we must pause and realize that the form (8.4-11) already suggests that this expression makes sense only when used as the integrand of an integral. The delta function is highly singular and, strictly speaking, defined only when it appears as part of an integrand. The expression (8.4-15) thus can similarly be used only as a factor of an integrand. A brief study shows that the factors of the cosine function in (8.4-15) remain finite for all values of ρ and ρ'. Since the argument of the cosine function tends towards infinity, this term can never give a finite contribution if it appears in the integrand of an integral. We thus conclude that the cosine terms give no contribution and can be omitted. The sine terms can be expressed by delta functions with the help of the following representation of the delta function

$$\delta(x) = \lim_{K \to \infty} \frac{\sin Kx}{\pi x} \tag{8.4-16}$$

Both ρ and ρ' are positive quantities. The term with $\sin (\rho + \rho')\infty$ is thus zero, and we obtain (C_e is taken to be real)

$$I = \frac{\pi}{2} C_e^2 \left(\cos^2 \sigma d + \frac{\sigma^2}{\rho^2} \sin^2 \sigma d \right) \delta(\rho - \rho') \tag{8.4-17}$$

Combining the equations (8.4-11) and (8.4-17) allows us to express the amplitude coefficient C_e in terms of the power P

$$C_e = \left[\frac{2\rho^2 \omega \mu_0 P}{\pi \beta(\rho^2 \cos^2 \sigma d + \sigma^2 \sin^2 \sigma d)} \right]^{1/2} \tag{8.4-18}$$

We thus have not only determined the even TE radiation modes but have also proved their orthogonality among each other. We do not repeat the detailed derivation of the amplitude coefficients and orthogonality relations for the other radiation modes, but state only the end results.

Odd TE Radiation Modes

The fields for $|x| < d$ are given by

$$E_y = C_0 \sin \sigma x \tag{8.4-19}$$

$$H_z = \frac{i\sigma}{\omega\mu_0} C_0 \cos \sigma x \tag{8.4-20}$$

On the outside of the slab, $|x| > d$, we have

$$E_y = \frac{x}{|x|} \left(D_0 e^{-i\rho|x|} + F_0 e^{i\rho|x|} \right) \tag{8.4-21}$$

$$H_z = \frac{\rho}{\omega\mu_0} \left(D_0 e^{-i\rho|x|} - F_0 e^{i\rho|x|} \right) \tag{8.4-22}$$

The relation between the constants C_0, D_0, and F_0 is given by

$$D_0 = \frac{1}{2} C_0 e^{i\rho d} \left(\sin \sigma d + i \frac{\sigma}{\rho} \cos \sigma d \right) \tag{8.4-23}$$

and

$$F_0 = D_0^* \tag{8.4-24}$$

The amplitude coefficient is related to the power carried by the mode. (The term "power carried by the mode" is used loosely. P is actually the power that results if we integrate the power in the field over an infinitesimal interval of ρ.)

$$C_0 = \left[\frac{2\rho^2 \omega\mu_0 P}{\pi\beta(\rho^2 \sin^2 \sigma d + \sigma^2 \cos^2 \sigma d)} \right]^{1/2} \tag{8.4-25}$$

Even TM Radiation Modes

The field for $|x| < d$ is

$$H_y = C_e \cos \sigma x \tag{8.4-26}$$

$$E_z = \frac{i\sigma}{n_1^2 \omega\varepsilon_0} C_e \sin \sigma x \tag{8.4-27}$$

The field outside the slab at $|x| > d$ is

$$H_y = D_e e^{-i\rho|x|} + F_e e^{i\rho|x|} \tag{8.4-28}$$

$$E_z = -\frac{x}{|x|} \frac{\rho}{n_2^2 \omega\varepsilon_0} \left(D_e e^{-i\rho|x|} - F_e e^{i\rho|x|} \right) \tag{8.4-29}$$

The relation between the constants is given by

$$F_e = D_e^*$$

(8.4-30)

and

$$D_e = \frac{1}{2} C_e e^{i\rho d} \left(\cos \sigma d - i \left(\frac{n_2}{n_1} \right)^2 \frac{\sigma}{\rho} \sin \sigma d \right)$$

(8.4-31)

and C_e is related to the power by

$$C_e = \left[\frac{2n_1^2 \rho^2 \omega \varepsilon_0 P}{\pi \beta \left(\frac{n_1^2}{n_2^2} \rho^2 \cos^2 \sigma d + \frac{n_2^2}{n_1^2} \sigma^2 \sin^2 \sigma d \right)} \right]^{1/2}$$

(8.4-32)

Odd TM Radiation Modes

The field inside the slab, $|x| < d$, is

$$H_y = C_0 \sin \sigma x$$

(8.4-33)

$$E_z = -\frac{i\sigma}{n_1^2 \omega \varepsilon_0} C_0 \cos \sigma x$$

(8.4-34)

Outside the slab, we have

$$H_y = \frac{x}{|x|} \left(D_0 e^{-i\rho|x|} + F_0 e^{i\rho|x|} \right)$$

(8.4-35)

$$E_z = -\frac{\rho}{n_2^2 \omega \varepsilon_0} \left[D_0 e^{-i\rho|x|} - F_0 e^{i\rho|x|} \right]$$

(8.4-36)

with

$$F_0 = D_0^*$$

(8.4-37)

$$D_0 = \frac{1}{2} C_0 e^{i\rho d} \left[\sin \sigma d + i \frac{n_2^2}{n_1^2} \frac{\sigma}{\rho} \cos \sigma d \right]$$

(8.4-38)

and

$$C_0 = \left[\frac{2n_1^2 \rho^2 \omega \varepsilon_0 P}{\pi \beta \left(\frac{n_1^2}{n_2^2} \sigma^2 \sin^2 \sigma d + \frac{n_2^2}{n_1^2} \sigma^2 \cos^2 \sigma d \right)} \right]^{1/2}$$

(8.4-39)

Radiation Modes of the Optical Fiber

The radiation modes of the dielectric slab with the imposed restriction (8.3-1) are quite simple. The corresponding modes of the optical fiber are expressed in terms of Bessel functions instead of the simpler circular functions of the slab. The general radiation modes of the optical fiber are very much more complicated. For the special cases $v = 0$ (which corresponds to $(\partial/\partial\phi = 0)$, these modes are given in reference 99. For the special case $v = 1$, the complicated radiation modes are given in reference 96.

The radiation modes of the round optical fiber can be expressed by Bessel and Neumann functions. There arises an interesting mathematical complication that is worth discussing. We have seen that the boundary conditions at the radius $r = a$ of the fiber core provide us with four equations for the determination of the amplitude coefficients of the modes. The radiation modes carry six undetermined coefficients. Two stem from the field expressions inside the rod, and four from those outside the rod. The two additional constants (there are only four in the case of guided modes) stem from the fact that the radiation modes contain incoming as well as outgoing traveling waves. One of the six coefficients can be related to the power carried by the mode. There is, however, still one more undetermined coefficient than there are determining equations. The physical reason for this mathematical puzzle can be explained in the following way. We avoided a similar problem in our derivation of the radiation modes of the slab waveguide by separating them into even and odd modes. Such a separation is not possible in the case of the round rod. There are, however, also two independent modes intermixed that need to be taken apart. Each mode contains an undetermined amplitude coefficient. If two modes are superimposed, there are then two undetermined coefficients. This is just what happens to the radiation modes of the optical fiber. We can separate these modes by the following procedure. We introduce an arbitrary linear relation between the two undetermined coefficients and obtain one radiation mode. We label the remaining amplitude coefficient A_1. A different mode results by using a different linear relation between the two undetermined coefficients. We label the one remaining coefficient A_2. We now have two arbitrary modes 1 and 2. These modes must be orthogonalized in order to obtain a complete orthogonal system of modes. The orthogonalization is still arbitrary to some extent. This arbitrariness always exist when one wants to form two orthogonal vectors out of two linearly independent but nonorthogonal vectors. Two vectors can always be combined in an infinite number of ways into two new orthogonal vectors. The remaining ambiguity can be used to simplify the relations between the two undetermined coefficients. The result of this orthogonalization procedure of the radiation modes of the optical fiber can be seen in reference 96. The resulting equations are very complicated and consequently unsuitable for a simple explanation of the physical processes that lead to radiation from dielectric waveguides.

The Range of the Guided and Radiation Modes

We have discovered that dielectric waveguides possess a discrete spectrum of guided modes and a continuous spectrum of radiation modes. We must, however, investigate the range of possible eigenvalues (propagation constants) of the two sets of modes. This discussion can be based on the constants γ and ρ. We know that guided modes exist for real values of γ. The condition that γ of (8.3-14) as well as κ of (8.3-11) are both real quantities limits the range of β to the following interval

$$n_2 k_0 < |\beta_g| < n_1 k \qquad (8.4\text{-}40)$$

The range indicated by (8.4-40) outlines the region that the discrete values of the propagation constant of the guided modes can occupy.

Radiation modes are obtained for real values of ρ of (8.4-6). The range of possible β values of the radiation modes is composed of two parts. For real values of β, we obtain from (8.4-6)

$$0 \leq |\beta_r| < n_2 k_0 \qquad (8.4\text{-}41)$$

This range does not overlap with the range of the propagation constants of the guide modes. Both ranges are mutually exclusive, but they meet at $n_2 k_0$. There is, however, also another possible range of values of β_r that results in positive values of ρ. This range is given by imaginary values of β

$$\beta_r = -i\,|\beta_r| \qquad (8.4\text{-}42a)$$

with

$$0 < |\beta_r| < \infty \qquad (8.4\text{-}42b)$$

This range of β values corresponds to evanescent modes. We have encountered evanescent modes in Sections 1.3 and 1.6. We are now finding out that evanescent waves are also possible as radiation modes of dielectric waveguides.

The ranges of the possible values of the propagation constants of guided modes and of radiation modes are shown in Figure 8.4.1. The ranges of guided and radiation modes indicated in Figure 8.4.1 are the same for the round optical fiber and for the slab waveguide.

The evanescent modes of the continuous spectrum of modes are necessary

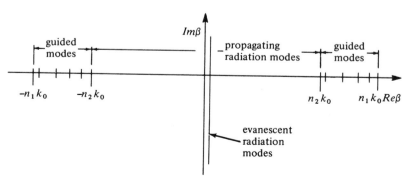

Figure 8.4.1 Spectral β range of guided and radiation modes of dielectric waveguides.

to express the fine details of the field shape close to the surface of the waveguide. However, they do not carry power away from the guide, and are thus not very important for the study of radiation losses of the guided modes of dielectric waveguides.

8.5 ORTHOGONALITY RELATIONS

The modes of the slab waveguide as well as the round optical fiber modes are orthogonal among each other.[97] Instead of proving the orthogonality relations by direct calculation, we provide a general proof that holds for all systems of cylindrical symmetry. The term "cylindrical symmetry" implies only that the modes do not change (except for phase) along the axis of the system.

Using the time dependence (8.3-7), we can write Maxwell's equations (1.2-1) and (1.2-2) as follows

$$\nabla \times \mathbf{H} = i\omega\varepsilon\mathbf{E} \tag{8.5-1}$$

and

$$\nabla \times \mathbf{E} = -i\omega\mu_0\mathbf{H} \tag{8.5-2}$$

We assume that the electric and magnetic fields belong to individual modes of the structure, which we label with indices v and μ. We apply (8.5-1) to mode v, and multiply the complex conjugate of this equation with $\mathbf{E}_\mu$. Equation (8.5-2) is applied to mode μ and multiplied with $\mathbf{H}_v^*$. Subtraction of the two resulting equations and integration leads to the expression

$$i\omega \int_{-\infty}^{\infty} (\varepsilon\mathbf{E}_\mu \cdot \mathbf{E}_v^* - \mu_0\mathbf{H}_\mu \cdot \mathbf{H}_v^*)dxdy = -\int_{-\infty}^{\infty} [\mathbf{E}_\mu \cdot (\nabla \times \mathbf{H}_v^*)$$
$$-\mathbf{H}_v^* \cdot (\nabla \times \mathbf{E}_\mu)]dxdy \tag{8.5-3}$$

This relation can also be written in the following way

$$i\omega \int_{-\infty}^{\infty} (\varepsilon\mathbf{E}_\mu \cdot \mathbf{E}_v^* - \mu_0\mathbf{H}_\mu \cdot \mathbf{H}_v^*)dxdy = -\int_{-\infty}^{\infty} \nabla \cdot (\mathbf{H}_v^* \times \mathbf{E}_\mu)dxdy \tag{8.5-4}$$

The z dependence of the modes is given by (8.3-7), so that (8.5-4) can be expressed as

$$i\omega \int_{-\infty}^{\infty} (\varepsilon\mathbf{E}_\mu \cdot \mathbf{E}_v^* - \mu_0\mathbf{H}_\mu \cdot \mathbf{H}_v^*)dxdy = i(\beta_\mu - \beta_v)\int_{-\infty}^{\infty} (\mathbf{H}_v^* \times \mathbf{E}_\mu)_z dxdy$$
$$-\int_{-\infty}^{\infty} \nabla_t \cdot (\mathbf{H}_v^* \times \mathbf{E}_\mu)dxdy \tag{8.5-5}$$

The operator ∇_t is the transverse part of the operator ∇. The integration is extended over the infinite cross section in the x, y plane. We can apply the divergence theorem to the second integral on the right-hand side of (8.5-5)

$$\int_{-\infty}^{\infty} \nabla_t \cdot (\mathbf{H}^* \times \mathbf{E}_\mu) dx dy = \int_C (\mathbf{H}_\nu^* \times \mathbf{E}_\mu) \cdot \mathbf{n} \, ds \qquad (8.5\text{-}6)$$

The line integral on the right-hand side is extended over an infinitely large circle C whose outward normal is $\mathbf{n}$. If one or both of the modes are guided modes of the waveguide, its field strength at infinity vanishes, and the integral on the right-hand side of (8.5-6) is zero. However, the integral vanishes also if both modes are radiation modes of the structure. This fact can be made plausible in the following way. The radiation modes are oscillatory as functions of x and y. We can argue that the orthogonality relations for radiation modes are only useful if they appear under an integral with respect to one of the mode labels ν or μ. (The mode labels for radiation modes are continuous variables.) For $\nu \neq \mu$, an integration over an infinitesimal region of ν or μ must result in the vanishing of any oscillatory expression that contains infinitely large factors (x and y) in its argument. The same reasoning allows us to set the delta function (8.4-16) equal to zero for $x \neq 0$. Thus we see that the integral on the right-hand side of (8.5-6) vanishes for all modes as long as $\nu \neq \mu$. We have now derived the important relation

$$(\beta_\nu - \beta_\mu) \int_{-\infty}^{\infty} (\mathbf{E}_\mu \times \mathbf{H}_\nu^*)_z dx dy = \omega \int_{-\infty}^{\infty} (\varepsilon \mathbf{E}_\mu \cdot \mathbf{E}_\nu^* - \mu_0 \mathbf{H}_\mu \cdot \mathbf{H}_\nu^*) dx dy \quad (8.5\text{-}7)$$

The subscript z indicates the z component of the vector product. We take the complex conjugate of (8.5-7), interchange the indices ν and μ, and subtract the resulting equation from the original equation (8.5-7). This procedure results in

$$(\beta_\nu - \beta_\mu) \int_{-\infty}^{\infty} \left[(\mathbf{E}_\mu \times \mathbf{H}_\nu^*)_z + (\mathbf{E}_\nu^* \times \mathbf{H}_\mu)_z \right] dx dy = 0 \qquad (8.5\text{-}8)$$

If $\beta_\nu \neq \beta_\mu$, we also have

$$\int_{-\infty}^{\infty} \left[(\mathbf{E}_\mu \times \mathbf{H}_\nu^*)_z + (\mathbf{E}_\nu^* \times \mathbf{H}_\mu)_z \right] dx dy = 0 \qquad (8.5\text{-}9)$$

It is apparent from inspection of (1.4-16) through (1.4-23) that Maxwell's equations are invariant with respect to the following transformation

$$\left. \begin{array}{l} \beta \rightarrow -\beta \\ E_z \rightarrow -E_z \\ H_z \rightarrow +H_z \\ E_x \rightarrow +E_x \\ E_y \rightarrow +E_y \\ H_x \rightarrow -H_x \\ H_y \rightarrow -H_y \end{array} \right\} \qquad (8.5\text{-}10)$$

The transformation (8.5-10) changes any given solution of Maxwell's equations into another solution of Maxwell's equations. Our relation (8.5-9) holds for any two mode solutions of Maxwell's equations. We subject the mode μ (but not the mode ν) to the transformation (8.5-10). Since (8.5-9) holds for any two mode solutions, it also holds for the new combination of old mode ν and transformed mode μ. We thus have

$$\int_{-\infty}^{\infty} [(\mathbf{E}_\mu \times \mathbf{H}_\nu^*)_z - (\mathbf{E}_\nu^* \times \mathbf{H}_\mu)_z] dx dy = 0 \qquad (8.5\text{-}11)$$

By adding (8.5-11) to (8.5-9), we obtain

$$\int_{-\infty}^{\infty} (\mathbf{E}_\mu \times \mathbf{H}_\nu^*)_z dx dy = 0 \qquad \text{for} \qquad \nu \neq \mu \qquad (8.5\text{-}12)$$

Note that the step from (8.5-9) to (8.5-12) does not work for $\beta_\mu = -\beta_\nu$. Equation (8.5-12) is the desired orthogonality relation between any two modes of any lossless dielectric waveguide. If the waveguide is lossy, the eigenvalues β_ν and β_μ become complex, and our derivation cannot be carried through in its present form. Equation (8.5-12) holds for any two modes regardless of whether mode ν and μ are both guided modes, both radiation modes, or whether one of the modes is a guided mode while the other one is a radiation mode. Equation (8.5-12) implies also that the right-hand side of (8.5-7) vanishes for $\nu \neq \mu$ as well as $\nu = \mu$. However, a mode is not orthogonal to its own counterpart traveling in the opposite direction ($\beta_\mu = -\beta_\nu$).

We can immediately apply (8.5-12) to derive a few useful relations for the slab waveguide modes. Equation (8.3-17) can be extended, so that we have for any two guided TE modes regardless of whether they are even or odd or mixed

$$\frac{\beta_\nu}{2\omega\mu_0} \int_{-\infty}^{\infty} E_{\nu y} E_{\mu y}^* \, dx = P\delta_{\nu\mu} \qquad (8.5\text{-}13)$$

$\delta_{\nu\mu}$ is the Kronecker delta symbol, which is zero for $\nu \neq \mu$ and unity for $\nu = \mu$. The orthogonality relation for even or odd TE radiation modes was already stated in (8.4-11).

For guided TM modes, we obtain as an extension of (8.3-39)

$$\frac{\beta_\nu}{2\omega\varepsilon_0} \int_{-\infty}^{\infty} \frac{1}{n^2} H_{\nu y} H_{\mu y}^* \, dx = P\delta_{\nu\mu} \qquad (8.5\text{-}14)$$

The orthogonality relation for TM radiation modes is

$$\frac{\beta}{2\omega\varepsilon_0} \int_{-\infty}^{\infty} \frac{1}{n^2} H_y(\rho) H_y^*(\rho') = P\delta(\rho - \rho') \qquad (8.5\text{-}15)$$

The orthogonality relations (8.5-13) and (8.5-14) hold also if one of the modes is a radiation mode while the other one is a guided mode. The right-hand side of both equations vanishes in this case.

Any arbitrary field distribution of the slab waveguide can be expressed in terms of the orthogonal modes of the guide. It is thus possible to write

$$E_y = \sum_\nu c_\nu E_{\nu y} + \sum \int_0^\infty q(\rho) E_y(\rho) d\rho \qquad (8.5\text{-}16)$$

The first sum extends over all even and odd TE modes, while the combination of sum and integral extends over all radiation modes. The summation symbol indicates that both the even and odd modes must be used for this expansion. The other field components can be expanded similarly. The TM modes, instead of the TE modes, are needed for the components H_y, E_x, and E_z.

The expansion coefficients can be obtained very easily with the help of the orthogonality relations. For the special case (8.5-16), we obtain with the help of (8.5-13)

$$c_\nu = \frac{\beta_\nu}{2\omega\mu_0 P} \int_{-\infty}^\infty E_y E_{\nu y}^* dx \qquad (8.5\text{-}17)$$

Similarly, we obtain with the help of (8.4-11)

$$q(\rho) = \frac{\beta}{2\omega\mu_0 P} \int_{-\infty}^\infty E_y E_y^*(\rho) dx \qquad (8.5\text{-}18)$$

In general, the expansion coefficients c_ν and $q(\rho)$ are functions of z.

For the H_y component, we obtain an expansion in terms of TM modes.

$$H_y = \sum_\nu d_\nu H_{\nu y} + \sum \int_0^\infty p(\rho) H_y(\rho) d\rho \qquad (8.5\text{-}19)$$

with

$$d_\nu = \frac{\beta_\nu}{2\omega\varepsilon_0 P} \int_{-\infty}^\infty \frac{1}{n^2} H_y H_{\nu y}^* dx \qquad (8.5\text{-}20)$$

and

$$p(\rho) = \frac{\beta}{2\omega\varepsilon_0 P} \int_{-\infty}^\infty \frac{1}{n^2} H_y H_y^*(\rho) dx \qquad (8.5\text{-}21)$$

The power P was assumed to be the same for all the modes. It is convenient to use the unit of power for P—for example, 1 w. The actual amount of power that each mode contributes to the field expansion (8.5-16) or (8.5-19) is determined by the values of the expansion coefficients.

For a field consisting only of the field components H_z, H_x, and E_y, the total power carried by the field is given by the expression

$$P_t = P\left\{ \sum_\nu |c_\nu|^2 + \sum \int_0^\infty |q(\rho)|^2 d\rho \right\} \qquad (8.5\text{-}22)$$

This expression follows from the first or second part of (8.3-17) with the help of (8.3-2) and the orthogonality relations (8.5-13) and (8.4-11).

The expansion of arbitrary fields in terms of TE or TM modes is possible only for the fields that also obey the restriction (8.3-1). More general fields can be expanded only in terms of all the modes of the slab waveguide. The expressions (8.5-16) and (8.5-19) are of course still formally correct. It is only necessary to consider the summations extended over all possible modes of the guide.

The possibility of expanding arbitrary fields in terms of modes of the structure is extremely useful for many field problems. We shall treat problems of mode conversion and radiation losses by this normal mode method. The problem of solving Maxwell's equations for arbitrary fields is reduced to systems of ordinary differential equations for the expansion coefficients c_ν and $q(\rho)$. Even though exact solutions of these equation systems are as complicated as the original problem, they are often more convenient for perturbation solutions.

8.6 USEFUL APPROXIMATIONS

The properties of the guided modes are determined by the values of their propagation constants. Since the propagation constant is obtained as a solution of a transcendental eigenvalue equation, it can usually not be obtained in explicit form. However, for certain cases of practical interest, useful approximate solutions of the eigenvalue equations can be obtained. The present section is devoted to the derivation of some approximate solutions of the eigenvalue equations (8.2-70) of the round optical fiber and (8.3-16) and (8.3-23) for the TE modes of the slab waveguide. The corresponding results for TM modes are stated without derivation.

We begin with a study of the slab waveguide. The even TE modes are obtained from the equation

$$\tan \kappa d = \frac{\gamma}{\kappa} \qquad (8.6\text{-}1)$$

while the propagation constants of the odd TE modes follow from the equation

$$\tan \kappa d = -\frac{\kappa}{\gamma} \qquad (8.6\text{-}2)$$

Introducing the abbreviation

$$V = \sqrt{n_1^2 - n_2^2}\, k_0 d \qquad (8.6\text{-}3)$$

with k_0 of (8.3-6), we obtain from (8.3-11) and (8.3-14) the following relation between the constants κ and γ

$$(\kappa d)^2 + (\gamma d)^2 = V^2 \qquad (8.6\text{-}4)$$

At cutoff, we have

$$\gamma d = 0 \qquad (8.6\text{-}5)$$

From (8.6-4), we see that the cutoff values of κd and V are identical

$$(\kappa d)_c = V_c = v\frac{\pi}{2} \qquad v = 0, 1, 2, 3 \ldots \qquad (8.6\text{-}6)$$

The last part of (8.6-6) is obtained with the help of Figure 8.3.1 as a consequence of the eigenvalue equations (8.6-1) and (8.6-2). Even values of v belong to even TE modes, while odd values of v belong to odd modes. Expressing κd in the form

$$\kappa d = v\frac{\pi}{2} + \eta \qquad (8.6\text{-}7)$$

with $\eta \ll 1$, we obtain from (8.6-1) for even TE modes (v even)

$$\gamma d = v\frac{\pi}{2}\,\eta \qquad (8.6\text{-}8)$$

From (8.6-4) and (8.6-6), we find that to first order in η

$$\eta = V - v\frac{\pi}{2} \qquad (8.6\text{-}9)$$

We thus obtain as the approximate solution for even TE modes of the slab waveguide close to cutoff

$$\gamma d = v\frac{\pi}{2}\left(V - v\frac{\pi}{2}\right) \qquad (8.6\text{-}10)$$

However, using odd values for v, we again obtain (8.6-8) from the eigenvalue equation (8.6-2) for odd TE modes. The near cutoff solution (8.6-10) is thus valid for both even and odd TE modes of the slab waveguide. Even modes are obtained for even values of v, while odd modes are associated with odd values of v.

The only exception to this rule is the lowest order even TE mode, whose cutoff frequency is 0. For $V_c = 0$, our derivation fails. Near cutoff, we have $\kappa d \ll 1$ for the lowest order even mode, so that we obtain from (8.6-1)

$$\gamma d = (\kappa d)^2 \qquad (8.6\text{-}11)$$

If we replace κd with the help of (8.6-4), we obtain a quadratic equation for γd with the solution

$$\gamma d = \frac{1}{2}\left(\sqrt{4V^2 + 1} - 1\right) \qquad (8.6\text{-}12)$$

The propagation constant is obtained from (8.3-14). With the help of (8.6-10), we obtain near cutoff, for the even and odd TE modes with the exception of the lowest order mode, the approximation

$$\beta d = \sqrt{(n_2 k_0 d)^2 + v^2 \frac{\pi^2}{4}\left(V - v\frac{\pi}{2}\right)^2} \qquad (8.6\text{-}13)$$

For the lowest order even TE mode, we obtain from (8.3-14) and (8.6-12)

$$\beta d = \sqrt{(n_2 k_0 d)^2 + \frac{1}{2}\left(1 + 2V^2 - \sqrt{1 + 4V^2}\right)} \qquad (8.6\text{-}14)$$

The parameter V is given by (8.6-3). It is also possible to obtain approximate solutions of the eigenvalue equations at the other extreme, far from cutoff. For its derivation, we follow a method that has been successfully used by A. W. Snyder.[90,92]

We consider κd as a function of V, and take the V derivative of (8.6-1)

$$\frac{1}{\cos^2 \kappa d}\frac{\partial(\kappa d)}{\partial V} = -\frac{\gamma d}{(\kappa d)^2}\frac{\partial(\kappa d)}{\partial V} + \frac{V - (\kappa d)\dfrac{\partial(\kappa d)}{\partial V}}{(\kappa d)(\gamma d)} \qquad (8.6\text{-}15)$$

We used (8.6-4) to express γd in terms of κd and V. From (8.6-1) and (8.6-4), we obtain

$$\cos^2 \kappa d = \frac{(\kappa d)^2}{V^2} = \frac{\kappa^2}{(n_1^2 - n_2^2)k_0^2} \qquad (8.6\text{-}16)$$

This allows us to express (8.6-15) in the following form

$$\frac{\partial(\kappa d)}{\partial V} = \frac{\kappa d}{V(1 + \gamma d)} \qquad (8.6\text{-}17)$$

Far from cutoff, we have $\gamma d \gg 1$ and $V \gg 1$, so that we obtain from (8.6-4) the approximation $\gamma d = V$. Equation (8.6-17) can thus be approximated as

$$\frac{\partial(\kappa d)}{\partial V} = \frac{\kappa d}{V(1 + V)} \qquad (8.6\text{-}18)$$

For $V \to \infty$, we see from Figure 8.3.1 that for the even TE modes

$$\kappa d = (v + 1)\frac{\pi}{2} \qquad (8.6\text{-}19)$$

with $v = 0, 2, 4 \ldots$ We thus find as a solution of the differential equation (8.6-18)

$$\kappa d = (v + 1) \frac{\pi}{2} \frac{V}{1 + V} \tag{8.6-20}$$

Equation (8.6-20) is not only the solution of the eigenvalue equation (8.6-1) for the even TE modes of the slab far from cutoff, but it can be shown by direct substitution into (8.6-2) that it also approximately satisfies (8.6-2) for the odd TE modes if $V \gg 1$. Again we obtain the values for the even modes for even integer values of v and the odd modes for odd integer v. The propagation constant β is of course obtained from (8.3-11)

$$\beta = \sqrt{n_1^2 k_0^2 - \kappa^2} = \sqrt{n_2^2 k_0^2 + \gamma^2} \tag{8.6-21}$$

Figure 8.6.1 shows a comparison of the approximate values of κd for the lowest order even TE mode obtained from (8.6-4) and (8.6-12) for small V values and from (8.6-20) with $v = 0$ for large values of V. Also shown in the figure is the solution of the exact equation (8.6-1). It is apparent that the approximate values of κd represent the exact values quite well in their region of applicability.

We present without derivation the results of corresponding calculations for the even and odd TM modes of the slab waveguide obtained from (8.3-38) and (8.3-45). Near cutoff, we obtain

$$\gamma d = \frac{n_2^2}{n_1^2} v \frac{\pi}{2} \left(V - v \frac{\pi}{2} \right) \tag{8.6-22}$$

for the even TM modes with even integer values of $v(v \neq 0)$ and for the odd TM modes with odd integer values of v. For the lowest order even TM mode, we obtain

$$\gamma d = \frac{1}{2} \frac{n_1^2}{n_2^2} \left(\sqrt{4 \frac{n_2^4}{n_1^4} V^2 + 1} - 1 \right) \tag{8.6-23}$$

Far from cutoff, we have

$$\kappa d = (v + 1) \frac{\pi}{2} \frac{V}{\left(\frac{n_2^2}{n_1^2} \right) + V} \tag{8.6-24}$$

with even integer v for even TM modes and with odd integer v for odd TM modes of the slab waveguide.

Similar approximations for γd near cutoff and for κd far from cutoff can also be derived for the modes of the round optical fiber. It is convenient to

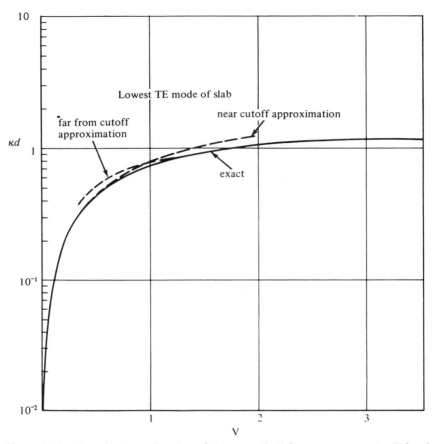

Figure 8.6.1   Plot of κd as a function of the normalized frequency parameter V for the lowest order TE mode of the slab waveguide (solid line). The approximate solutions are shown as dotted lines.

use the eigenvalue equation in the form (8.2-70). Near cutoff, where $\gamma d \ll 1$, (8.2-70) can be approximated by (8.2-76) with H^- given by (8.2-75) for $v = 2, 3, 4 \ldots$. We use the approximation*

$$\kappa a = V - \frac{(a\gamma)^2}{2V_c} \tag{8.6-25}$$

V_c is the value of V at cutoff. We obtain $V_c = \kappa_c a$ from (8.2-94) for EH modes and from (8.2-95) for HE modes. By using expansions of the Bessel functions, with the help of (8.6-25) in terms of $(a\gamma)^2/2V_c$, we find from (8.2-75) and (8.2-76) the following approximation for both the EH and HE modes of the round fiber near their respective cutoffs

*The definitions (8.6-3) and (8.6-4) apply also to fibers with $d = a$ indicating the fiber core radius.

$$(\gamma a)^2 =$$

$$\frac{V_c J_v(V)\left[\dfrac{V}{v-1}J_v(V) - (\varepsilon+1)J_{v-1}(V)\right]}{\left\{\left[\dfrac{v-2}{2(v-1)}(\varepsilon+1) - 1\right]J_v^2(V_c) + \left[\dfrac{2\varepsilon}{V_c} + \dfrac{V_c}{v-1}\left(1 + \dfrac{\varepsilon+1}{4v}\right)\right]J_v(V_c)J_{v-1}(V_c)\right.}$$
$$\left. - \dfrac{2\varepsilon + v(\varepsilon+1)}{2v}J_{v-1}^2(V_c)\right\}$$

$$(8.6\text{-}26)$$

We used the abbreviation

$$\varepsilon = \frac{n_1^2}{n_2^2} \tag{8.6-27}$$

Equation (8.6-26) holds for both EH and HE modes for $v \geq 2$. Note that the Bessel functions in the numerator depend on V, while the Bessel functions in the denominator depend on V_c. One of the two terms in the numerator is small when V is near cutoff for EH modes according to (8.2-94) or for HE modes according to (8.2-95). Equation (8.6-26) does not hold for $v = 0$. The important approximation for $v = 1$ is obtained by using (8.2-74) in (8.2-76). We obtain, near cutoff

$$\frac{\varepsilon + 1}{\kappa a}\frac{J_0(\kappa a)}{J_1(\kappa a)} + 2\ln\frac{\Gamma \gamma a}{2} = 0 \tag{8.6-28}$$

with $\Gamma = 1.781672$. Near cutoff, we can use the approximation

$$\kappa a \approx V \tag{8.6-29}$$

We thus obtain for the $EH_{1\mu}$ and $HE_{1\mu}$ modes the approximation

$$\gamma a = 1.122 \exp\left\{-\frac{\varepsilon+1}{2V}\frac{J_0(V)}{J_1(V)}\right\} \tag{8.6-30}$$

How well this approximation works is shown in Figure 8.6.2 for the HE_{11} mode ($V_c = 0$) for the case $\varepsilon - 1 \ll 1$. Even for $V = 1$, the approximation cannot be distinguished from the exact solution on the scale of the figure. Figure 8.6.2 shows how very rapidly γa decreases for decreasing values of V. It is apparent that the notion that the HE_{11} mode is guided even for zero frequencies is really quite academic. Below $V = 0.5$, the mode can hardly be considered as guided, since its field extends undiminished for thousands of core radii outside the fiber core. Equation (8.6-30) can be further approximated by using the approximations for small arguments (8.2-82) and (8.2-83) of the Bessel functions. However, with these approximations, the agreement with the exact solution is not nearly as good.

Unfortunately, the same method does not work for the case $v = 0$ applying

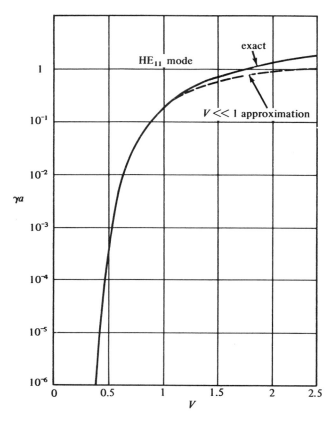

Figure 8.6.2 Plot of γa as a function of V for the HE_{11} mode of the round optical fiber. The dotted line is the approximate solution.

to the TE and TM modes. Using similar approximations, we obtain from (8.2-70) as the near cutoff approximation

$$\text{for TM modes} \quad (\gamma a)^2 = \frac{-V_c(V - V_c)}{\varepsilon \ln \dfrac{\Gamma \gamma a}{2}} \tag{8.6-31}$$

Equation (8.6-31) applies to TE modes if we set $\varepsilon = 1$.

Even though this equation is simpler than the original eigenvalue equation (8.2-70), it is still in implicit form and requires further approximate techniques to obtain γa as a function of V. However, (8.6-31) makes it easy to calculate $V - V_c$ as a function of γa, so that a graphic solution of the equation is simple. For the TE and TM mode solution of (8.6-31), V_c is obtained from (8.2-92).

For the TE modes of the slab waveguide and the TE modes of the round fiber, it is possible to express κa and γa (or κd and γd) as functions of V without any need for specifying the value of $\varepsilon = n_1^2/n_2^2$ separately. For all other modes, we need to know ε as well as V in order to obtain solutions for κa and γa. However, for many cases of practical interest, the ratio n_1/n_2 is so nearly unity that a good approximation is obtained by setting $\varepsilon = 1$ in (8.2-70). This approximation is particularly useful for most cladded optical fibers, whose core to cladding index ratios are typically so nearly unity that $(n_1/n_2) - 1 < 0.01$. For the important cases of this kind, we can approximate (8.2-70) by setting $\varepsilon = 1$ and obtain immediately, with the help of (8.2-56) through (8.2-59), the two eigenvalue equations

$$\text{for HE modes} \qquad \frac{J_{\nu-1}(\kappa a)}{\kappa a J_\nu(\kappa a)} = \frac{H_{\nu-1}^{(1)}(i\gamma a)}{i\gamma a H_\nu^{(1)}(i\gamma a)} \tag{8.6-32}$$

and

$$\text{for EH modes} \qquad \frac{J_{\nu+1}(\kappa a)}{\kappa a J_\nu(\kappa a)} = \frac{H_{\nu+1}^{(1)}(i\gamma a)}{i\gamma a H_\nu^{(1)}(i\gamma a)} \tag{8.6-33}$$

With the help of (8.2-64), the cutoff conditions (8.2-95) can be simplified for this special case $(n_1 \approx n_2)$ to read

$$\text{for HE modes} \qquad J_{\nu-2}(V_c) = 0 \tag{8.6-33}$$

while the cutoff condition (8.2-94) remains unchanged

$$\text{for EH modes} \qquad J_\nu(V_c) = 0 \tag{8.6-34}$$

Far from cutoff $(\gamma a \to \infty)$, we find from (8.6-32) and (8.6-33) the conditions

$$\text{for HE modes} \qquad J_{\nu-1}(\kappa a) = 0 \qquad \text{for} \qquad V \to \infty \tag{8.6-35}$$

and

$$\text{for EH modes} \qquad J_{\nu+1}(\kappa a) = 0 \qquad \text{for} \qquad V \to \infty \tag{8.6-36}$$

The approximate eigenvalue equations (8.6-32) and (8.6-33) are much more convenient for obtaining solutions than the original eigenvalue equation (8.2-70). Furthermore, we already know the approximations near cutoff (8.6-26), (8.6-30), and (8.6-31). However, we can also obtain very useful approximations for the region far from cutoff. We use again the method first employed by A. W. Snyder.[90,92] Starting with HE modes, we assume that $V \gg 1$ and $\gamma a \gg 1$, so that we can write, with the help of (8.2-54)

$$\frac{H_{\nu-1}^{(1)}(i\gamma a)}{H_\nu^{(1)}(i\gamma a)} = i \qquad \text{for} \qquad V \to \infty \tag{8.6-37}$$

We now obtain (8.6-32) in the approximate form

$$\gamma a J_{\nu-1}(\kappa a) = \kappa a J_\nu(\kappa a) \tag{8.6-38}$$

Next, we form the derivative of (8.6-38) with respect to V. After differentiating, we set approximately $\gamma a = V$ and eliminate the Bessel functions from the equation with the help of (8.6-38). We thus obtain

$$\frac{d(\kappa a)}{dV} = \frac{\kappa a}{V[V - 2(v - 1)]} \tag{8.6-39}$$

The solution of this differential equation is the desired approximation

for HE modes $v \neq 1$ $\kappa a = (\kappa a)_\infty \left[1 - \frac{2(v - 1)}{V}\right]^{1/[2(v-1)]}$ for $V \gg 1$

$$\tag{8.6-40}$$

The value for $(\kappa a)_\infty$ is the root of Equation (8.6-35). Roots of Bessel functions are tabulated in books presenting tables of these functions.[8,11]

The solution for $v = 1$ follows either by taking the limit $v \to 1$ in (8.6-40) or by integrating (8.6-39), with the result

for $HE_{1\mu}$ $\kappa a = (\kappa a)_\infty e^{-1/V}$ for $V \gg 1$ (8.6-41)

A similar procedure allows us to obtain the approximate solution of (8.6-33), with the result

for EH modes $\kappa a = (\kappa a)_\infty \left[1 - \frac{2(v + 1)}{V}\right]^{1/[2(v+1)]}$ for $V \gg 1$

$$\tag{8.6-42}$$

This equation holds for all positive integer values of v. The constant $(\kappa a)_\infty$ is obtained from (8.6-36) as the root of the Bessel function.

Figure 8.6.3 illustrates the accuracy of the approximation (8.6-41) for the important HE_{11} mode. The solid curve is the exact solution of (8.6-32), while the dotted curve shows the approximation for $V \gg 1$. Also shown as the dash-dotted curve are the κa values that can be obtained from (8.6-30) with the help of (8.6-4). Even though the approximations are of course most accurate near their respective limits, the approximation of κa which they afford does not deviate more than 7 percent at $V = 1.75$, where both approximations depart by nearly the same amount from the exact curve. For many purposes, the approximations near cutoff and far from cutoff provide solutions of the eigenvalue equation that are of sufficient accuracy.

Far from cutoff and for very small values of $n_1 - n_2$, it is possible to give simple approximations for the electromagnetic field components of the round fiber modes. These approximations were first derived by A. W. Snyder.[90,92] We assume that the waveguide is operated so far from cutoff that we can approximate $\beta = n_1 k$ and can assume $\gamma a \gg 1$. The eigenvalue equations (8.6-32) and (8.6-33) can thus be written in the form

for HE modes $\gamma a J_{v-1}(\kappa a) = \kappa a J_v(\kappa a)$ (8.6-43)

for EH modes $\gamma a J_{v+1}(\kappa a) = -\kappa a J_v(\kappa a)$ (8.6-44)

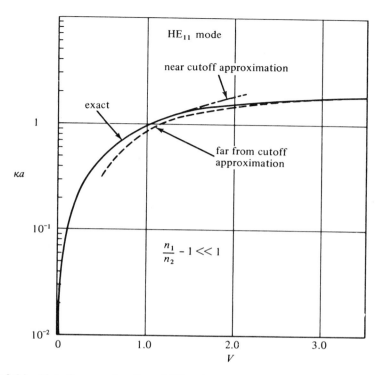

Figure 8.6.3 Plot of κa as a function of V for the HE_{11} mode of the round optical fiber. The dotted lines are approproximate solutions.

where the approximation (8.2-54) has been used. Using the functional relation[11,61]

$$J_v'(\kappa a) = -\frac{v}{\kappa a}J_v(\kappa a) + J_{v-1}(\kappa a) = \frac{v}{\kappa a}J_v(\kappa a) - J_{v+1}(\kappa a) \quad (8.6\text{-}45)$$

and the approximations mentioned above, we obtain the following approximation for (8.2-46)

$$B = \mp i\frac{|v|}{v}n\sqrt{\frac{\varepsilon_0}{\mu_0}}A \qquad (8.6\text{-}46)$$

The upper sign holds for HE modes, while the lower sign holds for EH modes. The refractive index n stands for $n_1 \approx n_2$. Using (8.6-45) once more, we obtain, with the help of (8.6-46) from (8.2-25) through (8.2-30), for $r \leq a^*$

$$E_z = AJ_v(\kappa r)e^{iv\phi} = \frac{1}{2}F_{z|v|}^{(1)}e^{iv\phi} \qquad (8.6\text{-}47)$$

---

* The factor $\exp[i(\omega t - \beta z)]$ has been omitted.

$$H_z = \mp i \frac{|v|}{v} n \sqrt{\frac{\varepsilon_0}{\mu_0}} A J_v(\kappa r) e^{iv\phi} = \mp i \frac{|v|}{2v} G^{(1)}_{z|v|} e^{iv\phi} \tag{8.6-48}$$

$$E_r = \mp i \frac{nk_0 A}{\kappa} J_{v \mp 1}(\kappa r) e^{iv\phi} = \mp \frac{i}{2} F^{(1)}_{|v| \mp 1} e^{iv\phi} \tag{8.6-49}$$

$$E_\phi = \frac{|v|}{v} \frac{nk_0 A}{\kappa} J_{v \mp 1}(\kappa r) e^{iv\phi} = \frac{|v|}{2v} F^{(1)}_{|v| \mp 1} e^{iv\phi} \tag{8.6-50}$$

$$H_r = -\frac{|v|}{v} \sqrt{\frac{\varepsilon_0}{\mu_0}} \frac{n^2 k_0 A}{\kappa} J_{v \mp 1}(\kappa r) e^{iv\phi} = -\frac{|v|}{2v} G^{(1)}_{|v| \mp 1} e^{iv\phi} \tag{8.6-51}$$

$$H_\phi = \mp i \sqrt{\frac{\varepsilon_0}{\mu_0}} \frac{n^2 k_0 A}{\kappa} J_{v \mp 1}(\kappa r) e^{iv\phi} = \mp \frac{i}{2} G^{(1)}_{|v| \mp 1} e^{iv\phi} \tag{8.6-52}$$

From (8.2-33) through (8.2-38), we obtain the field expressions for $r > a$

$$E_z = A \sqrt{\frac{a}{r}} J_v(\kappa a) e^{-\gamma(r-a)} e^{iv\phi} = \frac{1}{2} F^{(2)}_{z|v|} e^{iv\phi} \tag{8.6-53}$$

$$H_z = \mp i \frac{|v|}{v} n \sqrt{\frac{\varepsilon_0}{\mu_0}} A \sqrt{\frac{a}{r}} J_v(\kappa a) e^{-\gamma(r-a)} e^{iv\phi} = \mp i \frac{|v|}{2v} G^{(2)}_{z|v|} e^{iv\phi} \tag{8.6-54}$$

$$E_r = \mp i \frac{nk_0 A}{\kappa} \sqrt{\frac{a}{r}} J_{v \mp 1}(\kappa a) e^{-\gamma(r-a)} e^{iv\phi} = \mp \frac{i}{2} F^{(2)}_{|v| \mp 1} e^{iv\phi} \tag{8.6-55}$$

$$E_\phi = \frac{|v|}{v} \frac{nk_0 A}{\kappa} \sqrt{\frac{a}{r}} J_{v \mp 1}(\kappa a) e^{-\gamma(r-a)} e^{iv\phi} = \frac{|v|}{2v} F^{(2)}_{|v| \mp 1} e^{iv\phi} \tag{8.6-56}$$

$$H_r = -\frac{|v|}{v} \sqrt{\frac{\varepsilon_0}{\mu_0}} \frac{n^2 k_0 A}{\kappa} \sqrt{\frac{a}{r}} J_{v \mp 1}(\kappa a) e^{-\gamma(r-a)} e^{iv\phi} = -\frac{|v|}{2v} G^{(2)}_{|v| \mp 1} e^{iv\phi} \tag{8.6-57}$$

$$H_\phi = \mp i \sqrt{\frac{\varepsilon_0}{\mu_0}} \frac{n^2 k_0 A}{\kappa} \sqrt{\frac{a}{r}} J_{v \mp 1}(\kappa a) e^{-\gamma(r-a)} e^{iv\phi} = \mp \frac{i}{2} G^{(2)}_{|v| \mp 1} e^{iv\phi} \tag{8.6-58}$$

In order to derive these equations, we used (8.2-44) and (8.2-45) to express C and D in terms of A and B. All Hankel function were approximated with the help of (8.2-54), and the approximate eigenvalue equations (8.6-43) and (8.6-44) were used to express $J_v(\kappa a)$ in terms of $J_{v \mp 1}$. The upper signs refer again to the HE modes, while the lower signs belong to EH modes. The absolute value signs occur in the equations for those v values that stem from the functional relations (8.6-45), and indicate the order number of a Bessel function. It was pointed out in Section 8.2, below Equation (8.2-49), that the v appearing in the exponential function $\exp(iv\phi)$ can also be replaced with its negative value without affecting the order number v of the Bessel functions. When we want to change the exponent of $\exp(iv\phi)$ to $\exp(-iv\phi)$, we replace v with its negative value except in the order number of the Bessel functions and except where it is written explicitly with an absolute value sign. By replacing v with $-v$ in this manner in (8.6-47) through (8.6-58) and by adding and subtracting the new equations from the old ones, we obtain the electromagnetic field components of the fiber modes in the following form

$$E_z = F_{zv} \begin{Bmatrix} \cos v\phi \\ i \sin v\phi \end{Bmatrix} \tag{8.6-59}$$

$$H_z = \mp G_{zv} \begin{Bmatrix} -\sin v\phi \\ i \cos v\phi \end{Bmatrix} \tag{8.6-60}$$

$$E_r = \mp F_{v\mp1} \begin{Bmatrix} i \cos v\phi \\ -\sin v\phi \end{Bmatrix} \tag{8.6-61}$$

$$E_\phi = F_{v\mp1} \begin{Bmatrix} i \sin v\phi \\ \cos v\phi \end{Bmatrix} \tag{8.6-62}$$

$$H_r = -G_{v\mp1} \begin{Bmatrix} i \sin v\phi \\ \cos v\phi \end{Bmatrix} \tag{8.6-63}$$

$$H_\phi = \mp G_{v\mp1} \begin{Bmatrix} i \cos v\phi \\ -\sin v\phi \end{Bmatrix} \tag{8.6-64}$$

(The expressions in brackets are multiplied with F and G.) The two possible signs and the two sets of sin or cos functions are to be used independently of each other. The two signs belong to the HE modes (upper sign) and to the EH modes (lower sign), as before. The two sets of circular functions express two different polarizations of the vector field. Each polarization can be assumed by either set of modes (HE or EH). The real notation for the ϕ dependence is more immediately meaningful than the complex notation employed earlier. The special form of the F and G functions, as defined by (8.6-47) through (8.6-58), holds of course only for the far from cutoff approximation for small values of $n_1 - n_2$. However, the form of Equations (8.6-59) through (8.6-64) is generally correct and is independent of any approximation. We shall use this type of notation in Section 10.4 for the description of the HE_{11} mode.

The approximate field expressions derived above are useful for finding a simple expression for the field in Cartesian coordinates. Using the inverse of the transformation (8.2-3) and (8.2-4)

$$F_x = F_r \cos \phi - F_\phi \sin \phi \tag{8.6-65}$$

and

$$F_y = F_r \sin \phi + F_\phi \cos \phi \tag{8.6-66}$$

we can immediately express the transverse field components (8.6-61) through (8.6-64) as follows

$$E_x = \mp F_{v \mp 1} \begin{Bmatrix} i \cos (v \mp 1)\phi \\ - \sin (v \mp 1)\phi \end{Bmatrix} \tag{8.6-67}$$

$$E_y = F_{v \mp 1} \begin{Bmatrix} i \sin (v \mp 1)\phi \\ \cos (v \mp 1)\phi \end{Bmatrix} \tag{8.6-68}$$

$$H_x = -G_{v \mp 1} \begin{Bmatrix} i \sin (v \mp 1)\phi \\ \cos (v \mp 1)\phi \end{Bmatrix} \tag{8.6-69}$$

$$H_y = \mp G_{v \mp 1} \begin{Bmatrix} i \cos (v \mp 1)\phi \\ - \sin (v \mp 1)\phi \end{Bmatrix} \tag{8.6-70}$$

The upper and lower signs indicate, as always, the HE and EH modes. It is immediately apparent from these equations that the $HE_{1\mu}$ modes have only one transverse E and H component. Using the upper set of signs for the HE modes and taking $v = 1$, we see that, for the upper set of circular functions, only the E_x and H_y components of the $HE_{1\mu}$ modes exist in this approximation. The lower set of circular functions describe a polarization that is rotated by $90°$ since, with its use, we find that only E_y and H_x are nonvanishing. The field of the HE_{11} mode is thus particularly simple far from cutoff and for waveguides with small values of $n_1 - n_2$.

It is possible to form combinations of HE and EH modes that are simpler than the fields (8.6-67) through (8.6-70). In order to accomplish this, it is important to realize that the HE modes of order $v = v' + 1$ are degenerate with the EH modes of order $v = v' - 1$. This degeneracy is not exact. It holds approximately for small values of $n_1 - n_2$ but is independent of the value of γa. By inverting (8.6-32), replacing v with $v' + 1$, and using (8.2-64) to eliminate $J_{v'+1}$ from the equation, we obtain

$$\kappa a \frac{J_{v'-1}(\kappa a)}{J_{v'}(\kappa a)} = i\gamma a \frac{H_{v'-1}^{(1)}(i\gamma a)}{H_{v'}^{(1)}(i\gamma a)} \tag{8.6-71}$$

Exactly the same equation is obtained if we invert (8.6-33) and replace v with $v' - 1$. To the approximation implicit in the form of the eigenvalue equations (8.6-32) and 8.6-33), HE modes of order $v = v' + 1$ have the same propagation constant as the EH modes of order $v = v' - 1$. The same relationship is also apparent in (8.6-40) and (8.6-42).

We now use the lower set of circular functions in (8.6-67) through (8.6-70) for HE modes and EH modes, and form the following combinations

$$(E_x)_{HEv'+1} + (E_x)_{EHv'-1} = 0 \tag{8.6-72}$$

$$(E_y)_{HEv'+1} + (E_y)_{EHv'-1} = 2F_{v'} \cos v'\phi \tag{8.6-73}$$

$$(H_x)_{HEv'+1} + (H_x)_{EHv'-1} = -2G_{v'} \cos v'\phi \tag{8.6-74}$$

$$(H_y)_{HEv'+1} + (H_y)_{EHv'-1} = 0 \tag{8.6-75}$$

These new modes have only the nonvanishing components E_z, E_y, H_z, and H_x, and are thus much simpler in structure than the original HE and EH modes. Subtraction of the HE and EH modes leads to a field that is polarized at 90° to the superposition field expressed by (8.6-72) through (8.6-75). However, since the degeneracy between the modes is not exact, the two modes travel with slightly different phase velocities, and the linearly polarized field becomes elliptically polarized. However, after a distance $\Lambda = 2\pi N/ (\beta_2 - \beta_1)$ with arbitrary integer N (β_2 and β_1 are the slightly different propagation constants of the original EH and HE modes), the superposition field again appears in the linearly polarized form (8.6–72) through (8.6–75).

8.7 GAUSSIAN APPROXIMATION OF THE FUNDAMENTAL MODE

In Section 8.2, we gave an exact description of the modes of round optical fibers in terms of Bessel and Hankel functions. In Section 8.6, we derived approximate expressions for the propagation parameters of these modes and also explained how a simplified approximate description is possible by using superpositions of HE and EH modes. In this section, we go one step further and approximate the guided mode of a single mode fiber (one supporting only one mode, albeit in two possible orthogonal polarizations) by a Gaussian function.[122] This simplification of the description of the field of a single mode fiber is made possible by two observations.

We have seen at the end of Section 8.6 that the guided modes of a fiber with small refractive index difference, $(n_1 - n_2)/n_1 \ll 1$, have one predominant transverse field component, for example E_y. By comparison, E_x and the longitudinal component E_z are very much smaller. This observation allows us to treat the guided mode field as a scalar quantity which need not be described by the full set of Maxwell's equations but can be treated as the solution of the scalar wave equation, (3.2-1)

$$\nabla^2\phi + n^2 k_0^2 \phi = 0 \tag{8.7-1}$$

The function ϕ stands for the dominant transverse electric field component, $k_0 = 2\pi/\lambda$ (λ = vacuum wavelength of light) is the free space propagation constant and $n(x,y)$ is the refractive index of the fiber that, usually, does not depend on z, the coordinate that coincides with the direction of the fiber axis.

The second observation involves the shape of the predominant transverse field component. It can be shown[122] that the field distributions of the dominant guided modes of fibers with a wide class of refractive index distributions have very nearly Gaussian shapes. This is true not only for the HE_{11} mode of the step-index fiber (a fiber with constant core refractive index) but also for the fields of fibers with arbitrary convex refractive index distributions. Therefore, it is reasonable to express the predominant electric field component of the single guided mode as

$$\phi = \left(\frac{2}{\pi}\right)^{1/2} \frac{1}{w} \exp \left(-r^2/w^2\right)e^{-i\beta z} \qquad (8.7\text{-}2)$$

In this expression $r^2 = x^2 + y^2$, w is the width parameter and β the propagation constant of the guided mode field. The factor in front of the exponential function is arbitrary, it was chosen for normalization purposes. Once we accept that (8.7-2) has the correct shape, only two unknown parameters, β and w, remain to be determined with the help of a variational principle.

To lay the groundwork for the determination of the two unknown parameters, we now digress for a discussion of the required variational expression. We claim that the solutions of the wave equation (8.7-1) are functions that minimize the integral

$$J = \int_v \left[(\nabla\phi) \cdot (\nabla\phi^*) - n^2 k_0^2 \phi\phi^*\right]dV = \min \qquad (8.7\text{-}3)$$

(the asterisk indicates complex conjugation) where the integration range extends over a large cylinder with the fiber located at its axis. The length L of the cylinder is arbitrary, its radius is assumed to tend towards infinity. The ∇ operator was defined in Section 1.2 and the other functions and parameter are familiar from our discussion of the wave equation (8.7-1). Application of the well known variational calculus[123] proves that the wave equation (8.7-1) is the Euler equation of the variational expression (8.7-3). This means that the functions that minimize J satisfy the wave equation.

The minimum value of J is actually zero. To show this we rewrite (8.7-3) by performing a partial integration

$$J = \int_s \phi^*(\nabla\phi) \cdot d\mathbf{s} - \int_V \left[\nabla^2\phi + n^2 k_0^2 \phi\right]\phi^* \, dV \qquad (8.7\text{-}4)$$

The surface element $d\mathbf{s}$ is a vector pointing towards the outside of the cylinder in a direction normal to it. The first integral in (8.7-4) extends over the surface of the large cylinder. For a guided mode, the function ϕ vanishes on the curved cylindrical surface with infinite radius. The guided mode field can be written in the form

$$\phi = \hat{\phi}(x,y)e^{-i\beta z} \qquad (8.7\text{-}5)$$

Since the z-dependence is limited to the exponential function in (8.7-5), the integrand of the surface integral in (8.7-4) is independent of z. Thus, the contributions to the integral from the two flat end surfaces of the cylinder are independent of its length and are equal and opposite in value. This shows that the entire surface integral vanishes. However, the contribution of the volume integral in (8.7-4) is also zero if the function ϕ is a solution of the wave equation. This concludes the proof that $J = 0$ if ϕ is a legitimate guided mode field.

Finally, we modify the variational expression (8.7-3) by substitution of (8.7-5). We consider the volume integral as consisting of an integral over the infinite cross section of the fiber that, in turn, is integrated over the length coordinate z. Since the integrand is independent of z, integration over this variable simply multiplies the remaining integral over the cross section by the length L of the cylinder. Dividing by L and writing J instead of J/L, we obtain

$$J = \int_{-\infty}^{\infty}\int \{(\nabla_t \hat{\phi}) \cdot (\nabla_t \hat{\phi}^*) - [n^2(x,y)k_0^2 - \beta^2]\hat{\phi}\hat{\phi}^*\}\, dx\, dy \quad (8.7\text{-}6)$$

The differential operator ∇_t is the transverse part (the x and y derivatives) of ∇. Equation (8.7-6) is the desired variational expression that we need to determine β and w in (8.7-2).

Because we have established that $J = 0$ for solutions of the wave equation, we immediately find an expression for the propagation constant by solving (8.7-6) for β^2.

$$\beta^2 = \frac{\int_{-\infty}^{\infty}\int [n^2 k_0^2 \hat{\phi}\hat{\phi}^* - (\nabla_t \hat{\phi}) \cdot (\nabla_t \hat{\phi}^*)]\, dx\, dy}{\int_{-\infty}^{\infty}\int \hat{\phi}\hat{\phi}^*\, dx\, dy} \quad (8.7\text{-}7)$$

If we knew the function ϕ, its propagation constant β could be computed from (8.7-7). However, even if we do not know the mode function exactly, (8.7-7) provides us with a good approximation of the propagation constant. Using techniques of the variational calculus it can be shown that (8.7-7) remains unchanged to first order if the exact mode function $\hat{\phi}$ in (8.7-7) is replaced by a slightly perturbed function. An integral expression with this remarkable property is called a stationary value.[124] So, even if we do not know the exact form of the mode function we can still obtain a very good approximation for its propagation constant by using in (8.7-7) a function that approximates the exact mode function reasonably well. We substitute the Gaussian approximation (8.7-2) into (8.7-7) and obtain

$$\beta^2 = \left\{\frac{4k_0^2}{w^2}\int_0^{\infty} rn^2(r)\exp(-2r^2/w^2)\, dr\right\} - \frac{2}{w^2} \quad (8.7\text{-}8)$$

The normalization of (8.7-2) was chosen to make the denominator of (8.7-7) unity. The stationary expression (8.7-8) was derived from (8.7-7) under the assumption that the refractive index distribution depends only on the radial coordinate r, a condition that is satisfied by most commonly used optical fibers.

It remains to find a suitable value for the width parameter w which, until now, has remained unknown. To find w we remember that β^2 of (8.7-7) and (8.7-8) is a stationary value. If (8.7-2) were a mode function with the correct value for w, β^2 would not change if w were varied slightly, thus, its derivative would vanish

$$\frac{d\beta^2}{dw} = 0 \qquad (8.7\text{-}9)$$

This condition allows us to determine w. Differentiation of (8.7-8) and substitution into (8.7-9) yields the following implicit equation for w

$$1 + 2k_0^2 \int_0^\infty r\left(\frac{2r^2}{w^2} - 1\right)n^2(r)\, \exp(-2r^2/w^2)\, dr = 0 \qquad (8.7\text{-}10)$$

To find the Gaussian approximation for the dominant mode of a single mode fiber thus requires us to solve (8.7-10) for w and to compute the propagation constant from (8.7-8).

The equations (8.7-8) and (8.7-10) have been derived in reference 122 by direct substitution of (8.7-2) into the wave equation (8.7-1). Equation (8.7-10) was written in reference 122 in a different form, but the two expressions are fully equivalent. The derivation presented here is more general and its validity is more immediately apparent.

We apply our simplified theory to single mode fibers whose cores have refractive index profiles belonging to an important special class—the power

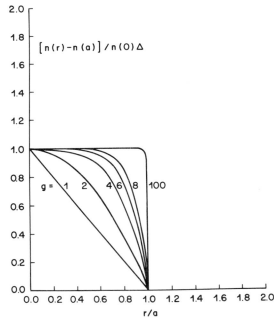

Figure 8.7.1   Normalized power law refractive index profiles for several values of the exponent g. (From D. Marcuse, Gaussian Approximation of the Fundamental Modes of Graded-Index Fibers, *J. Opt. Soc. Am*, Vol. 68, No. 1, January 1978, pp. 103–109. Reprinted with permission.)

y the equation

$$) = \begin{cases} n_1^2[1 - 2(r/a)^g\Delta] & \text{for } r < a \\ n_2^2 = n_1^2(1 - 2\Delta) & \text{for } r > a \end{cases} \quad (8.7\text{-}11)$$

core radius, n_1 the refractive index value at $r = 0$, n_2 the
e cladding—the material surrounding the core, g is the
er law profile and Δ is defined as

$$\Delta = \frac{n_1^2 - n_2^2}{2n_1^2} \quad (8.7\text{-}12)$$

profile defined by (8.7-11) describes a fiber with power
$r < a$ and an infinitely extended cladding with constant
refractive index n_2. Figure 8.7.1. shows the normalized shape of the power law

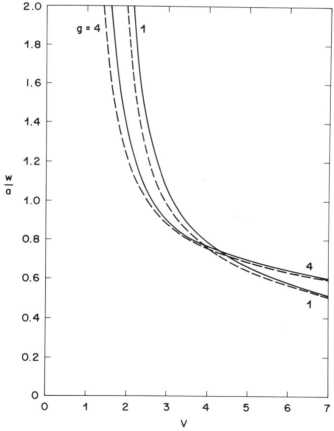

Figure 8.7.2 Comparison of w/a, obtained from numerical integration of the wave equation and subsequent optimization of its width (solid lines), with the approximation obtained from (8.7-10) (dotted lines) for $g = 1$ and 4. (From D. Marcuse, Gaussian Approximation of the Funda-mental Modes of Graded-Index Fibers, J. Opt. Soc. Am, Vol. 68, No. 1, January 1978, pp. 103–109. Reprinted with permission.)

profile of the core for several values of the exponent g; $g = \infty$ corresponds to the step-index profile with constant refractive index n_1 inside the core.

If we set $g = 2$ and let the power law extend to infinity, instead of using $n(r) = n_2$ for $r > a$, (8.7-11) describes the square law medium whose modes were discussed in Section 7.3. When we use the infinitely extended square law medium in (8.7-10), we obtain for the Gaussian beam width parameter

$$w^2 = \frac{a\sqrt{2}}{n_1 k_0 \sqrt{\Delta}} \tag{8.7-13}$$

and for the propagation constant

$$\beta^2 = n_1^2 k_0^2 \left(1 - \frac{2\sqrt{2\Delta}}{n_1 k_0 a}\right) \tag{8.7-14}$$

These solutions are indeed identical with (7.3-21) and (7.3-22) (with $p = q = 0$). The parameters of the power law medium (8.7-11) were defined

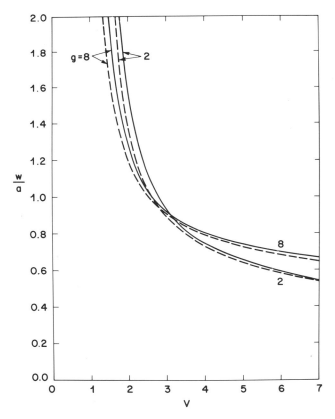

Figure 8.7.3   Same as Figure 8.7.2 with $g = 2$ and 8. (From D. Marcuse, Gaussian Approximation of the Fundamental Modes of Graded-Index Fibers, *J. Opt. Soc. Am*, Vol. 68, No. 1, January 1978, pp. 103–109. Reprinted with permission.)

differently than those of the square law medium (7.3-3) to conform with the standard notation used in the literature.

Next, we show graphs of numerical solutions of (8.7-10) and 8.7-8). For general values of g, and in particular for profiles with constant refractive index values of the cladding region, the integrals in (8.7-8) and (8.7-10) cannot be solved analytically, however computer solutions are easily obtained. For comparison, the wave equation was also solved by a direct numerical method[122] which yields values for the field distribution and the propagation constant that are not influenced by the prior assumption of a Gaussian shape. The numerical solutions for the modal fields were used to define the Gaussian beam width parameter by a fitting procedure.[122] This independent method makes it possible to assess the accuracy of the Gaussian approximation used here. Figures 8.7.2 through 8.7.4 show solutions of (8.7-10) for several values of g as dotted curves. The independent variable V, used in the figures, is defined as

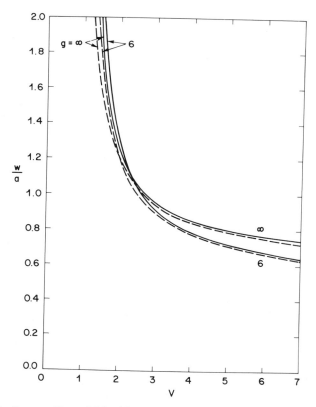

Figure 8.7.4   Same as Figure 8.7.2 with $g = 6$ and ∞. (From D. Marcuse, Gaussian Approximation of the Fundamental Modes of Graded-Index Fibers, *J. Opt. Soc. Am*, Vol. 68, No. 1, January 1978, pp. 103–109. Reprinted with permission.)

$$V = k_0 a (n_1^2 - n_2^2)^{1/2} = n_1 k_0 a \sqrt{2\Delta} \qquad (8.7\text{-}15)$$

The solid curves were obtained by the above mentioned matching procedure and are probably more accurate than the dotted curves computed from our theory. However, the comparison shows how well this simple theory works. The reader may wonder why the dotted and solid curves for $g = 2$ in Figure 8.7.3. do not coincide, since we have just shown that our solution becomes exact for $g = 2$. However, there really is no discrepancy because Figures 8.7.2. through 8.7.4 were computed for the truncated power law profile (8.7-11), while our solution becomes exact for $g = 2$ only if the power law is allowed to extend to infinity.

The propagation constant β varies only between the values $n_1 k_0$ and $n_2 k_0$. For this reason we use the related quantity

$$\kappa = (n_1^2 k_0^2 - \beta^2)^{1/2} \qquad (8.7\text{-}16)$$

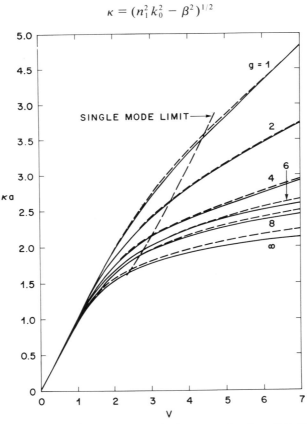

Figure 8.7.5 The propagation parameter κa is shown as a function of V. The solid lines are obtained from the numerical solution of the wave equation, the dotted lines are the result of solving (8.7-8) and (8.7-10). The diagonal dotted line indicates the location of the cutoff of the second mode. (From D. Marcuse, Gaussian Approximation of the Fundamental Modes of Graded-Index Fibers, *J. Opt. Soc. Am*, Vol. 68, No. 1, January 1978, pp. 103–109. Reprinted with permission.)

For small values of Δ, the dimensionless quantity κa, plotted in Figure 8.7.5, is a universal function of V. The dotted curves in Figure 8.7.5 were again computed from (8.7-10) and (8.7-8), while the solid curves were obtained from the 'exact' numerical solution of the wave equation.

The Gaussian approximation for the solutions of single mode fibers is very much simpler than the exact solution, or even than the simplified solution discussed in Section 8.6. For many purposes it may be of sufficient accuracy. The Gaussian approximation is well suited for describing the field inside the fiber core and for computing approximate values of the propagation constant. However, it is a poor approximation for the evanescent field in the cladding far from the core. We know that the exact cladding field is represented by modified Hankel functions which, asymptotically, become exponential functions of r—not Gaussian functions. For applications requiring knowledge of the cladding field, such as computation of the mode loss due to a lossy jacket or the properties of a directional coupler, the Gaussian approximation cannot be used.[125]

CHAPTER

9

DIELECTRIC WAVEGUIDES WITH IMPERFECTIONS

9.1 INTRODUCTION

Dielectric waveguides are multimode structures. Even if only a single guided mode can exist, there is always the continuous spectrum of radiation modes. If the geometry of the waveguide is perfect and if we can neglect losses in the dielectric material itself, the guided modes travel without change and without loss. Studying the modes of a perfect waveguide is an important first step of determining its properties. In order to be able to evaluate the performance of a realistic waveguide, it is necessary to study its behavior if departures from the perfect geometry occur. It is impossible to build dielectric waveguides (or any other structure) to such a perfection that accidental departures from perfect geometry are unobservable.

The imperfections of dielectric waveguides that must be considered are losses of the dielectric material, departures from perfect straightness, in-homogeneities of the dielectric material, and departures of the core cladding interface from a perfect plane in slab waveguides or from perfect circular cylinders in round optical fibers. It would be too tedious to describe the effect of all these imperfections in this book. The losses introduced by dissipative dielectric materials cause mode losses that are similar to the bulk loss of the material. This is particularly true if the losses of the core and the cladding are nearly identical. Of all the imperfections mentioned, we shall discuss the influence of boundary deflections and of waveguide curvature in this chapter.

9.2 SLAB WAVEGUIDE WITH IMPERFECT BOUNDARY

The modes of the slab waveguide were derived in the previous chapter by solving the field equations inside and outside the waveguide and by determining the amplitude and propagation constants by satisfying boundary conditions. This method distracts from the fact that the field solutions of the slab waveguide are really obtained from (8.3-2) through (8.3-4). If we consider the refractive index as a continuous variable and if we were able to solve the reduced wave equation (8.3-4) for this index distribution, there would be no need to bother with boundary conditions. It would be possible to approximate the index distribution by a smoothed function, solve the wave equation (8.3-4), and subsequently allow the index distribution to become as abrupt as desired. It is well to remember that the mode problem of the square law medium was treated entirely as a solution of the reduced wave equation. The only boundary conditions that we encountered were conditions at infinity.

Our treatment of the slab waveguide with imperfect walls is based on finding solutions of the reduced wave equation throughout all space without any need for satisfying boundary conditions except those at infinity. We have seen that any arbitrary field distribution can be expressed as a superposition of modes of the perfect waveguide. It is thus also possible to express the field of the imperfect waveguide in terms of the modes of the ideal guide. This field expansion is substituted into the reduced wave equation, and the expansion coefficients are (approximately) determined so that the field distribution becomes a solution of (8.3-4).

Instead of expanding the field of the imperfect waveguide in terms of the modes of the perfectly straight guide, it is possible to use so-called local normal modes.[113] Local normal modes resemble the mode solutions of the perfect waveguide. In fact, they are identical in form with the exact mode solutions of the perfect guide. However, the guide dimensions appearing as parameters in the mode expressions and the eigenvalue equation are now allowed to be functions of the length dimension z to conform exactly to the actual shape of the imperfect waveguide. The local normal modes thus satisfy the boundary conditions at the deformed core-cladding interface of the imperfect waveguide, and they are also mutually orthogonal to each other; but they do not satisfiy the wave equation or Maxwell's equations. The expansion coefficients for the general field of the imperfect waveguide must be chosen so that the total field becomes a solution of Maxwell's equations. We do not use local normal modes in this book, but employ, instead, an expansion of the general field in terms of the modes of the perfect waveguide. This procedure appears simpler for our purposes.

We require of the departures of the waveguide walls from perfect geometry that they obey the restriction (8.3-1). This limits the boundary imperfections to changes in slab thickness that do not vary in the y direction. The departures of the waveguide boundary are supposed to be very slight, so that they can

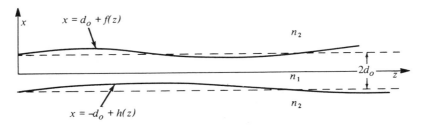

Figure 9.2.1   Slab waveguide with imperfect core-cladding boundary.

be treated by perturbation theory.[84,89,98] A slab waveguide with imperfect walls is shown in Figure 9.2.1. The wall departures appear in the figure vastly exaggerated. The index distribution of the waveguide with imperfect walls can be indicated as follows

$$n^2(x, z) = n_o^2(x) + \eta(x, z) \tag{9.2-1}$$

The index distribution n_o describes the perfect slab waveguide

$$n_o(x, z) = \begin{cases} n_2 & \text{for } |x| > d_0 \\ n_1 & \text{for } |x| < d_0 \end{cases} \tag{9.2-2}$$

If the waveguide were perfect, we would have $\eta = 0$. For the imperfect waveguide, we obtain the value $\eta = n_1^2 - n_2^2$ for the narrow regions where the guide wall protrudes over the boundary of the perfect wall and $\eta = n_2^2 - n_1^2$ for those regions where the core of the slab has receded from its perfect position. The values of η are given by the following equation

$$\eta = \begin{cases} 0 & \text{for } x > d_0 + f(z) \\ n_1^2 - n_2^2 & \text{for } d_0 < x < d_0 + f(z) \\ 0 & \text{for } -d_0 + h(z) < x < d_0 \\ -(n_1^2 - n_2^2) & \text{for } -d_0 < x \le -d_0 + h(z) \\ 0 & \text{for } -\infty < x < -d_0 \end{cases} \begin{array}{c} \text{for } f > 0 \\ h > 0 \end{array} \tag{9.2-3}$$

Similar equations hold for those z values where $f(z) < 0$, $h(z) > 0$, and for all other possible combinations. The difference $n_1^2 - n_2^2$ must not necessarily be small in order for a perturbation theory to be applicable. It is sufficient that the region over which η is different from zero is very narrow.

We substitute (8.5-16) and (9.2-1) into (8.3-4) and obtain

$$\sum_v \left\{ \frac{\partial^2 c_v}{\partial z^2} - 2i\beta_v \frac{\partial c_v}{\partial z} + c_v k_o^2 \eta \right\} E_{vy}$$

$$+ \sum_v \int_0^\infty \left\{ \frac{\partial^2 q(\rho)}{\partial z^2} - 2i\beta \frac{\partial q(\rho)}{\partial z} + q(\rho) k_0^2 \eta \right\} E_y(\rho) d\rho = 0 \tag{9.2-4}$$

The fact that the modes E_{vy} and $E_y(\rho)$ satisfy the wave equation for the ideal slab waveguide was used to obtain (9.2-4). It was also tacitly assumed that the field of the imperfect waveguide has only the three field components H_x, H_z, and E_y. Once E_y of (8.5-16) has been determined, the magnetic field components can be obtained with the help of (8.3-2) and (8.3-3). We want to solve the following problem. The lowest order TE mode of the perfect waveguide is incident on an imperfect section of waveguide. How does the power carried by the incident TE mode of the perfect guide get converted to other guided modes and to radiation?

Equation (9.2-4) contains only derivatives with respect to the z coordinate. This equation does, however, still depend on the x coordinate through the x dependence of η and of the normal modes. In order to obtain a system of coupled differential equations, we multiply (9.2-4) by

$$\frac{\beta_\mu}{2\omega\mu_0} E^*_{\mu y} \tag{9.2-5}$$

(μ is an integer; μ_0 is the magnetic permeability) and integrate over the entire cross section of the guide. Using the orthogonality relation (8.5-13), we obtain

$$\frac{\partial^2 c_\mu}{\partial z^2} - 2i\beta_\mu \frac{\partial c_\mu}{\partial z} + \sum_\nu c_\nu F_{\nu\mu}(z) + \sum \int_0^\infty q(\rho)G_\mu(\rho, z)d\rho = 0 \tag{9.2-6}$$

with

$$F_{\nu\mu}(z) = \frac{\beta_\mu k_0^2}{2\omega\mu_0 P} \int_{-\infty}^\infty E^*_{\mu y}\eta(x, z)E_{\nu y}\, dx \tag{9.2-7}$$

and

$$G_\mu(\rho) = \frac{\beta_\mu k_0^2}{2\omega\mu_0 P} \int_{-\infty}^\infty E^*_{\mu y}\eta(x, z)E_y(\rho)dx \tag{9.2-8}$$

Similarly, we obtain

$$\frac{\partial^2 q(\rho')}{\partial z^2} - 2i\beta' \frac{\partial q(\rho')}{\partial z} + \sum_\nu c_\nu \bar{F}_\nu(\rho', z) + \sum \int_0^\infty q(\rho)\bar{G}(\rho, \rho')d\rho = 0 \tag{9.2-9}$$

with

$$\bar{F}_\nu(\rho', z) = \frac{\beta' k_0^2}{2\omega\mu_0 P} \int_{-\infty}^\infty E^*_y(\rho')\eta(x, z)E_{\nu y}\, dx \tag{9.2-10}$$

and

$$\bar{G}(\rho, \rho') = \frac{\beta' k_0^2}{2\omega\mu_0 P} \int_{-\infty}^\infty E^*_y(\rho')\eta(x, z)E_y(\rho)dx \tag{9.2-11}$$

The dependence of the propagation constant β of the radiation modes on ρ' is indicated by the prime. It is apparent that the existence of the perturbation η couples the expansion coefficients c_ν and $q(\rho)$ to each other. The coupled differential-integral equations (9.2-6) and (9.2-9) are very complicated indeed.

Before we attempt to solve the equation systems let us take a look at the equations that result if $\eta = 0$. In this case, we obtain the simple uncoupled equations

$$\frac{\partial^2 c_\mu}{\partial z^2} - 2i\beta_\mu \frac{\partial c_\mu}{\partial z} = 0 \tag{9.2-12}$$

and

$$\frac{\partial^2 q(\rho')}{\partial z^2} - 2i\beta' \frac{\partial q(\rho')}{\partial z} = 0 \tag{9.2-13}$$

Since both equations are of exactly the same form, it is sufficient to concentrate on (9.2-12). The solution of this equation is

$$c_\mu = A + B e^{2i\beta_\mu z} \tag{9.2-14}$$

The mode solutions that entered the series expansion (8.5-16) had the time- and z-dependence (8.3-7). Since β_μ has always been taken to be positive, all modes inherently travel in the positive z direction. In the expansion, however, these modes appear multiplied with c_ν. The space- and time-dependence of the combination $c_\nu E_{\nu y}$ is given by the expression

$$c_\nu e^{i(\omega t - \beta_\nu z)} = A e^{i(\omega t - \beta_\nu z)} + B e^{i(\omega t + \beta_\nu z)} \tag{9.2-17}$$

It is thus apparent that, even though we started with modes traveling only in the positive z direction, we now obtain modes traveling in positive as well as negative z direction. The solutions of the wave equation have thus restored the generality.

Let us next consider the inhomogeneous differential equations

$$\frac{\partial^2 c_\mu}{\partial z^2} - 2i\beta_\mu \frac{\partial c_\mu}{\partial z} = \phi_\mu(z) \tag{9.2-18}$$

and

$$\frac{\partial^2 q(\rho')}{\partial z^2} - 2i\beta' \frac{\partial q(\rho')}{\partial z} = \psi_{\rho'}(z) \tag{9.2-19}$$

Equation (9.2-18) has the solution

$$c_\mu(z) = A_\mu + B_\mu e^{2i\beta_\mu z}$$
$$+ \frac{1}{2i\beta_\mu}\left\{ -\int_0^z \phi_\mu(\xi)d\xi + e^{2i\beta_\mu z}\int_0^z e^{-2i\beta_\mu \xi}\phi_\mu(\xi)d\xi \right\} \tag{9.2-20}$$

It is advantageous to separate the coefficient c_μ into two parts. The first part

$$c_\mu^{(+)} = A_\mu - \frac{1}{2i\beta_\mu}\int_0^z \phi_\mu(\xi)d\xi \tag{9.2-21}$$

contributes to waves traveling in the positive z direction. The second part

$$c_\mu^{(-)} = \left[B_\mu + \frac{1}{2i\beta_\mu} \int_0^z e^{-2i\beta_\mu\xi} \phi_\mu(\xi)d\xi \right] e^{2i\beta_\mu z} \qquad (9.2\text{-}22)$$

contributes to waves traveling in the negative z direction. It may appear as though this separation of the coefficient into positive and negative parts depends not only on the factors in front of the integrals but also on the result of the integration. This is true, generally speaking. However, it turns out that for our application the separation into components contributing to waves traveling in the positive and negative z direction is meaningful.

The coefficients A_μ and B_μ must be determined from initial conditions. In order to have a definite physical problem in mind, we assume that the lowest order even TE mode is incident from the perfect slab waveguide on a waveguide section of length L that contains wall imperfections. The imperfect piece of waveguide is again followed by a perfect waveguide of infinite length. This model allows us to calculate the loss contribution of an imperfect piece of waveguide of length L. Our assumption enables us to formulate definite initial conditions. We assume, as stated, that, at $z = 0$, only the lowest order even TE mode travels to the right. This allows us to postulate that the amplitude c_o of this mode must be unity

$$c_0^{(+)}(0) = 1 \qquad (9.2\text{-}23)$$

while the forward-traveling part of all other modes must vanish

$$c_\mu^{(+)}(0) = 0 \qquad \text{for} \qquad \mu \neq 0 \qquad (9.2\text{-}24)$$

Similarly, we postulate that no mode appears at $z = L$ traveling in negative z direction

$$c_\mu^{(-)}(L) = 0 \qquad \text{for all } \mu \qquad (9.2\text{-}25)$$

The initial conditions (9.2-23) and (9.2-24) lead to

$$A_\mu = \delta_{o\mu} \qquad (9.2\text{-}26)$$

The condition (9.2-25) leads to

$$B_\mu = -\frac{1}{2i\beta_\mu} \int_0^L e^{-2i\beta_\mu\xi} \phi_\mu(\xi)d\xi \qquad (9.2\text{-}27)$$

Corresponding equations hold also for the solutions of (9.2-19).

Equation (8.5-22) gives the total power carried by all modes. If we exclude the power of the incident TE mode and extend (8.5-22) to hold for the power flowing through the infinite plane at $z = 0$ to the left as well as the power that flows through the infinite plane at $z = L$ to the right, we obtain the amount of power that the incident mode loses to forward and backward guided modes as well as forward and backward scattered radiation. The relative power loss from the incident TE mode can thus be expressed by the following equation

$$\frac{\Delta P}{P} = \sum_{v=0}^{\infty} [(1 - \delta_{ov}) |c_v^{+}(L)|^2 + |c^{(-)}(0)|^2]$$

$$+ \sum_{1}^{2} \int_0^{\infty} [|q^{(+)}(\rho, L)|^2 + |q^{(-)}(\rho, 0)|^2] d\rho \qquad (9.2\text{-}28)$$

With the help of (9.2-21), (9.2-22), (9.2-26), (9.2-27), we can write the equation for the relative power loss as follows

$$\frac{\Delta P}{P} = \sum_{v=0}^{\infty} \left\{ (1 - \delta_{ov}) \left| \frac{1}{2i\beta_v} \int_0^L \phi_v(\xi) d\xi \right|^2 + \left| \frac{1}{2i\beta_v} \int_0^L e^{-2i\beta_v \xi} \phi_v(\xi) d\xi \right|^2 \right\}$$

$$+ \sum_{1}^{2} \int_0^{\infty} \left\{ \left| \frac{1}{2i\beta(\rho)} \int_0^L \psi_\rho(\xi) d\xi \right|^2 + \left| \frac{1}{2i\beta(\rho)} \int_0^L e^{-2i\beta(\rho)\xi} \psi_\rho(\xi) d\xi \right|^2 \right\} d\rho$$

$$(9.2\text{-}29)$$

The sum from one to two in front of the integral reminds us to add the loss contribution of the even and odd radiation modes. The sum over the guided modes is assumed to contain even and odd modes.

Our theory is of use only if we can derive expressions for the functions $\phi_v(z)$ and $\psi_\rho(z)$. We accomplish this by a perturbation theory. The perturbation theory assumes that the amount of power carried by the incident lowest order even TE mode is not depleted very much by the section of imperfect waveguide. In other words, $c_o^{(+)}(z)$ is assumed to remain almost unity. This seemingly very restrictive assumption will later be relaxed for certain important cases. Since the coefficient of the incident mode does not change very much from its initial value unity, we can also assume that the coefficients of the unwanted modes do not build up to large values. These perturbation assumptions are reasonable only if the expressions (9.2-7), (9.2-8), (9.2-10), and (9.2-11) are small. The products of the expansion coefficients with these small quantities are of second order and can be neglected. The only term in the sums appearing in (9.2-6) and (9.2-9) that must be considered is the products of $c_o = 1$ with the small quantities $F_{v\mu}(z)$ etc. Considering these perturbation assumptions, we obtain by comparison of (9.2-6) with (9.2-18)

$$\phi_\mu(z) = -c_o F_{o\mu} = -\frac{\beta_\mu k_o^2}{2\omega\mu_o P} \int_{-\infty}^{\infty} E_{\mu y}^* \eta E_{oy} dx \qquad (9.2\text{-}30)$$

and similarly

$$\psi_{\rho'}(z) = -c_o \bar{F}_v(\rho', z) = -\frac{\beta(\rho') k_o^2}{2\omega\mu_o P} \int_{-\infty}^{\infty} E_y^*(\rho') \eta E_{oy} dx \qquad (9.2\text{-}31)$$

The region over which $\eta(x)$ contributes to the integrals is assumed to be very narrow. This assumption is in agreement with our perturbation postulates, and means physically that the waveguide boundaries depart only very slightly from their ideal shape. Since η is different from zero only over such a narrow region, we can safely assume that the field components appearing under the integrals remain constant over the x range that contributes to the integrals. We replace the fields with their values at $x = \pm d$. The expression (9.2-30) becomes, with the help of (8.3-9), (8.3-18) and (9.2-3)

$$\phi_{\mu e}(z) = -(n_1^2 - n_2^2) \frac{\beta_\mu k_o^2 \cos \kappa_o d_o \cos \kappa_\mu d_o}{\sqrt{\left(\beta_o d_o + \dfrac{\beta_o}{\gamma_o}\right)\left(\beta_\mu d_o + \dfrac{\beta_\mu}{\gamma_\mu}\right)}} [f(z) - h(z)]e^{i(\beta_\mu - \beta_o)z}$$

(9.2-32)

for the case that μ indicates an even TE mode, and

$$\phi_{\mu o}(z) = -(n_1^2 - n_2^2) \frac{\beta_\mu k_o^2 \cos \kappa_o d_o \sin \kappa_\mu d_o}{\sqrt{\left(\beta_o d_o + \dfrac{\beta_o}{\gamma_o}\right)\left(\beta_\mu d_o + \dfrac{\beta_\mu}{\gamma_\mu}\right)}} [f(z) + h(z)]e^{i(\beta_\mu - \beta_o)z}$$

(9.2-33)

for the case that μ indicates an odd TE mode. The z-dependent part of the factor (8.3-7) has been reinstated into the equations. The function (9.2-31) can be treated very similarly, with the result

$$\psi_{\rho e} = -(n_1^2 - n_2^2) \frac{\sqrt{\beta(\rho)}\, k_o^2 \rho \cos \kappa_o d_o \cos \sigma d_o}{\sqrt{\pi\left(\beta_o d_o + \dfrac{\beta_o}{\gamma_o}\right)(\rho^2 \cos^2 \sigma d_o + \sigma^2 \sin^2 \sigma d_o)}}$$
$$\cdot [f(z) - h(z)]e^{i[\beta(\rho) - \beta_o]z}$$

(9.2-34)

for the even radiation modes, and

$$\psi_{\rho o} = -(n_1^2 - n_2^2) \frac{\sqrt{\beta(\rho)}\, k_o^2 \rho \cos \kappa_o d_o \sin \sigma d_o}{\sqrt{\pi\left(\beta_o d_o + \dfrac{\beta_o}{\gamma_o}\right)(\rho^2 \sin^2 \sigma d_o + \sigma^2 \cos^2 \sigma d_o)}}$$
$$\cdot [f(z) + h(z)]e^{i[\beta(\rho) - \beta_o]z}$$

(9.2-35)

for the odd radiation modes. The eigenvalues β_μ with their related parameters κ_μ and γ_μ are solutions of the eigenvalue equation (8.3-16) if they belong to even modes and of (8.3-23) if they belong to odd modes. A distinction in notation for the two types of eigenvalue was avoided for reasons of simplicity. The functions $f(z)$ and $h(z)$ enter the equations via (9.2-3). These functions describe the deflections of the upper and lower boundaries of the slab waveguide.

It is interesting to consider the coefficients that enter the equation (9.2-28) for the relative power loss. From (9.2-21), (9.2-26), and (9.2-32), we obtain for the amplitudes of the even guided TE modes

$$c_{\mu e}^{(+)}(L) = (n_1^2 - n_2^2) \frac{L k_o^2 \cos \kappa_o d \cos \kappa_\mu d_o}{2i\sqrt{\left(\beta_o d_o + \dfrac{\beta_o}{\gamma_o}\right)\left(\beta_\mu d_o + \dfrac{\beta_\mu}{\gamma_\mu}\right)}} [F(\beta_o - \beta_\mu) - H(\beta_o - \beta_\mu)]$$

(9.2-36)

with

$$F(\beta_o - \beta_\mu) = \frac{1}{L} \int_0^L f(z) e^{-i(\beta_o - \beta_\mu)z} dz \qquad (9.2\text{-}37)$$

and

$$H(\beta_o - \beta_\mu) = \frac{1}{L} \int_0^L h(z) e^{-i(\beta_o - \beta_\mu)z} dz \qquad (9.2\text{-}38)$$

The coefficient for the odd guided TE mode is

$$c_{\mu o}^{(+)}(L) = (n_1^2 - n_2^2) \frac{Lk_o^2 \cos \kappa_o d_o \sin \kappa_\mu d_o}{2i \sqrt{\left(\beta_o d_o + \dfrac{\beta_o}{\gamma_o}\right) \left(\beta_\mu d_o + \dfrac{\beta_\mu}{\gamma_\mu}\right)}} [F(\beta_o - \beta_\mu) + H(\beta_o - \beta_\mu)] \qquad (9.2\text{-}39)$$

The coefficient for the even radiation modes is

$$q_e^{(+)}(\rho,L) = \frac{n_1^2 - n_2^2}{2i} \frac{L\rho k_o^2 \cos \kappa_o d_o \cos \sigma d_o}{\sqrt{\pi\beta\left(\beta_o d_o + \dfrac{\beta_o}{\gamma_o}\right)(\rho^2 \cos^2 \sigma d_o + \sigma^2 \sin^2 \sigma d_o)}}$$

$$\cdot [F(\beta_o - \beta) - H(\beta_o - \beta)] \qquad (9.2\text{-}40)$$

while

$$q_o^{(+)}(\rho,L) = \frac{n_1^2 - n_2^2}{2i} \frac{L\rho k_o^2 \cos \kappa_o d_o \sin \sigma d_o}{\sqrt{\pi\beta\left(\beta_o d_o + \dfrac{\beta_o}{\gamma_o}\right)(\rho^2 \sin^2 \sigma d_o + \sigma^2 \cos^2 \sigma d_o)}}$$

$$\cdot [F(\beta_o - \beta) + H(\beta_o - \beta)] \qquad (9.2\text{-}41)$$

is the coefficient of the odd radiation modes. ($\cos \kappa_o d$ can be expressed in terms of κ and k_o with the help of (8.6-16).) The corresponding coefficients $c_\mu^{(-)}(0)$ etc. are obtained from the coefficients stated above simply by replacing β_μ with $-\beta_\mu$ or by replacing β with $-\beta$ in the F and H functions.

What is noteworthy about these equations is their dependence on F and H, which are simply the Fourier components of the wall distortion functions f and h. It is important to note that these Fourier transforms are taken over the same range over which the losses are considered. Our equations hold for the losses of a slab waveguide of length L. This same length L is also used as the base period for the Fourier coefficients. This length dependence of the Fourier transforms has important consequences for the loss behavior of guides with systemic wall deflections and with random wall perturbations.

We have now collected all the equations for calculating the losses that the lowest order even TE mode incurs as it travels through a section of slab waveguide with wall perturbations.

9.3 SLAB WAVEGUIDE WITH SINUSOIDAL WALL PERTURBATIONS

The perturbation theory of the previous section shall now be applied to a specific case. We assume that the wall distortion functions are of the following form[98]

$$f(z) = a \sin \theta z \qquad (9.3\text{-}1)$$

and

$$h(z) = -a \sin(\theta z + \phi) \qquad (9.3\text{-}2)$$

If

$$\phi = 0 \qquad (9.3\text{-}3)$$

(9.3-1) and (9.3-2) describe a slab with periodically varying thickness. If

$$\phi = \pi \qquad (9.3\text{-}4)$$

the slab waveguide has a constant thickness but periodically changing direction. The Fourier coefficients of these wall distortion functions follow from (9.2-37)

$$F(\beta_0 - \beta_\mu) = \frac{a}{iL} e^{i[\theta - (\beta_o - \beta_\mu)]L/2} \frac{\sin[\theta - (\beta_o - \beta_\mu)]\frac{L}{2}}{\theta - (\beta_o - \beta_\mu)} \qquad (9.3\text{-}5)$$

and from (8.6-38)

$$H(\beta_o - \beta_\mu) = -\frac{a}{iL} e^{i\{[\theta - (\beta_o - \beta_\mu)]L/2 + \phi\}} \frac{\sin[\theta - (\beta_o - \beta_\mu)]\frac{L}{2}}{\theta - (\beta_o - \beta_\mu)} \qquad (9.3\text{-}6)$$

Terms with $\theta + (\beta_o - \beta_\mu)$ were neglected in these two equations. Since β_o belongs to the lowest order mode, its value is larger than that of any other mode, and we have $\beta_o - \beta_\mu > 0$. We assume that L is much larger than the wavelength of the radiation in the waveguide. It is apparent that the Fourier coefficients have large values only if the denominators in (9.3-5) and (9.3-6) vanish. For all other values, the Fourier coefficients are negligibly small. It is for this reason that the terms with $\theta + (\beta_o - \beta_\mu)$ could be neglected, since this term cannot vanish. Vanishingly small Fourier coefficients result in vanishingly small mode amplitudes. This discussion allows us to formulate the following important result. Two modes in the waveguide are coupled by a sinusoidal wall deflection only if the mechanical frequency θ of the wall perturbation is related to the propagation constants of the modes by the following relation

$$\theta = \beta_o - \beta_\mu \qquad (9.3\text{-}7)$$

This statement holds for coupling to radiation modes as well as to spurious guided modes. We conclude, therefore, that a sinusoidal wall perturbation couples the incident mode either to one other guided mode or to a narrow region of the modes of the continuous spectrum.

Let us first assume that (9.3-7) is satisfied for a guided spurious mode. From (9.3-5) and (9.3-6), we obtain, with (9.3-7) and $\phi = 0$

$$F(\beta_o - \beta_\mu) = \frac{a}{2i} \tag{9.3-8}$$

and

$$H(\beta_o - \beta_\mu) = -\frac{a}{2i} \tag{9.3-9}$$

We see from (9.2-39) that in this case there is no coupling to odd modes. A symmetrical thickness variation of the slab waveguide couples the symmetric incident modes only to other symmetric modes. Had we used $\phi = \pi$, the sign of (9.3-9) would have been positive and coupling could occur only to odd modes.

We use our equations to calculate from (9.2-28) the relative power loss of the incident mode to the one spurious mode to which it is coupled ($\phi = 0$)

$$\left(\frac{\Delta P}{P}\right)_g = a^2 L^2 \frac{[(n_1^2 - n_2^2)k_o^2 \cos \kappa_o d \cos \kappa_\mu d]^2}{4\left(\beta_o d_o + \dfrac{\beta_o}{\gamma_o}\right)\left(\beta_\mu d_o + \dfrac{\beta_\mu}{\gamma_\mu}\right)} \tag{9.3-10}$$

The coupling to the radiation modes follows the same law. Thickness variations couple only to even radiation modes, while periodic changes of the direction of the guide couple only to odd modes. The power loss to the radiation modes cannot be obtained quite as easily, since even a strictly sinusoidal wall perturbation couples to infinitely many radiation modes. If the relation (9.3-7) is satisfied for the radiation modes, we can allow θ to fall anywhere inside a range of values. This range is determined by the range of values of the radiation modes. Since backward scattered power as well as forward scattered power must be considered, the possible range of the radiation modes that are of interest is given by

$$-n_2 k_o < \beta < n_2 k_o \tag{9.3-11}$$

The range of mechanical frequencies θ that causes coupling to radiation modes is thus given by

$$\beta_o - n_2 k_o < \theta < \beta_o + n_2 k_o \tag{9.3-12}$$

Sinusoidal wall perturbations whose mechanical frequencies fall outside the range (9.3-12) do not cause power loss by radiation. This result holds, strictly speaking, only in the limit of infinitely long waveguide sections. It is, however, correct to a good approximation for long (compared to the wavelength) but finite sections of guide.

The formula (9.2-28) for power loss caused by radiation can be written in a more convenient form. The integral in (9.2-28) not only extends over the range of propagating radiation modes but also includes evanescent radiation modes. These modes do not carry power, so that it is sufficient to limit the integration range to propagating radiation modes. We convert the integration over the variable ρ to an integration over β. From (8.4-6), we obtain

$$d\rho = -\frac{\beta}{\rho} d\beta \tag{9.3-13}$$

The integrals extend from $\beta = 0$ to $\beta = n_2 k_o$. We have seen that the coefficient $q^{(-)}(\rho)$ is obtained from $q^{(+)}(\rho)$ by changing the propagation constant that appears in the Fourier coefficients-from β to $-\beta$. We can thus combine the two integrals appearing in (9.2-28) to one integral by extending the integration range from $-n_2 k_o$ to $+k_o n_2$. The relative power loss caused by radiation alone can thus be written

$$\frac{\Delta P}{P} = \int_{-n_2 k_o}^{n_2 k_o} [\,|q_e|^2 + |q_o|^2\,] \frac{|\beta|}{\rho} d\beta \tag{9.3-14}$$

The superscripts $+$ or $-$ are no longer necessary, and have been dropped. The sum in front of the integral of (9.2-28) has been explicitly taken care of by including the even and odd modes under the integral sign.

In the special case $\phi = 0$, only the expansion coefficients q_e for the even modes exist. We obtain, using (9.3-5) and (9.3-6)

$$\frac{\Delta P}{P} = \frac{a^2(n_1^2 - n_2^2)^2 k_o^4 \cos^2 \kappa_o d_o}{\pi\left(\beta_o d_o + \dfrac{\beta_o}{\gamma_o}\right)} \int_{-n_2 k_o}^{n_2 k_o} \frac{\rho \sin^2 \Gamma \dfrac{L}{2} \cos^2 \sigma d_o}{\Gamma^2(\rho^2 \cos^2 \sigma d_o + \sigma^2 \sin^2 \sigma d_o)} d\beta \tag{9.3-15}$$

with

$$\Gamma = \theta - (\beta_o - \beta) \tag{9.3-16}$$

This integral is very hard to solve for small values of L. However, if L is much larger than the wavelength of the radiation, we can easily find a good approximation. We have seen in (8.4-16) that for large values of L

$$\delta(\Gamma) \approx \frac{1}{\pi} \frac{\sin \Gamma \dfrac{L}{2}}{\Gamma} \tag{9.3-17}$$

What actually appears in the integrand is the square of the delta function. However, the square of the delta function behaves very similar to the delta function itself. It allows us to take the factors of the delta function out of the integral, since the delta function contributes to the integration only over an infinitesimal range. The resulting integral is

$$\int_{-n_2 k_o}^{n_2 k_o} \frac{\sin^2 \Gamma \frac{L}{2}}{\Gamma^2} \, d\beta \approx \int_{-\infty}^{\infty} \frac{\sin^2 \Gamma \frac{L}{2}}{\Gamma^2} \, d\Gamma = \frac{\pi L}{2} \qquad (9.3\text{-}18)$$

The change of the integration interval to infinite limits has a negligible effect, since the integrand contributes to the integral only in the immediate vicinity of $\Gamma = 0$.

The relative radiation loss caused by the periodic thickness variation of the slab waveguide is thus (α is the amplitude loss coefficient)

$$\frac{\Delta P}{P} = 2\alpha L = L \frac{a^2 (n_1^2 - n_2^2)^2 k_o^4 \rho_\theta \cos^2 \kappa_o d_o \cos^2 \sigma_\theta d_o}{2 \left(\beta_o d_o + \dfrac{\beta_o}{\gamma_o} \right) (\rho_\theta^2 \cos^2 \sigma_\theta d_o + \sigma_\theta^2 \sin^2 \sigma_\theta d_o)} \qquad (9.3\text{-}19)$$

with

$$\rho_\theta = \sqrt{n_2 k_o^2 - (\beta_o - \theta)^2} \qquad (9.3\text{-}20)$$

and

$$\sigma_\theta = \sqrt{n_1^2 k_o^2 - (\beta_o - \theta)^2} \qquad (9.3\text{-}21)$$

There is a very interesting difference between the power loss to one guided mode, (9.3-10), and the power loss to the continuum of radiation modes. The loss caused by the guided mode is proportional to L^2, while the loss to the continuous spectrum of radiation modes is proportional to L. Both loss figures apply only as long as $\Delta P/P \ll 1$.

We shall later investigate the coupling to the guided mode in more detail. The radiation loss can be generalized by the following argument. The radiation caused by the wall perturbation leaves the waveguide and disappears into space. We shall calculate the radiation far field pattern later in this section. When we have lost a certain small amount of power, we find that the incident mode has not changed except for the fact that its amplitude is slightly reduced. The converted power is no longer traveling along with the incident mode, since it has radiated away into space. This is not true for the coupling to the guided mode. There the converted power exists in the guide and keeps on interacting with the incident mode power. In case of radiation, the incident mode simply suffers loss but does not interact appreciably with the converted power. This argument allows us to extend the range of our results out of the region of the perturbation theory. Applying (9.3-19) repeatedly, each time with the reduced value of the incident mode power, allows us to obtain the power at the end of a waveguide of length D as

$$P = P_o e^{-2\alpha D} \qquad (9.3\text{-}22)$$

The attenuation coefficient α is defined by (9.3-19). Equation (9.3-22) holds for small values of $2\alpha\lambda$ and improves in accuracy as the angle increases at which the radiation leaves the waveguide. (λ = wavelength)

We obtain the radiation angle from an analysis of the far field radiation pattern.[100] We keep the assumption (9.3-3) and obtain for the expansion coefficient $q_e^{(+)}$ from (9.2-40), (9.3-5), and (9.3-6) the following expression.

$$q_e^{(+)}(\rho,L) = -(n_1^2 - n_2^2) \frac{a\rho k_o^2 \cos \kappa_o d_o \cos \sigma d_o \, e^{i\Gamma L/2}}{\sqrt{\pi\beta\left(\beta_o d_o + \dfrac{\beta_o}{\gamma_o}\right)}\, (\rho^2 \cos^2 \sigma d_o + \sigma^2 \sin^2 \sigma d_o)}$$

$$\cdot \frac{\sin \Gamma \dfrac{L}{2}}{\Gamma} \tag{9.3-23}$$

with Γ given by (9.3-16). The electric radiation field is obtained from (8.5-16)

$$E_y = \int_0^\infty q_e^{(+)}(\rho,L)E_y(\rho,z)d\rho \tag{9.3-24}$$

Equations (9.3-23) and (9.3-24) describe the radiation field of the slab with sinusoidally varying thickness at a point x, z in space. The radiation is generated by the imperfect portion of slab waveguide of length L. We are interested only in the far field, so that we assume

$$L \ll z \tag{9.3-25}$$

The field of the radiation mode $E_y(\rho)$ outside the slab is given by (8.4-4), (8.4-9), (8.4-10), and (8.4-18). These equations can be combined to yield

$$E_y(\rho) = \frac{\rho\sqrt{2\omega\mu_o P}\left[\cos \sigma d_o \cos \rho(|x| - d_o) - \dfrac{\sigma}{\rho} \sin \sigma d_o \sin \rho(|x| - d_o)\right]}{\sqrt{\pi\beta(\rho^2 \cos^2 \sigma d_o + \sigma^2 \sin^2 \sigma d_o)}} e^{-i\beta z} \tag{9.3-26}$$

The radiation field is thus obtained from (9.3-23), (9.3-24), and (9.3-26)

$$E_y = -\frac{ak_o^2\sqrt{2\omega\mu_o P}\,(n_1^2 - n_2^2)\cos \kappa_o d_o}{\pi\sqrt{\beta_o d_o + \dfrac{\beta_o}{\gamma_o}}}$$

$$\cdot \int_0^\infty \frac{e^{i\Gamma L/2}\rho^2 \cos \sigma d_o \left[\cos \sigma d_o \cos \rho(|x| - d_o) - \dfrac{\sigma}{\rho} \sin \sigma d_o \sin \rho(|x| - d_o)\right]}{\beta(\rho^2 \cos^2 \sigma d_o + \sigma^2 \sin^2 \sigma d_o)}$$

$$\cdot \frac{\sin \Gamma \dfrac{L}{2}}{\Gamma} e^{-i\beta z}d\rho \tag{9.3-27}$$

This integral is far too complicated to hope for a general solution. However, it is possible to obtain a far field approximation by the method of stationary phase.[9] This useful approximate method for solving integrals with rapidly oscillating functions in their integrands was introduced in Section 2.3. We assume that both x and z become very much larger than L. This causes the functions that contain x and z in their arguments to vary much more rapidly than the functions with L and d_o in their arguments. The main contribution to the integral comes from the region where the arguments of the rapidly oscillating functions assume stationary values. We thus must examine expressions of the form

$$\rho x + \beta z = f(\rho) \tag{9.3-28}$$

Using (8.4-6), we obtain

$$\frac{df}{d\rho} = x - \frac{\rho}{\beta} z \tag{9.3-29}$$

and

$$\frac{d^2 f}{d\rho^2} = -\frac{n_2^2 k_o^2}{\beta^3} z \tag{9.3-30}$$

The requirement of stationary phase $df/d\rho = 0$ leads to the condition

$$\frac{\rho_s}{\beta_s} = \frac{x}{z} = \tan \alpha_s \tag{9.3-31}$$

By expanding $f(\rho)$ in Taylor series around the stationary point, we obtain

$$\int_0^\infty e^{-i(\rho x + \beta z)} d\rho \approx e^{-i(\rho_s x + \beta_s z)} \int_{-\infty}^\infty e^{i(n_o^2 k_o^2 / 2\beta_s^3)z(\rho - \rho_s)^2} d(\rho - \rho_s)$$

$$= (1 + i)\sqrt{\pi} \, \frac{\beta_s^{3/2}}{n_2 k_o \sqrt{z}} e^{-i(\rho_s x + \beta_s z)} \tag{9.3-32}$$

Equation (2.3-23) was used to evaluate the integral. The integral (9.3-32) is the only integration that needs to be performed in (9.3-27). We decompose the sine and cosine functions that contain x into exponential functions. Only that part of the decomposition that contains functions of the form (9.3-28) in their argument needs to be considered. The exponential functions with arguments $\beta z - \rho x$ give no appreciable contributions as long as x and z are both larger than zero. The factors multiplying the exponential functions with stationary phase can be considered as constants at the point $\rho = \rho_s$, and can be taken out of the integral. The remaining integration has been performed in (9.3-32), so that we obtain

$$E_y = -\frac{a(1 + i)k_o\rho_s\sqrt{\beta_s}\sqrt{2\omega\mu_o}P(n_1^2 - n_2^2)\cos\kappa_o d_o\cos\sigma_s d_o e^{i\rho_s d_o}e^{i\Gamma_s L/2}}{2n_2\sqrt{\pi}\sqrt{\left(\beta_o d_o + \dfrac{\beta_o}{\gamma_o}\right)(\rho_s\cos\sigma_s d_o + i\sigma_s\sin\sigma_s d_o)}}$$

$$\cdot\frac{\sin\Gamma_s\dfrac{L}{2}}{\Gamma_s}\frac{e^{-in_2k_o(x\sin\alpha_s + z\cos\alpha_s)}}{\sqrt{z}} \tag{9.3-33}$$

with

$$\rho_s = n_2 k_o\sin\alpha_s \tag{9.3-34}$$

and

$$\beta_s = n_2 k_o\cos\alpha_s \tag{9.3-35}$$

Γ_s is obtained from (9.3-16) by replacing β with β_s. The expression for the far field reveals several interesting features. We see that the field travels approximately like a plane wave in the direction of α_s, the angle at which the field point x, z appears from the source of the radiation.* Since we are in the far field, the source of finite length appears like a point source at the coordinate origin. The last factor of the radiation field (9.3-33) is, however, a cylindrical wave. This fact becomes more apparent when we introduce, with the help of (9.3-31)

$$x = r\sin\alpha_s \tag{9.3-36}$$

and

$$z = r\cos\alpha_s \tag{9.3-37}$$

and write

$$\sqrt{\beta_s}\frac{e^{-in_2k_o(x\sin\alpha_s + z\cos\alpha_s)}}{\sqrt{z}} = \sqrt{n_2 k_o}\frac{e^{-in_2k_o r}}{\sqrt{r}} \tag{9.3-38}$$

The appearance of a cylindrical, instead of a spherical, wave is of course caused by the two-dimensional geometry of the slab. The most important factor in (9.3-33) is

$$G = \frac{\sin\Gamma_s\dfrac{L}{2}}{\Gamma_s} \tag{9.3-39}$$

We know that this factor approaches a delta function for increasing values of L. It is only necessary that L be much larger than the wavelength λ for G/π

* We see in this example that the superposition of the standing wave radiation modes results in a traveling wave.

to be a good approximation of a delta function. This requirement does not contradict the far field requirement (9.3-25). The factor (9.3-39) determines the antenna pattern of the slab waveguide with sinusoidal wall perturbations. The main lobe of this antenna pattern becomes narrower with increasing values of L. The direction of the main lobe is given by

$$\Gamma_s = \theta - (\beta_o - \beta_s) = 0 \qquad (9.3\text{-}40)$$

Appreciable radiation can therefore be found only in the direction given by (9.3-35)

$$\cos \alpha_s = \frac{\beta_o - \theta}{n_2 k_o} \qquad (9.3\text{-}41)$$

The radiation is directed perpendicular to the surface of the waveguide for $\theta = \beta_o$, and it is directed tangential to the waveguide for $\theta = \beta_o - n_2 k_o$. In other words, the radiation is perpendicular when $\beta_s = 0$, and it becomes more nearly tangential with increasing values of β_s. As $\beta_s = n_2 k_o$, the radiation lobe points parallel to the surface of the waveguide. Larger values of the propagation constant (of radiation modes) are impossible. For $\beta_s > n_2 k_o$, we obtain a guided mode. This picture shows very clearly how the radiation field becomes more and more forward-directed and, as it reaches the limit of tangential radiation, merges with the guided mode. When we talk of the direction of the radiation lobe as if it could be changed at will, we assume that we can vary the mechanical frequency θ continuously. The approximation (9.3-22) becomes increasingly more precise as the angle α_s determined by (9.3-41) increases.

The reader may wonder why we devote so much space to the discussion of the slab waveguide with strictly sinusoidal wall perturbations. After all, it is highly unlikely that such a case would ever occur in practice unless very special care is taken to produce such a guide. The value of our discussion becomes immediately apparent when we realize that any arbitrary wall perturbation can be expressed as a Fourier series or integral expansion in terms of sinusoidal functions. We have seen that a given sinusoidal wall perturbation couples the incident guided mode only to one other guided mode or only to a very narrow region of the continuous spectrum of radiation modes. Our discussion is thus applicable to each Fourier component of the series expansion of an arbitrary wall distortion function. In order to obtain the radiation losses for an arbitrary wall deformation, we would introduce the Fourier component $F^2 d\theta$ instead of the amplitude a^2 and integrate (9.3-19) over the θ interval of (9.3-12). This procedure follows, in fact, directly from substituting (9.2-40) and (9.2-41) into (9.3-14). With $\Gamma = 0$, we obtain from (9.3-16) $d\beta = -d\theta$. The quantity $|F|^2$ (and of course similarly $|H|^2$) is called the "power spectrum" of the wall distortion function. This term is borrowed from the Fourier analysis of electrical signals. The absolute square values of the Fourier expansion coefficients of a generally time-varying signal are indeed proportional to the power density at the given frequency. Limiting, for simplicity, the wall distortion to one side of the waveguide only, $h(z) = 0$, we obtain from (9.2-40), (9.2-41), and (9.3-14)

$$\frac{1}{L}\frac{\Delta P}{P} = \int_{-n_2 k_o}^{n_2 k_o} I(\beta)\, |F(\beta_o - \beta)|^2 L\, d\beta \qquad (9.3\text{-}42)$$

with

$$I(\beta) = (n_1^2 - n_2^2)^2 \frac{\rho k_o^4 \cos^2 \kappa_o d_o}{4\pi\left(\beta_o d_o + \dfrac{\beta_o}{\gamma_o}\right)} \left[\frac{\cos^2 \sigma d_o}{\rho^2 \cos^2 \sigma d_o + \sigma^2 \sin^2 \sigma d_o}\right.$$

$$\left. + \frac{\sin^2 \sigma d_o}{\rho^2 \sin^2 \sigma d_o + \sigma^2 \cos^2 \sigma d_o}\right] \qquad (9.3\text{-}43)$$

The understanding that we gained from the discussion of the effect of a strictly sinusoidal wall perturbation helps us to gain an intuitive feeling for the effect of a more complicated wall perturbation that is expressed as a Fourier integral. If the spectrum of the wall distortion function were known, we could calculate the radiation loss from (9.3-42). In general, the "power spectrum" of the wall distortion function is not known, and it is necessary to study certain models in order to obtain an insight into the losses that are to be expected. Instead of studying models based on different "power spectra," we shall use a statistical approach in the next section.

Starting out with a perturbation theory, we have seen that it is possible to calculate the radiation loss suffered by a guided mode even for long waveguides. We are about to show that a similar approach can be used for the interaction between different guided modes. It is of course not possible simply to extend the loss formula (9.3-10) to (9.3-22). The loss formula (9.3-10) does not depend linearly on L, so that it is not possible to define a loss per unit length. The physical argument used to obtain (9.3-22) from (9.3-19) is not applicable to guided modes, because the power that is converted from the incident mode to a spurious mode remains in the guide and keeps on interacting with the original mode. There is a certain unique feature about the coupling of guided modes by a strictly sinusoidal wall perturbation that can be utilized to obtain a solution far exceeding the applicability of perturbation theory. We have seen (compare [9.3-7]) that a sinusoidal wall perturbation effectively couples only two guided modes. We therefore ignore the possible existence of all the other guided modes and the radiation field, and pretend that only the two guided modes exist, coupled by the sinusoidal wall perturbation. Designating the amplitude coefficients of these two guide modes by c_0 and c_1, we obtain from (9.2-6)

$$\frac{\partial^2 c_0}{\partial z^2} - 2i\beta_o \frac{\partial c_0}{\partial z} = -c_0 F_{00} - c_1 F_{10} \qquad (9.3\text{-}44)$$

and

$$\frac{\partial^2 c_1}{\partial z^2} - 2i\beta_1 \frac{\partial c_1}{\partial z} = -c_0 F_{01} - c_1 F_{11} \tag{9.3-45}$$

This equation system is of course applicable only to the special case of a sinusoidally varying wall distortion function. The diagonal terms F_{00} and F_{11} are therefore themselves sinusoidally varying with z. They represent a sinusoidally changing correction of the propagation constant with zero average value. Since the wall perturbation is considered to be only very slight, we can safely neglect these diagonal terms. We assume that both modes travel in the positive z direction. In the absence of coupling, both c_0 and c_1 would be constant. For the weak coupling that we consider, the changes of c_0 and c_1 are only very slight over the distance of one wavelength. It is thus possible to neglect the second derivative of the coefficients c_0 and c_1 compared to the much larger product of the first derivative with 2β. After we neglect the small terms, we obtain the following simpler equation system (compare the coupled wave equations [2.7-15] and [2.7-16])

$$\frac{\partial c_0}{\partial z} = -\kappa c_1 \tag{9.3-46}$$

and

$$\frac{\partial c_1}{\partial z} = \kappa c_0. \tag{9.3-47}$$

The coupling coefficient of (9.3-47) can of course be derived from (9.2-7). However, it is simpler to use (9.2-36), (9.3-8), and (9.3-9) with $L = z$. Remembering that c_0 is unity in (9.2-36), we obtain by differentiation with respect to z and comparison with (9.3-47)

$$\kappa = -(n_1^2 - n_2^2) \frac{ak_o^2 \cos \kappa_o d_o \cos \kappa_1 d_o}{2\sqrt{\left(\beta_o d_o + \frac{\beta_o}{\gamma_o}\right)\left(\beta_1 d_o + \frac{\beta_1}{\gamma_1}\right)}} \tag{9.3-48}$$

The $\cos \kappa d_o$ terms could be eliminated with the help of (8.6-16). The coupling coefficient is real. The coupling coefficient for (9.3-46) is obtained by interchanging the role of c_o and c_1 in (9.2-36). The coefficient (9.3-48) is invariant with respect to an interchange of the indices 0 and 1. The change in sign that is indicated in (9.3-46) can be obtained by taking a closer look at the Fourier coefficient (9.2-37) and (9.3-8). In deriving (9.3-8), we neglected a term of the form

$$\frac{\sin x \dfrac{L}{2}}{x} \tag{9.3-49}$$

with $x = \theta + (\beta_0 - \beta_1)$, because x could not vanish. However, after we interchange the indices 0 and 1, it is this neglected term that becomes important, while the term with $x = \theta - (\beta_0 - \beta_1)$ now turns into a form that cannot vanish. The two terms have opposite signs, because they originated from the decomposition of the function $\sin \theta z$ into exponential components. This discussion explains the different signs in (9.3-46) and (9.3-47).

By eliminating c_1 from (9.3-46) and (9.3-47), we obtain

$$\frac{d^2 c_0}{dz^2} + \kappa^2 c_0 = 0 \tag{9.3-50}$$

and similarly

$$\frac{d^2 c_1}{dz^2} + \kappa^2 c_1 = 0 \tag{9.3-51}$$

Since c_0 was assumed to be the amplitude coefficient of the incident mode with $c_0 = 1$ at $z = 0$, we obtain the solution

$$c_0 = \cos \kappa z \tag{9.3-52}$$

and

$$c_1 = \sin \kappa z \tag{9.3-53}$$

For small values of κz, we obtain from (9.3-53), with $z = L$

$$\frac{\Delta P}{P} = |c_1|^2 = \kappa^2 L^2 \tag{9.3-54}$$

in complete agreement with (9.3-10).

The solutions (9.3-52) and (9.3-53) far exceed the applicability of perturbation theory. They are an excellent description of the power interplay between the two guided modes that are coupled together by the sinusoidal wall perturbation. It is now clear why the approximate expression for the relative power loss (9.3-10) was dependent on L^2 instead of being simply proportional to L.

The two guided modes can exchange their power completely. We see from (9.3-52) and (9.3-53) that mode 1 grows at the expense of mode 0. After a distance

$$D = \frac{\pi}{2\kappa} \tag{9.3-55}$$

the incident mode is completely exhausted, while all the power has been transferred to mode 1.

The power interplay between two modes coupled by a sinusoidal wall perturbation cannot be extended to random shapes of the wall distortion function. Even though it is only one Fourier component of the wall distortion function that is responsible for the coupling of the two modes, our mechanism fails if the sinusoidal function (9.3-1) is only one of many Fourier components of the total wall distortion function. There are two reasons why the simple coupling theory breaks down in this more complicated case. First of all, there are now more modes coupled to the incident guided mode. Not only does the mode lose power by radiation, caused by those Fourier components that couple to the continuous spectrum, but coupling to other guided modes becomes also possible. It is thus no longer admissible to consider only two guided modes, but the other competing processes must be taken into account. However, even if the radiation losses were small and only one other guided mode existed, we would no longer be able to use the simple coupling theory. To understand this, we must realize that the direction of power transfer depends critically on the phase relationship between the two coupled modes and the phase of the sinusoidal wall function. The Fourier component that enters the theory is a function of z, since it is obtained by integration to just the point z at which the two modes are being considered. Not only would the amplitude a of the Fourier component change its value with increasing values of z, but, more importantly, its phase would not have the constant value required for continuous energy interchange. For a random function $f(z)$, the phase of the Fourier component would change randomly as a function of z. This would cause the direction of power transfer between the two modes to change back and forth randomly, so that no complete power transfer would be possible. If the two modes ever do exchange their power, we must consider this as purely accidental.

The theory of coupling between two guided modes caused by a precisely sinusoidal wall perturbation does not have a direct application to the study of mode losses caused by random wall perturbations. However, our study of strictly sinusoidal wall perturbations has important applications for the construction of mode couplers. This mechanism makes it possible to transfer power between two guided modes.

The theory of mode coupling by sinusoidal wall perturbations and of the radiation loss and its directional properties has been verified by an experiment at millimeter wave frequencies.[99] All the aspects of the theory of sinusoidal wall perturbations could be confirmed in excellent agreement with the perturbation theory. The experiment was conducted not with a slab waveguide but with a round Teflon rod that was excited in the circular electric TE_{01} mode. This choice of waveguide was dictated by experimental convenience. The slab waveguide is a very simple device from the point of view of its mathematical treatment. However, it is hard to excite the TE and TM modes of the slab waveguide, since this requires a field that has no variation in y direction from $-\infty$ to $+\infty$. The circular symmetric modes of the round waveguide, on the other hand, are easy to excite. Furthermore, they are quite similar to the TE and TM modes of the slab, with the only exception that the lowest order TE mode of the round waveguide is not a dominant mode. It cannot exist at arbitrarily low frequencies or on an arbitrarily thin rod. The

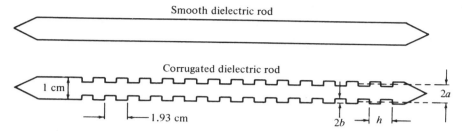

Figure 9.3.1   The smooth and corrugated Teflon rods used in the microwave experiment. (From D. Marcuse and R. M. Derosier, Mode Conversion Caused by Diameter Changes of a Round Dielectric Waveguide, *B.S.T.J.*, Vol. **48**, No. 10, December 1969. Copyright 1969, The American Telephone and Telegraph Co., reprinted by permission.)

experiments were compared with a theory of the TE modes of the round rod, not with our slab waveguide theory. The general features and even the numbers of the two theories are very nearly the same.

The theory of the TE modes of the round dielectric rod and its experimental verification are described in reference 99. We discuss here only the results of the experiment.

Instead of sinusoidal diameter changes, the Teflon rod was made with periodic grooves for reasons of machining simplicity. The grooves were only 76 and 230 microns deep, so that it would have been very difficult to machine sinusoidal diameter changes. Furthermore, there is no need to reproduce the shape of sinusoidal diameter changes precisely. We have seen that a Fourier component works just as well as a true sinusoidal wall perturbation provided that its amplitude and phase are independent of z. The two rods used in the experiment are shown in Figure 9.3-1.

The loss measurements were based on a comparison of the losses of a smooth and a corrugated rod. The rods were suspended in air with the help of fine nylon threads that did not interfere noticeably with the wave propagation. The TE_{01} mode was excited at approximately 50 GHz with the help of a klystron and a rectangular-to-round waveguide transducer. The transducer was constructed to excite predominantly the TE_{01} mode of the round, hollow metallic waveguide. To suppress any unwanted modes, a piece of round helical waveguide was inserted between the transducer and the dielectric waveguide. The experimental arrangement is shown schematically in Figure 9.3.2. The grooves were spaced so that condition (9.3-7) was satisfied for the TE_{01} and TE_{02} modes. The rod diameter was chosen so that the TE_{02} mode could propagate at frequencies above approximately 50 GHz but was cut off below that frequency. It was thus possible to observe the coupling phenomenon between the two guided modes by operating above 50 GHz and to observe the coupling to the radiation modes below 50 GHz. We have seen that coupling to guided modes and coupling to radiation modes are

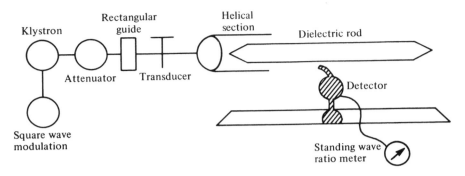

Figure 9.3.2   Block diagram of the microwave experiment. (From D. Marcuse and R. M. Derosier, Mode Conversion Caused by Diameter Changes of a Round Dielectric Waveguide, *B.S.T.J.*, Vol. **48**, No. 10, December 1969. Copyright 1969, The American Telephone and Telegraph Co., printed by permission.)

mutually exclusive if the wall perturbation is strictly sinusoidal. The periodic grooves approximate the true sinusoidal condition sufficiently closely. The round rod could support other modes. We know that the dominant HE_{11} mode can always exist on a dielectric rod without suffering a cutoff frequency. However, this mode was not excited by the mode launcher. Of the family of TE_{0n} modes, only TE_{01} and TE_{02} were possible.

It is very easy to detect the power that travels in the TE_{02} mode without interference from the TE_{01} mode existing simultaneously on the rod. The TE_{02} mode was always quite close to its cutoff frequency, and consequently extended much further out into the space around the rod than the more tightly guided TE_{01} mode. The detector consisted simply of an L-shaped piece of rectangular waveguide that could be moved parallel to the rod at variable distances from its surface, as shown in Figure 9.3-2. A few millimeters from the rod, no TE_{01} power could be detected while the TE_{02} mode was still strongly in evidence. The millimeter wave power picked up by the L-shaped piece of waveguide was detected with a single diode detector. The klystron power was square wave modulated, and this modulation was amplified in a low-frequency amplifier connected to the diode detector. The coupling between the TE_{01} mode of the metallic waveguide and the corresponding-mode of the dielectric rod was accomplished simply by inserting the dielectric rod into the round metallic waveguide. This arrangement did excite the TE_{01} mode of the dielectric waveguide very strongly, but excited only a small residual amount of the TE_{02} mode. The existence of a small amount of TE_{02} power and other unwanted modes stemming from the source contributed to a slight interference in some of the measured results, which was not serious however.

The build-up of TE_{02} power as a result of the periodic wall perturbation is shown in Figure 9.3.3. The grooves were only 76 μ deep in this case, which

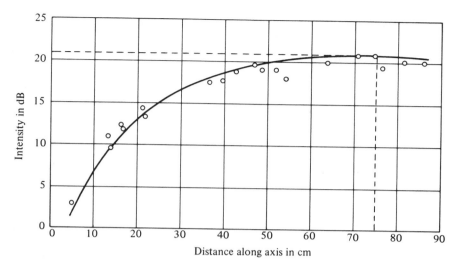

Figure 9.3.3   Buildup of TE_{02} power as a result of coupling between the TE_{01} and TE_{02} modes. Groove depth $= 7.6 \times 10^{-3}$ cm. (From D. Marcuse and R. M. Derosier, Mode Conversion Caused by Diameter Changes of a Round Dielectric Waveguide, *B.S.T.J.* Vol. **48**, No. 10, December 1969. Copyright 1969, The American Telephone and Telegraph Co., reprinted by permission.)

made them hardly visible. Yet all the power is converted from the incident TE_{01} mode to the TE_{02} mode over a distance only of 75 cm. The theoretical prediction based on (9.3-55), with κ obtained not from (9.3-48) but from the corresponding theory of the round rod,[99] was $D = 80$ cm. The slight discrepancy between theory and experiment can be attributed to machining tolerances of ± 10 μ. The slight scatter in the data is caused by interference of the TE_{02} mode power with another unwanted mode. In order to confirm that it was indeed the TE_{02} mode whose power buildup was detected, we probed the radiation field at the end of the smooth rod and at the end of the corrugated rod. The result is shown in Figure 9.3.4. The transverse mode pattern obtained at the end of the smooth rod (solid line) shows the characteristic shape of the TE_{01} mode with only a slight admixture of TE_{02} mode power, which was inadvertently produced by the mode launcher. When the smooth rod was replaced with the corrugated rod, the TE_{02} mode pattern (dotted line) appeared. The corrugated rod was inserted into the metallic waveguide to such an extent that the remaining free end protruded from the metallic waveguide corresponded to the complete mode conversion length D. The figure shows that mode conversion was practically complete. The slight asymmetry and distortion of the TE_{02} transverse mode profile are again attributable to the presence of a small amount of unwanted mode power produced by the launcher.

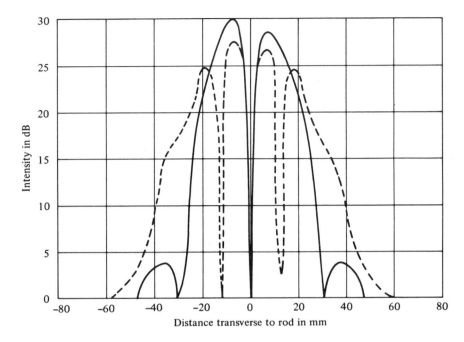

Figure 9.3.4   Transverse field patterns at the end of the smooth rod (solid line) and at the end of the corrugated rod (dotted line). The solid line corresponds to the TE_{01} mode with only a slight admixture of TE_{02} power. The dotted line corresponds to the TE_{02} mode. (From D. Marcuse and R. M. Derosier, Mode Conversion Caused by Diameter Changes of a Round Dielectric Waveguide, *B.S.T.J.*, Vol. **48**, No. 10, December 1969. Copyright 1969, The American Telephone and Telegraph Co., reprinted by permission.)

The radiation loss predicted for the grooves with 76 μ depth was too slight to be observable. It was thus necessary to deepen the grooves. The rod with 230 μ deep grooves was again used to study the mode conversion between the two guided modes. The result is shown in Figure 9.3.5. The resulting complete conversion length of 25 cm is again in good agreement with theory. In fact, it is remarkable that the coupling coefficient obtained from perturbation theory correctly predicts the mode conversion phenomenon of such strength. Complete mode conversion is achieved (and correctly predicted) over a distance of only 45 wavelengths. Figure 9.3.5 also demonstrates convincingly the validity of the assumption that only two modes are affected by the periodic wall perturbation. The mode coupling phenomenon goes through two complete cycles without any indication that power has been lost to radiation.

Lowering the frequency below the cutoff of the TE_{02} mode makes TE_{01} the dominant mode. There are of course other modes possible, notably

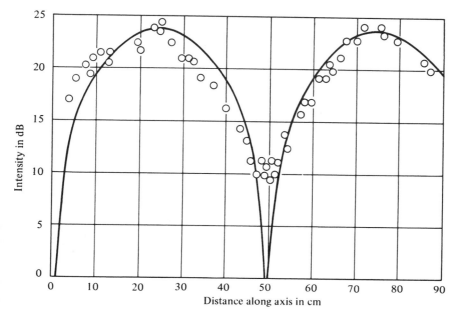

Figure 9.3.5 Buildup of the TE_{02} mode along the corrugated rod. The energy exchange goes through two complete cycles. Groove depth $= 2.3 \times 10^{-2}$ cm. (From D. Marcuse and R. M. Derosier, Mode Conversion Caused by Diameter Changes of a Round Dielectric Waveguide, *B.S.T.J.*, Vol. **48**, No. 10, December 1969. Copyright 1969, The American Telephone and Telegraph Co., reprinted by permission.)

HE_{11}, but these modes are not coupled to TE_{01} by the circular symmetry of the structure. Below the cutoff of TE_{02}, the incident TE_{01} mode can couple only to the radiation modes. The radiation loss of the TE_{01} mode on the rod with 230 μ deep grooves is shown as a function of frequency in Figure 9.3.6. The dots are the measured points. The solid line is the theoretical curve computed under the assumption that the effective rod radius that has to be used in the theory is the average value of the high and low portions of the grooves. There is some doubt as to what value should be used for the radius of the rod. The two dotted lines show the theoretical result if the high points or the low points are used as the guide radii. Almost all the experimental points fall between the two dotted lines, showing the close agreement of theory and experiment. The experimental accuracy is no better than approximately 1 dB. The high losses that are obtained as the frequency approaches the cut-off frequency of the TE_{02} mode exceed the applicability of the simple perturbation theory by far. The experiment justifies our extension (9.3-22) to arbitrary loss values.

The angle of the radiation lobe is shown in Figure 9.3.7. The experimental points again fall very closely to the theoretical curve.

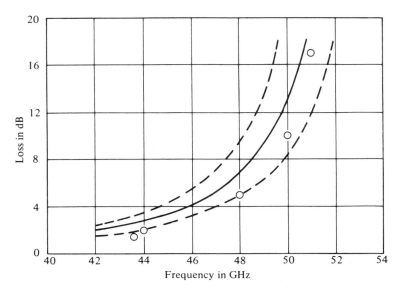

Figure 9.3.6   Radiation loss as a function of frequency. Dotted lines represent theoretical loss assuming that the guide radius is either the maximum or the minimum value on top of the grooves or in the grooves. The solid curve shows the theoretical loss based on the average radius of the corrugated rod. The dots are the measured points. Groove depth = 2.3×10^{-2} cm. (From D. Marcuse and R. M. Derosier, Mode Conversion Caused by Diameter Changes of a Round Dielectric Waveguide, *B.S.T.J.*, Vol. **48**, No. 10, December 1969. Copyright 1969. The American Telephone and Telegraph Co., reprinted by permission.)

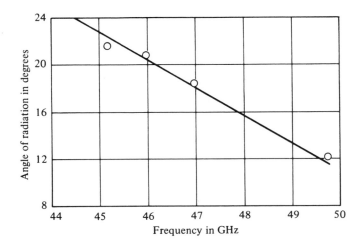

Figure 9.3.7   The angle of the far field radiation lobe as a function of frequency. The solid line represents the theory. The dots are measured points. (From D. Marcuse and R. M. Derosier, Mode Conversion Caused by Diameter Changes of a Round Dielectric Waveguide, *B.S.T.J*, Vol. **48**, No. 10, December 1969. Copyright 1969, The American Telephone and Telegraph Co., reprinted by permission.)

We conclude this section with a few numerical predictions of radiation and mode conversion effects applicable to the optical region. We have demonstrated the validity of the perturbation theory and its extensions at millimeter wave frequencies. However, it is more important for the purpose of this book to know the implications of the theory for light transmission through optical fibers. We assume that the free space wavelength of the light guided by the slab waveguide is

$$\lambda = 1\mu \tag{9.3-87}$$

This infrared frequency was chosen simply to use a round number.

We begin by asking: what amplitude a of the sinusoidal wall perturbations or a Fourier component of an arbitrary wall perturbation causes 10 percent of the power to be lost to the next guided mode over a distance of 1 cm? ($\Delta P/P = 0.1$, $L = 1$ cm.) We use $k_o d = 3.0$. With (9.3-87), this leads to a half width of the slab of $d = 0.477\ \mu$. Under these conditions, there are three TE modes possible. Their propagation constants are obtained from (8.3-16) with $n_1 = 1.5$ and $n_2 = 1$.

$$\left.\begin{array}{l} \beta_o d = 4.336 \quad \text{even mode} \\ \beta_1 d = 3.831 \quad \text{odd mode} \\ \beta_2 d = 3.051 \quad \text{even mode} \end{array}\right\} \tag{9.3-88}$$

From (9.3-10), we calculate the required value of the amplitude of the sinusoidal perturbation, coupling mode 0 to mode 2 ($\phi = 0$, thickness variation)

$$a = 2.5\ 10^{-5}\ \mu = 0.25\ \text{Å} \tag{9.3-89}$$

This extraordinarily small number indicates the extreme tolerance requirements that mode coupling imposes on the waveguide dimensions. For $n_1 = 1.01$ and $n_2 = 1$, we obtain for $k_o d = 23$, the guide half width of $d = 3.66\ \mu$, and the propagation constants of the three possible TE modes

$$\left.\begin{array}{l} \beta_o = 23.199 \quad \text{even mode} \\ \beta_1 = 23.112 \quad \text{odd mode} \\ \beta_2 = 23.002 \quad \text{even mode} \end{array}\right\} \tag{9.3-90}$$

The required amplitude of the sinusoidal wall thickness perturbation ($\phi = 0$) that causes 10 percent loss in $L = 1$ cm by coupling the mode 0 to the even mode 2 is

$$a = 1.17\ 10^{-2}\ \mu = 117\ \text{Å} \tag{9.3-91}$$

Both guides were designed to allow three guided modes to propagate. Their dimensions at $\lambda = 1\ \mu$ are, therefore quite different. Under these conditions, the guide with the slight index difference is far less sensitive to thickness variation than the guide with the high index difference. However, even this second example, which is representative of typical cladded optical fibers, shows that the tolerance requirements are quite stringent. It is apparent from these examples that there is virtually no hope of operating a multimode optical fiber with only one guided mode. The inavoidable wall perturbations and other influences cause the incident single mode power to spread very rapidly over all possible guided modes. However, it is possible to operate a fiber under conditions that ensure single guided mode operation. If the guide radius or thickness is below the cutoff value of any but the dominant mode, only this mode can propagate. The random wall perturbations can then no longer couple this mode to guided modes, but they can still cause radiation losses.

We estimate the radiation losses for single guided mode operation. We use (9.3-87) and $n_1 = 1.01$, $n_2 = 1$, $k_o d = 15$, so that we obtain $d = 2.39$ μ. We ask for the amplitude of the sinusoidal wall thickness perturbations ($\phi = 0$), which causes 10 percent power to be lossed in $L = 1$ cm. From (9.3-19), we calculate

$$a = 5.46 \ 10^{-2} \ \mu = 546 \ \text{Å} \tag{9.3-92}$$

The power loss caused by radiation is thus not quite as rapid as that caused by coupling to other guided modes. However, we must remember that the amplitude (9.3-92) belongs only to one Fourier component. Whereas only one specific Fourier component can cause loss to any given guided mode, a large range of mechanical frequencies contributes to radiation loss. We must add the losses caused by all Fourier components to obtain the overall radiation loss. Instead of pursuing this method, we turn to a statistical discussion of radiation losses caused by random wall perturbations.

Incidentally, almost the same numbers obtained from the slab waveguide theory are calculated from the theory of the TE modes in the round dielectric fiber (reference 99). This agreement of the result of different models shows that results based on the simple slab waveguide model are representative for other types of waveguide and, in particular, for the round optical fiber.

9.4 RANDOM WALL PERTURBATIONS

Our general formulas of power loss caused by mode conversion to guided modes and to radiation modes can be used to calculate the average losses that can be expected from a slab waveguide with random wall perturbations.

Both walls of the slab contribute to the loss. We assume in this section that one wall of the waveguide is perfect, $h(z) = 0$. The results obtained on this basis certainly give the right order of magnitude of the power losses. Furthermore, it is easy to account for the random wall perturbations of the other wall. If both walls contain a broad spectrum of mechanical frequencies of equal "power" distribution, the waveguide losses are twice as high if both sides of the slab are distorted. For a narrow spectrum of wall distortion with perfect correlation between the two walls, the losses can at the most be four times as high as those calculated under the assumption of one perfectly straight wall.

We begin by considering the mode conversion losses to other guided modes. The loss contribution of each guided mode is, according to (9.2-28), given by the absolute square value of its expansion coefficient. The average loss contribution of a given even guided mode is thus, according to (9.2-36)

$$< |c_{\mu e}|^2 > = (n_1^2 - n_2^2)^2 \frac{L^2 k_o^4 \cos^2 \kappa_o d_o \cos^2 \kappa_\mu d_o}{4 \left(\beta_o d_o + \dfrac{\beta_o}{\gamma_o} \right) \left(\beta_\mu d_o + \dfrac{\beta_\mu}{\gamma_\mu} \right)} < |F|^2 > \qquad (9.4\text{-}1)$$

The symbol $< \; >$ indicates an ensemble average. This means that we consider many statistically identical systems and their mode conversion losses. The average over the loss values obtained from all the systems is given by (9.4-1). Many systems are ergodic, which means that the average obtained from measuring different portions of one and the same waveguide would be the same as the average taken over measurements obtained from many similar waveguides. For an ergodic system, the ensemble average can be replaced by an average over loss measurements taken at different points along the guide.

The quantity of interest in (9.4-1) is the ensemble average of the Fourier coefficient

$$< |F|^2 > = \frac{1}{L^2} \int_0^L dz \int_0^L dz' < f(z)f(z') > e^{- i(\beta_o - \beta_\mu)(z - z')} \qquad (9.4\text{-}2)$$

The average value of the product of the function $f(z)$ with itself is called the autocorrelation function of $f(z)$. The autocorrelation function is independent of the position z for a stationary random process. It depends only on the difference $z - z'$.

$$R(z - z') = < f(z)f(z') > \qquad (9.4\text{-}3)$$

The integral can be simplified by introducing the new variables

$$\left. \begin{array}{l} u = z - z' \\ v = z' \end{array} \right\} \qquad (9.4\text{-}4)$$

The integration range and the lines of constant values for the new variables are shown in Figure 9.4.1.

The integral (9.4-2) appears now in the form

$$< |F|^2 > = \frac{1}{L^2} \int_0^L du \; R(u)e^{- i(\beta_o - \beta_\mu)u} \int_0^{L-u} dv$$

$$+ \frac{1}{L^2} \int_{-L}^0 du \; R(u)e^{- i(\beta_o - \beta_\mu)u} \int_{-u}^L dv \qquad (9.4\text{-}5)$$

The new limits on the integrals are obtained from Figure 9.4-1. The line $u = 0$ forms the diagonal of the square that runs from the coordinate origin to its upper right corner. Below this diagonal, u runs through the values from $u = 0$ to $u = L$ that are indicated on the first integral. The corresponding

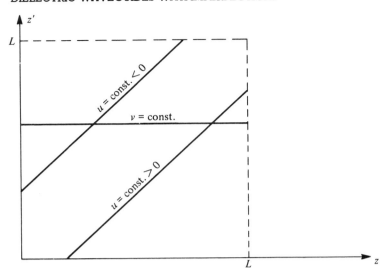

Figure 9.4.1   Conversion from z, z' coordinates to the coordinates u and v.

range for v is obtained as this variable starts out at the bottom of the square with $v = 0$ and rises to the point where the line $u = \text{const}$ intersects the vertical $z = L$. At this latter point, we have $v = z' = L - u$. The second double integral covers the region of the square above the diagonal. In the upper left corner, we have $u = -L$, so that the range of u values covers the interval from $u = -L$, to $u = 0$, as indicated on the second double integral. The range of v values begins at the intersection of the line $u = \text{const}$ with the vertical $z = 0$. At that point, we have $v = -u$. The v values increase until the line $u = \text{const}$ reaches the top of the square with $v = L$.

The v integrals can be performed immediately, with the result

$$< |F|^2 > = \frac{1}{L^2} \int_{-L}^{L} (L - |u|)R(u)e^{-i(\beta_o - \beta_\mu)u}du \qquad (9.4\text{-}6)$$

Equation (9.4-6) is still exact. It is possible, however, to introduce a useful approximation. For truly random processes, $R(u)$ must decrease to zero as the distance u between the two points z and z' on the waveguide increases. If this were not the case, there would be a systematic deformation of the guide walls contrary to our assumption. We assume that L is much larger than the distance over which $R(u)$ gives an appreciable contribution. It is thus possible to neglect $|u|$ compared to L. Furthermore, since $R(u)$ vanishes for large values of u, it makes no difference if we increase the integration range to ∞. We thus obtain the approximation

$$< |F|^2 > = \frac{1}{L} \int_{-\infty}^{\infty} R(u)e^{-i(\beta_o - \beta_\mu)u}du \qquad (9.4\text{-}7)$$

The average value of the "power spectrum" $|F|^2$ of the wall distortion function is thus proportional to the Fourier transform of the correlation function.

The expression (9.4-7) is inversely proportional to L. It is thus apparent that the relative power loss (9.4-1) to the guided mode μ becomes proportional to L, and it is possible to define a loss per unit length. This is a very interesting result. We have seen in Section 9.3 that the expression for the relative power loss, (9.3-10), caused by a sinusoidal wall distortion was proportional to L^2. The average loss caused by a random wall distortion is, however, proportional to L. The nature of the loss process has changed completely. This result is in agreement with our discussion of the loss contributions of the different Fourier components of the wall distortion function. It was pointed out in this discussion in Section 9.3 that total power exchange between only two guided modes is no longer possible if the wall distortion function varies randomly, because the phase of the Fourier transform becomes a random variable. Our present result proves that total power exchange can no longer result, even though it is only one Fourier component of the wall distortion function that is responsible for the coupling between two guided modes. The power loss to a given guided mode becomes a true loss per unit length. The Fourier component of a random function is inversely proportional to the square root of length, while the Fourier component of a strictly periodic function is independent of length.

If we want to obtain more than this very general result, we must know the correlation function of the wall distortion. This function is no easier to obtain than the " power spectrum " of the wall distortion function. In fact, we have seen that one is the Fourier transform of the other. However, we know certain general features of the correlation function, so that it is possible to use model assumptions and study their effect on the power loss. The correlation function must vanish for large values of u. It also must be a symmetrical function of u, since it should not matter whether we correlate the point z' to the right or to the left of z with z. A particularly simple correlation function is the following

$$R(u) = A^2 e^{-(|u|/B)} \tag{9.4-8}$$

The constant A is the rms deviation of the wall from perfect straightness, since $R(0) = \langle f^2(z) \rangle = A^2$. The constant B determines the rate of decay of $R(u)$ with increasing u, and is thus called the correlation length.

From (9.4-7) and (9.4-8), we obtain

$$\langle |F|^2 \rangle = \frac{2A^2}{BL} \frac{1}{(\beta_o - \beta_\mu)^2 + \dfrac{1}{B^2}} \tag{9.4-9}$$

The power loss to the μth even guided mode is thus, from (9.4-1) and (9.4-9)

$$\langle |c_{\mu e}|^2 \rangle = \frac{LA^2(n_1^2 - n_2^2)^2 k_o^4 \cos^2 \kappa_o d_o \cos^2 \kappa_\mu d_o}{2B\left(\beta_o d_o + \dfrac{\beta_o}{\gamma_o}\right)\left(\beta_\mu d_o + \dfrac{\beta_\mu}{\gamma_\mu}\right)\left[(\beta_o - \beta_\mu)^2 + \dfrac{1}{B^2}\right]} \tag{9.4-10}$$

The loss to odd guided modes is obtained by replacing $\cos \kappa_\mu d_o$ with $\sin \kappa_\mu d_o$. The total loss caused by all guided modes is the sum of the loss contributions of all even and odd guided modes.

The relative average radiation loss is obtained by taking an ensemble average of (9.3-42) and using (9.3-43) and (9.4-9)

$$\left\langle \frac{\Delta P}{P} \right\rangle = 2\alpha L = L \frac{A^2(n_1^2 - n_2^2)^2 k_o^4 \cos^2 \kappa_o d_o}{2\pi B \left(\beta_o d_o + \dfrac{\beta_o}{\gamma_o} \right)}$$

$$\cdot \int_{-n_2 k_o}^{n_2 k_o} \frac{\rho}{(\beta_o - \beta)^2 + \dfrac{1}{B^2}} \left[\frac{\cos^2 \sigma d_o}{\rho^2 \cos^2 \sigma d_o + \sigma^2 \sin^2 \sigma d_o} \right.$$

$$\left. + \frac{\sin^2 \sigma d_o}{\rho^2 \sin^2 \sigma d_o + \sigma^2 \cos^2 \sigma d_o} \right] d\beta \qquad (9.4\text{-}11)$$

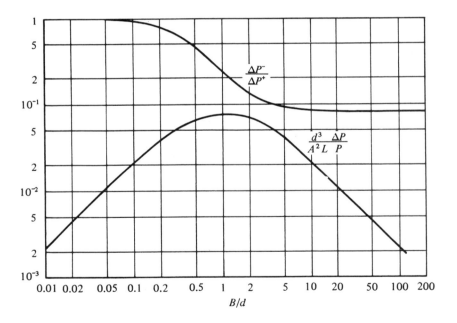

Figure 9.4.2   Normalized radiation loss $(d^3/A^2L)(\Delta P/P)$ and ratio of backward to forward scattered power $\Delta P^-/\Delta P^+$ as functions of the normalized correlation length B/d for $n_1 = 1.5$, $n_2 = 1.0$, $kd = 1.3$. Single guided mode operation. ($d =$ half width of the core, $A =$ rms deviation of one side of the core-cladding interface, $L =$ length of waveguide section contributing to loss, $n_1 =$ core index, $n_2 =$ cladding index, $k =$ free space propagation constant.) (From D. Marcuse, Mode Conversion Caused by Surface Imperfections of a Dielectric Slab Waveguide, B.S.T.J., Vol. 48, No. 10, December 1969. Copyright 1969, The American Telephone and Telegraph Co., reprinted by permission.)

Equation (9.4-11) is the definition of the power loss per unit length 2α, which determines the total loss, according to (9.3-22) (α is the amplitude loss coefficient.)

Equation (9.4-11) must be evaluated by numerical methods. We discuss two different cases. We begin by assuming

$$n_1 = 1.5 \qquad n_2 = 1 \qquad (9.4\text{-}12)$$

It is convenient to introduce a dimensionless quantity

$$\frac{d_o^3}{A^2 L} \frac{\Delta P}{P} \qquad (9.4\text{-}13)$$

This normalized relative power loss has the advantage of eliminating the rms deviation A from the equation. The normalized loss as a function of the normalized correlation length B/d_o is shown in Figures 9.4.2, 9.4.3, and 9.4.4 for several values of the parameter $k_o d_o$.

Also shown in these figures is the ratio of backward to forward scattered power, $\Delta P^-/\Delta P^+$. The number of possible guided modes is increasing. Figure 9.4.2 holds for the case that only the incident TE mode can propagate.

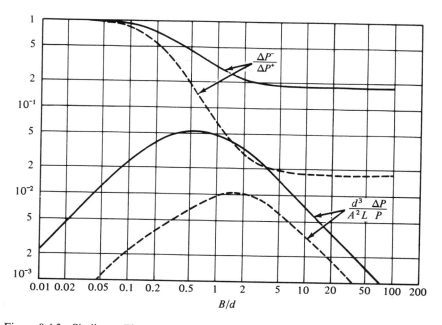

Figure 9.4.3 Similar to Figure 9.4.2. Two guided modes can propagate. $n_1 = 1.5$, $n_2 = 1$, $kd = 1.8$. The dotted lines indicate power lost to the unwanted guided mode. The solid lines indicate radiation loss. (From D. Marcuse, Mode Conversion Caused by Surface Imperfections of a Dielectric Slab Waveguide, *B.S.T.J.*, Vol. **48**, No. 10, December 1969. Copyright 1969, The American Telephone and Telegraph Co., reprinted by permission.)

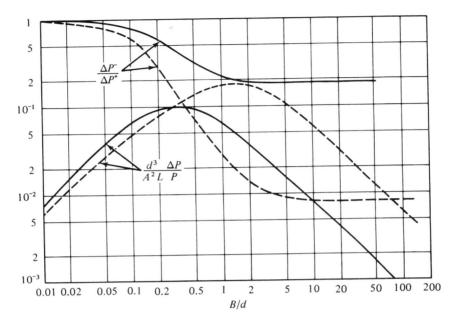

Figure 9.4.4   Similar to Figure 9.4.2. Three guided modes can propagate. $n_1 = 1.5$, $n_2 = 1$, $kd = 3$. The dotted lines indicate power lost to the two unwanted guided modes. The solid lines indicate radiation loss. (From D. Marcuse, Mode Conversion Caused by Surface Imperfections of a Dielectric Slab Waveguide, B.S.T.J., Vol. 48, No. 10, December 1969. Copyright 1969, The American Telephone and Telegraph Co., reprinted by permission.)

This guide is operated with a dominant mode. The next two figures hold for two- and three-mode operation. The relative power loss caused by radiation is shown as the solid line, while the relative power loss to the unwanted guided modes is plotted as a dotted line.

The case of low index difference often encountered in optical fibers

$$n_1 = 1.01 \qquad n_2 = 1 \qquad (9.4\text{-}14)$$

is shown in Figures 9.4.5, 9.4.6, and 9.4.7.

All these figures have the same characteristic features in common. The normalized relative power loss has a maximum near $B/d_o = 1$. The power loss is thus worst if the correlation length is approximately equal to the half width of the guide or, since the guide half width has the same order of magnitude as the wavelength, if the correlation length is approximately equal to the wavelength. The ratio of backward to forward scattered power $\Delta P^-/\Delta P^+$ becomes unity for small values of B and decreases to a finite value (not shown in all figures) for large values of B. This behavior is typical for light scattering from small particles. We might think of the correlation length as defining the

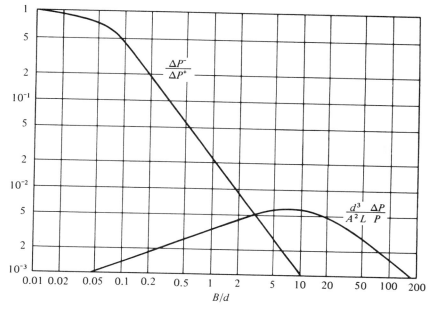

Figure 9.4.5   Similar to Figure 9.4.2. The parameters $n_1 = 1.01$, $n_2 = 1$, $kd = 8$ ensure single-mode operation. (From D. Marcuse, Mode Conversion Caused by Surface Imperfections of a Dielectric Slab Waveguide, *B.S.T.J.*, Vol. **48**, No. 10, December 1969. Copyright 1969, The American Telephone and Telegraph Co., reprinted by permission.)

particle size of the scatterers on the surface of the waveguide. Short correlation length indicates a small particle size, while long correlation length corresponds to large particle size. These particles must not be taken literally. The term is simply meant to indicate the region of surface perturbation whose deflections are correlated. Scattering by particles that are much smaller than the wavelength is well known to be isotropic. Large particles, on the other hand, scatter predominantly in forward direction.

The shape of the normalized relative power loss as a function of correlation length depends on the assumed form of the correlation function. Our curves hold of course for the exponential correlation function (9.4-8). Other functions and superpositions of different functions could (and should) be used. Several other functions have been tried. They all were monotonically decreasing functions symmetrical in u. It was found that the shape of the relative power loss curves to the left of the maximum as well as the position and height of the maximum do not depend on the form of the correlation function. No matter what particular form the correlation functions assumes, the peak of the loss curve and the maximum loss value remain the same. The shape of the loss curve to the right of the loss peak is very strongly dependent on the shape of the correlation function.

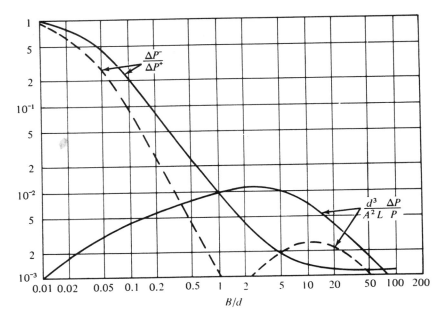

Figure 9.4.6   Similar to Figure 9.4.2. Two guided modes can propagate. $n_1 = 1.01$, $n_2 = 1$, $kd = 15$. The meaning of the dotted and solid lines is explained in Figure 9.4.3. (From D. Marcuse, Mode Conversion Caused by Surface Imperfections of a Dielectric Slab Waveguide, B.S.T.J., Vol. **48**, No. 10, December 1969. Copyright 1969, The American Telephone and Telegraph Co., reprinted by permission.)

Similar curves for the relative power loss have been calculated for the TE and TM modes[99] as well as for the dominant HE_{11} mode of the round waveguide.[96] The shape of these curves agrees remarkably well with those shown here for the slab waveguide. The loss peaks for the curves of the round waveguide are approximately four times higher than the loss peaks appearing in these curves. The explanation is simple. Our curves hold for the case that only one of the two sides of the slab has random imperfections. If both walls were deformed in a perfectly correlated way, the losses would be four times as high. This is just the case that applies to diameter changes of the round dielectric rod. The agreement between the slab theory and the much more complicated theory of the round rod is excellent.

We conclude this section with a few numerical examples to gain an insight into the losses that can be expected from random wall perturbations. Let us assume that the refractive index is given by (9.4-12). We assume that $k_o d_o = 1.3$, so that only the dominant guided mode can propagate. Furthermore, we assume that the correlation function assumes its worst possible value. Any other assumption would depend on the shape of the correlation

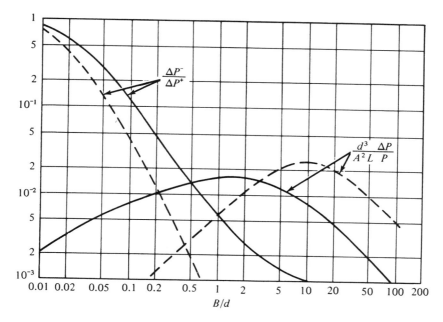

Figure 9.4.7 Similar to Figure 9.4.2. There guided modes can propagate. $n_1 = 1.01$, $n_2 = 1$, $kd = 23$. (From D. Marcuse, Mode Conversion Caused by Surface Imperfections of a Dielectric Slab Waveguide, *B.S.T.J.*, Vol. **48**, No. 10, December 1969. Copyright 1969, The American Telephone and Telegraph Co., reprinted by permission.)

function and would thus not be too meaningful. We have $d_o = 0.207 \mu$ for $\lambda = 1 \mu$. The normalized loss at the peak of Figure 9.4.2 is

$$\frac{d^3}{A^2 L} \frac{\Delta P}{P} = 7 \cdot 10^{-2} \qquad (9.4\text{-}15)$$

We ask again: what value of the rms deviation A is required to cause 10 percent radiation loss at a distance of $L = 1$ cm? We obtain

$$A = 1.12 \ 10^{-3} \mu = 11.2 \ \text{Å} \qquad (9.4\text{-}16)$$

The relative tolerance requirement is

$$\frac{A}{d} = 5.4 \ 10^{-3} = 0.54 \ \text{percent} \qquad (9.4\text{-}17)$$

Figure 9.4.3 shows that the radiation loss outweighs the loss to the one unwanted guided mode. For this example, we have, with $k_o d_o = 1.8$ and with $\lambda = 1 \mu$, a guide half width of $d = 0.286 \mu$. The normalized relative radiation loss of 0.05 at $B/d_o = 1$ causes 10 percent radiation loss at $L = 1$ cm if

$$A = 2.16 \ 10^{-3} \mu = 21.6 \ \text{Å} \qquad (9.4\text{-}18)$$

The relative tolerance requirement is

$$\frac{A}{d_o} = 7.6 \ 10^{-3} = 0.76 \text{ percent} \tag{9.4-19}$$

The addition of the guided mode loss does not change these figures appreciably.

The radiation loss of Figure 9.4.4 is slightly less than the loss to the two unwanted guided modes. With $k_o d_o = 3$, we have again for $\lambda = 1 \ \mu$ the half width $d_o = 0.477 \ \mu$. The combined normalized radiation and mode conversion loss at $B/d_o = 1$ is 0.25. The rms deviation responsible for 10 percent loss at $L = 1$ cm is

$$A = 2.08 \ 10^{-3} = 20.8 \text{ Å} \tag{9.4-20}$$

The relative tolerance requirements are

$$\frac{A}{d_o} = 4.35 \ 10^{-3} = 0.44 \text{ percent} \tag{9.4-21}$$

These comparisons show that the tolerance requirements are nearly the same in all three cases.

Next, we consider the case of a slab waveguide with 1 percent index difference, (9.4-14). For the dominant mode case, Figure 9.4.5 with $k_o d_o = 8$, corresponding to $d_o = 1.28 \ \mu$ and a loss peak of $6 \ 10^{-3}$ located at $B/d = 10$, we obtain

$$A = 5.9 \ 10^{-2} \ \mu = 590 \text{ Å} \tag{9.4-22}$$

for the usual requirements. The relative tolerance requirement is

$$\frac{A}{d_o} = 4.6 \ 10^{-2} = 4.6 \text{ percent} \tag{9.4-23}$$

In case that two unwanted guided modes can propagate in addition to the incident mode, we have, form Figure 9.4.7, with $k_o d_o = 23$, $d_o = 3.66 \ \mu$, a total value of radiation plus mode conversion loss of $3.4.10^{-2}$, so that the tolerance requirement that follows from the usual condition is

$$A = 0.12 \ \mu \tag{9.4-24}$$

or the relative tolerance requirement

$$\frac{A}{d_o} = 3.3 \ 10^{-2} = 3.3 \text{ percent} \tag{9.4-25}$$

The relative tolerance requirements for these low index guides are again nearly the same for the case of a dominant mode guide and for the case of a guide that can support three guided modes. The tolerance requirements of the low index guide are much more relaxed as compared to those of the high index guide. This comparison shows that the high index guide is more susceptible to losses caused by random wall perturbation than the low index guide.

As a final example, let us consider the case of a fiber guide for long-distance light transmission. We again consider the low index guide (9.4.14), and assume that it is operated under dominant mode conditions corresponding to Figure 9.4.5. The only difference between our present case and the one leading to (9.4-22) is the requirement that the radiation losses are 10 dB/km. Under these conditions, we must require that the rms deviation of the wall is in the order of

$$A = 9 \ 10^{-4} \ \mu = 9 \ \text{Å} \tag{9.4-26}$$

The tolerance requirements for long-distance transmission are extremely stringent indeed, even for the less critical low index guide. Our estimate may well be too pessimistic, however, since we have assumed that the waveguide has wall imperfections whose correlation length is in the order of its half width. It may well be that the correlation length of practical drawn glass fibers are far longer and consequently lead to far less radiation and mode conversion loss. One way of estimating the correlation length is to observe the ratio of forward to backward scattered light. If the ratio is near unity, we know that the correlation length is indeed short. Predominantly forward scattering indicates a long correlation length. Some preliminary measurements have resulted in scattering losses of the order of 40 dB/km.[105] This result indicates perhaps that the tolerances of the guide were indeed extremely small if the correlation length happened to be in the order of the guide width. More likely, however, the correlation length is actually much longer. In that case, our model is not too instructive, since the shape of the correlation function critically influences the loss value. It would be necessary to know more about the "power spectrum" of the wall distortion function or, alternately, about the correlation function in order to be able to make theoretical loss predictions.

9.5 STEPS AND TAPERS OF THE SLAB WAVEGUIDE

In Section 9.2, we derived the expansion coefficients of the field expansion (8.5-16) directly from the requirement that the electric field component must be a solution of the wave equation. We applied this method only to fields of the transverse electric type. The same method can of course be used to handle all other types of fields. However, the practical procedure for determining the amplitude coefficients is much more complicated in the general case. To illustrate this point, let us consider the wave equation for the H_y component of the transverse magnetic fields. The form (8.3-33) for the wave equation of TM modes is correct only in a homogeneous dielectric medium. We could use this equation in Section 8.3 since we solved the wave equation separately in each medium and matched the two solutions at the dielectric interfaces $x = \pm d$.

The method of Section 9.2 does not require any boundary conditions, but it requires the wave equation in a form that holds everywhere, even at the dielectric interface. The proper form of the reduced wave equation for $\mathbf{H}$ is

$$\nabla^2\mathbf{H} + \frac{1}{\varepsilon}(\nabla\varepsilon) \times (\nabla \times \mathbf{H}) + \omega^2\varepsilon\mu_o\mathbf{H} = 0 \qquad (9.5\text{-}1)$$

The derivation of this equation is left to the reader as an exercise. For a transverse magnetic field whose only magnetic field component is H_y and for a medium whose dielectric constant varies only in x direction, (9.5-1) reduces to

$$\nabla^2 H_y - \frac{1}{\varepsilon}\frac{\partial\varepsilon}{\partial x}\frac{\partial H_y}{\partial x} + \omega^2\varepsilon\mu_o\,H_y = 0 \qquad (9.5\text{-}2)$$

At the dielectric interface, the x derivative of the dielectric constant contributes a delta function, while $\partial H_y/\partial x$ has a discontinuity at the same point. The wave equation for the E_y component of the transverse electric fields does not have this complication, since the only derivative of ε that occurs in (1.3-4) is $\partial\varepsilon/\partial y$. This derivative vanishes because of our condition (8.3-1). It is possible to handle Equation (9.5-2) by the method of Section 9.2. This procedure becomes much more complicated, however. It is necessary to approximate ε by a steep but continuous function. Another possible approach to this problem is the use of local normal modes, which are mentioned briefly at the beginning of Section 9.2. There are simpler methods that can be used. The discussion of one such method is the object of this section.

It is possible to approximate any arbitrary deformation of the interface of the slab by a succession of steps,[101] as shown in Figure 9.5.1. In the limit of infinitely many steps and vanishing step height, the succession of steps is capable of approximating any arbitrary function. If the mode conversion of one step is known, the contribution to mode conversion from more complicated shapes can be synthesized by superposition of the fields that are generated at each individual step. This method is simple and remarkably successful. It is applicable to any type of field, even to the complicated fields of the round optical fiber.[96] The validity of this method has been verified by

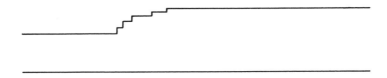

Figure 9.5.1   An arbitrarily deformed core-cladding interface can be approximated by a series of steps.

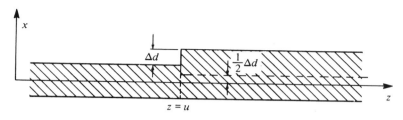

Figure 9.5.2 Geometry of a single step in the clore-cladding interface of a slab waveguide.

experiments. Furthermore, it leads to the same results as the method of Section 9.2. We shall apply the method of successive steps to the TE and TM modes of the slab waveguide and discuss its applications to other types of guides. The step method is also only an approximation. We shall point out the approximations as they arise.

We begin with a discussion of one individual step, as shown in Figure 9.5.2. We assume that the lowest order even TE mode is travelling on the guide. At the location of the step, this mode loses power by mode conversion to other guided modes and to the radiation field. For simplicity, we limit the discussion to the case of a dominant mode waveguide, which means that only one guided mode can exist on the guide before as well as after the step. The field on the guide can be calculated by expansion into normal modes. The expansion coefficients are determined by requiring the tangential electric and magnetic field components E_y and H_x to be continuous at the discontinuity at $z = u$. This condition leads to the equation for the E_y components

$$E_1^{(i)} + a_r E_1^{(r)} + \int_0^\infty q_e^{(-)}(\rho) E_{e1}^{(r)}(\rho) d\rho + \int_0^\infty q_o^{(-)}(\rho) E_{o1}^{(r)}(\rho) d\rho$$

$$= c_t E_2^{(t)} + \int_0^\infty q_e^{(+)}(\rho) E_{e2}^{(t)}(\rho) d\rho + \int_0^\infty q_o^{(+)}(\rho) E_{o2}^{(t)}(\rho) d\rho \quad (9.5\text{-}3)$$

and for the H_x components

$$H_1^{(i)} + a_r H_1^{(r)} + \int_0^\infty q_e^{(-)}(\rho) H_{e1}^{(r)}(\rho) d\rho + \int_0^\infty q_o^{(-)}(\rho) H_{o1}^{(r)}(\rho) d\rho$$

$$= c_t H_2^{(t)} + \int_0^\infty q_e^{(+)}(\rho) H_{e2}^{(t)}(\rho) d\rho + \int_0^\infty q_o^{(+)}(\rho) H_{o2}^{(t)}(\rho) d\rho \quad (9.5\text{-}4)$$

The expressions on the left-hand side of the equations represent the fields immediately to the left of the step. This field is a superposition of the incident wave (superscript i), a reflected guided wave (superscript r), and reflected even and odd radiation modes. The constant a_r is the reflection coefficient of the guided mode. The coefficients $q_e^{(-)}$ and $q_o^{(-)}$ are the amplitudes of the reflected even and odd radiation modes. The subscript 1 attached to the field components indicates the fields on the waveguide to the left of the step. The expressions on the right-hand side of the equations represent the transmitted guided mode and forward scattered radiation modes. The superscript t indicates transmitted modes traveling in the positive z direction. The constant c_t is the transmission coefficient of the guided mode; $q_e^{(+)}$ and $q_o^{(+)}$ are the amplitudes of the forward traveling even and odd radiation modes. In Section 8.6, the expansion coefficients were determined from the requirement that the field must satisfy the wave equation. In this section, we determine the expansion coefficients from the requirement of continuity of the transverse field components at the position of the step.

The modes of the waveguide to the left of the step are mutually orthogonal among each other. The same is true for the modes of the waveguide to the right of the step. However, the modes of the waveguide to the left of the step are not orthogonal to the waveguide modes to the right of the step, because of the different size of the waveguides. However, for small step height, there is an approximate orthogonality of the modes on one side of the step with the modes on the other side. We isolate the expansion coefficient $q_e^{(+)}$, for example, by using the orthogonality of the modes to the right of the step. The coefficients $q_e^{(-)}$ and $q_o^{(-)}$ are small. We can neglect the difference between the radiation modes on either side of the step, but keep in mind that $E^{(t)}(\rho)$ depends on $\exp(-i\beta z)$ while $E^{(r)}(\rho)$ depends on $\exp(i\beta z)$. We achieve the determination of $q_e^{(+)}$ by neglecting the amplitude reflection coefficient a_r (which is negligibly small for small step height), multiplication of (9.5-3) by

$$\frac{\beta}{2\omega\mu_o} E_{e2}^{(t)*}(\rho') \tag{9.5-5}$$

and by subsequent integration over x. We thus obtain, with the help of (8.4-11), at $z = u$

$$q_e^{(+)}(\rho) = \frac{\beta}{2\omega\mu_o P} \int_{-\infty}^{\infty} E_1^{(i)} E_{e2}^{(t)*}(\rho)dx + q_e^{(-)}(\rho)e^{2i\beta u} \tag{9.5-6}$$

We can also obtain $q_e^{(+)}$ from (9.5-4). We express the H_x components appearing in this equation in terms of E_y with the help of (8.3-2)

$$\beta_o E_1^{(i)} - a_r \beta_o E_1^{(r)} - \int_0^\infty \beta q_e^{(-)}(\rho)E_{e1}^{(r)}(\rho)d\rho - \int_0^\infty \beta q_o^{(-)}(\rho)E_{o1}^{(r)}(\rho)d\rho$$

$$= \beta_o c_t E_2^{(t)} + \int_0^\infty \beta q_e^{(+)}(\rho)E_{e2}^{(t)}(\rho)d\rho + \int_0^\infty \beta q_o^{(+)}(\rho)E_{o2}^{(t)}(\rho)d\rho \tag{9.5-7}$$

Multiplication with

$$\frac{1}{2\omega\mu_o} E_{2e}^{(t)*}(\rho') \tag{9.5-8}$$

and integration over x yields, with the help of (8.4-11), at $z = u$

$$q_e^{(+)}(\rho) = \frac{\beta_o}{2\omega\mu_o P} \int_{-\infty}^{\infty} E_1^{(i)} E_{e2}^{(t)*}(\rho) dx - q_e^{(-)}(\rho) e^{2i\beta u} \qquad (9.5-9)$$

The reflection coefficient a_r was again neglected. By adding (9.5-6) and (9.5-9), $q_e^{(-)}(\rho)$ drops out, and we obtain

$$q_e^{(+)}(\rho) = \frac{\beta_o + \beta}{4\omega\mu_o P} \int_{-\infty}^{\infty} E_1^{(i)} E_{e2}^{(t)*}(\rho) dx \qquad \text{at} \qquad z = u \qquad (9.5-10)$$

Subtraction of (9.5-6) and (9.5-9) yields an expression for $q_e^{(-)}(\rho)$ that is very similar to (9.5-10), except that β is replaced with $-\beta$ and $E^{(t)*}(\rho)$ with $E^{(r)*}(\rho)$. (Explain why.) The expression for the expansion coefficient of the odd radiation modes follows from (9.5-10) by replacing E_{e2} with E_{o2}.

We assume that the step height Δd is much smaller than the half width of the slab d. The y component of the electric field of the radiation modes of the waveguide to the right of the step is (see [8.4-1] through [8.4-10])

$$E_{e2}^{(t)*}(\rho, x') = e^{i\beta z} \begin{cases} C_e \cos \sigma x' & \text{for} \quad |x'| < d_2 \\ C_e \left[\cos \sigma d_2 \cos \rho(|x'| - d_2) \right. \\ \qquad \left. - \frac{\sigma}{\rho} \sin \sigma d_2 \sin \rho(|x'| - d_2) \right] & \text{for} \quad |x'| > d_2 \end{cases}$$

$$(9.5-11)$$

The coordinate x' is referred to the dotted line in Figure 9.5.2. We expand this expression in a power series in Δd, keeping only the first two terms

$$E_{e2}^{(t)*}(\rho) = E_{e1}^{(t)*}(\rho, x') + \frac{1}{2} \left(\frac{\partial E_{e2}^*(\rho, x')}{\partial d_2} \right)_{d_2 = d_1} \Delta d \qquad (9.5-12)$$

with

$$d_2 = d_1 + \frac{1}{2} \Delta d \qquad (9.5-13)$$

The coordinate x' is related to x by

$$x' = x - \frac{1}{2} \Delta d \qquad (9.5-14)$$

The second term on the right of (9.5-12) is already small of first order. The difference between x and x' is also a first order term. Replacing x' with x in this term causes only a change to second order in Δd, which is negligible in a first order theory. We thus obtain from (9.5-12)

$$E_{e2}^{(t)*}(\rho) = E_{e1}^{(t)*}(\rho,x) - \frac{1}{2}\left(\frac{\partial E_{e1}^{(t)*}(\rho,x')}{\partial x'}\right)_{x'=x} \Delta d$$

$$+ \frac{1}{2}\left(\frac{\partial E_{e2}^{(t)*}(\rho,x)}{\partial d_2}\right)_{d_2=d_1} \Delta d \quad (9.5\text{-}15)$$

The first term of (9.5-15) does not contribute to (9.5-10), because of mode orthogonality. We thus obtain

$$q_e^{(+)}(\rho) = \Delta d \frac{\beta_o + \beta}{8\omega\mu_o P} \int_{-\infty}^{\infty} E_1^{(i)} \left[\left(\frac{\partial E_{e2}^{(t)*}}{\partial d_2}\right)_{d_2=d_1} - \left(\frac{\partial E_{e1}^{(t)*}}{\partial x'}\right)_{x'=x}\right] dx \quad (9.5\text{-}16)$$

The incident mode is supposed to be an even function of x. The radiation mode $E_e(\rho)$ is also an even function. The derivative $\partial E_{e1}(\rho,x')/\partial x'$ of an even function must be an odd function. The product of the guided mode with the x' derivative of the radiation mode is an odd function, so that its contribution to the symmetrical integral vanishes. We are thus left with

$$q_e^{(+)}(\rho) = \Delta d \frac{\beta_o + \beta}{8\omega\mu_o P} \int_{-\infty}^{\infty} E_1^{(i)} \left(\frac{\partial E_{e2}^{(t)*}}{\partial d_2}\right)_{d_2=d_1} dx \quad (9.5\text{-}17)$$

The amplitude C_e does depend on d, according to (8.4-18). However, whether we write (9.5-11) with C_e or its derivative, this field component is still orthogonal to the incident guided mode if it is taken at $d_2 = d_1$, with $x' = x$. The term containing the derivative of C_e does not contribute to (9.5-17), and we can treat C_e as though it were independent of d_2. The propagation constant β as well as σ and ρ do not depend on d_2. The derivative of the radiation mode is thus simply

$$\left(\frac{\partial E_{e2}^{(t)*}}{\partial d_2}\right)_{d_2=d_1} = e^{i\beta z} \begin{cases} 0 \quad \text{for} \quad |x| < d \\ \\ -C_e \dfrac{n_1^2 - n_2^2}{\rho} k_o^2 \cos \sigma d_1 \sin \rho(|x| - d_1) \\ \\ \qquad\qquad\qquad \text{for} \quad |x| > d \end{cases}$$

$$(9.5\text{-}18)$$

This simple expression is caused by a cancellation of terms, and is typical of the TE modes of the slab waveguide. The corresponding expressions for the HE$_{11}$ mode of the round fiber are far more complicated.[96] The expression

$$\sigma^2 - \rho^2 = (n_1^2 - n_2^2)k_o^2 \quad (9.5\text{-}19)$$

was used to simplify (9.5-18) even further.

The equations (8.3-12), (8.3-18), and (9.5-18) are used to obtain from (9.5-17)

$$q_e(\rho,u) = -\frac{(n_1^2 - n_2^2)k_o^2\rho \cos \kappa d \cos \sigma d\, e^{-i(\beta_o - \beta)u}\Delta d}{2(\beta_o - \beta)\sqrt{\pi|\beta|\left(\beta_o d_o + \dfrac{\beta_o}{\gamma_o}\right)(\rho^2 \cos^2 \sigma d + \sigma^2 \sin^2\sigma d)}}$$

(9.5-20)

The superscript + was dropped from (9.5-20), since both expressions $q_e^{(+)}$ and $q_e^{(-)}$ differ only in the sign of the propagation constant β of the radiation mode; β_o is the propagation constant of the incident guided mode.

We derived the amplitude of the radiation mode that is excited at a step of the slab waveguide in order to utilize the result to calculate the radiation losses caused by more general wall perturbations. In order to achieve this goal, we consider Δd as an infinitesimal increment in a series of steps approximating a function $f(z)$. We thus use the relation

$$\Delta d = \frac{df}{dz}\, dz$$

(9.5-21)

Each step of the succession generates radiation modes. Each step is located at a different position $z = u$. At a given point far from the disturbed region, all the radiation modes that are excited add up, so that we have for a given radiation mode designated by ρ ($q = \hat{q}du$)

$$\bar{E}_e(\rho,x,z) = \int_0^L \hat{q}_e(\rho,u)E_e(\rho,x,z)du = E_e(\rho,x,z)\int_0^L \hat{q}_e(\rho,u)du$$

(9.5-22)

The dash over the field component on the left indicates the superposition field of all the contributions from all the steps. The mode expression can be taken out of the integral, since it does not depend in any way on the position u of each step. This consideration shows that the effective amplitude of the radiation mode, labeled ρ, is the integral over (9.5-20). Writing z instead of u, we obtain

$$q_e(\rho) = -\frac{n_1^2 - n_2^2}{2\sqrt{\pi|\beta|}}\rho k_o^2 \int_0^L \frac{\cos \kappa d \cos \sigma d\, e^{-i\int_o^z(\beta_o - \beta)du}\left(\dfrac{df}{dz} - \dfrac{dh}{dz}\right)}{(\beta_o - \beta)\sqrt{\left(\beta_o d + \dfrac{\beta_o}{\gamma}\right)(\rho^2 \cos^2 \sigma d + \sigma^2 \sin^2 \sigma d)}}\, dz$$

(9.5-23)

with

$$d = d_o + f(z)$$

(9.5-24)

The integral $-i\int_o^z(\beta_o - \beta)du$ is necessary since β_o is a function of z and the incremental phase change is $-i(\beta_o - \beta)du$. In a similar way, we obtain the corresponding mode amplitude for the odd radiation modes

$$q_o(\rho) = - \frac{n_1^2 - n_2^2}{2\sqrt{\pi|\beta|}} \rho k_o^2 \int_0^L \frac{\cos \kappa d \sin \sigma d \; e^{-i\int_0^z(\beta_o - \beta)du}\left(\frac{df}{dz} + \frac{dh}{dz}\right)}{(\beta_o - \beta)\sqrt{\left(\beta_o d_o + \frac{\beta_o}{\gamma}\right)(\rho^2 \sin^2 \sigma d + \sigma^2 \cos^2 \sigma d)}} dz$$

(9.5-25)

The calculations can also be carried out for the TM modes with no additional complication. Writing p instead of q for the TM radiation mode amplitudes excited by the guided lowest order even TM mode, we obtain for the even radiation modes

$$p_e(\rho) = - \frac{n_1^2 - n_2^2}{2\sqrt{\pi|\beta|\,n_2^2}} \times$$

$$\rho \int_0^L \frac{\sqrt{\gamma}\,(\beta_o \beta \cos \sigma d + \gamma\sigma \sin \sigma d)\cos \kappa d \; e^{-i\int_0^z(\beta_o - \beta)du}\left(\frac{df}{dz} - \frac{dh}{dz}\right)}{(\beta_o - \beta)\sqrt{\beta_o\left(\gamma d + (n_1 n_2)^2 \dfrac{\kappa^2 + \gamma^2}{n_2^4\kappa^2 + n_1^4\gamma^2}\right)\left(\dfrac{n_1^2}{n_2^2}\rho^2 \cos^2 \sigma d + \dfrac{n_2^2}{n_1^2}\sigma^2 \sin^2 \sigma d\right)}} dz$$

(9.5-26)

and for the odd TM radiation modes

$$p_o(\rho) = - \frac{n_1^2 - n_2^2}{2\sqrt{\pi|\beta|\,n_2^2}} \times$$

$$\rho \int_0^L \frac{\sqrt{\gamma}\,(\beta_o \beta \sin \sigma d - \gamma\sigma \cos \sigma d)\cos \kappa d \; e^{-i\int_0^z(\beta_o - \beta)du}\left(\frac{df}{dz} + \frac{dh}{dz}\right)}{(\beta_o - \beta)\sqrt{\beta_o\left(\gamma d + (n_1 n_2)^2 \dfrac{\kappa^2 + \gamma^2}{n_2^4\kappa^2 + n_1^4\gamma^2}\right)\left(\dfrac{n_1^2}{n_2^2}\rho^2 \sin^2 \sigma d + \dfrac{n_2^2}{n_1^2}\sigma^2 \cos^2 \sigma d\right)}} dz$$

(9.5-27)

We have added the wall distortion function $h(z)$ of the lower wall of the slab to all these equations. Our theory thus holds for arbitrary deformations of both walls provided that the waveguide can support only the lowest order TE or lowest order TM mode everywhere. If the guide size increases to the point where additional guided modes can propagate, these modes must also be taken into account.

Not only do our Equations (9.5-23) through (9.5-27) hold for very slight wall distortions of the type discussed in the previous sections but they also hold approximately for larger deflections of the walls. In fact, these equations have been used to predict the radiation losses of fairly large abrupt steps by treating them as abrupt tapers. Good experimental agreement has been obtained even in these cases.[96] The propagation constant β_o of the incident guided mode and the parameters κ and γ are all functions of z, since they depend on the width of the waveguide. Our theory is applicable in those cases where the radiation leaves the tapered waveguide sufficiently rapidly so that it cannot convert back to the guided mode. This condition is satisfied to a good approximation for gentle, long tapers as well as for abrupt tapers. The applicability for the latter has been experimentally verified.[96]

A comparison of (for example) (9.2-40) with (9.5-23) does not seem to lead to immediate agreement between these two equations. However, it is easy to demonstrate that these two equations are indeed identical in the case of a slightly perturbed guide for which (9.2-40) was derived. The equations of Section 9.2 do not hold for large deviations of the waveguide walls. The theory of our present section has thus a wider range of applicability.

Let us assume that the walls deviate only very slightly from their perfect shape. This assumption causes d to be approximately equal to d_o and also makes β_o, κ, and γ constants. It is then possible to take all these expressions out of the integral of (9.5-23), so that only the following integration remains

$$I = \int_0^L e^{-i(\beta_o - \beta)z} \frac{d}{dz}[f(z) - h(z)]dz \tag{9.5-28}$$

By performing a partial integration, we obtain

$$I = [f(L) - h(L)]e^{-i(\beta_o - \beta)L} - [f(0) - h(0)]$$
$$+ i(\beta_o - \beta)\int_0^L [f(z) - h(z)]e^{-i(\beta_o - \beta)z}dz \tag{9.5-29}$$

It was assumed in Section 9.2 that the waveguide is perfect at $z = 0$ and at $z = L$. We thus have $f(L) = 0$, $f(0) = 0$, $h(L) = 0$, and $h(0) = 0$. The terms in the first two brackets give no contribution. We obtain, using (9.2-37) and (9.2-38)

$$I = i(\beta_o - \beta)L[F(\beta_o - \beta) - H(\beta_o - \beta)] \tag{9.5-30}$$

Substitution of (9.5-30) into (9.5-23) results in (9.2-40).

This result is satisfying for more than one reason. It does prove that the theory of radiation losses of wall perturbations, based on synthesizing an arbitrary function as a succession of steps, leads to the same results as the method of Section 9.2. We have thus also extended the theory of that section to TM modes by obtaining (9.5-26) and (9.5-27). The derivation of the corresponding equations with the methods of section 9.2 would have been considerably more difficult. Knowing that the radiation loss theory, based on the loss of an individual step, works allows us to handle far more complicated problems in the same way. It is thus possible to calculate the radiation losses of the HE_{11} mode for random wall perturbations as well as for steps and tapers[96]. Finally, we have obtained results whose applicability is wider than those of the previous sections. Our Equations (9.5-23) through (9.5-27) hold approximately for all types of tapers provided that the waveguide can support only the dominant mode. An extension to the case of many guided modes would be difficult, since these modes would travel on the guide once they are excited and would be able to keep on interacting with the original incident mode, making a simple perturbation theory useless.[91] The advantage of radiation losses is that the converted power is lost from the guide, so that the incident mode simply suffers attenuation.

We conclude this section with a discussion of the implications of the theory for the radiation losses of random wall perturbations applied to TM modes. We also discuss the results of similar calculations applied to the dominant HE_{11} mode of the round optical fiber.

By using the same procedure that led from (9.5-23) to (9.2-40), we can derive the TM radiation mode amplitudes that correspond to (9.2-40) and (9.2-41). These expressions can then be used to obtain the relative radiation loss caused by random wall imperfections. The equivalent of (9.4-11) for TM modes is

$$
\left\langle \frac{\Delta P}{P} \right\rangle = \frac{A^2 \gamma L (n_1^2 - n_2^2)^2}{2\pi B \beta_o n_2^4} \int_{-n_2 k_o}^{n_2 k_o} \left\{ \frac{\rho \cos^2 \kappa d}{\left[(\beta_o - \beta)^2 + \frac{1}{B^2}\right]\left[\gamma d + n_1^2 n_2^2 \frac{\kappa^2 + \gamma^2}{n_2^4 \kappa^2 + n_1^4 \gamma^2}\right]} \right.
$$

$$
\cdot \left[\frac{(\beta_o \beta \cos \sigma d + \gamma \sigma \sin \sigma d)^2}{\frac{n_1^2}{n_2^2} \rho^2 \cos^2 \sigma d + \frac{n_2^2}{n_1^2} \sigma^2 \sin^2 \sigma d} \right.
$$

$$
\left. \left. + \frac{(\beta_o \beta \sin \sigma d - \gamma \sigma \cos \sigma d)^2}{\frac{n_1^2}{n_2^2} \rho^2 \sin^2 \sigma d + \frac{n_2^2}{n_1^2} \sigma^2 \cos^2 \sigma d} \right] \right\} d\beta
$$

$$
(9.5\text{-}31)
$$

A numerical evaluation of this equation is shown in Figures 9.5.3 and 9.5.4.

The loss of the TM mode is shown as a solid line in these figures, while the losses for the TE modes have been drawn as dotted lines for comparison. Figure 9.5.3 holds for $n_1 = 1.5$, $n_2 = 1$, $k_o d = 1.3$, while Figure 9.5.4 holds for $n_1 = 1.01$, $n_2 = 1$ and $k_o d = 8$. It is remarkable how nearly identical the losses of the TE and TM modes are. It appears therefore, that the radiation losses caused by random wall imperfections are nearly independent of the polarization of the incident mode.

A corresponding theory for radiation losses from wall imperfections has been worked out for the dominant HE_{11} mode of the round dielectric waveguide (optical fiber).[96] This theory was also based on the loss calculations for one individual step. The wall perturbations are restricted to symmetrical diameter changes of the rod. The resulting theory is extremely complicated. We can do no more than quote some results of this theory. The losses of the

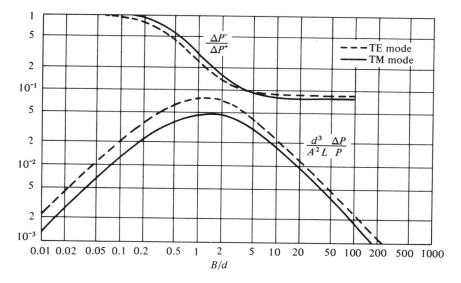

Figure 9.5.3 Comparison of TM mode (solid line) and TE mode (dotted line) radiation loss of the slab waveguide caused by random distortion of one side of the core-cladding interface. $n_1 = 1.5$, $n_2 = 1$, $kd = 1.3$. (From D. Marcuse, Radiation Losses of Tapered Dielectric Slab Waveguides, *B.S.T.J.*, Vol. **49**, No. 2, February 1970. Copyright 1970, Telegraph Co., reprinted by permission.)

HE_{11} mode caused by random wall perturbations with an exponential correlation function (9.4-8) are shown in Figures 9.5.5 and 9.5.6. The solid lines in each figure represent the losses of the HE_{11} mode. Also shown as a dotted line are the losses of the circular electric TE_{01} mode. This latter mode corresponds to the TE mode of the dielectric slab in some respects except that it does have a cut-off frequency. Figure 9.5.5 holds for $n_1 = 1.432$, $n_2 = 1$. The values of $k_o a$ are indicated as parameters at each curve. Figure 9.5.6 holds for $n_1 = 1.01$, $n_2 = 1$. The agreement between the losses of the HE_{11} mode and the TE_{01} mode is again quite close. A comparison of these curves and the loss curves for the modes of the slab waveguide shows that the losses of the modes of the round rod appear to be four times as high as those of the slab waveguide. The reason for this difference has been explained earlier. It is caused by the assumption that the slab waveguide has one perfect wall. The round rod with diameter changes corresponds to a slab with symmetrical thickness changes. This requires that both walls of the slab be deformed simultaneously with perfect correlation between both sides. The losses for this case are obtained from Figures 9.4.2 through 9.4.7 and 9.5.3 and 9.5.4 by multiplication with four. When this adjustment is made, the agreement between the loss calculations for all cases is very close. This

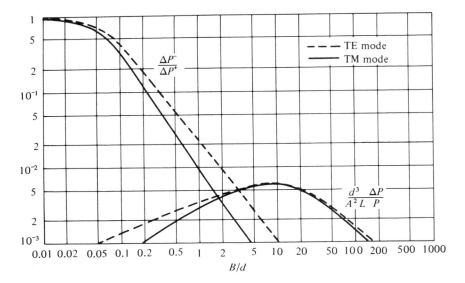

Figure 9.5.4   Similar to Figure 9.5.3, with $n_1 = 1.01$, $n_2 = 1$, $kd = 8$. (From D. Marcuse, Radiation Losses of Tapered Dielectric Slab Waveguides, *B.S.T.J.*, Vol. **49**, No. 2, February 1970. Copyright 1970, The American Telephone and Telegraph Co., reprinted by permission.)

result indicates again that the mode losses caused by random wall perturbations of dielectric waveguides are nearly independent of the type of waveguide and of the propagating guided mode.

The theory of this section can be applied to steps and tapers of dielectric waveguides. The losses of abrupt steps can be calculated by two different methods. We can use our expressions (9.5-23) through (9.5-27) to calculate the relative power losses of a steep taper, with the help of (9.2-28). The second way to calculate the radiation losses of the step uses a more direct method. We can use (9.5-3) and (9.5-4) to calculate approximate values of the guided mode transmission coefficient c_t and of the reflection coefficient a_r. The losses caused by radiation can then be obtained from the expression

$$\frac{\Delta P}{P} = 1 - |c_t|^2 - |a_r|^2 \qquad (9.5\text{-}32)$$

This expression is based on the fact that the radiation losses must be the difference between the total power P minus the transmitted power $|c_t|^2 P$ and minus the reflected power $|a_r|^2 P$. Both methods of calculation are approximate, so that exact agreement cannot be expected. The agreement is surprisingly close, however, even for steps with a two-to-one ratio of the slab width. Figure 9.5.7 shows a comparison of TE mode radiation losses of a step in the slab waveguide.[101]

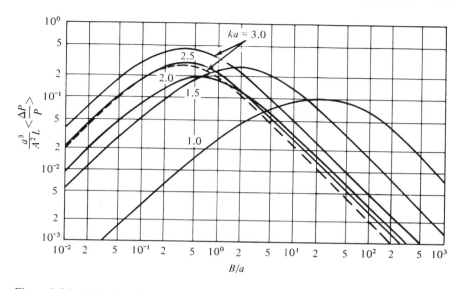

Figure 9.5.5   Radiation losses caused by random changes of the core radius of a round optical fiber. The solid lines apply to the HE_{11} mode. The dotted line shows the TE_{01} mode for comparison. $n_1 = 1.432$, $n_2 = 1$, the ka values are indicated as parameters on the curves. B is the correlation length.

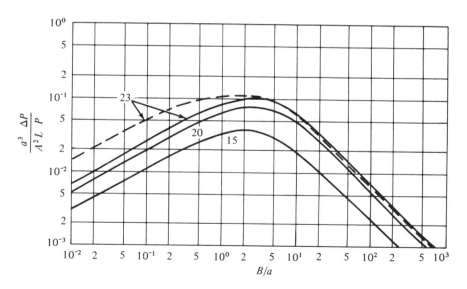

Figure 9.5.6   Similar to Figure 9.5.5. $n_1 = 1.01$, $n_2 = 1$. The numbers indicated are the values of ka. (From D Marcuse, Radiation Losses of Tapered Dielectric Slab Waveguides, B.S.T.J., Vol. 49, No. 2, February 1970. Copyright 1970, The American Telephone and Telegraph Co., reprinted by permission.)

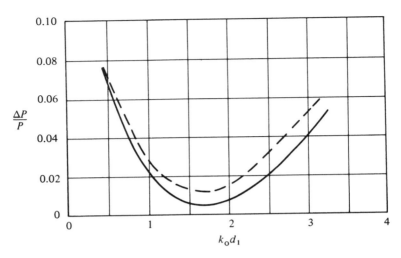

Figure 9.5.7   Relative radiation loss of the lowest order TE mode of the slab waveguide caused by a single step. The solid and dotted lines are different approximate methods of computing the loss. The ratio of the two core half widths before and after the step is d_2/d_1 = 0.5; $n_1 = 1.432$, $n_2 = 1$. (From D. Marcuse, Radiation Losses of Tapered Dielectric Slab Waveguides, *B.S.T.J.*, Vol. **49**, No. 2, February 1970. Copyright 1970, The American Telephone and Telegraph Co., reprinted by permission.)

Figure 9.5.7 was drawn for $n_1 = 1.432$, $n_2 = 1$, and for a step ratio of $d_2/d_1 = 0.5$. The product $k_o d_1$, the independent variable in Figure 9.5.7, uses the half width d_1 of the larger section of waveguide. It is surprising how low the radiation losses of the dominant TE mode of the slab waveguide can be. The losses of the TM modes are somewhat higher.[101] The solid line represents the result of the radiation loss method, while the dotted line was computed from (9.5-32). The agreement between these two methods of calculation is quite good considering the size of the step.

Even better agreement is obtained for the case $n_1 = 1.01$, $n_2 = 1$, and d_2/d_1 = 0.5. The results of calculations for this case are shown in Figure 9.5.8. The solid lines were again obtained from the radiation loss theory, while the dotted-lines represent the result of (9.5-32). The losses of the TE and TM modes are very nearly the same for this low index guide.

The step losses of the slab waveguide are much lower than the step losses of the round dielectric waveguide. The theory of the HE_{11} mode of reference 96 was used to calculate the losses of abrupt steps of the round dielectric waveguide. The results of this calculation are shown in Figures 9.5.9 and 9.5.10.

Figure 9.5.9 shows the losses of the HE_{11} mode of the round wave-guide and, for purposes of comparison, the losses of the TE mode and TM mode of the dielectric slab for $n_1 = 1.432$, $n_2 = 1$, and $a_2/a_1 = 0.5$. The radius of the

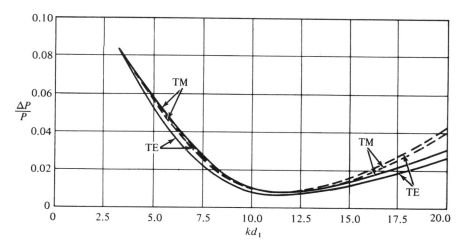

Figure 9.5.8 Single-step radiation losses of the lowest order TE and TM modes of the slab waveguide. $d_2/d_1 = 0.5$, $n_1 = 1.01$, $n_2 = 1$. The dotted and solid lines apply to different approximations. (From D. Marcuse, Radiation Losses of Tapered Dielectric Slab Waveguides, *B.S.T.J.*, Vol. **49**, No. 2, February 1970. Copyright 1970, The American Telephone and Telegraph Co., reprinted by permission.)

round guide is given by a; for the slab waveguide, we have used $a = d$ in these figures. The two methods of calculation are again indicated as the solid line for the radiation loss method and as the dotted line for the method of (9.5-32). The agreement between the two methods is again very good. The figure shows clearly how much higher the losses of the HE_{11} mode of the round rod are compared to the TE and TM mode losses of the slab waveguide. Figure 9.5.10 holds for the case $n_1 = 1.01$, $n_2 = 1$. It shows that the losses of the slab waveguide modes approach the losses of the round waveguide mode for larger values of $k_o a_1$.

The step loss theory was verified by a microwave experiment.[96] The experiment used a resonant cavity with a smooth Teflon rod, $n_1 = 1.432$, $n_2 = 1$, for calibration purposes and teflon rods with two different steps. The results of these step loss measurements are entered as crosses in Figure 9.5-11. The curve was drawn for both analytical approaches. The agreement is so close in this case that the two curves nearly coincide. The agreement between the measured points and the theoretical curve is also quite good. For a comparison of (9.5-9) and (9.5-11), note that $ka_2 = (a_2/a_1)ka_1$.

In an attempt to check the loss prediction of the slab waveguide theory, the resonant cavity setup was used to measure the radiation losses of a two-to-one step in a rectangular dielectric waveguide.[96] It is very hard to realize a true slab waveguide. The rectangular dielectric waveguide had a two-to-one ratio of its wide side to its narrow side. The step reduced the narrow side

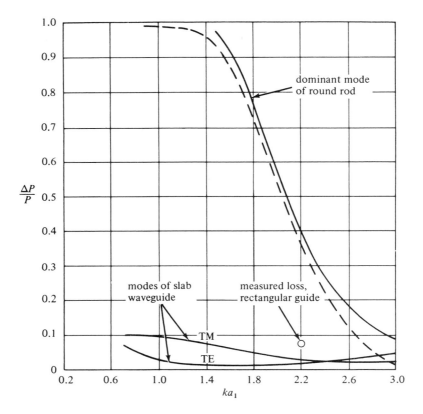

Figure 9.5.9   Comparison of the radiation losses caused by a single step in the core width of the slab waveguide and the core diameter of the round optical fiber. The step height is $d_2/d_1 = 0.5$, $n_1 = 1.432$, $n_2 = 1$. The solid and dotted lines refer to two different methods of approximation. The circle indicates the result of a step loss measurement of a rectangular dielectric waveguide. (From D. Marcuse, Radiation Losses of the Dominant Mode in Round Dielectric Waveguides, *B.S.T.J.*, Vol. **49**, No. 8, October 1970. Copyright 1970, The American Telephone and Telegraph Co., reprinted by permission.)

of the waveguide without changing its wide side. This rectangular dielectric waveguide is an approximation to the slab waveguide. The result of the measurement of the rectangular waveguide is shown as the circle in Figures 9.5.9 and 9.5.11. The loss of this type of waveguide is obviously much less than that of the round waveguide, but it is also considerably higher than the predicted loss of the slab waveguide. This result indicates that the rectangular dielectric waveguide is not as tolerant of steps as the slab waveguide. However, it approximates the slab waveguide, as is evident by the much lower loss compared with the round waveguide.

The radiation loss can be reduced to any desired amount if the step or

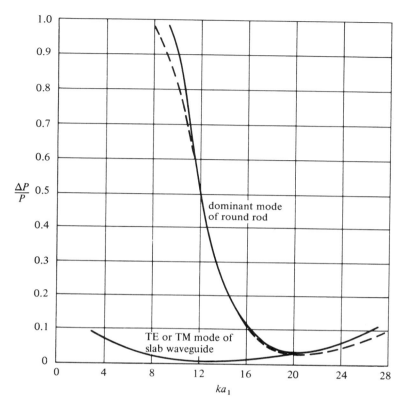

Figure 9.5.10 Similar to Figure 9.5.9, with $d_2/d_1 = 0.5$, $n_1 = 1.01$ $n_2 = 1$. (From D. Marcuse, Radiation Losses of the Dominant Mode in Round Dielectric Waveguides, *B.S.T.J.*, Vol. 49, No. 8, October 1970. Copyright 1970, The American Telephone and Telegraph Co., reprinted by permission.)

steep taper is stretched out into a gradual taper. It is informative to consider the mechanism of the loss reduction for long tapers. The losses of tapers are determined by the amplitudes of the radiation modes (9.5-23) through (9.5-27). If the taper is so short that $\int_0^z (\beta_o - \beta)du \ll 1$, and if we ignore the z dependence of the other factors in the integrand, we find that the integral is approximately proportional to $f(L) - f(0) \pm (h(L) - h(0))$. This means that for short tapers it is only the height of the step that determines the radiation loss. The losses of abrupt steps and short tapers are identical. This is the reason why our approximate theory of the radiation loss of tapers works even for abrupt steps. As the taper becomes longer, so that $\int_0^z (\beta_o - \beta)du$ is no longer smaller than unity, another effect comes into play. The exponential function, with exponent $-i\int_0^z (\beta_o - \beta)du$, is an oscillatory function (more accurately speaking, its real and imaginary parts are oscillatory functions).

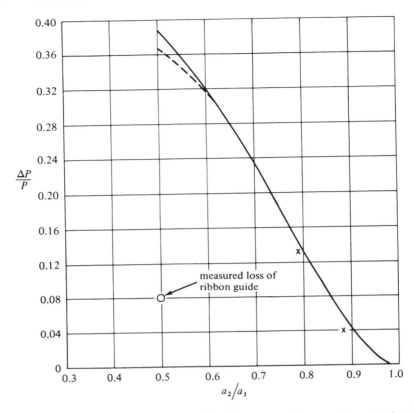

Figure 9.5.11   Single-step radiation loss of the HE_{11} mode of the round fiber as a function of the step height ratio a_2/a_1. The crosses indicate measured values corresponding to the theoretical parameters. The circle represents the result of a measurement on a rectangular dielectric waveguide. $n_1 = 1.432$, $n_2 = 1$, $ka_2 = 1.1$. (From D. Marcuse, Radiation Losses of the Dominant Mode in Round Dielectric Waveguides, *B.S.T.J.*, Vol. 49, No. 8, October 1970. Copyright 1970, The American Telephone and Telegraph Co., reprinted by permission.)

Rapidly oscillating functions appearing in the integrands of integrals have the tendency to cancel any contribution to the integral. This tendency is demonstrated most dramatically in the example of the delta function (8.4-16). This function contributes to the integral only at its pole, while the rapid oscillations of the sine function cancel all contributions of any other function that appears multiplied with it in the integrand. The exponential function in the integrals (9.5-23) through (9.5-27) does not vary extremely rapidly. However, if the value of the derivatives of the functions $f(z)$ and $h(z)$ are small and do not change rapidly, the cancelling influence of the exponential function becomes pronounced. In fact, for infinitely long tapers, the integrals approach the value zero.

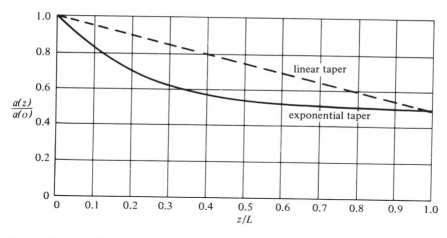

Figure 9.5.12   Profiles of a linear (dotted line) and an exponential taper (solid line). (From D. Marcuse, Radiation Losses of the Dominant Mode in Round Dielectric Waveguides, *B.S.T.J.*, Vol. **49**, No. 8, October 1970. Copyright 1970, The American Telephone and Telegraph Co., reprinted by permission.)

The factor $\int_0^z (\beta_o - \beta)du$ in the exponent of the exponential function depends on z. The propagation constant β_o is larger in the wider portion of the taper, and decreases on its narrower portion. A large value of this factor leads to more effective cancellation by the oscillatory exponential function. The contribution to the integral stems, therefore, more from the narrow portion of a linear taper than from its wider part. It appears that an advantage could be gained by shaping the taper in such a way that its contribution to the radiation loss is nearly equally distributed over its entire length. An analysis shows that a taper with an exponential shape would tend to spread the radiation loss more evenly over its entire length than a linear taper. The profiles of a linear taper and of an exponential taper are shown in Figure 9.5.12. The discontinuity of the first derivative of the exponential taper at $z = 0$ does not contribute to the integral. It makes no difference, therefore, whether we smooth this function to give it a continuous first derivative or whether we leave its shape as shown in the figure. The radiation losses of the HE_{11} mode of the round waveguide are shown for both tapers in Figure 9.5.13 as a function of taper length. This figure shows several interesting features. The radiation loss is constant from $L/d = 0$ to $L/d = 1$. Tapers of such a short length are just as lossy as abrupt steps. As the length of the taper is increased, the losses decrease rapidly. A linear taper with $L/d = 400$ has only 1 percent of the loss of an abrupt step. The figure also shows that our expectation of finding lower loss for the exponential taper was correct. The gain is considerable. However, an increase in the length of the linear taper compensates for any advantage the exponential taper may offer. The

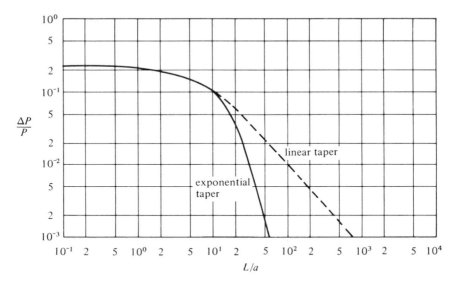

Figure 9.5.13   Relative radiation losses of a linear taper (dotted line) and an exponential taper (solid line) as a function of taper length to core radius ratio L/a. The ratio of the radius of the core on the smaller end of the taper to its value at the wider end is $a_2/a_1 = 0.5$; $n_1 = 1.432$, $n_2 = 1$, $ka_1 = 2.5$. (From D. Marcuse, Radiation Losses of the Dominant Mode in Round Dielectric Waveguides, *B.S.T.J.*, Vol. **49**, No. 8, October 1970. Copyright 1970, The American Telephone and Telegraph Co., reprinted by permission.)

overall length of these tapers need not be very great at all. For $\lambda = 1 \mu$, we have a guide radius of approximately 1μ. A taper with $L/d = 1000$ is thus only 1 mm long.

9.6 BENDING LOSSES

Dielectric waveguides cannot guide electromagnetic energy around bends without losing power by radiation.[114] In practice, this bending loss may be negligibly small if the radius of curvature of the bend is sufficiently large. However, radiation is always emitted by a bent dielectric waveguide, and bending losses can become large if the guide is bent abruptly.[106]

A general discussion of bending losses is difficult. We thus follow our usual practice and study the physical principle involved in bending losses by considering the slab waveguide.

The reason for the occurrence of bending losses can be explained as follows. A portion of the energy of dielectric waveguides travels outside of the core

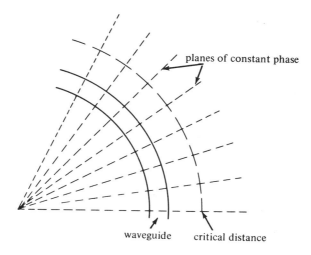

planes of constant phase

waveguide   critical distance

Figure 9.6.1   Curved slab waveguide with (approximate) planes of constant phase. The critical distance corresponds to the region where the speed of the phase fronts exceeds the speed of light.

region. Far from cutoff, the fraction of energy that finds itself outside the waveguide core may be very small, but it is present nevertheless. The field outside the waveguide core is decaying exponentially in transverse direction to the waveguide axis. Mathematically, some finite amount of field energy can be found even at great distances. As the waveguide is bent, the planes of constant phase have the tendency to assume the shape that is schematically shown in Figure 9.6.1. Close to the waveguide axis, the planes of constant phase (or phase fronts) move at the velocity that is typical for the propagation of the particular mode in the straight waveguide. This guided mode velocity is slower than the velocity of plane waves in the medium outside the waveguide core. We can picture the motion of the phase fronts as if they were pivoted at the center of curvature and rotate around this center. It is thus apparent that the phase front velocity is smaller than the phase velocity of the mode in the straight guide on the side of the waveguide that is directed toward the center of curvature, while the phase front velocity is higher on the opposite side. At the critical distance, the phase front velocity equals the plane wave velocity in the medium outside the waveguide. At this point, the exponentially decaying guided field is in trouble. Electromagnetic fields cannot be made to move faster than the characteristic velocity in the particular medium. The field circumvents this problem by detaching itself from the guided mode field and radiating into space. This explains, in a crude, qualitative way, why bent dielectric waveguides must lose power by radiation.

Our mathematical treatment of the problem of bending losses is based on the realization that, for very slight bends, the field distribution and phase velocity near the waveguide core must be very nearly the same as for the straight waveguide. However, we also know the exact form that the wave field must assume because of its cylindrical symmetry with the center of curvature as the center of a cylindrical coordinate system. We use the coordinate system shown in Figure 9.6.2.

The y-axis (this axis is commonly called the z-axis) of the cylindrical coordinate system is directed out of the plane of the drawing. The solution of Maxwell's equations in cylindrical polar coordinates is given as follows

$$E_y = B H_v^{(2)} (n_2 k_o r) e^{i(\omega t - v\phi)} \tag{9.6-1}$$

$$H_r = B \frac{\omega \varepsilon_o}{k_o^2} \frac{v}{r} H_v^{(2)}(n_2 k_o r) e^{i(\omega t - v\phi)} \tag{9.6-2}$$

$$H_\phi = -iB \frac{n_2 \omega \varepsilon_o}{k_o} H_v^{(2)'}(n_2 k_o r) e^{i(\omega t - v\phi)} \tag{9.6-3}$$

we have $k_o = \omega \sqrt{\varepsilon_o \mu_o}$, and assume that there is no y dependence. This field solution holds only for $r \geq b$ outside the waveguide core. In the core, the field is a superposition of Hankel functions of the first and second kind, while the field outside the core for $r \leq a$ is given by the Bessel function $J_v(n_2 k_o r)$. The prime on the Hankel function in (9.6-3) indicates its derivative with respect to the argument $n_2 k_o r$. The choice of functions is dictated by the requirement that the field must decay exponentially in the direction of increasing r for $r > b$ and must become a traveling wave moving in the direction of positive r for $r \gg b$. The Hankel function of the second kind

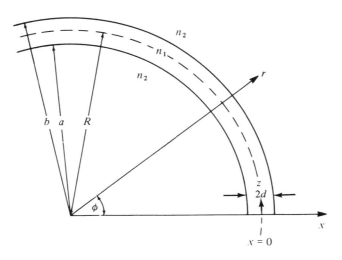

Figure 9.6.2   Coordinate system used to describe the radiation losses of a bent slab waveguide.

satisfies both of these requirements. Since no such restriction exists inside the waveguide core, both Hankel functions of the first and second kind must be used, or, alternately, we could use the Bessel and Neumann functions. For $r < a$, we must use a cylinder function that does not have a singularity at $r = 0$. The only function that satisfies this requirement is $J_\nu(n_2 k_o r)$. The order number ν of the cylinder function need not be an integer in this problem, since we need not require that the functions be periodic in the variable ϕ with the period 2π. We do not need the field solutions inside the fiber core and for $r < a$ for our present purpose.

The field solution (9.6-1) through (9.6-3) is required by the geometry and by Maxwell's equations. Assuming that the field has no y dependence and that the curved slab is infinitely extended in y direction, the field can have no other form. However, we also know that in the limit $R \to \infty$ the field near the core must assume the form (8.3-12) of the symmetric TE mode of the straight slab waveguide. This requirement allows us to determine the constant B. Once this parameter is determined, we can use the field equations to determine the power flow in radial direction at infinite distance from the guide and use this information to determine the loss of the guided mode.

The exact approach to the solution of this problem would of course involve the determination of the unknown amplitude coefficients from the boundary conditions. We would then obtain an eigenvalue equation for the determination of ν. This equation would not have real solutions. The complex value of ν would determine the mode losses. This latter procedure leads to the precise solution of the problem.[106,114] However, it is much more complicated than our approximate approach. We circumvent the necessity for computing ν from the eigenvalue equation by assuming that the propagation constant of the wave in the bent waveguide is very nearly identical to the propagation constant of the corresponding mode in the straight guide. By comparison with the modes of the straight guide, we obtain the relation

$$\nu\phi = \beta z \qquad (9.6\text{-}4)$$

The z-axis is now the arc length along the bent center line of the waveguide, $z = R\phi$. The order number of the Bessel function is thus given as

$$\nu = \beta R \qquad (9.6\text{-}5)$$

The propagation constant β is assumed to be identical to that of the symmetric TE mode of the straight slab waveguide. It is obtained as a solution of the eigenvalue equation (8.3-16).

In order to be able to determine the constant B, we need an approximation of the Hankel function. It is apparent from (9.6-5) that ν is of the same order of magnitude as the argument $n_2 k_o r$ of the Bessel function. Using the x coordinate shown in Figure 9.6.2, we obtain the relation

$$r = R + x \qquad (9.6\text{-}6)$$

The ratio of order number to argument is thus

$$\cosh \alpha = \frac{v}{n_2 k_o r} = \frac{\beta}{n_2 k_o} \frac{1}{1 + \frac{x}{R}} \approx \frac{\beta}{n_2 k_o}\left(1 - \frac{x}{R}\right) \tag{9.6-7}$$

The propagation constant of the mode is larger than the propagation constant of plane waves in the medium outside the core

$$\beta > n_2 k_o \tag{9.6-8}$$

For $x = 0$, we see that the ratio (9.6-7) is larger than unity. However, as x increases, (9.6-7) must eventually become smaller than unity. However, for large values of R and for small values of x, we can assume that the ratio (9.6-7) is larger than unity.* The approximation of the Hankel function that is valid in this case can be found in books on functions[61]

$$H_v^{(2)}(n_2 k_o r) = + i \frac{e^{v(\alpha - \tanh \alpha)}}{\sqrt{\frac{\pi}{2} v \tanh \alpha}} \tag{9.6-9}$$

The argument α of the hyperbolic tangent function is defined by (9.6-7). We thus obtain

$$u = \tanh \alpha = \frac{\sqrt{\cosh^2 \alpha - 1}}{\cosh \alpha} = \frac{\sqrt{v^2 - (n_2 k_o r)^2}}{v} \tag{9.6-10}$$

It is sufficient to approximate $v \tanh \alpha$ in the denominator of (9.6-9) as follows

$$v \tanh \alpha = \sqrt{v^2 - (n_2 k_o R)^2} = R\sqrt{\beta^2 - (n_2 k_o)^2}$$

so that we obtain from (8.3-14)

$$v \tanh \alpha = \gamma R \tag{9.6-11}$$

The argument of the exponential function in (9.6-9) must be approximated more precisely. We use the relation

$$\alpha = \tanh^{-1} u = \frac{1}{2} \ln\left(\frac{1 + u}{1 - u}\right) = u + \frac{1}{3}u^3 + \frac{1}{5}u^5 + \frac{1}{7}u^7 + \dots \tag{9.6-12}$$

with u defined by (9.6-10). According to (9.6-7), $\cosh \alpha$ is not too different from unity, so that $\tanh \alpha = u$ is a small quantity justifying the expansion

---

* The conditions $\beta/\left[n_2 k_o\left(1 + \frac{d}{R}\right)\right] > 1$ and $\beta/\left[n_1 k_o\left(1 - \frac{d}{R}\right)\right] < 1$ are both required for the validity of our approximate theory. If one or both of these conditions are violated, our theory fails.

on the right-hand side of (9.6-12). We can now write

$$\alpha - \tanh \alpha = \frac{1}{3} u^3 + \frac{1}{5} u^5 + \frac{1}{7} u^7 + \cdots \tag{9.6-13}$$

With the help of (9.6-5), (9.6-6), and (9.6-10), we obtain

$$u^p = \left[\frac{\sqrt{\beta^2 - (n_2 k_o r/R)^2}}{\beta} \right]^p \approx \left[\frac{\sqrt{\gamma^2 - 2(n_2 k_o)^2 x/R}}{\beta} \right]^p$$

$$\approx \left(\frac{\gamma}{\beta} \right)^p \left(1 - p \frac{(n_2 k_o)^2}{\gamma^2} \frac{x}{R} \right) \tag{9.6-14}$$

Substitution into (9.6-13) yields

$$\alpha - \tanh \alpha = \frac{1}{3} \left(\frac{\gamma}{\beta} \right)^3 + \frac{1}{5} \left(\frac{\gamma}{\beta} \right)^5 + \frac{1}{7} \left(\frac{\gamma}{\beta} \right)^7 + \cdots$$

$$- \left(\frac{n_2 k_o}{\beta} \right)^2 \frac{\gamma}{\beta} \frac{x}{R} \left\{ 1 + \left(\frac{\gamma}{\beta} \right)^2 + \left(\frac{\gamma}{\beta} \right)^4 + \left(\frac{\gamma}{\beta} \right)^6 + \cdots \right\} \tag{9.6-15}$$

The first infinite series on the right-hand side of (9.6-15) can again be expressed by (9.6-12), while the expression in the bracket is a simple geometrical series. We can thus write

$$\alpha - \tanh \alpha = \left(\tanh^{-1} \frac{\gamma}{\beta} \right) - \frac{\gamma}{\beta} - \frac{(n_2 k_o)^2}{\beta^3 R} \frac{\gamma x}{1 - \left(\frac{\gamma}{\beta} \right)^2} \tag{9.6-16}$$

Using the identity

$$1 - \left(\frac{\gamma}{\beta} \right)^2 = \frac{(n_2 k_o)^2}{\beta^2} \tag{9.6-17}$$

we finally obtain from (9.6-5), (9.6-9), (9.6-11), (9.6-16), and (9.6-17)

$$H_\nu^{(2)} (n_2 k_o r) = + \frac{i}{\sqrt{\frac{\pi}{2} \gamma R}} \exp \left(\beta R \tanh^{-1} \frac{\gamma}{\beta} - \gamma R \right) e^{-\gamma x} \tag{9.6-18}$$

It is apparent that the electromagnetic field in the form (9.6-1), if approximated by (9.6-18), has the x dependence of the field (8.3-12). By comparing the two field expressions, we obtain the amplitude coefficient B in the form

$$B = - i \sqrt{\frac{\pi}{2} \gamma R} \sqrt{\frac{2\omega \mu_o P}{\beta d + \frac{\beta}{\gamma}}} \cos \kappa d \, e^{\gamma d} e^{-(\beta \tanh^{-1}(\gamma/\beta) - \gamma)R} \tag{9.6-19}$$

Computing the bending loss is now an easy matter. The power loss coefficient is the ratio of the power contribution of the unit length of guide to the radiation field at infinity divided by the power P flowing in the waveguide.

$$2\alpha = \frac{(LS_r)_{r \to \infty}}{P} \qquad (9.6\text{-}20)$$

Since α is the amplitude attenuation coefficient, the power loss coefficient is 2α. L is the length of the arc that is obtained by extending the two radii leading to the end points of an arc of length unity at the waveguide axis to the distance r at which S_r is computed. S_r is the r component of the power flow density vector (Poynting vector). We have

$$L = \frac{r}{R} \qquad (9.6\text{-}21)$$

and

$$S_r = -\frac{1}{2} E_y H_\phi^* = -i|B|^2 \frac{n_2 \omega \varepsilon_o}{2k_o} H_v^{(2)}(n_2 k_o r)[H_v^{(2)'}(n_2 k_o r)]^* \qquad (9.6\text{-}22)$$

The asterisk indicates complex conjugation.

Since the argument is now much larger than the index, we need the approximation of the Hankel function for large argument[11, 61]

$$H_v^{(2)}(n_2 k_o r) = \frac{\sqrt{2}}{\sqrt{\pi n_2 k_o r}} e^{-i n_2 k_o r} e^{i(\pi/4)} e^{i v(\pi/2)} \qquad (9.6\text{-}23)$$

Substitution of (9.6-19), (9.6-21), (9.6-22), and (9.6-23) into (9.6-20) gives the final result

$$2\alpha = \frac{\gamma^2}{\beta(1 + \gamma d)} \frac{\kappa^2}{(n_1^2 - n_2^2)k_o^2} e^{2\gamma d} e^{-2(\beta \tanh^{-1}(\gamma/\beta) - \gamma)R} \qquad (9.6\text{-}24)$$

To express (9.6-24) in this form, we used the relation

$$\kappa^2 + \gamma^2 = (n_1^2 - n_2^2)k_o^2 \qquad (9.6\text{-}25)$$

and also the expression (8.6-16)

$$\cos^2 \kappa d = \frac{\kappa^2}{\kappa^2 + \gamma^2} \qquad (9.6\text{-}26)$$

For small values of γ/β, we can use the simpler approximation

$$2\alpha = \frac{\gamma^2}{\beta(1 + \gamma d)} \frac{\kappa^2}{(n_1^2 - n_2^2)k_o^2} e^{2\gamma d} e^{-(2/3)(\gamma^3/\beta^2)R} \qquad (9.6\text{-}27)$$

The reader may be concerned that we have computed the power loss at a position z on the waveguide by computing the radiated power at an infinite distance from the waveguide, for surely the power flowing at a given z_1 and $x \to \infty$ must have been contributed at a different position z_2 a long distance

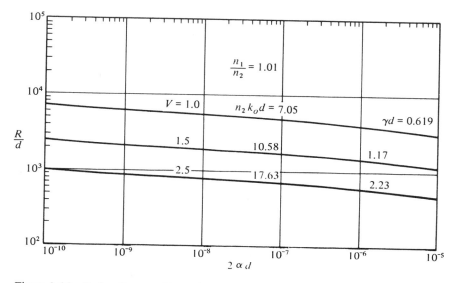

Figure 9.6.3 Ratio of waveguide radius of curvature R to slab half width d that causes a radiation loss 2α. The core to cladding index ratio is $n_1/n_2 = 1.01$.

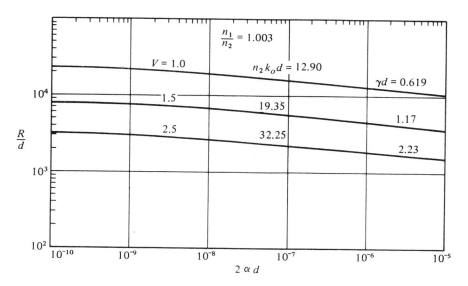

Figure 9.6.4 Similar to Figure 9.6.3, except that $n_1/n_2 = 1.003$.

before the point z_1. The justification for our procedure follows from our assumption that the amplitude of the guided wave is not changed, even though we use it to compute the power loss. This assumption of constant amplitude is typical for any first order perturbation theory. Since the radial power flow is independent of z (or, equivalently, of the angle ϕ), it does not matter which element of the waveguide actually contributed to. The relative power loss per unit length is constant and independent of the position on the waveguide.

The present approximate method of computing the bending loss is in very good agreement with Marcatili's result.[106] He used an approximate solution of the complex eigenvalue equation to compute the bending loss. His expressions are much more complicated than the result of our theory. However, the numerical values of the loss computed from both theories are in excellent agreement. Our theory is not restricted to small values of $n_1 - n_2$.

The values of γ (or β) must be obtained from the solution of the eigenvalue equation (8.3-16) or from the approximations (8.6-12) and (8.6-20).

We have treated only the simple case of the symmetric TE mode of the slab waveguide. A theory of the bending losses of the HE_{11} mode of the round fiber is much more complicated. It is tempting to try to use the slab theory to obtain at least an estimate of the bending losses to be expected for the HE_{11} mode of the round fiber. For any such attempt, one should use the values of κ, γ, and β of the theory of the fiber instead of obtaining these values with $a = d$ from the eigenvalue equation of the slab waveguide.*

Plots of the normalized radius of curvature R/d as a function of the bending loss $2\alpha d$ are shown in Figures 9.6.3 and 9.6.4. The refractive index ratio of core index to the index of the surrounding medium is $n_1/n_2 = 1.01$ in Figure 9.6.3 and $n_1/n_2 = 1.003$ in Figure 9.6.4. We have plotted the inverse of the function (9.6-27) by using the approximate solution (8.6-12) in order to show the radius of curvature that is permissible if a certain acceptable bending loss is prescribed. It is remarkable how little R/d changes with $2\alpha d$. A slight change of the radius of curvature causes a change of the loss coefficient by many orders of magnitude. Typical operating conditions may involve an index ratio of $n_1/n_2 = 1.003$, a slab half width of $d = 1\ \mu$, and $n_2 kd = 12.9$. The bending losses vary from $2\alpha d = 10^{-10}$ or $2\alpha = 0.434$ dB/km to $2\alpha d = 10^{-5}$ or $2\alpha = 43.4$ dB/m as the radius of curvature varies from $R = 2.2$ cm to 1 cm.

---

* The slab waveguide formula (9.6-24) is in good agreement with experimental results obtained for the HE_{11} mode.[25]

CHAPTER

10

COUPLING BETWEEN DIELECTRIC WAVEGUIDES

10.1 INTRODUCTION

We have demonstrated in Sections 2.7 and 9.3 that two waves can be coupled so effectively that complete transfer of power from one wave to the other is possible. This type of coupling was caused by a sinusoidal imperfection of the core-cladding boundaries of the waveguide or by a sinusoidal variation of the index of refraction. In the present chapter, we deal with coupling of guided modes of two different waveguides. Coupling between different waveguides can also be caused by imperfections in the waveguide geometry or by inhomogeneities of the dielectric medium of the waveguide. Power in one of the guides is scattered by an imperfection, and radiates into the space surrounding the guide. If a perfect waveguide were placed into this radiation field, the latter would simply penetrate through it and be partially reflected by it, but no power would be coupled to any of the guided modes of the second waveguide. However, if the second waveguide also has imperfections, some of the radiated power can be scattered into its guided modes.[104] We call this type of coupling scattering coupling. Complete power exchange does not result from this mechanism.

However, two dielectric waveguides can exchange power by quite a different mechanism, even if both of them are perfect.[103] We have seen in the preceding chapter that the electromagnetic field of guided modes in dielectric waveguides decays exponentially in transverse direction to the waveguide axis. If two dielectric waveguides are placed alongside each other, some of the field of one guide reaches the other guide. The presence of the second guide distorts the field of the guided mode of the first guide. This distorted field can be expressed as a superposition of the mode fields of each guide plus a small additional field. Since the mode field of the second guide is required to express the distortion of the field of the first guide, it is clear that the two modes interact. We show in this chapter that complete transfer of power is again possible by this type of coupling mechanism. It is required only that the phase velocities of the modes in either guide are identical in the absence of coupling. Modes with different phase velocities cannot interact very effectively, and only very little power can be exchanged.

Mode coupling between two different dielectric waveguides has two important applications. An undesirable feature of waveguide coupling is cross talk between two transmission lines.[84,85,86,87] For communications purposes, it is desirable to be able to bunch many optical fibers into one cable. However, because of their close proximity, the individual waveguides can exchange power, so that some part of the signal that is being transmitted in one guide can enter a neighboring guide and interfere with the signal that is being transmitted there. Both types of coupling are undesirable in their ability to cause cross talk. However, waveguide coupling offers the possibility to feed power from one guide into another in a controlled way. It can thus be used to excite a waveguide. An important example of this type is the prism coupler that is used to excite guided modes in thin dielectric films.[107,108] Even though the prism coupler uses unguided radiation from a laser to excite guided modes, it is closely related to the coupling mechanism between perfect waveguides. In both cases, it is "frustrated total internal reflection" that is the cause of coupling. We shall not discuss the prism coupler in detail. However, it is interesting to mention it briefly. Figure 10.1.1 shows its principle.

A laser beam is incident on the lower face of the prism shown in the figure. In the absence of the waveguide, it would be totally reflected at the glass-air interface. However, we know from the discussion in Section 1.6 that an

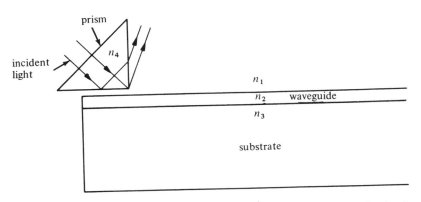

Figure 10.1.1   Schematic of the prism coupler. Frustrated total internal reflection is used to excite guided modes in thin dielectric films.

exponentially decaying field reaches into the air space under the prism. The thin film waveguide is placed in close proximity to the prism, so that the evanescent field can still reach it with some intensity. The refractive index of the waveguide material is higher than that of the prism, $n_2 > n_4$, so that if the waveguide medium were infinitely extended, a traveling wave would be excited by the evanescent field carrying power into the medium. Even though the medium with index n_2 is only a thin film, considerable power can be transferred from the incident field to the guided mode of the thin film if the component of the propagation vector of the incident field in the direction of the waveguide axis matches the propagation constant of the guided mode. Approximately 80 percent of power can be transferred in this way. It has been shown that more than 90 percent of power can be transferred if the air gap between the prism and the thin film is tapered.[109,121]

Coupling by evanescent fields has the advantage of being unidirectional. Waveguide coupling can thus be used to transfer power from one guide so that it travels in the other guide in the same direction. A device of this kind is known as a directional coupler.

10.2 COUPLED WAVE EQUATIONS

We have encountered coupled wave equations[88] of a special type in Section 9.3, Equations (9.3-46) and (9.3-47), and in Section 2.7, Equations (2.7-15) and (2.7-16). Even though coupled wave equations are quite plausible and could be written down simply by intuition, we present an approximate derivation of coupled wave equations for the case of two arbitrary dielectric waveguides. Each waveguide consists of a dielectric medium with a maximum of the refractive index in the region of the waveguide. The refractive index is inhomogeneous only in a narrow region, and assumes a constant value outside the guide. The distributions of the square of the refractive indices of the two guides are shown in Figure 10.2.1 under the assumption that each exists by itself. The coordinate x is meant to indicate a suitable transverse direction. The problem is not limited to two dimensions. If both waveguides exist side by side, the square of the refractive index of the medium in which both guides are present can be expressed as

$$n^2 = (n_1^2 - n_3^2) + (n_2^2 - n_3^2) + n_3^2 \tag{10.2-1}$$

n_1^2 is the index distribution shown in the top part of Figure 10.2.1, and n_2^2 is the distribution shown in the lower part of the figure. The square of the index n_3^2 is the constant value outside the region of both guides. Since $n_2^2 - n_3^2 = 0$ outside the region of the second guide, is clear that (10.2-1) describes the square of the refractive index distribution correctly in the vicinity and inside the first waveguide. Since $n_1^2 - n_3^2 = 0$ outside the first

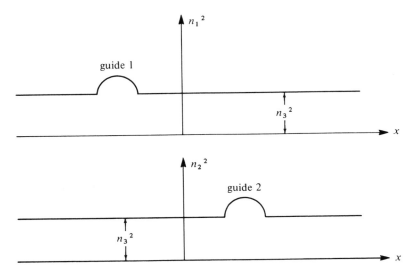

Figure 10.2.1   Transverse distributions of the refractive index defining the two waveguides, each in the absence of the other. (From D. Marcuse, The Coupling of Degenerate Modes in Two Parallel Dielectric Waveguides *B.S.T.J.*, Vol. 6, 60. N, July–August 1971. Copyright 1971, The American Telephone and Telegraph Co., reprinted by permission.)

waveguide, (10.2-1) is the correct index distribution everywhere provided that the two waveguides do not overlap.

We indicate the electromagnetic field of each waveguide in the absence of the other by subscripts 1 and 2 attached to the field quantities. We thus write the electric field of each guide as

$$\mathbf{E}_v = \hat{\mathbf{E}}_v e^{i(\omega t - \beta_v z)} \quad v = 1, 2 \tag{10.2-2}$$

and the magnetic field as

$$\mathbf{H}_v = \hat{\mathbf{H}}_v e^{i(\omega t - \beta_v z)} \quad v = 1, 2 \tag{10.2-3}$$

These fields satisfy Maxwell's equations in the following form

$$\mathbf{V}_t \times \mathbf{H}_v - i\beta_v(\mathbf{z} \times \mathbf{H}_v) - i\omega\varepsilon_0 n_v^2 \mathbf{E}_v = 0 \tag{10.2-4}$$

$$\mathbf{V}_t \times \mathbf{E}_v - i\beta_v(\mathbf{z} \times \mathbf{E}_v) + i\omega\mu_0 \mathbf{H}_v = 0 \tag{10.2-5}$$

with $v = 1$ or 2. $\mathbf{z}$ is a unit vector in z direction, $\mathbf{V}_t$ is the transverse part of the $\mathbf{V}$ vector operator.

When both guides are placed near each other, the total field can approximately be expressed as a superposition of the unperturbed fields of each guide. However, since we anticipate that the two guides influence each other, we must provide for the possibility that the field amplitudes may change with distance. We thus express the total field in the form

$$\mathbf{E} = A_1(z)\mathbf{E}_1 + A_2(z)\mathbf{E}_2 \tag{10.2-6}$$

and

$$\mathbf{H} = A_1(z)\mathbf{H}_1 + A_2(z)\mathbf{H}_2 \tag{10.2-7}$$

These field expressions are not exact. Additional small terms are required to express the fields exactly. The total electric field $\mathbf{E}$ and magnetic field $\mathbf{H}$ satisfy Maxwell's equations

$$\nabla \times \mathbf{H} = i\omega\varepsilon_0 n^2 \mathbf{E} \tag{10.2-8}$$

and

$$\nabla \times \mathbf{E} = -i\omega\mu_0\mathbf{H} \tag{10.2-9}$$

with n^2 of (10.2-1). Substitution of (10.2-6) and (10.2-7) into (10.2-8) and (10.2-9) leads to

$$A_1\left[\nabla_t \times \mathbf{H}_1 - i\beta_1(\mathbf{z} \times \mathbf{H}_1)\right] + \frac{\partial A_1}{\partial z}(\mathbf{z} \times \mathbf{H}_1) - i\omega\varepsilon_0 n^2 A_1 \mathbf{E}_1$$

$$+ A_2\left[\nabla_t \times \mathbf{H}_2 - i\beta_2(\mathbf{z} \times \mathbf{H}_2)\right] + \frac{\partial A_2}{\partial z}(\mathbf{z} \times \mathbf{H}_2) - i\omega\varepsilon_0 n^2 A_2 \mathbf{E}_2 = 0$$

$$\tag{10.2-10}$$

and

$$A_1\left[\nabla_t \times \mathbf{E}_1 - i\beta_1(\mathbf{z} \times \mathbf{E}_1)\right] + \frac{\partial A_1}{\partial z}(\mathbf{z} \times \mathbf{E}_1) + i\omega\mu_0 A_1 \mathbf{H}_1$$

$$+ A_2\left[\nabla_t \times \mathbf{E}_2 - i\beta_2(\mathbf{z} \times \mathbf{E}_2)\right] + \frac{\partial A_2}{\partial z}(\mathbf{z} \times \mathbf{E}_2) + i\omega\mu_0 A_2 \mathbf{H}_2 = 0$$

$$\tag{10.2-11}$$

By using (10.2-1), (10.2-4), and (10.2-5), we can simplify these equations, so that they assume the form

$$\frac{\partial A_1}{\partial z}(\mathbf{z} \times \mathbf{H}_1) - i\omega\varepsilon_0\left(n_2^2 - n_3^2\right)A_1 \mathbf{E}_1 + \frac{\partial A_2}{\partial z}(\mathbf{z} \times \mathbf{H}_2)$$

$$- i\omega\varepsilon_0(n_1^2 - n_3^2) A_2 \mathbf{E}_2 = 0 \tag{10.2-12}$$

and

$$\frac{\partial A_1}{\partial z}(\mathbf{z} \times \mathbf{E}_1) + \frac{\partial A_2}{\partial z}(\mathbf{z} \times \mathbf{E}_2) = 0 \tag{10.2-13}$$

As the next step in the derivation of the coupled wave equations, we take the scalar product of (10.2-12) with $\mathbf{E}_1^-$ and a similar product of (10.2-13) with $\mathbf{H}_1^-$ and substract the two equations. The superscripts $-$ indicate that the radian frequency ω and the propagation constants β have been replaced by their negative values. This procedure is necessary to remove the time dependence from the equations. If we were to limit ourselves to real values of the refractive indices, we could have used the complex conjugate quantities instead of $\mathbf{E}_1^-$ and $\mathbf{H}_1^-$. However, we want to allow for the possibility that the refractive indices may be complex, meaning that the media composing the guides as well as the surrounding medium may be lossy. Finally, we integrate the resulting equation over the infinite cross section.

$$
\int_{-\infty}^{\infty}\int \left\{ \frac{\partial A_1}{\partial z}\left[\mathbf{E}_1^- \cdot (\mathbf{z} \times \mathbf{H}_1) - \mathbf{H}_1^- \cdot (\mathbf{z} \times \mathbf{E}_1) \right] \right.
$$

$$
+ \frac{\partial A_2}{\partial z}\left[\mathbf{E}_1^- \cdot (\mathbf{z} \times \mathbf{H}_2) - \mathbf{H}_1^- \cdot (\mathbf{z} \times \mathbf{E}_2) \right] - i\omega\varepsilon_0(n_2^2 - n_3^2)A_1\,\mathbf{E}_1^- \cdot \mathbf{E}_1
$$

$$
\left. - i\omega\varepsilon_0(n_1^2 - n_3^2)A_2\mathbf{E}_1^- \cdot \mathbf{E}_2 \right\} dxdy = 0 \tag{10.2-14}
$$

We can simplify this equation by neglecting small terms. The expression $n_2^2 - n_3^2$ vanishes outside the region of the second waveguide. However, in the region where this term makes its contribution, the field of the first guide is already quite weak. Since it is the square of $\mathbf{E}_1$ that appears multiplied with $n_2^2 - n_3^2$, the entire term is small of second order and can be neglected. The product $\mathbf{E}_1^- \cdot (\mathbf{z} \times \mathbf{H}_1)$ is a zero order term as far as its magnitude is concerned. The term $(n_1^2 - n_3^2)\mathbf{E}_1^- \cdot \mathbf{E}_2$ is small of first order, since the difference of the squares of the refractive indices is non-vanishing only in the region of the first guide where the field of the second guide, $\mathbf{E}_2$, is small. This comparison of orders of magnitude suggests that the derivatives $\partial A/\partial z$ are quantities that are themselves of first order. Since the product $\mathbf{E}_1^- \cdot (\mathbf{z} \times \mathbf{H}_2)$ is a first order term, because the fields of different waveguides overlap only slightly, we see that the term with $\partial A_2/\partial z$ is also small of second order and can thus be neglected. Keeping only first order terms and rearranging the equation results in

$$
\frac{\partial A_1}{\partial z} = ic_1 A_2 e^{i(\beta_1 - \beta_2)z} \tag{10.2-15}
$$

We obtain similarly, by multiplication of (10.2-12) and (10.2-13) with $\mathbf{E}_2^-$ and $\mathbf{H}_2^-$, the equation

$$
\frac{\partial A_2}{\partial z} = ic_2 A_1 e^{-i(\beta_1 - \beta_2)z} \tag{10.2-16}
$$

with

$$
c_1 = -\omega\varepsilon_0 \frac{\displaystyle\int_{-\infty}^{\infty}\int (n_1^2 - n_3^2)\hat{\mathbf{E}}_1^- \cdot \hat{\mathbf{E}}_2\,dxdy}{\displaystyle\int_{-\infty}^{\infty}\int \mathbf{z}\cdot(\hat{\mathbf{E}}_1^- \times \hat{\mathbf{H}}_1 + \hat{\mathbf{E}}_1 \times \hat{\mathbf{H}}_1^-)\,dxdy} \tag{10.2-17}
$$

and

$$c_2 = -\omega\varepsilon_0 \frac{\displaystyle\int_{-\infty}^{\infty}\int (n_2^2 - n_3^2)\hat{\mathbf{E}}_2^- \cdot \hat{\mathbf{E}}_1\,dx\,dy}{\displaystyle\int_{-\infty}^{\infty}\int \mathbf{z}\cdot(\hat{\mathbf{E}}_2^- \times \hat{\mathbf{H}}_2 + \hat{\mathbf{E}}_2 \times \hat{\mathbf{H}}_2^-)\,dx\,dy} \qquad (10.2\text{-}18)$$

The coupling coefficients (10.2-17) and (10.2-18) are independent of z. To indicate that the term $\exp(\pm i\beta_v z)$ has been removed, we use the notation $\hat{\mathbf{E}}_1$, $\hat{\mathbf{E}}_1^-$, etc., introduced in (10.2-2) and (10.2-3). The reader is reminded that the quantities with negative superscripts result from the ordinary quantities by changing the sign of ω and β.

In going from (10.2-14) to (10.2-15) through (10.2-18), we used the fact that the field amplitudes A_1 and A_2 do not depend on the transverse coordinates x and y, so that they could be taken out of the integrals. The terms in the mixed scalar-vector products were rearranged using a well-known vector identity.

The coupled wave equations (10.2-15) and (10.2-16) are not exact, as our derivation has shown. In addition to neglecting certain second order terms, we have limited ourselves to only two modes. Even if the two guides, which are coupled together by being close to each other, support only one guided mode, there is still the possibility that coupling to radiation modes could also occur. If the guides are multimode, all the modes would be coupled together to some extent.[85,86,87] Limiting the coupled wave equations to only two modes thus also constitutes an approximation. However, we shall see shortly that only modes with identical phase propagation constants can exchange significant amounts of energy. Limiting the coupled wave equations to only two modes is thus really a very good approximation, and allows us to study the energy exchange between the two modes to a high accuracy. We already encountered a similar situation in Section 9.3.

The coupled wave equations are often written in a slightly different form. By introducing the wave amplitudes

$$a_v = A_v e^{-i\beta_v z} \quad v = 1, 2 \qquad (10.2\text{-}19)$$

we can write (10.2-15) and (10.2-16) in the familiar form[88]

$$\frac{\partial a_1}{\partial z} = -i\beta_1 a_1 + i c_1 a_2 \qquad (10.2\text{-}20)$$

and

$$\frac{\partial a_2}{\partial z} = -i\beta_2 a_2 + i c_2 a_1 \qquad (10.2\text{-}21)$$

These equations have such a clear intuitive meaning that it would have been possible to write them down without derivation. However, our derivation has the advantage that the values of the coupling coefficients are now explicitly known. Since the index distributions $n_1(x,y)$ and $n_2(x,y)$ as well as the constant value of the surrounding medium n_3 are allowed to be complex, the coupling coefficients c_1 and c_2 must, in general, also be complex quantities. In case of lossless media, the refractive index distributions are real, and Equations (10.2-17) and (10.2-18) can be simplified. For real values of n_1, n_2, and n_3, we could have used the complex conjugate quantities $\mathbf{E}_1^*$ instead of $\mathbf{E}_1^-$. (The derivation presented here does not make it compellingly clear that $\mathbf{E}_1^-$, instead of $\mathbf{E}_1^*$, must be used. This fact comes out more clearly in the derivation presented in reference 103, because $\mathbf{E}_1^-$ must be a solution of Maxwell's equations, with ω and β replaced by their negative counterparts, while n_1^2 remains in its original form without becoming n_1^{2*}.) The expressions in the denominator can then be interpreted as $4P$, with P being the power of the mode in guide 1 (for $A_1 = 1$). If, in addition, the two waveguides are identical (n_1 having the same distribution centered in guide 1 as n_2 has centered in guide 2), we obtain, instead of (10.2-17) and (10.2-18)

$$c_2 = -\frac{\omega\varepsilon_0}{4P} \int_{-\infty}^{\infty}\int (n_2^2 - n_3^2)\hat{\mathbf{E}}_2^* \cdot \hat{\mathbf{E}}_1\, dx\, dy \qquad (10.2\text{-}22)$$

and

$$c_1 = c_2 \qquad (10.2\text{-}23)$$

It is easy to show that for real values of c_1 and c_2 the condition (10.2-23) is required for conservation of power between the two modes. With the help of (10.2-20) and (10.2-21), we obtain the expression (Re indicates the real part)

$$\frac{d}{dz}(|a_1|^2 + |a_2|^2) = 2\,\mathrm{Re}\,\{i(c_2 - c_1^*)a_1 a_2^*\}$$

provided that β_1 and β_2 are real. The left-hand side is the z derivative of the total power carried in both guides. This quantity must vanish if power is conserved. Because $a_1(0)$ and $a_2(0)$ can be chosen arbitrarily, we thus obtain the following condition as a requirement for conservation of power

$$c_1 = c_2^* \qquad (10.2\text{-}24)$$

Equation (10.2-24) is identical to (10.2-23) for real values of c_1 and c_2.

It is easy to see that substantial amounts of power can be transferred only if $\beta_1 = \beta_2$. Let us assume that $A_2 = 0$ at $z = 0$. We then obtain from (10.2-16)

$$A_2(L) = ic_2 \int_0^L A_1(z)e^{-i(\beta_1 - \beta_2)z}\, dz \qquad (10.2\text{-}25)$$

If $\beta_1 - \beta_2 \neq 0$, the function $A_1(z)$ is multiplied by $\cos(\beta_1 - \beta_2)$ z and $\sin$ $(\beta_1 - \beta_2)z$. Both functions oscillate rapidly and prevent any appreciable buildup of the integral (10.2-25). However, if $\beta_1 - \beta_2 = 0$, the integral grows proportional to L (at least initially when $A_1(z)$ has not yet changed very much). This consideration shows that $A_2(z)$ can obtain appreciable values only if the propagation constants of both modes are identical. This is the reason that a coupled wave theory, considering only two modes, works even in the case of multimode waveguides, since considerable exchange of power occurs only between modes with equal phase velocity. If, however, the coupling coefficients are not constant but arbitrary functions of z, a two-mode theory would be invalid for multimode waveguides.

The coupled wave equations with constant coupling coefficients have the following simple solution for $\beta_1 = \beta_2 = \beta$

$$a_1(z) = \frac{1}{2}\left\{a_1(0)[e^{i\Delta\beta z} + e^{-i\Delta\beta z}] + \sqrt{\frac{c_1}{c_2}}\, a_2(0)[e^{i\Delta\beta z} - e^{-i\Delta\beta z}]\right\}e^{-i\beta z} \quad (10.2\text{-}26)$$

and

$$a_2(z) = \frac{1}{2}\left\{a_2(0)[e^{i\Delta\beta z} + e^{-i\Delta\beta z}] + \sqrt{\frac{c_2}{c_1}}\, a_1(0)[e^{i\Delta\beta z} - e^{-i\Delta\beta z}]\right\}e^{-i\beta z} \quad (10.2\text{-}27)$$

with

$$\Delta\beta = \sqrt{c_1 c_2} \quad (10.2\text{-}28)$$

The terms in this solution have been grouped to display clearly the initial values of a_1 and a_2 at $z = 0$. By regrouping terms, it becomes apparent that the solutions consist of a superposition of two new modes with phase constants

$$\beta_+ = \beta + \Delta\beta \quad (10.2\text{-}29)$$

and

$$\beta_- = \beta - \Delta\beta \quad (10.2\text{-}30)$$

The two coupled waveguides thus possess normal modes with the altered phase propagation constants (10.2-29) and (10.2-30). If the coupling coefficients are real, $\Delta\beta$ is real. Taking for simplicity $a_2(0) = 0$, we obtain from (10.2-26) and (10.2-27), using (10.2-24)

$$a_1(z) = a_1(0)\cos(\Delta\beta z)e^{-i\beta z} \quad (10.2\text{-}31)$$

and

$$a_2(z) = ia_1(0)\sin(\Delta\beta z)e^{-i\beta z} \quad (10.2\text{-}32)$$

This solution shows clearly that for real values of $\beta_1 = \beta_2$ energy is continuously exchanged between the two waveguides. If, however, $\Delta\beta$ is not real, we see that the factors $\exp(-i\Delta\beta z)$ decreases while $\exp(+i\Delta\beta z)$ grows (or vice versa). For sufficiently large values of z, one of the two factors becomes negligible, and we have

$$a_1(z) = \frac{1}{2} a_1(0)e^{-i(\beta + \Delta\beta)z} \tag{10.2-33}$$

and

$$a_2(z) = \frac{1}{2} \sqrt{\frac{c_2}{c_1}} a_1(0)e^{-i(\beta + \Delta\beta)z} \tag{10.2-34}$$

provided that $a_2(0) = 0$. Even though both mode amplitudes are decreasing, we see that, in case of complex $\Delta\beta$, both waveguides carry equal amounts of power (c_1/c_2 is still not too different from unity) after a sufficiently large distance even though only one of the guides was excited initially. The periodic exchange of power is dying out, and the power in both guides (whatever remains of it) equalizes.

For the purpose of studying cross talk, it is sufficient to consider only the transfer of a small amount of energy, since cross talk is an undesirable effect and should be kept as small as possible. We assume once more that initially all the power is in guide 1, $a_2(0) = 0$, and that $\Delta\beta z \ll 1$ at the end of the waveguides at $z = L$. From (10.2-27), we obtain

$$\frac{a_2(L)}{a_1(0)e^{-i\beta L}} = i \sqrt{\frac{c_2}{c_1}} \Delta\beta L \tag{10.2-35}$$

From (10.2-26), we obtain to the same approximation

$$a_1(L) = a_1(0)e^{-i\beta L} \tag{10.2-36}$$

The power flowing in each guide is proportional to the square of the field amplitude. We thus get for the power ratio at the end of the two guides

$$C = \frac{|a_2(L)|^2}{|a_1(L)|^2} = \left|\frac{c_2}{c_1}\right| (|\Delta\beta| L)^2 \tag{10.2-37}$$

For most practical purposes, we can assume that (10.2-24) holds, at least approximately, so that we obtain finally to a good approximation

$$C = |\Delta\beta|^2 L^2 \tag{10.2-38}$$

It is important to remember that (10.2-38) holds only as long as $C \ll 1$.

In summary, we saw in this section that two dielectric waveguides influence each other if they are placed sufficiently close to each other. The coupling between the guides can be expressed by two coupled wave equations with a constant coupling coefficient. The coupling is caused by the fact that the field tail of the guided mode in one guide reaches to the vicinity of the other guide. Our coupling theory holds for any type of dielectric waveguide regardless of the distribution of refractive index that produces the guiding action. The two guides and the modes carried by them need not be identical, and it is permissible for the dielectric media of the guides and for the surrounding medium to have losses. The only requirement is that the propagation constants of the two modes be nearly identical. If only one mode can exist on each guide, this requirement too can be relaxed. However, modes with different propagation constants do not couple effectively by means of a constant coupling coefficients. The only case of real interest is thus the degenerate case with both modes having equal phase velocity.

10.3 COUPLED SLAB WAVEGUIDES

The general coupling theory shall now be applied to two special cases. In this section, we consider coupling of two identical parallel slab waveguides, as shown in Figure 10.3.1. The coupling coefficients for the problem of two identical modes being coupled can be solved by applying (10.2-22) of the preceding section. Use of (10.2-22) instead of (10.2-17) is permissible if we assume that the core and cladding indices n_c and n_m are real numbers. The surrounding medium is allowed to have a complex refractive index n_3, but in order for (10.2-22) to be approximately correct, we must assume that the cladding is sufficiently thick so that the mode losses caused by the lossy surrounding medium are slight. These assumptions are reasonable since waveguides are useful only if they guide energy with low losses.

The coupling problem can also be solved more directly. We can start from first principles and solve the problem of guided modes of the combined structure, as shown in Figure 10.3.1. This is a formidable problem for cladded waveguides. However, in the limiting case $n_m = n_3$, it is easy to solve the mode problem of the combined structure. The reader is encouraged to try this procedure as an exercise. With reasonable approximations, this direct approach yields Equation (10.3-12), to be derived in this section.

To save space, we shall only outline the details of the calculation of the coupling problem of two cladded slab waveguides. Even with the help of (10.2-22), the problem is difficult to solve rigorously (rigorously in the sense of accepting (10.2-22) as rigorous). However, if we assume that the mode fields have decayed to small values at the boundary of the cladding, $x = D$, we can simplify the theory considerably. We start with the even TE modes of the slab wave guide. The only electric field component can be expressed as

$$\hat{\mathbf{E}}_y = \begin{cases} A \cos \kappa x & 0 \leq x \leq d \\ B e^{\gamma x} + C e^{-\gamma x} & d \leq x \leq D \\ F e^{-\rho x} & D \leq x < \infty \end{cases} \qquad (10.3\text{-}1)$$

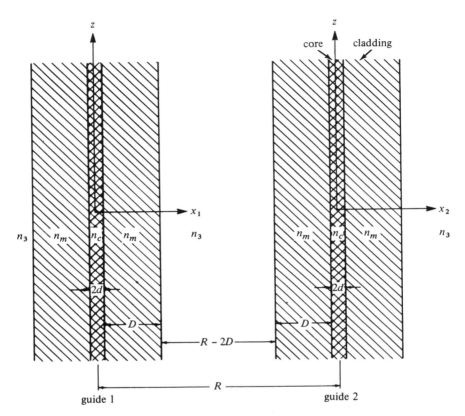

Figure 10.3.1 Geometry of two coupled cladded slab waveguides. (From D. Marcuse, The Coupling of Degenerate Modes in Two Parallel Dielectric Waveguides, *B.S.T.J.*, Vol. **6**, No. 6, July–August 1971. Copyright 1971, The American Telephone and Telegraph Co., reprinted by permission.)

For $x < 0$, $\hat{E}_y(x)$ is assumed to be continued as an even function. The parameters occuring in (10.3-1) are given by

$$k_0 = \omega \sqrt{\varepsilon_0 \mu_0} \tag{10.3-2}$$

$$\kappa = \sqrt{n_c^2 k_0^2 - \beta^2} \tag{10.3-3}$$

$$\gamma = \sqrt{\beta^2 - n_m^2 k_0^2} \tag{10.3-4}$$

$$\rho = \sqrt{\beta^2 - n_3^2 k_0^2} \tag{10.3-5}$$

The relation between $\hat{E}_y$ and E_y is given by (10.2-2). The magnetic field components are obtained from (10.3-1) with the help of (8.3-2) and (8.3-3). We now approximate the theory by assuming that

$$e^{-\gamma(D-d)} \ll 1 \qquad (10.3\text{-}6)$$

Neglecting squares of this small quantity, we obtain*

$$B = 0 \qquad (10.3\text{-}7)$$

$$C = Ae^{\gamma d} \cos \kappa d \qquad (10.3\text{-}8)$$

$$F = \frac{2\kappa\gamma A}{(\gamma + \rho)\sqrt{\kappa^2 + \gamma^2}} e^{\rho D} e^{-\gamma(D-d)} \qquad (10.3\text{-}9)$$

The coefficient A can be expressed by the power P carried by the mode

$$A = \sqrt{\frac{2\omega\mu_0 P}{\beta d + \dfrac{\beta}{\gamma}}} \qquad (10.3\text{-}10)$$

The eigenvalue equation of the mode of the cladded slab waveguide differs from (8.3-16) only by a term proportional to the square of (10.3-6). In the spirit of our approximation, we thus use (8.3-16) unaltered. The mode field (10.3-1) is expressed in a coordinate system that is centered at the core of each waveguide.

Since, according to (10.2-23) and (10.2-28), $\Delta\beta = c_1$, we obtain by straight-forward evaluation of the integral from (10.2-22)

$$\Delta\beta = \frac{4\kappa^2\gamma^3\rho}{\beta(1 + \gamma d)(\gamma + \rho)^2(\kappa^2 + \gamma^2)} e^{-2\gamma(D-d)} e^{-\rho(R-2D)} \qquad (10.3\text{-}11)$$

The refractive indices of core and cladding were assumed to be real, but n_3 and consequently ρ are allowed to be complex quantities. This enables us to study the decoupling of the two waveguides that is caused by making the surrounding medium lossy.

In the special case that the cladding and surrounding medium have identical refractive indices, $n_m = n_3$, we obtain $\gamma = \rho$, and (10.3-11) specializes to the following form

$$\Delta\beta = \frac{\kappa^2\gamma^2}{\beta(1 + \gamma d)(\kappa^2 + \gamma^2)} e^{-\gamma(R-2d)} \qquad (10.3\text{-}12)$$

As mentioned earlier, this expression can be derived directly by solving the problem of the normal modes of the combined structure composed of both slab waveguides. The normal modes of the combined structure have slightly different propagation constants according to (10.2-29) and (10.2-30). The direct calculation of these propagation constants results in Equation (10.3-12). The same expression can also be found in Kapany's book.[79]

Since it is possible to reduce the coupling between the two waveguides

---

* The approximation $B = 0$ was made after F had been calculated.

by adding loss to the surrounding medium, it is important to know the effect of this external loss on the loss of the guided modes. The guided mode loss can be calculated by a simple perturbation technique that is similar to the one employed in Section 9.6 for the calculation of the bending losses. Since the technique is quite straightforward, we content ourselves with a verbal description and stating the end result.

The field expressions (10.3-1) allow us to calculate the x component of the Poynting vector S_x at the cladding boundary at $x = D$. The power loss (α is the amplitude attenuation coefficient) is thus given by

$$2\alpha = \frac{2S_x}{P} \tag{10.3-13}$$

The factor 2 on the left-hand side of this equation is required, since 2α is the power loss; the factor 2 on the right-hand side accounts for the fact that power is flowing away on either side of the waveguide. The presence of the neighboring guide is ignored in this calculation. Its effect is slight, since the power loss is caused by radiation which penetrates the second guide except for some small reflection. However, even the reflected power does not return to the mode in the first guide. The simple calculation results in

$$2\alpha = \frac{8\kappa^2 \gamma^3 Im(\rho)e^{-2\gamma(D-d)}}{\beta(1 + \gamma d)(\kappa^2 + \gamma^2)|\gamma + \rho|^2} \tag{10.3-14}$$

The symbol $Im(—)$ indicates the imaginary part.

Equation (10.3-14) reveals several interesting features of the mode attenuation. It shows that the attenuation decreases exponentially with increasing cladding thickness. The mode loss can thus be made arbitrarily small by making the cladding sufficiently thick. The mode attenuation is, furthermore, dependent on the imaginary part of ρ. If ρ is real, the mode loss vanishes. It is of course important to remember that we assumed that the materials of core and cladding are lossless. Losses in these materials cause additional mode losses. The parameter ρ can be real when n_3 is real and when, in addition, $\beta > n_3 k_0$. However, if

$$\beta < n_3 k_0 \tag{10.3-15}$$

then ρ is imaginary and the mode suffers loss even though n_3 is real. This is another example of "frustrated" total internal reflection. The exponential field decay caused by total internal reflection at the core-cladding interface converts itself to a radiation field if the refractive index of the surrounding medium is large enough. For small values of n_3, mode loss is caused by losses in the surrounding medium.

Analogous expressions can of course be derived for the TM modes of the slab waveguides. TE modes and TM modes do not couple, since the scalar product of their electric fields vanishes. The change of the propagation constant caused by coupling of two symmetric TM modes of the slab waveguides is

$$\Delta\beta = \frac{n_c^2 n_m^2 \kappa^2 \gamma^2 [2e^{(\gamma-\rho)(D-d)} - 1]e^{-2\gamma(D-d)} e^{-\rho(R-2D)}}{\beta[(n_m^4 \kappa^2 + n_c^4 \gamma^2)\gamma d + n_c^2 n_m^2(\kappa^2 + \gamma^2)]} \qquad (10.3\text{-}16)$$

The coupling of TE modes and TM modes is apparently not the same. Only in the limit $n_c \approx n_m \approx n_3$ do (10.3-11) and (10.3-16) result in the same expressions.

The losses to the TM modes caused by the surrounding medium are given by

$$2\alpha = \frac{8n_c^2 n_m^4 |n_3|^4 \kappa^2 \gamma^3 Im\left(\dfrac{\rho}{n_3^2}\right)e^{-2\gamma(D-d)}}{\beta[\gamma d(n_m^4 \kappa^2 + n_c^4 \gamma^2) + n_c^2 n_m^2(\kappa^2 + \gamma^2)]\,|n_m^2\rho + n_3^2\gamma|^2} \qquad (10.3\text{-}17)$$

10.4 COUPLING OF HE$_{11}$ MODES OF ROUND FIBERS

In complete analogy to the slab waveguide case discussed in the preceding section, we now briefly discuss the results of coupling between the HE$_{11}$ modes of two round optical fibers.

The fiber geometry is shown in Figure 10.4.1.

The modes can be expressed by the equations for the longitudinal electric and magnetic z components inside the core

$$\left.\begin{array}{l} \hat{E}_z = AJ_1(\kappa r)\cos\phi \\ \hat{H}_z = BJ_1(\kappa r)\sin\phi \end{array}\right\} 0 \leq r \leq a \qquad (10.4\text{-}1)$$

In the cladding, we have

$$\left.\begin{array}{l} \hat{E}_z = [CH_1^{(1)}(i\gamma r) + DH_1^{(2)}(i\gamma r)]\cos\phi \\ \hat{H}_z = [FH_1^{(1)}(i\gamma r) + GH_1^{(2)}(i\gamma r)]\sin\phi \end{array}\right\} a \leq r \leq b \qquad (10.4\text{-}2)$$

and, finally, the field outside the cladding in the surrounding medium is given by

$$\left.\begin{array}{l} \hat{E}_z = MH_1^{(1)}(i\rho r)\cos\phi \\ \hat{H}_z = NH_1^{(1)}(i\rho r)\sin\phi \end{array}\right\} b \leq r < \infty \qquad (10.4\text{-}3)$$

The other field components are obtained from (8.2-7) through (8.2-10), with κ^2 standing for either κ^2, $-\gamma^2$, or $-\rho^2$. To the same approximation used in the slab waveguide case, the coefficients B, C, and F are given by

$$B = \sqrt{\frac{\varepsilon_0}{\mu_0}}\,\frac{a^2 k_0 \kappa^2 \gamma^2}{\beta(\kappa^2 + \gamma^2)}\left\{\frac{n_c^2}{\kappa a}\left[\frac{1}{\kappa a} - \frac{J_0(\kappa a)}{J_1(\kappa a)}\right] + \frac{n_m^2}{\gamma a}\left[\frac{1}{\gamma a} - \frac{iH_0^{(1)}(i\gamma a)}{H_1^{(1)}(i\gamma a)}\right]\right\}A$$

$$(10.4\text{-}4)$$

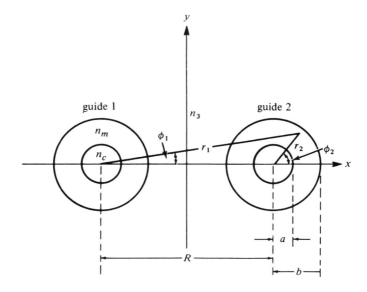

Figure 10.4.1   Geometry of two coupled cladded round optical fibers. (From D. Marcuse, The Coupling of Degenerate Modes in Two Parallel Dielectric Waveguides, *B.S.T.J.*, **6**, No. 6, July–August 1971. Copyright 1971, The American Telephone and Telegraph Co., reprinted by permission.)

and

$$\frac{C}{A} = \frac{F}{B} = \frac{J_1(\kappa a)}{H_1^{(1)}(i\gamma a)} \tag{10.4-5}$$

To the approximation that the field intensity is weak at the interface between the cladding and the surrounding medium, we have*

$$D = G = 0 \tag{10.4-6}$$

The coefficients M and N are approximated by replacing the Hankel functions by their approximations for large argument

$$M = \frac{\left(\frac{\rho}{\gamma}\right)^{3/2} e^{(\rho-\gamma)b} \left\{ C \left[\left(\frac{\beta}{b}\right)^2 \left(1 - \frac{\gamma^2}{\rho^2}\right)^2 \left(\frac{\gamma}{\rho} - 1\right) \right. \right.}{\left[\frac{\beta}{b}\left(1 - \frac{\gamma^2}{\rho^2}\right)\right]^2 - k_0^2\gamma^2 \left(1 + \frac{\gamma}{\rho}\right)\left(n_m^2 + n_3^2\frac{\gamma}{\rho}\right)}$$

$$\frac{\left. \left. - k_0^2\gamma^2\left(1 + \frac{\gamma}{\rho}\right)\left\{n_m^2 + n_3^2\frac{\gamma^2}{\rho^2} + \frac{\gamma}{\rho}(n_m^2 - n_3^2)\right\}\right] + 2\omega\mu_0 F\gamma\frac{\beta}{b}\left(1 - \frac{\gamma^2}{\rho^2}\right)\right\}}{} \tag{10.4-7}$$

* Just as in the slab case, $D = G = 0$ can be taken only after M and N have been computed.

and

$$N = - \sqrt{\frac{\rho}{\gamma}} e^{(\rho - \gamma)b} \frac{2Fk_0^2\gamma^2 \left(n_m^2 + n_3^2 \frac{\gamma}{\rho}\right) + 2\omega n_m^2 \varepsilon_0 \gamma \frac{\beta}{b} \left(\frac{\gamma^2}{\rho^2} - 1\right)C}{\left[\frac{\beta}{b}\left(1 - \frac{\gamma^2}{\rho^2}\right)\right]^2 - k_0^2\gamma^2\left(1 + \frac{\gamma}{\rho}\right)\left(n_m^2 + n_3^2\frac{\gamma}{\rho}\right)}$$

$$(10.4\text{-}8)$$

The coefficient A, finally, is related to the power carried by the mode

$$P = \frac{\pi}{4} \sqrt{\frac{\varepsilon_0}{\mu_0}} A^2 \left\{ \frac{k_0\beta}{\kappa^4} \left[(a\kappa)^2 (J_0^2(\kappa a) + J_1^2(\kappa a)) - 2J_1^2(\kappa a) \right] \left(n_c^2 + \frac{\mu_0}{\varepsilon_0} \frac{B^2}{A^2} \right) \right.$$
$$+ \frac{k_0\beta}{\gamma^4} \left[(a\gamma)^2 \left(1 - \left(\frac{iH_0^{(1)}(i\gamma a)}{H_1^{(1)}(i\gamma a)} \right)^2 \right) + 2 \right] J_1^2(\kappa a) \left(n_m^2 + \frac{\mu_0}{\varepsilon_0} \frac{B^2}{A^2} \right)$$
$$\left. + 2 \sqrt{\frac{\mu_0}{\varepsilon_0}} \frac{B}{A} \left[\frac{\beta_0^2 + n_c^2 k_0^2}{\kappa^4} - \frac{\beta_0^2 + n_m^2 k_0^2}{\gamma^4} \right] J_1^2(\kappa a) \right\}$$

$$(10.4\text{-}9)$$

The parameters κ, γ, and ρ are again given by (10.3-3), (10.3-4), and (10.3-5). The eigenvalue equation (8.2-49) is approximately applicable, since the eigenvalue equation of the cladded HE$_{11}$ mode differs from the eigenvalue equation (8.2-49) with infinite cladding only by a term that is the square of (10.3-6).

The evaluation of the integral in (10.2-22) is not as easy as it was for the slab waveguide. In fact, the integral cannot be solved exactly. Using a coordinate system that is centered on the axis of the second guide, we can write (10.2-22) in the form

$$\Delta\beta = - \frac{\omega\varepsilon_0}{4P} \int_0^b r_2 dr_2 \int_0^{2\pi} (n_2^2 - n_3^2)\hat{\mathbf{E}}_2^* \cdot \hat{\mathbf{E}}_1 d\phi_2 \qquad (10.4\text{-}10)$$

Since the field of each guide is expressed in a coordinate system centered at the axis of this guide, we must transform the fields to a common coordinate system before the scalar product of the electric field vectors can be taken. We write the electric field components in the coordinate system of each guide in the form (compare [8.6-59], [8.6-61], and [8.6-62], the difference of a factor i is caused by the definition of the F functions)

$$\left. \begin{array}{l} \hat{E}_{zv} = F_{zv}(r_v)\cos\phi_v \\ \hat{E}_{rv} = F_{rv}(r_v)\cos\phi_v \\ \hat{E}_{\phi v} = F_{\phi v}(r_v)\sin\phi_v \end{array} \right\} \qquad (10.4\text{-}11)$$

The index v assumes the values 1 or 2, and indicates the waveguide to which the field component or coordinate bearing that label belongs. The purpose of the notation (10.4-11) is to display explicitly the r- and ϕ-dependence of each field component in a general way. It can be shown that the scalar product of the two field vectors in the relative orientation of the two guides, as indicated by Figure 10.4.1, can be expressed as follows

$$\hat{\mathbf{E}}_2^* \cdot \hat{\mathbf{E}}_1 = F_{z1}(r_1)F_{z2}^*(r_2)\cos\phi_1\cos\phi_2$$
$$+ [F_{r1}(r_1)F_{r2}^*(r_2)\cos\phi_1\cos\phi_2$$
$$+ F_{\phi 1}(r_1)F_{\phi 2}^*(r_2)\sin\phi_1\sin\phi_2]\cos(\phi_2 - \phi_1)$$
$$+ [F_{\phi 1}(r_1)F_{r2}^*(r_2)\sin\phi_1\cos\phi_2$$
$$- F_{r1}(r_1)F_{\phi 2}^*(r_2)\cos\phi_1\sin\phi_2]\sin(\phi_2 - \phi_1) \qquad (10.4\text{-}12)$$

In order to be able to solve the integral in (10.4-10), we assume that the two waveguides are far apart. It turns out that the resulting solution to our problem works even when the two guides are quite close together. However, the approximations that are necessary to make the integral tractable are based on the assumption that $R \gg a$. A first consequence of this assumption is the realization that the angle ϕ_1 never deviates far from zero. Stated more specifically, we should say that only those regions of the range of integration contribute appreciably to the integral where ϕ_1 is quite small. The reason for this statement is the close concentration of the guided mode around the core of the waveguide. Even if the cladding is wide, the region close to the waveguide core contributes more significantly to the integral because of the high field intensity to be found there. We thus boldly set $\phi_1 = 0$. This assumption reduces the scaler product (10.4-12) to the simpler form

$$\hat{\mathbf{E}}_2^* \cdot \hat{\mathbf{E}}_1 = F_{z1}(r_1)F_{z2}^*(r_2)\cos\phi_2$$
$$+ F_{r1}(r_1)[F_{r2}^*(r_2)\cos^2\phi_2 - F_{\phi 2}^*(r_2)\sin^2\phi_2] \qquad (10.4\text{-}13)$$

As the next step in approximating the integral we, use the following approximation for the radius r_1

$$r_1 = \sqrt{R^2 + r_2^2 + 2r_2R\cos\phi_2} \approx R + r_2\cos\phi_2 \qquad (10.4\text{-}14)$$

The functions $F(r_1)$ appearing in (10.4-13) according to (10.4-3) and (8.2-7) are the Hankel function of first order and its derivative. Since ρr_1 is very large, we can use the approximation of the Hankel functions for large arguments together with the approximation (10.4-14)

$$H_1^{(1)}(i\rho r_1) \approx -\sqrt{\frac{2}{\pi\rho r_1}}e^{-\rho r_1} \approx -\sqrt{\frac{2}{\pi\rho R}}e^{-\rho R}e^{-\rho r_2\cos\phi_2} \qquad (10.4\text{-}15)$$

$$H_1^{(1)'}(i\rho r_1) \approx iH_1^{(1)}(i\rho r_1) \qquad (10.4\text{-}16)$$

Having approximated the scalar product of the electric field vectors and the Hankel functions as indicated, it is now possible to perform the ϕ_2 integration in (10.4-10). The following integrals are encountered

$$\int_0^{2\pi} e^{-\rho r_2 \cos \phi_2} \cos \phi_2 d\phi_2 = 2\pi i J_1(i\rho r_2) \qquad (10.4\text{-}17)$$

$$\int_0^{2\pi} e^{-\rho r_2 \cos \phi_2} \cos^2 \phi_2 d\phi_2 = 2\pi \left\{ J_0(i\rho r_2) + \frac{i}{\rho r_2} J_1(i\rho r_2) \right\} \qquad (10.4\text{-}18)$$

and

$$\int_0^{2\pi} e^{-\rho r_2 \cos \phi_2} \sin^2 \phi_2 d\phi_2 = -\frac{2\pi i}{\rho r_2} J_1(i\rho r_2) \qquad (10.4\text{-}19)$$

The remaining integration involves only products of cylinder functions and, though tedious, presents no further difficulties. We thus obtain as a result of an approximate evaluation of (10.4-10)

$$\Delta \beta_h = \sqrt{\frac{\varepsilon_0}{\mu_0}} \frac{A^2}{2P} \frac{M}{A} \frac{1}{k_0} \sqrt{\frac{2\pi}{\rho R}} e^{-\rho R}$$

$$\cdot \left\{ \frac{e^{(\rho-\gamma)b} J_1(\kappa a)}{\pi \sqrt{\rho \gamma} H_1^{(1)}(i\gamma a)} \left[\left(1 + \frac{\beta^2}{\gamma \rho} \right)(\gamma + \rho) + \frac{\beta^2}{b\gamma^2 \rho^2} (\rho^2 - \gamma^2) \left(1 - \frac{k_0}{\beta} \sqrt{\frac{\mu_0}{\varepsilon_0}} \frac{B}{A} \right) \right] \right.$$

$$+ i J_1(i\rho a) \left[a \left(\frac{\beta^2}{\kappa} - \kappa \right) J_0(\kappa a) + a \left(\frac{\beta^2}{\gamma} + \gamma \right) \frac{i H_0^{(1)}(i\gamma a)}{H_1^{(1)}(i\gamma a)} J_1(\kappa a) \right]$$

$$\left. - \frac{\beta^2}{\kappa^2 \gamma^2} (\kappa^2 + \gamma^2) \left(1 - \frac{k_0}{\beta} \sqrt{\frac{\mu_0}{\varepsilon_0}} \frac{B}{A} \right) J_1(\kappa a) \right] \right\} \qquad (10.4\text{-}20)$$

The subscript h on $\Delta \beta$ serves as a reminder that the electric field vectors of the two modes of the guides shown in Figure 10.4.1 are polarized horizontally. This means that near the axis of the waveguides the electric field vectors are parallel to the line joining the centers of the two guides. In the opposite case of vertical polarization, the guides are still oriented as shown in Figure 10.4.1, but the electric field vectors are now oriented perpendicular (or vertical) to the line joining the waveguide centers. The coupling of the vertically polarized modes is slightly different from that of the horizontally polarized modes. A similar calculation leads to the result

$$\Delta \beta_v = \frac{A^2}{2P} \frac{N}{A} \frac{1}{\rho} \sqrt{\frac{2\pi}{\rho R}} e^{-\rho R}$$

$$\cdot \left\{ \frac{e^{(\rho-\gamma)b}}{\pi \gamma b \sqrt{\rho \gamma}} \frac{J_1(\kappa a)}{H_1^{(1)}(i\gamma a)} \left[k_0 \sqrt{\frac{\mu_0}{\varepsilon_0}} \frac{B}{A} \left((\gamma + \rho) b + \frac{\rho^2 - \gamma^2}{\gamma \rho} \right) - \frac{\beta}{\gamma \rho} (\rho^2 - \gamma^2) \right] \right.$$

$$+ i\rho J_1(i\rho a) \left[k_0 a \sqrt{\frac{\mu_0}{\varepsilon_0}} \frac{B}{A} \left(\frac{1}{k_0} J_0(\kappa a) + \frac{1}{\gamma} \frac{i H_0^{(1)}(i\gamma a)}{H_1^{(1)}(i\gamma a)} J_1(\kappa a) \right) \right.$$

$$\left. \left. + \frac{\beta}{\kappa^2 \gamma^2} (\kappa^2 + \gamma^2) \left(1 - \frac{k_0}{\beta} \sqrt{\frac{\mu_0}{\varepsilon_0}} \frac{B}{A} \right) J_1(\kappa a) \right] \right\} \qquad (10.4\text{-}21)$$

n_c and n_m must be real, but n_3 is allowed to be complex in these equations. The coefficients A, B, M, and N have all been defined earlier in this section. The parameters k_0, κ, γ, and ρ are defined by (10.3-2) through (10.3-5).

A vertically polarized mode does not couple with a horizontally polarized mode. The two-mode theory is thus applicable.

In spite of the apparently crude approximations, Equations (10.4-20) and (10.4-21) express the coupling between two HE_{11} modes of cladded round optical fibers very well. A. L. Jones[84] has published his results of a HE_{11} mode coupling theory, and has shown comparison between his theoretical results with measurements by Bracey et al.[110] His theory is in excellent agreement with the experimental results. However, his theory does not include the case of cladded fibers in a lossy environment. Numerical curves computed for the special case $n_m = n_3$ and $\rho = \gamma$ have been compared with Jones' curves. The agreement is excellent, even for the case that the two fiber cores touch each other. For the special case of uncladded fibers, Jones' theory is thus in complete agreement with the theory presented here, and both theories agree well with Bracey's measurements.[110]

It is again desirable to know the HE_{11} mode loss that is caused either by the surrounding medium having an index large enough for (10.3-15) to be satisfied or by a lossy surrounding medium. Using the approach outlined in Section 10.3, we obtain without difficulty

$$2\alpha = \sqrt{\frac{\varepsilon_0}{\mu_0}} \frac{e^{-2\rho_r b}}{|\rho| P} \left\{ \frac{\beta}{b} \frac{1}{|\rho|^4} Im\left[\sqrt{\frac{\mu_0}{\varepsilon_0}} MN^*(\rho^{*2} - \rho^2) \right] \right.$$

$$\left. - k_0 |M|^2 Im\left(\frac{n_3^2}{\rho}\right) - \frac{\mu_0}{\varepsilon_0} k_0 |N|^2 Im\left(\frac{1}{\rho}\right) \right\} \quad (10.4\text{-}22)$$

The symbol $Im(\ )$ indicates the imaginary part; ρ_r is the real part of ρ. The mode factor P cancels with the same factor that is contained in M^2 and N^2 via their dependence on A^2.

10.5 CROSS TALK

The results of the preceding two sections enable us to discuss the cross talk problem of cladded slab waveguides and cladded round optical fibers.[103] Figure 10.5.1 shows the product of $|\Delta\beta| L$ as a function of the loss in the surrounding medium for the slab waveguide case. The loss in the surrounding medium is expressed as the loss that a plane wave would suffer as it travels perpendicularly to the waveguide axis from the cladding boundary of the first guide to the cladding boundary of the second guide. This loss is plotted in db. The solid curves of Figure 10.5.1 hold for the TM mode, while the dotted curves apply to the TE mode. TM mode and TE mode correspond

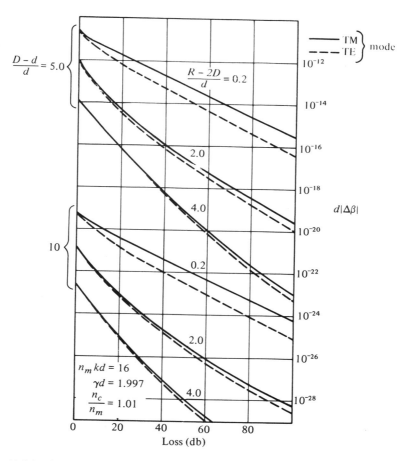

Figure 10.5.1 Crosstalk parameter $|\Delta\beta|d$ of two coupled cladded slab waveguides as a function of the plane wave loss between the cladding boundaries of the two guides. The parameter $(D-d)/d$ describes the normalized cladding thickness; the parameter $(R-2D)/d$ describes the normalized cladding separation between the two guides. The dotted lines apply to TE modes (corresponding to vertical polarization of the fiber case); the solid lines apply to TM modes (horizontal polarization). $n_c/n_m = 1.01$, $n_m kd = 16$, $\gamma d = 1.997$. (From D. Marcuse, The Coupling of Degenerate Modes in Two Parallel Dielectric Waveguides, B.S.T.J., Vol. 6, No. 6, July–August 1971. Copyright 1971, The American Telephone and Telegraph Co., reprinted by permission.)

in their polarizations to the case of horizontal and vertical polarization in the round fiber case. The curves are further identified with two parameters. We used two values of the relative cladding thickness $(D - d)/d = 5$ and 10. It is apparent how effectively an increase in cladding thickness reduces the coupling between the two guides. The second parameter is the relative spacing $(R - 2D)/d$ between the two guides. The values 0.2, 2.0, and 4.0 are used. The decrease in coupling with increasing values of this parameter is again attributable to the increase in distance between the waveguide cores. The loss plotted on the abscissa takes the varying distance into account. The loss is always the actual loss encountered in going from one guide to the other adjusted for the proper waveguide separation. Figure 10.5.1 shows the reduction in coupling with increasing loss of the surrounding medium. The real part of the refractive index of the surrounding medium, n_3, was taken to be equal to the index n_m of the cladding.

Figure 10.5.2 shows similar data for the cladded round fiber. It is apparent from both figures that horizontally polarized modes couple more strongly than vertically polarized modes. It is also apparent that the slab waveguides are coupled less tightly than the round fibers. This difference in coupling strength is caused by the fact that $\gamma_{slab} > \gamma_{fiber}$. Larger γ values cause the field to be concentrated more tightly near the core, reducing the field that reaches over to the other guide.

Figure 10.5.3 is a plot of the mode losses as a function of the power loss $2ak_0 Im(n_3)$ in the surrounding medium normalized with respect to the waveguide core. The core radius a must be replaced by the slab half thickness d for the slab waveguide case. Similarly, the cladding thickness $(b - a)/a$ of the fiber stands also for the cladding thickness $(D - d)/d$. The slab waveguide losses are lower because $\gamma_{slab} > \gamma_{fiber}$. The most remarkable feature of Figure 10.5.3 is the fact that the mode losses depend only very slightly on the loss in the surrounding medium after an initial rapid increase in loss from zero values. As long as the surrounding medium has a few db loss per core radius, the mode loss is determined primarily by the cladding thickness. This behavior makes it easy to discuss the problems of mode losses and cross talk simultaneously. Let us consider an example. We assume a waveguide with $a = 1\mu$ ($d = 1\mu$) and postulate that we can tolerate a mode loss caused by the lossy surrounding medium of 1 db/km. This means that we must have $2a\alpha < 10^{-9}$ db. Of the curves shown in Figure 10.5.3, only the uppermost curve for the HE_{11} mode exceeds the tolerable mode loss. Next, let us assume that we want to tolerate a cross talk of $C = 10^{-6}$ for a guide length of 1 km. According to (10.2-38), this means that we must require $|\Delta\beta| a < 10^{-12}$. Except for the topmost curve, all curves in Figure 10.5.1 are thus acceptable. Of the curves in Figure 10.5.2, none of the upper set of curves for $(b - a)/a = 5$ is acceptable unless we provide high loss in the surrounding medium. This comparison of loss with cross talk suggests the following conclusion. If we want to uncouple two parallel dielectric waveguides by adding loss to the surrounding medium, we simultaneously increase the mode loss to unacceptably high levels. Or, in other words, if the cladding thickness is insufficient to provide acceptable mode isolation, it is also insufficient to protect the guided modes from the losses of the surrounding medium. This conclusion is based on more evidence than the two sample cases presented here.[103] We thus reach the conclusion that losses in the surrounding medium are unsuitable to decrease the cross

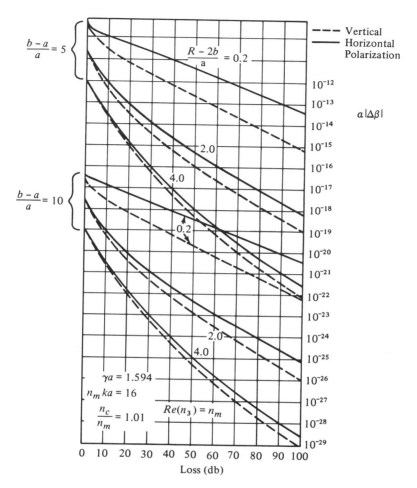

Figure 10.5.2 Coupling parameter $|\Delta\beta|/a$ of two cladded optical fibers as a function of the plane wave loss between the cladding boundaries of the two guides. The parameter $(b-a)/a$ describes the normalized cladding thickness, $(R-2b)/a$ describes the normalized separation between the claddings of the two guides. The dotted lines apply to vertical polarization; the solid lines apply to horizontal polarization. $n_c/n_m = 1.01$, $n_m ka = 16$, $\gamma a = 1.594$. (From D. Marcuse, The Coupling of Degenerate Modes in Two Parallel Dielectric Waveguides, *B.S.T.J.*, Vol. **6**, No. 6, July–August 1971. Copyright 1971, The American Telephone and Telegraph Co., reprinted by permission.)

talk between two dielectric waveguides. Either the cladding thickness is already sufficient for decoupling the two guides to make the loss in the surrounding medium unnecessary or the mode losses become excessive by adding loss to the surrounding medium.

We do not have the space to discuss scattering cross talk. The interested

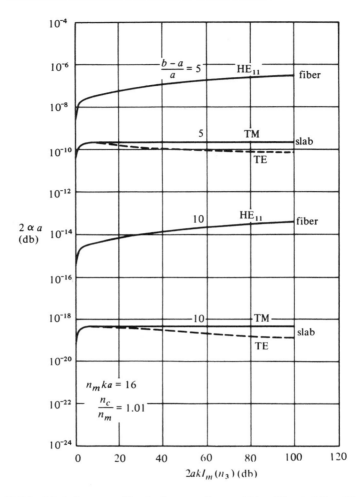

Figure 10.5.3   Mode loss caused by the lossy medium outside of the cladding. The abscissa represents the plane wave power loss in the outside medium over a distance equal to the core radius (or the slab core half width). The parameter $(b-a)/a$ represents the normalized width of the cladding. The solid lines apply to TM modes of the slab waveguide, while the dotted lines apply to TE modes. $n_c/n_m = 1.01$, $n_m ka = 16$ (also, $n_m kd = 16$). (From D. Marcuse, The Coupling of Degenerate Modes in Two Parallel Dielectric Waveguides, *B.S.T.J.*, Vol. **6**, No. 6, July–August 1971. Copyright 1971, The American Telephone and Telegraph Co., reprinted by permission.)

reader is referred to reference 104. However, the conclusion as to the desirability of losses in the surrounding medium holds also true for scattering cross talk. If scattering by waveguide imperfections is the source of cross talk, there is also a close link between the amount of cross talk provided by the scattering and the mode loss that is caused by the same scattering mechanism.

One can conclude again that either the cross talk is already acceptable or, if it is high, the scattering losses suffered by the guided modes are also so high as to be unacceptable. In principle, it is of course possible to suppress scattering cross talk by adding loss to the surrounding medium. This loss is not detrimental to the mode propagation if the cladding is thick enough. However, if the losses in the surrounding medium are needed to assure waveguide isolation, the waveguide is already so bad that the added scattering losses suffered by the guided modes are too high. The only exception to this case can arise if the scattering losses are not caused by random waveguide imperfection but by a systematic sinusoidal deformation that persists throughout the entire length of both guides. This type of systematic imperfection need not increase the mode losses excessively, but it can cause appreciable cross talk. In this case, a lossy surrounding medium would be desirable to uncouple the waveguides.

The data presented in this chapter can of course also be used to calculate the properties of directional couplers. The coupling equations (10.2-17) and (10.2-18) are sufficiently general to cover any type of dielectric waveguide.

CHAPTER

11

THEORY OF MULTIMODE FIBERS

11.1 INTRODUCTION

Optical fibers for communications purposes are grouped into two classes: multimode fibers and single mode fibers.[126] As the name suggests, multimode fibers have relatively large cores and large V-numbers (see 8.7-15) and can support many modes. Single mode fibers, on the other hand, have small cores and V-numbers around or below $V = 2.4$ so that they support only the HE_{11} mode that can exist in two mutually orthogonal polarizations. To understand why single mode operation requires $V < 2.4$, remember that cutoff is defined as $\gamma = 0$ (see (8.2-55)) so that according to (8.6-4) at cutoff, $V_c = \kappa_c a$. The cutoff values $V = V_c$ for a number of modes are listed in Table 8.2.1. The HE_{11} mode has no cutoff, $V_c = 0$, but the next smallest cutoff value is $V_c = 2.405$.

When we discuss the dispersive properties of optical fibers in Chapter 12 we shall see that multimode and single mode fibers behave very differently and hence serve different purposes. A pulse propagating in single mode fibers broadens only very slightly due to the dispersive properties of the fiber material and the inherent dispersion of the waveguiding process. In multimode fibers, the power launched initially in a short pulse is distributed over all (or at least very many) modes, each of which travels with a slightly different group velocity. The receiver at the fiber output thus receives many different, overlapping pulses each carried by a different mode. Thus, modal dispersion in multimode fibers causes a pulse to broaden much more than in single mode fibers. Consequently, multimode fibers have a much more limited information carrying capacity than single mode fibers. Much effort has been devoted to the design of refractive index profiles that minimize modal dispersion and improvements of several orders of magnitude over the step index profile (characterized by constant core index) have been achieved. However, complete compensation of modal dispersion is not possible so that even the best multimode fibers have less information carrying capacity than single mode fibers. The advantage of single mode fibers is even more dramatic since it is possible to find an operating wavelength at which their dispersion vanishes to first order.

Discussing channel capacity, it may appear as though multimode fibers cannot compete with single mode fibers for communications purposes. However, this is not the case, there are other considerations that make multimode fibers attractive. For example, the larger core diameter facilitates splicing of multimode fibers. Another advantage becomes apparent when we consider the type of light source required for each fiber type. Single mode fibers must be excited with a laser, since incoherent light cannot be used to excite a single mode. Multimode fibers can also be excited with lasers but they can also be operated with (incoherent) light emitting diodes that are cheaper and have longer lifetimes than lasers. Because they are easier to handle, to splice and can be used with cheaper light sources, multimode fibers are being used in applications where high channel capacity is not important. If only a few megahertz of signal bandwidth are to be transmitted, it is not necessary to employ a transmission medium capable of carrying thousands of megahertz.

In this chapter, we discuss light propagation in multimode fibers in terms of ray and wave optics. We derive the ray equation in cylindrical coordinates and use it to compute ray trajectories in multimode fibers with different index profiles. The wave optics treatment used in this chapter is really geometrical optics in disguise since we are dealing with the WKB method that is inherently no more accurate than ray optics, but it gives valuable insights into the waveguiding mechanism helping, for example, to understand leaky waves. The WKB method is also useful for treating modal dispersion.

11.2 THE RAY EQUATION IN CYLINDRICAL COORDINATES

Multimode fibers satisfy the conditions required for the applicability of geometrical optics: the transverse dimension of the fiber core and the length scale governing refractive index fluctuations are both large compared to the light wavelength. Geometrical optics is described by the ray equation. In Section 3.2 we expressed the ray equation either in vector form, (3.2-16),

$$\frac{d}{ds}\left(n\frac{d\mathbf{r}}{ds}\right) = \nabla n \tag{11.2-1}$$

or in Cartesian coordinates, (3.4-11), (3.4-12), and (3.4-16). Because of the geometry of optical fibers it is convenient to express the ray equation in cylindrical coordinates. The desired transformation can be performed in several ways. It is possible to use the transformation of Cartesian to cylindrical coordinates to achieve the corresponding conversion of the ray equation. A simpler, more direct approach uses vector analysis.

In preparation for the transformation of the wave equation we derive a few simple vector relations. First, we need to express the gradient operator in cylindrical coordinates by using the definition of the gradient in terms of the total differential of a function $F(x, y, z) = F(r, \phi, z)$. Applying the rules of differential calculus we have

$$dF = (\nabla F) \cdot ds = \frac{\partial F}{\partial x} dx + \frac{\partial F}{\partial y} dy + \frac{\partial F}{\partial z} dz$$
$$= \frac{\partial F}{\partial r} dr + \frac{\partial F}{\partial \phi} d\phi + \frac{\partial F}{\partial z} dz \tag{11.2-2}$$

The change dF of F is defined as the difference $F(x_2, y_2, z_2) - F(x_1, y_1, z_1)$. The vector ds points from x_1, y_1, z_1 to its infinitesimal neighbor x_2, y_2, z_2. We are free to choose the direction of ds at will. Letting $ds = e_r \, dr$ (with e_r indicating the unit vector in r-direction) we obtain from (11.2-2)

$$(\nabla F)_r = \frac{\partial F}{\partial r} \tag{11.2-3}$$

Next, we let $ds = e_\phi r d\phi$ and obtain from (11.2-2)

$$(\nabla F)_\phi = \frac{1}{r} \frac{\partial F}{\partial \phi} \tag{11.2-4}$$

Finally, we set $ds = e_z \, dz$ and obtain

$$(\nabla F)_z = \frac{\partial F}{\partial z} \tag{11.2-5}$$

Equations (11.2-3) through (11.2-5) define the r, ϕ, and z components of the gradient operator.

We also need expressions for the derivatives of the unit vectors e_r, e_ϕ, e_z with respect to the direction s along the ray path. The unit vector e_z always points in z-direction so that its derivative vanishes,

$$\frac{de_z}{ds} = 0 \tag{11.2-6}$$

The directions of the vectors e_r and e_ϕ change as we move along the ray. The condition for a unit vector is expressed as

$$e_r \cdot e_r = 1 \tag{11.2-7}$$

Taking its derivative results in

$$\mathbf{e}_r \cdot \frac{d\mathbf{e}_r}{ds} = 0 \qquad (11.2\text{-}8)$$

which shows that the derivative of a unit vector is orthogonal to the vector itself. The derivative of $\mathbf{e}_r$ cannot have a component in the z-direction because $\mathbf{e}_r$ (and also $\mathbf{e}_\phi$) always points parallel to the x, y plane as the endpoint of the vector $\mathbf{r}$ moves along the light ray. Consequently, $d\mathbf{e}_r/ds$ must point in the direction of $\mathbf{e}_\phi$. From simple geometry it is easy to see that its magnitude is $d\phi/ds$ so that we have

$$\frac{d\mathbf{e}_r}{ds} = \frac{d\phi}{ds} \mathbf{e}_\phi \qquad (11.2\text{-}9)$$

A similar agrument convinces us that

$$\frac{d\mathbf{e}_\phi}{ds} = -\frac{d\phi}{ds} \mathbf{e}_r \qquad (11.2\text{-}10)$$

We are now ready to express the ray equation in cylindrical coordinates. The vector $\mathbf{r}$ appearing in (11.2-1) can be expressed in r, ϕ, z, coordinates as

$$\mathbf{r} = r\mathbf{e}_r + z\mathbf{e}_z \qquad (11.2\text{-}11)$$

Its s-derivative is

$$\frac{d\mathbf{r}}{ds} = \frac{dr}{ds} \mathbf{e}_r + r\frac{d\phi}{ds} \mathbf{e}_\phi + \frac{dz}{ds} \mathbf{e}_z \qquad (11.2\text{-}12)$$

Multiplication with the refractive index distribution $n = n(r, \phi, z)$ and differentiation yields

$$\begin{aligned}
\frac{d}{ds}\left(n\frac{d\mathbf{r}}{ds}\right) = &\left[\frac{d}{ds}\left(n\frac{dr}{ds}\right) - rn\left(\frac{d\phi}{ds}\right)^2\right]\mathbf{e}_r \\
&+ \left[n\frac{dr}{ds}\frac{d\phi}{ds} + \frac{d}{ds}\left(rn\frac{d\phi}{ds}\right)\right]\mathbf{e}_\phi \\
&+ \frac{d}{ds}\left(n\frac{dz}{ds}\right)\mathbf{e}_z \qquad (11.2\text{-}13)
\end{aligned}$$

Using (11.2-3) through (11.2-5), the r, ϕ, and z components of the ray equation (11.2-1) assume the following form

$$\frac{d}{ds}\left(n\frac{dr}{ds}\right) - rn\left(\frac{d\phi}{ds}\right)^2 = \frac{\partial n}{\partial r} \qquad (11.2\text{-}14)$$

$$\frac{d}{ds}\left(r^2 n\frac{d\phi}{ds}\right) = \frac{\partial n}{\partial \phi} \qquad (11.2\text{-}15)$$

$$\frac{d}{ds}\left(n\frac{dz}{ds}\right) = \frac{\partial n}{\partial z} \qquad (11.2\text{-}16)$$

The left side of the ϕ component, (11.2-15), was rewritten as a total derivative with respect to s.

Written in cylindrical coordinates, the ray equations become nonlinear because of the appearance of squares and products of the functions $r(s)$, $\phi(s)$ and their derivatives. However, in spite of this complication the new form of the ray equations is more useful for fiber problems than their representation in Cartesian coordinates.

Most optical fibers have rotational symmetry and are (at least nominally) uniform along z. This observation allows us to write the refractive index distribution simply as

$$n = n(r) \qquad (11.2\text{-}17)$$

Consequently, the right hand sides of (11.2-15) and (11.2-16) vanish, making it possible to formulate conservation laws.[127] In Section 3.2 we introduced the unit vector (3.2-8)

$$\mathbf{u} = \frac{d\mathbf{r}}{ds} = \frac{dr}{ds}\,\mathbf{e}_r + r\,\frac{d\phi}{ds}\,\mathbf{e}_\phi + \frac{dz}{ds}\,\mathbf{e}_z \qquad (11.2\text{-}18)$$

The right hand side of this equation follows from (11.2-12). Multiplication with nk_0 yields the propagation vector of wave optics

$$\boldsymbol{\beta} = nk_0\mathbf{u} = nk_0\left[\frac{dr}{ds}\,\mathbf{e}_r + r\,\frac{d\phi}{ds}\,\mathbf{e}_\phi + \frac{dz}{ds}\,\mathbf{e}_z\right] \qquad (11.2\text{-}19)$$

To elucidate this important relationship let us digress for a discussion of the connection between rays and modes. A fiber mode is a traveling wave along the fiber axis. In r direction, the field forms a standing wave that may be expressed in terms of Bessel functions and their derivatives [see (8.2-25) through (8.2-30)]. These standing waves can be decomposed into traveling waves by expressing each Bessel function as a superposition of Hankel functions of the first and second kind. While Bessel and Neumann functions represent standing waves, Hankel functions represent waves traveling in positive and negative r direction. If necessary, a similar decomposition of standing waves into traveling waves can be done for the ϕ dependent part of the field. However, in Section 8.2, we expressed the propagation of the mode in ϕ direction already as traveling waves. A description in terms of standing waves (sine and cosine functions) was given in (8.6-59) through (8.6-64).

After the mode field is decomposed into overlapping traveling waves, we can define families of rays as orthogonal trajectories to the curved wave fronts. The unit vectors pointing in the direction of each ray are, by definition, parallel to the propagation vectors of the traveling waves. This relationship is expressed in (11.2-19). The proportionality constant is the magnitude $n(r)k_0$ of the propagation vector of a locally plane wave. The description of rays traveling in the fiber is thus intimately related to the fiber modes of wave optics.

According to (11.2-19), the propagation vector has the following components,

$$\beta_r = nk_0 \frac{dr}{ds} \qquad (11.2\text{-}20)$$

$$\beta_\phi = nk_0 r \frac{d\phi}{ds} \qquad (11.2\text{-}21)$$

$$\beta_z = nk_0 \frac{dz}{ds} \qquad (11.2\text{-}22)$$

The sum of the squares of these components is, of course,

$$\beta_r^2 + \beta_\phi^2 + \beta_z^2 = n^2 k_0^2 \qquad (11.2\text{-}23)$$

The refractive index distribution of optical fibers is usually independent of z [see (11.2-17)] so that the right hand side of (11.2-16) vanishes. Consequently, $n(r)dz/ds$ must be constant so that (11.2-22) becomes

$$\beta_z = nk_0 \frac{dz}{ds} = \beta \qquad (11.2\text{-}24)$$

This conservation law states that locally plane waves travel in the fiber in such a way that the z-components of their propagation vectors remain unchanged even though the refractive index distribution varies as a function of r. This conservation law permits the existence of modes; β_z is identical with the propagation constant β of guided modes [see (8.2-22)].

Next we show that the ϕ component of the wave vector also satisfies a conservation law and that it is related to the azimuthal mode number appearing in the mode field expressions (8.2-25) through (8.2-30). Because the refractive index distribution is independent of ϕ, the right hand side of (11.2-15) is zero so that $n(r)r^2 d\phi/ds$ must be constant. In terms of the component (11.2-21) of the propagation vector this conservation law assumes the form

$$r\beta_\phi = nk_0 r^2 \frac{d\phi}{ds} = \nu \qquad (11.2\text{-}25)$$

To relate the constant ν to the azimuthal mode number appearing in (8.2-25) through (8.2-30), we consider the expression for the locally plane wave that is associated with the ray in question,

$$\exp(i\omega t - i\boldsymbol{\beta} \cdot \mathbf{r}_c) \qquad (11.2\text{-}26)$$

The coordinate vector $\mathbf{r}_c = (x - x_0, y - y_0, z - z_0)$ has its origin at the point (x_0, y_0, z_0) that is located on the ray. The expression (11.2-26) is meaningful only in the immediate vicinity of (x_0, y_0, z_0). However, since we want to use $\boldsymbol{\beta}$ in its cylindrical coordinate representation (11.2-20) through 11.2-22), we must also use cylindrical coordinates for $\mathbf{r}_c$,

$$\mathbf{r}_c = \{r - r_0, r_0\,(\phi - \phi_0), z - z_0\} \tag{11.2-27}$$

The scalar product in (11.2-26) can now be written in the following form

$$\boldsymbol{\beta} \cdot \mathbf{r}_c = \beta_r r + \beta_\phi r_0 \phi + \beta_z z + \psi \tag{11.2-28}$$

The term ψ is independent of the coordinates r, ϕ, z and contributes an unimportant, constant phase to (11.2-26). The ϕ dependent term in (11.2-28) makes the following contributions to (11.2-26)

$$e^{-i\beta_\phi r_0 \phi} = e^{-i\nu\phi} \tag{11.2-29}$$

On the right hand side of this expression we have substituted (11.2-25), using $r = r_0$, because the r coordinate in (11.2-25) is meant to indicate the point on the ray that we are now (temporarily) designating as r_0. The importance of this exercise becomes apparent when we compare (11.2-29) with (8.2-21) and realize that the constant ν introduced in (11.2-25) is actually the azimuthal mode number.

The conservation law (11.2-25) corresponds to the law of conservation of angular momentum in classical mechanics, ν assumes the role of angular momentum. Conservation of angular momentum, or the fact that an azimuthal mode number can be defined, is attributable to the rotational symmetry of the refractive index distribution relative to the fiber axis.

Substitution of (11.2-25) in (11.2-14) simplifies the r component of the ray equation

$$\frac{d}{ds}\left(n\frac{dr}{ds}\right) - \frac{\nu^2}{nr^3 k_0^2} = \frac{\partial n}{\partial r} \tag{11.2-30}$$

$r(s)$ is now the only dependent variable in this equation. It is often more convenient to use z, the length measured along the fiber axis, instead of s, the distance measured along the ray, as the independent variable. The transformation of the s-derivative is accomplished with the help of (11.2-24).

$$\frac{d}{ds} = \frac{dz}{ds}\frac{d}{dz} = \frac{\beta}{nk_0}\frac{d}{dz} \tag{11.2-31}$$

The r component of the ray equation now reads

$$\frac{d^2 r}{dz^2} - \frac{\nu^2}{r^3\beta^2} = \frac{k_0^2}{2\beta^2}\frac{dn^2}{dr} \tag{11.2-32}$$

The partial derivative symbol $\partial n/\partial r$ was replaced by the derivative dn/dr since r is the only variable on which n depends.

A further simplification of the ray equation can be achieved by multiplying (11.2-32) with dr/dz

$$\frac{dr}{dz}\frac{d^2r}{dz^2} - \frac{v^2}{r^3\beta^2}\frac{dr}{dz} - \frac{k_0^2}{2\beta^2}\frac{dr}{dz}\frac{dn^2}{dr} = 0 \qquad (11.2\text{-}33)$$

because the equation can now be written as a derivative with respect to z

$$\frac{1}{2}\frac{d}{dz}\left\{\beta^2\left(\frac{dr}{dz}\right)^2 + \frac{v^2}{r^2} - n^2k_0^2\right\} = 0 \qquad (11.2\text{-}34)$$

and can be integrated to yield

$$\beta^2\left(\frac{dr}{dz}\right)^2 + \frac{v^2}{r^2} - n^2k_0^2 = -E^2 \qquad (11.2\text{-}35)$$

with $-E^2$ introduced as an integration constant. The value of E can be determined by using the fact that on each ray trajectory there are "turning points" $r = r_1$ defined by the condition $dr/ds = 0$ and, consequently, also $\beta_r = 0$.

At a turning point we obtain from (11.2-35)

$$E^2 = n^2(r_1)k_0^2 - \frac{v^2}{r_0^2} = n^2(r_1)k_0^2 - \beta_\phi^2 = \beta_z^2 = \beta^2 \qquad (11.2\text{-}36)$$

The right hand side of this equation was obtained by using (11.2-25) and (11.2-23) with $\beta_r = 0$. The constant E is thus identified as the z-component of the propagation vector or, equivalently, as the model propagation constant β.

Equation (11.2-35) can immediately be solved for the derivative of r

$$\frac{dr}{dz} = \frac{1}{\beta}\left[n^2(r)k_0^2 - \beta^2 - \frac{v^2}{r^2}\right]^{1/2} \qquad (11.2\text{-}37)$$

and can then be integrated to yield the formal solution

$$z - z_0 = \beta\int_{r_0}^{r}\frac{dr}{\left[n^2(r)k_0^2 - \beta^2 - \frac{v^2}{r^2}\right]^{1/2}} \qquad (11.2\text{-}38)$$

z_0 is a second integration constant representing z at $r = r_0$.

Instead of the desired solution $r = r(z)$ we have found a formal, exact solution for the inverse function $z = z(r)$.

In terms of ray optics, β and v are just integration constants, their identification as modal parameters exceeds the realm of pure ray optics and was borrowed from our knowledge of wave optics.

For a given refractive index distribution $n(r)$, (11.2-38) can be integrated to give the trajectory of a ray having $r = r_0$ at $z = z_0$ and turning points given by (11.2-36). The trajectory of a ray that is trapped in the fiber core oscillates between an upper and a lower turning point while the ray progresses in the direction of the positive (or negative) z axis. However, there is no lower turning point for $v = 0$, in this case the ray crosses the z axis and proceeds to the turning point on the opposite side.

The calculation of the ray trajectory proceeds in stages. First, we follow the ray from its starting point r_0, z_0 by letting the r variable (upper integration limit) increase and compute corresponding values for z from (11.2-38). At the turning point, the integrand becomes infinite but the integral itself remains finite. The coordinate, $r = r_1$, of the turning point is either the largest or smallest value of this segment of the ray trajectory, because at this point the integrand changes from real to imaginary values. On reaching the turning point, the progression of r must be reversed. If the turning point was reached through growing values of r, its values must now start to decrease; if they decreased before, they must now start to increase. The sign of the square root in (11.2-38) must be chosen so that $z - z_0$ is growing all along the ray trajectory. When the integration proceeds from smaller to larger values of r the square root is chosen to be positive, it is chosen negative when the upper limit of the integral decreases. The opposite choice of signs must, of course, be made if the ray is to progress in negative z direction.

So far we have assumed that the ray is trapped in the core. However, the ray may be able to leave the core if it starts with a sufficiently steep slope. In this case there are no turning points and we can follow the ray trajectory in a continuous procedure by letting r increase (or decrease) indefinitely. Whether or not a ray is trapped depends on the value of β. For a given value of β we compute the turning points from (11.2-36). If the larger turning point is smaller than the fiber core radius, $r_1 < a$, the ray is trapped. The ray may still be trapped if $r_1 = a$ and there is a refractive index discontinuity at $r = a$. If the refractive index is continuous at the core boundary, the ray leaves the core if $r_1 > a$ This criterion is not sufficient for deciding whether the corresponding mode is truly guided. Even when $r_1 < a$, the mode may not be guided. This case of "tunneling leaky modes" will be discussed in Section 11.7.

We have described how the $r - z$ coordinate pairs of the ray are being determined. To find the ϕ coordinate we change the s derivative in (11.2-25) to a z derivative with the help of (11.2-31)

$$r^2\beta \frac{d\phi}{dz} = \nu \qquad (11.2\text{-}39)$$

Integration of this expression yields

$$\phi - \phi_0 = \frac{\nu}{\beta} \int_{z_0}^{z} \frac{dz}{r^2} \qquad (11.2\text{-}40)$$

This integration along z requires knowledge of $r = r(z)$ which, of course, is in hand once the $r - z$ coordinate pairs for the ray have been determined. An expression more like (11.2-38) is obtained by changing integration variables with the help of (11.2-37), with the result

$$\phi - \phi_0 = \nu \int_{r_0}^{r} \frac{dr}{r^2 \left[n^2(r)k_0^2 - \beta^2 - \frac{\nu^2}{r^2} \right]^{1/2}} \tag{11.2-41}$$

Analytical solutions of (11.2-38) and (11.2-41) can only be obtained for special refractive index distributions. In general, it is necessary to employ numerical methods.

We mentioned in Section 11.1 that the channel capacity of multimode fibers is critically dependent on modal dispersion. Reduction of modal dispersion requires equalization of the group delays of all the modes. However, before equalization can be attempted we must develop the means for computing the group delay along each ray. For this reason we derive a formula that is, of course, also applicable to the group delay of the corresponding mode.

The transit time for light traveling along a ray of length s is given by (3.4-1) and (3.4-2)

$$\tau = \frac{1}{c} \int_{0}^{s} n \, ds \tag{11.2-42}$$

This expression for the group delay ignores material dispersion. In Section 12.5, the theory will be improved by including the dispersive properties of the fiber material in the expression for the group delay. Using (11.2-31) and (11.2-37) we can write (11.2-42) in the following form

$$\tau = \frac{k_0}{c\beta} \int_{z_0}^{z} n^2 \, dz = \frac{k_0}{c} \int_{r_0}^{r} \frac{n^2(r) \, dr}{\left[n^2(r)k_0^2 - \beta^2 - \frac{\nu^2}{r^2} \right]^{1/2}} \tag{11.2-43}$$

This formula is not yet sufficient for our purposes since it does not permit comparison of the transit times along different rays. A comparison is only possible if we have an expression for the transit time relative to a fixed length of fiber. For this reason we convert (11.2-43) to the transit time per unit fiber length by dividing (11.2-43) by (11.2-38), which represents the length of the ray measured not along its trajectory but along the fiber axis.

$$\hat{\tau} = \frac{k_0}{c\beta} \frac{\displaystyle\int_{r_0}^{r_1} \frac{n^2(r) \, dr}{\left[n^2(r)k_0^2 - \beta^2 - \frac{\nu^2}{r^2} \right]^{1/2}}}{\displaystyle\int_{r_0}^{r_1} \frac{dr}{\left[n^2(r)k_0^2 - \beta^2 - \frac{\nu^2}{r^2} \right]^{1/2}}} \tag{11.2-44}$$

Equation (11.2-44) represents the transit time (or group delay) per unit fiber length. The integrals in this expression are extended between the inner and outer turning points, r_0 and r_1. This is an important point! The group delay per unit fiber length is not independent of the ray section that is used for its computation. However, the distance between successive turning points of a given ray is invariant, that is it is the same no matter which section of the ray trajectory was used. For rays (or modes) with $\nu = 0$ there is no lower turning point, in this case we use $r_0 = 0$.

Equation (11.2-44) is very important for the design of optimum refractive index profiles for multimode fibers of maximum channel capacity. Its only shortcoming is the absence of material dispersion. An improved expression that includes material dispersion will be derived in Section 12.5.

11.3 RAY TRAJECTORIES IN THE SQUARE LAW FIBER

Exact solutions can be computed from (11.2-38) and (11.2-40) for all rays traveling in a fiber whose core has a parabolic (square law) refractive index profile. We are limiting the discussion to trapped rays that do not enter the cladding region of the fiber, thus it is sufficient to consider only the region $r < a$ of (8.7-11). However, we cannot solve the ray equations for arbitrary g values but only for $g = 2$,

$$n^2(r) = n_1^2 \left[1 - 2 \left(\frac{r}{a} \right)^2 \Delta \right] \tag{11.3-1}$$

Substitution of this refractive index distribution allows us to write (11.2-38) in the following form

$$z = \beta \int_{r_0}^{r} \frac{r \, dr}{[\kappa^2 r^2 - A r^4 - \nu^2]^{1/2}} \tag{11.3-2}$$

with the abbreviations

$$\kappa^2 = n_1^2 k_0^2 - \beta^2 \tag{11.3-3}$$

and

$$A = \left(\frac{n_1 k_0}{a} \right)^2 (2\Delta) \tag{11.3-4}$$

We shifted the coordinate origin so that $z_0 = 0$. In addition, we let the starting point of the ray trajectory coincide with the lower turning point r_0. From (11.2-36) and (11.3-1) we find the following expressions for the lower turning point

$$r_0 = \left\{ \frac{1}{2A} \left[\kappa^2 - (\kappa^4 - 4A\nu^2)^{1/2} \right] \right\}^{1/2} \tag{11.3-5}$$

and for the upper turning point

$$r_1 = \left\{ \frac{1}{2A} \left[\kappa^2 + (\kappa^4 - 4A\nu^2)^{1/2} \right] \right\}^{1/2} \tag{11.3-6}$$

From (11.3-5) we have $r_0 = 0$ for $\nu = 0$. The integral in (11.3-2) has the following solution

$$z = \frac{\beta}{2\sqrt{A}} \left\{ \arcsin \left[\frac{2Ar^2 - \kappa^2}{(\kappa^4 - 4A\nu^2)^{1/2}} \right] \right.$$

$$\left. -\arcsin \left[\frac{2Ar_0^2 - \kappa^2}{(\kappa^4 - 4A\nu^2)^{1/2}} \right] \right\} \tag{11.3-7}$$

Using (11.3-5) and (11.3-6) we find that the agrument of the second arcsine function is -1 and that

$$(\kappa^4 - 4A\nu^2)^{1/2} = A(r_1^2 - r_0^2) \tag{11.3-8}$$

Another abbreviation whose usefulness will soon become apparent is

$$\Omega = \frac{\sqrt{A}}{\beta} = \frac{n_1 k_0}{a\beta} \sqrt{2\Delta} \tag{11.3-9}$$

Equation (11.3-7) now becomes

$$2\Omega z = \frac{\pi}{2} + \arcsin \left(\frac{2Ar^2 - \kappa^2}{A(r_1^2 - r_0^2)} \right) \tag{11.3-10}$$

This equation can be solved for r,

$$r(z) = \frac{1}{\sqrt{2}} \left\{ (r_1^2 + r_0^2) - (r_1^2 - r_0^2) \cos(2\Omega z) \right\}^{1/2} \tag{11.3-11}$$

The relation

$$\kappa^2 = A(r_1^2 + r_0^2) \tag{11.3-12}$$

obtained from (11.3-5) and (11.3-6), was used to write (11.3-11) in this simple form. Regrouping of terms under the square root sign yields the desired solution $r = r(z)$ for rays in the square law fiber, with $r = r_0$ (lower turning point) at $z = 0$,

$$r(z) = \{ r_1^2 \sin^2(\Omega z) + r_0^2 \cos^2(\Omega z) \}^{1/2} \tag{11.3-13}$$

This form of the ray trajectory makes it immediately apparent that the ray oscillates between the upper and lower turning points r_0 and r_1. The ray trajectory assumes a particularly simple form for rays with azimuthal mode number $\nu = 0$, corresponding to $r_0 = 0$.

$$r(z) = r_1 \sin(\Omega z) \tag{11.3-14}$$

Because these rays cross the fiber axis and move in the meridional plane, they are called meridional rays. Equation (11.3-14) shows the utility of the abbreviation (11.3-9) that is now identified as the spatial angular frequency of meridional rays.

So far we have only described the radial part of the ray trajectory. The corresponding azimuthal function $\phi = \phi(z)$ follows from (11.2-40) with the help of (11.3-13)

$$\phi(z) = \arctan\left[\frac{r_1}{r_0} \tan (\Omega z)\right] \tag{11.3-15}$$

Once again we chose the coordinate origin to make $\phi_0 = 0$. With this choice we find from (11.3-15) that meridional rays ($r_0 = 0$) have constant values of $\phi(z) = \pi/2$. Equations (11.3-13) and (11.3-15) completely describe the ray trajectory in parametric form, $r = r(z)$, $\phi = \phi(z)$.

Another special class of rays can be defined with the help of (11.3-13) and (11.3-15). If the two turning points coincide, we have

$$r(z) = r_0 = r_1 \tag{11.3-16}$$

and

$$\phi(z) = \Omega z \tag{11.3-17}$$

In this case, the rays travel along helical paths with constant radius and the same angular spatial frequency Ω as meridional rays; they are, of course, called helical rays.

Equation (7.2-8) represents meridonal rays in the square law medium in the paraxial approximation. The spatial angular frequency of the paraxial ray is, according to (7.2-8)

$$\Omega_P = (1/a) \sqrt{2\Delta} \tag{11.3-18}$$

(we have converted the notation used in Section 7.2 to the notation used here). The paraxial angular spatial frequency Ω_P agrees with (11.3-9) if we approximate $\beta \approx n_1 k_0$.

With the help of the exact solution (11.3-13), the group delay of an arbitrary ray in the square law fiber can now easily be computed from the left hand side of (11.2-43). Substitution of (11.3-1) and (11.3-13) into (11.2-43) yields after integration

$$\tau = \frac{n_1^2 k_0^2 + \beta^2}{2c\, k_0\, \beta} z_1 \tag{11.3-19}$$

We have extended the integration interval from the lower turning point r_0 to the upper turning point r_1, or equivalently from $z = 0$ to $z = z_1$. The group delay per unit length (specific group delay) is obtained from (11.3-19) by division with z_1

$$\hat{\tau} = \frac{n_1^2 k_0^2 + \beta^2}{2c\, k_0\, \beta} \tag{11.3-20}$$

It is noteworthy that the specific group delay does not depend explicitly on the azimuthal mode number ν but only on the propagation constant β. All rays (or modes) with the same propagation constant β have the same specific group delay. We shall see in Section 11.6 that the modes of the square law fiber are degenerate, this means that many modes with different azimuthal and radial mode numbers can have the same propagation constant. According to (11.3-20) these modes also have the same specific group delay if material dispersion is ignored.

The propagation constant of the guided modes cannot be smaller than $n_2 k_0$, (n_2 = cladding index) and cannot exceed $n_1 k_0$. These extremes correspond to maximum and minimum specific group delays. The difference in arrival time of these slowest and fastest rays in a fiber of length L follows from (11.3-20)

$$\Delta\tau = L\Delta\hat{\tau} = L\,(\hat{\tau}_{max} - \hat{\tau}_{min}) \approx \frac{n_1 L}{2c}\,\Delta^2 \tag{11.3-21}$$

The approximation is valid for small values of Δ [defined by (8.7-12)]. Equation (11.3-21) describes modal dispersion, that is it gives the length of a pulse at $z = L$ which started at $z = 0$ with zero length. This expression shows that the square law fiber provides partial delay equalization, since the pulse width is only proportional to Δ^2. By comparison, the width of pulses traveling in the step-index fiber is directly proportional to Δ,

$$\Delta\tau = \frac{n_1 L}{c}\,\Delta \tag{11.3-22}$$

which, for $\Delta \ll 1$, is a much larger value. However, we shall see in section 11.6 that even less pulse spreading can be achieved in fibers with properly designed power law refractive index profiles.

11.4 RAY TRAJECTORIES IN HYPERBOLIC SECANT AND LORENTZIAN FIBERS

In Section 11.3, we found exact solutions for the trajectories and group delays of all rays trapped in the core of a square law fiber. In this section, we consider two additional refractive index distributions with interesting properties: the hyperbolic secant and the Lorentzian profiles. Neither of these profiles permits exact integration of the ray equation for all ray trajectories. But there are exact solutions for the meridional rays of the hyperbolic secant profile, all of which have identical group delays. However, non-meridional rays are not equalized. Fibers with Lorentzian refractive index profiles have perfect equalization for all helical rays but not for meridional or any other, more general rays. The hyperbolic secant and the Lorentzian refractive index profiles are closely related to the square law profile; to first order in Δ [see (8.7-12)] they are identical.

The hyperbolic secant function is defined as $1/\cosh x$, the corresponding refractive index profile inside the fiber core is described by the following equation

$$n(r) = \frac{n_1}{\cosh(\gamma r)} \qquad \text{for } r < a \qquad (11.4\text{-}1)$$

In the cladding the refractive index is constant, $n(r) = n_2$ for $r > a$, the value of n_2 determines the constant γ,

$$\cosh(\gamma a) = \frac{n_1}{n_2} \qquad (11.4\text{-}2)$$

As always, a is the core radius. The integral (11.2-38) cannot be solved for arbitrary valus of ν with $n(r)$ given by (11.4-1). However, for $\nu = 0$ an exact analytical solution is easily obtained. Substitution of (11.4-1) into (11.2-38) yields

$$z = \beta \int_0^r \frac{\cosh(\gamma r)\, dr}{[n_1^2 k_0^2 - \beta^2 \cosh^2(\gamma r)]^{1/2}} \qquad (11.4\text{-}3)$$

Because we are dealing with meridional rays ($\nu = 0$) that cross the axis, we may set the lower integration limit in (11.2-38) $r_0 = 0$. After the integration is carried out we have,

$$z = \frac{1}{\gamma} \arcsin\left[\frac{\beta}{\kappa}\sinh(\gamma r)\right] \qquad (11.4\text{-}4)$$

with

$$\kappa = (n_1^2 k_0^2 - \beta^2)^{1/2} \qquad (11.4\text{-}5)$$

By using the inverse of the arcsine and of the hyperbolic sine functions we obtain the desired expression for the trajectories $r = r(z)$ of meridional rays of the hyperbolic secant fiber

$$r(z) = \frac{1}{\gamma}\sinh^{-1}\left[\frac{\kappa}{\beta}\sin(\gamma z)\right] \qquad (11.4\text{-}6)$$

At $z = 0$ the ray crosses the axis. A full oscillation period is completed when $z = L$ satisfies

$$L = \frac{2\pi}{\gamma} \qquad (11.4\text{-}7)$$

The trajectories of meridional rays of the hyperbolic secant fiber are very similar to those of the square law medium but there is one important difference. According to (11.3-9), the lengths of the oscillation periods $2\pi/\Omega$ of the meridional rays of the square law medium depend on β and hence on the amplitudes of the ray trajectories. The length L of the trajectory of a meridional ray of the hyperbolic secant fiber is independent of its amplitude. (We designate the maximum displacement of the merdional ray from the fiber axis as its amplitude.) This fiber thus accomplishes perfect focusing for its meridional rays. Unfortunately, this is not true of all rays so that a short piece of fiber with hyperbolic secant index profile is not a perfect lens.

Next we consider the specific group delays of meridional rays of the hyperbolic secant fiber. From (11.2-43), (11.4-1) and (11.4-6) we have

$$\hat{\tau}_m = \frac{\tau_m}{z_1} = \frac{n_1^2 k_0}{c\beta z_1} \int_0^{z_1} \frac{dz}{1 + \dfrac{\kappa^2}{\beta^2} \sin^2(\gamma z)} \tag{11.4-8}$$

After the integral is evaluated this expression assumes the form

$$\hat{\tau}_m = \frac{n_1}{c\gamma z_1} \arctan\left[\frac{n_1 k_0}{\beta} \tan(\gamma z_1)\right] \tag{11.4-9}$$

We choose $z_1 = L/4$ (the same result is obtained for $z_1 = L/2$ or $z_1 = L$) and obtain

$$\hat{\tau}_m = \frac{n_1}{c} \tag{11.4-10}$$

This expression shows clearly that the specific group delays of all meridional rays of the hyperbolic secant fiber are identical, they are perfectly equalized.

It remains to show that the hyperbolic secant profile does not provide perfect equalization for all rays. In particular we consider helical rays that follow spiraling paths with constant radii $r(z) = r_h$. The specific group delays of helical rays follow directly from (11.2-43) and (11.4-1)

$$\hat{\tau}_h = \frac{k_0}{c\beta} \frac{n_1^2}{\cosh^2(\gamma r_h)} \tag{11.4-11}$$

This expression is of little use unless we can determine the radius r_h of each helical ray. All rays move between the two turning points r_1 and r_2 defined by (11.2-36). For helical rays $r_1 = r_2 = r_h$. The condition for the occurrence of a double root is that not only (11.2-36) is satisfied, but also that its r derivative vanishes,

$$n \frac{dn}{dr} k_0^2 + \frac{\nu^2}{r^3} = 0 \tag{11.4-12}$$

Thus, we have the two simultaneous equations

$$\frac{n_1^2 k_0^2}{\cosh^2(\gamma r_h)} - \beta^2 - \frac{\nu^2}{r_h^2} = 0 \qquad (11.4\text{-}13)$$

and

$$-\frac{n_1^2 k_0^2 \gamma r_h}{\cosh^3(\gamma r_h)} \sinh(\gamma r_h) + \frac{\nu^2}{r_h^2} = 0 \qquad (11.4\text{-}14)$$

which together, define r_h. By addition of (11.4-13) and (11.4-14) we can eliminate ν

$$\frac{n_1^2 k_0^2}{\cosh^2(\gamma r_h)} \left[1 - \frac{\gamma r_h \sinh(\gamma r_h)}{\cosh(\gamma r_h)} \right] - \beta^2 = 0 \qquad (11.4\text{-}15)$$

Equation (11.4-5) defines r_h in terms of β. Since an exact, explicit solution of this equation cannot be obtained we settle for an approximation. Assuming that $\gamma r_h \ll 1$ we use the series expansions of the hyperbolic functions up to terms of second order in γr_h and obtain from (11.4-15)

$$(\gamma r_h)^2 = \frac{n_1^2 k_0^2 - \beta^2}{n_1^2 k_0^2 + \beta^2} \qquad (11.4\text{-}16)$$

For small values of $n_1 - n_2$ (remember β is bounded by $n_1 k_0$ and $n_2 k_0$) γr_h is indeed small so that our initial assumption is justified. To the same approximation we finally obtain from (11.4-11) an approximate expression for the specific group delay

$$\hat{\tau}_h = \frac{n_1^2 k_0^2 + \beta^2}{2 c \beta k_0} \qquad (11.4\text{-}17)$$

This expression is identical to (11.3-20), the specific group delay for all the modes of square law medium, because to first order in Δ the hyperbolic secant profile is indistinguishable from the square law profile. However, even though (11.4-17) is not exact, its dependence on β is correctly described to first order of approximation, the exact expression cannot be independent of β. Thus, it is clear that the helical rays of the hyperbolic secant fiber do not have identical group delays independent of β. Therefore, modal dispersion in the hyperbolic secant fiber is of the same order of magnitude as in the square law fiber. Perfect equalization is achieved only for meridional rays, but these constitute only a minority of all possible rays.

Next, we turn our attention to fibers with Lorentzian refractive index profiles and show that, in this case, all helical rays are perfectly equalized. The Lorentzian refractive index profile is defined as follows:

$$n^2(r) = \frac{n_1^2}{1 + 2 \dfrac{n_1^2}{n_2^2} \left(\dfrac{r}{a} \right)^2 \Delta} \qquad \text{for } r < a \qquad (11.4\text{-}18)$$

with

$$\Delta = \frac{n_1^2 - n_2'^2}{2n_1^2} \tag{11.4-19}$$

Substitution of (11.4-18) into (11.2-38) leads to an elliptical integral. Because the general ray trajectories in Lorentzian fibers have no simple solutions we consider only helical rays that spiral around the fiber axis at constant radii r_h. As explained above, the radius r_h is computed from (11.2-36),

$$\frac{n_1^2 k_0^2}{1 + 2\dfrac{n_1^2}{n_2^2}\left(\dfrac{r_h}{a}\right)^2 \Delta} - \beta^2 - \frac{\nu^2}{r_h^2} = 0 \tag{11.4-20}$$

and its r derivative,

$$-\frac{n_1^2 k_0^2}{\left[1 + 2\dfrac{n_1^2}{n_2^2}\left(\dfrac{r_h}{a}\right)^2 \Delta\right]\left[1 + 2\dfrac{n_1^2}{n_2^2}\left(\dfrac{r_h}{a}\right)^2 \Delta\right]} 2\dfrac{n_1^2}{n_2^2}\left(\dfrac{r_h}{a}\right)^2 \Delta + \frac{\nu^2}{r_h^2} = 0 \tag{11.4-21}$$

which was multiplied by $r_h/2$. Addition of (11.4-20) and (11.4-21) removes ν and results in

$$\frac{n_1^2 k_0^2}{\left[1 + 2\dfrac{n_1^2}{n_2^2}\left(\dfrac{r_h}{a}\right)^2 \Delta\right]^2} - \beta^2 = 0 \tag{11.4-22}$$

The desired radius of the helical rays is obtained by solving (11.4-22),

$$r_h = a\frac{n_2}{n_1}\left[\frac{n_1 k_0 - \beta}{2\beta\Delta}\right]^{1/2} \tag{11.4-23}$$

As for all other refractive index profiles considered so far, r_h depends on β. We are most interested in calculating the specific group delays for the helical rays of the Lorentzian fiber. From (11.2-43) and (11.4-18), we obtain

$$\hat{\tau}_h = \frac{1}{z}\tau_h = \frac{n_1^2 k_0}{c\beta\left[1 + 2\dfrac{n_1^2}{n_2^2}\left(\dfrac{r_h}{a}\right)^2 \Delta\right]} \tag{11.4-24}$$

Replacing r_h by the expression (11.4-23), yields the desired result

$$\hat{\tau}_h = \frac{n_1}{c} \tag{11.4-25}$$

This exact result shows that the specific group delays of all helical rays are identical, independent of β. This means that the helical rays in the Lorentzian fiber are perfectly equalized. However, all other rays have slightly different specific group delays.

There is no refractive index profile that equalizes the group delays of all rays exactly. But perfect equalization of certain ray families can be achieved. We have seen that the hyperbolic secant fiber equalizes all meridional rays and that the Lorentzian fiber equalizes all helical rays. Modal dispersion is of the same order of magnitude in the square law, hyperbolic secant and Lorentzian fibers; the width of the output pulse is proportional to Δ^2 (for an input pulse of zero length). We show in Section 11.6 that there is an optimum value of the exponent g that minimizes modal dispersion in fibers with power law refractive index profiles.

11.5 THE WKB METHOD

Ray optics gives much valuable information about light propagation in multimode fibers. In the preceding sections of this chapter we have expressed the trajectories and group delays of light in terms of modal parameters—the propagation constant β and the azimuthal mode number ν. From the point of view of ray optics β and ν are just integration constants, their identification as modal parameters had to be made by borrowing modal concepts, using our knowledge of wave optics.

In this section, we develop an approximate description of the modes of multimode fibers from a wave optics point of view using a method that was originally developed by Wentzel, Kramers, and Brillouin for quantum mechanics. Because of the contribution of these early workers this analytical procedure for obtaining approximate solutions of differential equations is called the WKB method. Its results are surprisingly accurate and have helped in the design of multimode fibers with optimal refractive index profiles.[128−130] Even though the WKB method is based on wave optics, we shall see that the expressions for the group delay derived from this wave theory are identical with the corresponding expressions derived from ray optics. Therefore, the WKB method is no more accurate than ray theory. However, the WKB theory provides an approximate eigenvalue equation for the modal propagation constant β that remained undetermined by ray theory. Thus, it allows us to count the total number of guided modes and even provides insight into the interesting tunneling leaky modes.

In this section, we derive the WKB solution in considerable detail to allow the reader to follow the reasoning and to see clearly the mathematical approximations that need to be made. Readers not interested in the details of the derivation are advised to skip this section but to look at its principal result, the eigenvalue equation (11.5-45).

The starting point of the WKB theory is the scalar wave equation (8.7-1)

$$\nabla^2 E + n^2(r)k_0^2 E = 0 \qquad (11.5\text{-}1)$$

In the quest for modal solutions of the guided wave problem we postulate the trial solution

$$E = F(r) \cos(\nu\phi)e^{-i\beta z} \qquad (11.5-2)$$

introducing the modal propagation constant β and the azimuthal mode number ν from the start. Instead of the cosine function, $\sin(\nu\phi)$ could have been used. Solutions containing both functions are required for obtaining a complete set of guided modes. The azimuthal mode number ν must be an integer to ensure that E remains single valued as a function of ϕ. Substitution of (11.5-2) into (11.5-1) yields

$$\frac{d^2F}{dr^2} + \frac{1}{r}\frac{dF}{dr} + \left(n^2 k_0^2 - \beta^2 - \frac{\nu^2}{r^2}\right)F = 0 \qquad (11.5-3)$$

We encountered this differential equation in Section 8.2, Eq. (8.2-23), except there $n(r)$ was a piecewise constant function. The transformation[128]

$$r = ae^x \qquad \text{or} \qquad x = \ln(r/a) \qquad (11.5-4)$$

reduces (11.5-3) to the simpler form

$$\frac{d^2F}{dx^2} + G(x)F = 0 \qquad (11.5-5)$$

with

$$
\begin{aligned}
G &= (n^2 k_0^2 - \beta^2)a^2 e^{2x} - \nu^2 \\
&= (n^2 k_0^2 - \beta^2)r^2 - \nu^2 \qquad (11.5-6)
\end{aligned}
$$

Throughout this section we shall use the variables x and r interchangeably without changing the functional symbols G or F.

Equation (11.5-5) is reminiscent of the wave equation (8.3-8) of the slab waveguide. In the spirit of the WKB method we attempt to solve (11.5-5) by using the trial solution

$$F = A(x)e^{iS(x)} \qquad (11.5-7)$$

where A and S are assumed to be real functions of x. Substitution into (11.5-5) yields

$$\frac{d^2A}{dx^2} + 2i\frac{dA}{dx}\frac{dS}{dx} + iA\frac{d^2S}{dx^2} - A\left(\frac{dS}{dx}\right)^2 + GA = 0 \qquad (11.5-8)$$

So far, the theory has been exact (except that we started with the scalar wave equation, which is not an exact representation of Maxwell's theory). We are now introducing the first approximation when we neglect d^2A/dx^2. Setting the real and imaginary parts of (11.5-8) separately equal to zero results in

$$\frac{dS}{dx} = \sqrt{G} \tag{11.5-9}$$

and

$$2\frac{dA}{dx}\frac{dS}{dx} + A\frac{d^2S}{dx^2} = 0 \tag{11.5-10}$$

The derivation of (11.5-9) is analogous to the derivation of the eikonal equation (3.2-5) in Section 3.2. However, there we wrote $k_0 S$ instead of S and justified neglecting the second derivative of the amplitude function by comparing it to terms containing the large parameter k_0^2. In the present case, S is not as large as the corresponding function in Section 3.2, since it describes the radial variation of the field instead of the rapid variation in the principal direction of wave propagation. However, the WKB theory yields excellent solutions even though the insignificance of d^2A/dx^2 is not quite as apparent as that of the corresponding term in Section 3.2.

Equation (11.5-10) can be written in the form

$$2\left(\frac{dS}{dx}\right)^{1/2}\frac{d}{dx}\left[A\left(\frac{dS}{dx}\right)^{1/2}\right] = 0 \tag{11.5-11}$$

or, simpler,

$$\frac{d}{dx}(AG^{1/4}) = 0 \tag{11.5-12}$$

so that we have after integration

$$A(x) = \frac{A_1}{G^{1/4}} \tag{11.5-13}$$

A_1 is an integration constant.

Equation (11.5-9) can also be integrated yielding

$$S = S_0 + \psi \tag{11.5-14}$$

with

$$S_0 = \int_{x_0}^{x} G^{1/2}\,dx = \int_{r_0}^{r}\left[n^2(r)k_0^2 - \beta^2 - \frac{\nu^2}{r^2}\right]^{1/2}dr \tag{11.5-15}$$

For convenience, we used the lower turning point r_0 as the lower integration limit, ψ is an integration constant. The amplitude coefficient (11.5-13) can also be expressed more explicitly as

$$A(r) = \frac{A_1}{[(n^2(r)k_0^2 - \beta^2)r^2 - \nu^2]^{1/4}} \tag{11.5-16}$$

In (11.5-7) we could have used the exponential function with negative imaginary argument and would have obtained the same results. This means that there are two independent solutions which may be linearly superimposed to yield a more general solution of the form

$$E = \hat{A} \frac{\cos[S_0(r) + \psi] \cos(\nu\phi) e^{-i\beta z}}{[(n^2(r)k_0^2 - \beta^2)r^2 - \nu^2]^{1/4}} \tag{11.5-17}$$

$\hat{A}$ is an arbitrary amplitude. Equation (11.5-17) seems to be an approximate solution of the wave equation, however, something is still missing. We have, as yet, no way to determine the propagation constant β. The missing element is the requirement that our solution is indeed a guided mode whose amplitude remains finite at $r = 0$ and vanishes at $r = \infty$. In fact, the solution (11.5-17) fails at the turning points $r = r_0$ and $r = r_1$ defined by (11.2-36), because at these points it becomes infinite. In addition, it does not hold in the region outside the interval $r_0 < r < r_1$ because it is of the wrong form. Between the turning points the solution is oscillatory as indicated by the function $\cos[S(r) + \psi]$. But outside of this interval the solution must have an evanescent behavior. Thus, (11.5-17) cannot be used to describe the field outside of the interval between the turning points.

The failure of the present solution at the turning points was introduced when we neglected the term d^2A/dx^2 in (11.5-8). Instead of trying to improve on this approximation we start out anew and try to find a solution that holds exactly at those points where the other solution fails. We find an approximation to the differential equation (11.5-5) in the vicinity of r_0 and r_1 by expanding $G(x)$ in series at those points. Thus we have, for example, at $x = x_0$ (corresponding to $r = r_0$)

$$G = (x - x_0)K_0^2 \tag{11.5-18}$$

with

$$K_0^2 = \left(\frac{dG}{dX}\right)_{x=x_0} \tag{11.5-19}$$

Since $G(x_0) = 0$ and $G(x) > 0$ for $x > x_0$ the derivative in (11.5-19) is positive so that K_0^2 is indeed a positive quantity. The differential equation (11.5-5) now assumes the approximate form

$$\frac{d^2F_0}{dx^2} + (x - x_0)K_0^2 F_0 = 0 \tag{11.5-20}$$

that is valid in the immediate vicinity of $x = x_0$. The exact solution of this (approximate) differential equation is[8]

$$F_0 = \sqrt{(x - x_0)} \left\{ D_1 H_{1/3}^{(1)} \left[\frac{2k_0}{3} (x - x_0)^{3/2} \right] \right.$$
$$\left. + D_2 H_{1/3}^{(2)} \left[\frac{2k_0}{3} (x - x_0)^{3/2} \right] \right\} \tag{11.5-21}$$

D_1 and D_2 are integration constants, $H_{1/3}^{(1)}$ and $H_{1/3}^{(2)}$ are the Hankel functions of the first and second kind of order $1/3$. These solutions must, of course, be valid on both sides of x_0. For $x > x_0$ the arguments of the Hankel functions are real and positive. However, for $x < x_0$ we have

$$(x - x_0)^{3/2} = e^{i3\pi/2} |x - x_0|^{3/2} = -i |x - x_0|^{3/2} \qquad (11.5\text{-}22)$$

The angle of the argument of the Hankel functions thus becomes $3\pi/2$. For the further development of the theory we need the approximation of the Hankel functions for large arguments [see (11.5-27) and (11.5-28)] which, for $H_{1/3}^{(2)}(z)$, is valid only if the angle of z lies in the range between -2π and π exclusive of the range boundaries. Since the angle of (11.5-22) lies outside of this range we must use analytical continuation to transform $H_{1/3}^{(1)}(z)$ and $H_{1/3}^{(2)}(z)$ so that the angles of their arguments are brought back into the desired range. Analytical continuation of the Hankel functions allows us to express functions of negative (complex) argument as superpositions of Hankel functions of positive argument[8]

$$H_{1/3}^{(1)}(-z) = -e^{-i\pi/3} H_{1/3}^{(2)}(z) \qquad (11.5\text{-}23)$$

and

$$H_{1/3}^{(2)}(-z) = H_{1/3}^{(2)}(z) + e^{i\pi/3} H_{1/3}^{(1)}(z) \qquad (11.5\text{-}24)$$

Thus we obtain from (11.5-21) through (11.5-24) for the solutions for $x < x_0$

$$F_0 = i\sqrt{|x - x_0|} \left\{ D_2 e^{i\pi/3} H_{1/3}^{(1)} \left[i\,\frac{2}{3} K_0 |x - x_0|^{3/2} \right] \right.$$
$$\left. + (D_2 - e^{-i\pi/3} D_1) H_{1/3}^{(2)} \left[i\,\frac{2}{3} K_0 |x - x_0|^{3/2} \right] \right\} \qquad (11.5\text{-}25)$$

We may now use the large-argument approximation to conclude that, for increasing argument values, $H_{1/3}^{(1)}(iz)$ decreases while $H_{1/3}^{(2)}(iz)$ increases rapidly. However, outside of the turning point interval the field is known to be evanescent and rapidly decreasing so that the coefficient of $H_{1/3}^{(2)}(iz)$ must vanish. This consideration allows us to determine the relation between D_1 and D_2.

$$D_2 = e^{-i\pi/3} D_1 \qquad (11.5\text{-}26)$$

Thus, we have found a relation between the two amplitude coefficients by using a boundary condition. Since $x \rightarrow -\infty$ as $r \rightarrow 0$ we seem to be dealing with a boundary condition at infinity. However, since the approximation (11.5-18) holds only in the vicinity of x_0 it does not make sense to let the argument truly go to infinity. But note, that we are assuming that (11.5-18) may be used sufficiently far from the turning point to allow us to apply the large-argument approximation for the Hankel functions.

Next, we must relate the amplitude coefficient D_1 to the amplitude $\hat{A}$ of

(11.5-17). To do this we use (11.5-21) for $x > x_0$ and replace the Hankel functions by their large-argument approximations.

$$H_{1/3}^{(1)}(z) = \left(\frac{2}{\pi z}\right)^{1/2} \exp\left[i\left(z - \frac{5}{12}\pi\right)\right] \quad \text{for } -\pi < \arg z < 2\pi \quad (11.5\text{-}27)$$

and

$$H_{1/3}^{(2)}(z) = \left(\frac{2}{\pi z}\right)^{1/2} \exp\left[-i\left(z - \frac{5}{12}\pi\right)\right] \quad \text{for } -2\pi < \arg z < \pi \quad (11.5\text{-}28)$$

with (11.5-26) through (11.5-28) we obtain from (11.5-21)

$$F_0(x) = 2\left(\frac{3}{\pi K_0}\right)^{1/2} D_1 e^{-i\pi} \frac{\cos\left[\frac{2}{3}K_0(x - x_0)^{3/2} - \frac{\pi}{4}\right]}{(x - x_0)^{1/4}} \quad (11.5\text{-}29)$$

By substitution of (11.5-18) into (11.5-17) we find that the r dependent part of the WKB field solution may be expressed as

$$F(x) = \hat{A} \frac{\cos\left[\frac{2}{3}K_0(x - _0)^{3/2} + \psi\right]}{\sqrt{K_0}(x - x_0)^{1/4}} \quad (11.5\text{-}30)$$

The functional form of the two approximate solutions (11.5-29) and (11.5-30) is indeed identical! Comparing the two expressions we obtain a relation between the two amplitude coefficients

$$D_1 = \frac{1}{2}\sqrt{\frac{\pi}{3}} e^{i\pi/6} \hat{A} \quad (11.5\text{-}31)$$

and, more importantly, determine the phase

$$\psi = -\frac{\pi}{4} \quad (11.5\text{-}32)$$

Having extracted all the necessary information from the solution in the vicinity of the lower turning point x_0, we now proceed to a similar treatment of the solution in the vicinity of the upper turning point x_1. In the vicinity of $x = x_1$ we use

$$G = -K_1^2(x - x_1) \quad (11.5\text{-}33)$$

with

$$-K_1^2 = \left(\frac{dG}{dx}\right)_{x=x_1} \quad (11.5\text{-}34)$$

K_1^2 is again a positive constant because dG/dx is a negative at $x = x_1$. The differential equation (11.5-5) now assumes the form

$$\frac{d^2 F_1}{dx^2} - K_1^2(x - x_1) F_1 = 0 \tag{11.5-35}$$

and has the solution[8]

$$F_1 = \sqrt{x - x_1} \; D_3 \, H_{1/3}^{(1)} \left[i\frac{2}{3} K_1 (x - x_1)^{3/2} \right] \tag{11.5-36}$$

for $x > x_1$. The Hankel function $H_{1/3}^{(2)}$ was omitted in (11.5-36) since it grows with increasing $x - x_1$ while the physically acceptable solution must decrease. The boundary condition is thus already satisfied by (11.5-36). For $x < x_1$ we use

$$i(x - x_1)^{3/2} = e^{2i\pi} |x - x_1|^{3/2} \tag{11.5-37}$$

Because the large-argument approximation (11.5-27) holds if the angle of z lies inside the range between $-\pi$ and 2π, but not at 2π, we must again use an analytical continuation of $H_{1/3}^{(1)}$

$$H_{1/3}^{(1)} (e^{2i\pi} z) = -H_{1/3}^{(1)}(z) - e^{-i\pi/3} H_{1/3}^{(2)}(z) \tag{11.5-38}$$

With the large-argument approximations (11.5-27) and (11.5-28) and with the analytical continuation (11.5-38) the solution (11.5-36) becomes at $x < x_1$

$$F_1 = -2\sqrt{\frac{3}{\pi}} \, e^{i\pi/3} D_3 \frac{\cos\left[\frac{2}{3} K_1 |x - x_1|^{3/2} - \frac{\pi}{4}\right]}{\sqrt{K_1} |x - x_1|^{1/4}} \tag{11.5-39}$$

As before, this solution must be connected with the WKB solution (11.5-17). To accomplish this we write (11.5-14) in the form

$$S(x) = S_{01} + \int_{x_1}^{x} G^{1/2} \, dx - \frac{\pi}{4} \tag{11.5-40}$$

with

$$S_{01} = \int_{x_0}^{x_1} G^{1/2} \, dx = \int_{r_0}^{r_1} \left[n^2 k_0^2 - \beta^2 - \frac{\nu^2}{r^2} \right]^{1/2} dr \tag{11.5-41}$$

We separated the integral S_{01}, taken between the turning points, from the remaining integral to be able to use the approximation (11.5-33) to approximate S as follows (note: $x < x_1$)

$$S = S_{01} + \int_{x_1}^{x} [-K_1^2(x - x_1)]^{1/2} dx - \frac{\pi}{4}$$

$$= S_{01} + iK_1 \int_{x_1}^{x} [e^{i\pi} |x - x_1| x - x_1|]^{1/2} dx - \frac{\pi}{4}$$

$$= S_{01} - \frac{2}{3} K_1 |x - x_1|^{3/2} - \frac{\pi}{4} \qquad (11.5\text{-}42)$$

This derivation was written out in such detail to show why the minus sign appears in the final result. The radial function contained in (11.5-17) now assumes the approximate form

$$F = \hat{A} \; \frac{\cos\left[\frac{2}{3} K_1 |x - x_1|^{3/2} - S_{01} + \pi/4\right]}{i \sqrt{K_1} |x - x_1|^{1/4}} \qquad (11.5\text{-}43)$$

For convenience, we changed the sign of the argument of the cosine function for easier comparison with (11.5-39). Once more we are struck by the fact that the two approximations (11.5-39) and (11.5-43) have the same functional form even though very different methods were used to obtain them. The comparison of (11.5-39) and (11.5-43) does not only allow us to relate the amplitude coefficients $\hat{A}$ and D_3 to each other, but it does much more, it finally leads to the important eigenvalue equation for the determination of β. When we compare the cosine functions we must keep in mind that the functional form remains unchanged when an integral multiple of π is added to their arguments. Actually, this should already have been done when we compared the solutions at the lower turning point, but the two multiples of π would later have combined into one without changing the result. Comparison of the arguments of the cosine functions of (11.5-39) and (11.5-43) thus yields the result

$$-S_{01} + \frac{\pi}{4} + \mu\pi = -\frac{\pi}{4} \qquad (11.5\text{-}44)$$

which, by using (11.5-41), can also be expressed in the form

$$\int_{r_0}^{r_1} \left[n^2(r) k_0^2 - \beta^2 - \frac{\nu^2}{r^2}\right]^{1/2} dr = \left(\mu + \frac{1}{2}\right)\pi \qquad (11.5\text{-}45)$$

The integer μ is called the radial mode number. Equation (11.5-45) is the famous WKB eigenvalue equation for the propagation constant β of the guided modes.

Comparison if the amplitude coefficients yields

$$D_3 = \frac{i}{2} \sqrt{\frac{\pi}{3}} \; e^{-i\pi/3} \, \hat{A} \, \cos(\mu\pi) \qquad (11.5\text{-}46)$$

This concludes our derivation of the WKB solution of the modes of a multimode fiber. We have described the derivation in sufficient detail to allow the reader to follow each step of the complicated agrument. Most books on optical fibers state the eigenvalue equation (11.5-45) without proof by giving heuristic arguments that are not wholly satisfactory.[131,132] A full derivation of the WKB method is usually only found in books on quantum mechanics that do not deal with the cylindrical geometry used here.

Our treatment provides an approximate solution of the scalar wave equation for all values of the r variable. The solution is somewhat cumbersome since it consists of several expressions. Equation (11.5-17) describes the guided wave in its oscillatory region between the turning points. The complete set of guided modes is obtained by using in addition to (11.5-17) the solution that follows by replacing $\cos \nu \phi$ with $\sin \nu \phi$.

The solution in the immediate vicinity of the turning point x_0 is described by (11.5-21), (11.5-26) and (11.5-31) and requires the expressions for the analytical continuation (11.5-23) and (11.5-24) if we want to pass from one side of the turning point to the other. A similar set of equations describes the solution in the vicinity of the turning point $x = x_1$. The connection between the x and r variables is given by (11.5-4). Outside of the turning points for $x < x_0$ ($r < r_0$) and $x > x_1$ ($r > r_1$) the WKB solution holds again except that we now need $F(r)$ in the original form (11.5-7) since the field is not oscillatory (cannot be described by a sine or cosine function) but must be described by an exponential function with the sign of the exponent chosen such that the function decays as r moves away from the nearest turning point. Outside of the turning points (11.5-7) becomes a growing or decaying exponential function because $S(x)$ becomes purely imaginary. The amplitudes of these evanescent solutions can easily be found by comparison with the Hankel function solution applicable near the turning point. The details of these calculations are not given here since we shall not make any use of the field solution.[133]

The most important result of this section is the eigenvalue equation (11.5-45) that is extremely useful for computing the dispersive properties of the modes of a multimode fiber. It is important to remember that the eigenvalue equation is not exact. However, it is a surprisingly good approximation for all modes that are not too close to cutoff. However, it is not useful for describing the mode of a single-mode fiber.

11.6 APPLICATION OF THE WKB METHOD TO MULTIMODE FIBERS

We use the eigenvalue equation (11.5-45), derived in Section 11.5, to extract information about the properties of the modes in multimode fibers.

We begin by calculating the propagation constants of the modes of the square law fiber with (core) refractive index profile

$$n^2(r) = n_1^2 \left[1 - 2 \left(\frac{r}{a} \right)^2 \Delta \right] \qquad (11.6-1)$$

Substitution of (11.6-1) into (11.5-45) yields

$$\int_{r_0}^{r_1} \left[\kappa^2 - 2n_1^2 k_0^2 \left(\frac{r}{a} \right)^2 \Delta - \frac{\nu^2}{r^2} \right]^{1/2} dr = \left(\mu + \frac{1}{2} \right) \pi \qquad (11.6\text{-}2)$$

with

$$\kappa^2 = n_1^2 k_0^2 - \beta^2 \qquad (11.6\text{-}3)$$

The integral can be solved by a change of variables from r to $u = r^2$. The resulting integral is listed in tables of indefinite integrals.[61] Using the fact that the square root expression vanishes at $r = r_0$ and $r = r_1$, we find

$$\left[\frac{a\kappa^2}{4n_1 k_0 \sqrt{2\Delta}} - \frac{\nu}{2} \right] \pi = \left(\mu + \frac{1}{2} \right) \pi \qquad (11.6\text{-}4)$$

Solving this expression for β^2 results in

$$\beta^2 = n_1^2 k_0^2 \left[1 - \frac{2\sqrt{2\Delta}}{n_1 k_0 a} (2\mu + \nu + 1) \right] \qquad (11.6\text{-}5)$$

It is a remarkable fact that this solution for the propagation constant of the modes of the square law fiber is exact, even though it was derived from the approximate WKB eigenvalue equation (11.5-45). However, this statement requires a qualification. Equation (11.6-5) is the exact solution of the scalar wave equation for an infinitely extended square law medium. But the scalar wave equation (11.5-1) is not an exact representation of Maxwell's equations. Furthermore, real square law fibers can at best have a truncated square law distribution that merges into a constant refractive index in the cladding region. Thus, (11.6-5) cannot be exact for real fibers. That the cladding value of the refractive index cannot affect the WKB solution becomes clear when we remember that the integral in (11.6-2) only extends between the two turning points that, for a guided mode, must both lie within the core. The influence of the evanescent field, extending into the cladding, is neglected by the WKB eigenvalue equation. However, we might have expected an approximate result even for the solution of the scalar wave equation of an infinitely extended square law medium because the WKB solution involved a number of approximations.

If we convert the exact solution (7.3-22) of the wave equation for the infinitely extended square law medium to our present notation, we obtain

$$\beta^2 = n_1^2 k_0^2 \left[1 - \frac{2\sqrt{2\Delta}}{n_1 k_0 a} (p + q + 1) \right] \qquad (11.6\text{-}6)$$

This expression applies to modes that are expressed in rectangular Cartesian coordinates in terms of Hermite–Gaussian functions. The solution (11.6-5) belongs to modes expressed in cylindrical coordinates. The corresponding exact solutions would be expressed as Laguerre–Gaussian functions. These differences explain the different dependence on the radial and azimuthal mode numbers μ and ν in (11.6-5) and on the mode numbers p and q in (11.6-6).

Exact analytical solutions of the integral in (11.6-2) cannot be obtained for arbitrary refractive index distributions. However, a very interesting relationship exists[129] between the propagation constant β and the number of modes $M(\beta)$ that have propagation constants whose values are below that given by the argument of M. In preparation for the derivation of this relationship we consider the mode number space spanned by the radial and azimuthal mode numbers μ and ν as shown in Figure 11.6.1. Each mode of the fiber, belonging to a pair of mode numbers μ and ν, is represented as a point in mode number space. All modes whose representation points lie on the curve labeled β = const have the same propagation constant. The lines β = const are usually curved but, according to (11.6-5), they are straight lines in square law fibers. Mode number space contains guided and radiation modes. The boundary between the two types of modes is the curve $\beta = n_2 k_0$ [see (8.2-55)]. All representation points in mode number space lying below the line $\beta = n_2 k_0$ belong to guided modes, the region above this line is occupied by radiation modes and so called leaky modes.

If we want to count the total number of guided modes we must remember that each point in mode number space represents four modes, because for each pair of mode numbers μ and ν the corresponding mode field can have azimuthal mode dependence $\cos \nu\phi$ and $\sin \nu\phi$ and can exist in two mutually orthogonal polarizations. We say the modes are fourfold degenerate. The only

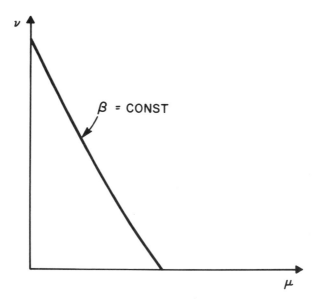

Figure 11.6.1   Mode number space μ, ν showing the curve β = const.

exception are the modes with $\nu = 0$ that are only doubly degenerate, because $\cos \nu\phi$ becomes unity while $\sin \nu\phi$ no longer exists. Since modes with $\nu = 0$ represent only a small minority we ignore this anomaly and assign four modes to the unit area of mode number space. Thus, the total number of modes with propagation constant less than β is[129]

$$M(\beta) = 4 \int_0^{\nu(\beta)} \mu \, d\nu \qquad (11.6\text{-}7)$$

The upper limit of the integral lies on the curve with constant β whose value is indicated by the arguments of M and ν. We substitute the value of μ from the eigenvalue equation (11.5-45) but ignore the additive constant $1/2$ on the assumption that most modes have large values of μ. The expression for the mode number now becomes

$$M(\beta) = \frac{4}{\pi} \int_0^{\nu(\beta)} \int_{r_0}^{r_1} \left[n^2(r)k_0^2 - \beta^2 - \frac{\nu^2}{r^2} \right]^{1/2} dr \, d\nu \qquad (11.6\text{-}8)$$

The ν integration of this double integral is clearly very much easier to perform than the r integration. For this reason we change the order of integration, integrating first over ν and than over r. The integral over ν extends from 0 to a maximum value that is defined as the point where the square root expression under the integral vanishes. The r integral also begins at $r = 0$ and extends to a point $r = R$ defined by

$$n(R)k_0 = \beta \qquad (11.6\text{-}9)$$

The integral (11.6-8) now assumes the form

$$M(\beta) = \frac{4}{\pi} \int_0^R \int_0^{r[n^2(r)k_0^2 - \beta^2]^{1/2}} \left[n^2(r)k_0^2 - \beta^2 - \frac{\nu^2}{r^2} \right]^{1/2} d\nu \, dr \qquad (11.6\text{-}10)$$

After the ν integration is carried out we are left with the expression

$$M(\beta) = \int_0^R [n^2(r)k_0^2 - \beta^2] r \, dr \qquad (11.6\text{-}11)$$

To make further progress we have to know the refractive index distribution $n(r)$. A particularly interesting refractive index distribution of great practical importance is the power law profile (8.7-11)[129]

$$n^2(r) = n_1^2 \left[1 - 2 \left(\frac{r}{a} \right)^g \Delta \right] \qquad \text{for } r < a \qquad (11.6\text{-}12)$$

Using (11.6-12), we can integrate (11.6-8) and obtain

$$M(\beta) = \frac{g}{g+2} (n_1 k_0 a)^2 \Delta \left[\frac{n_1^2 k_0^2 - \beta^2}{2n_1^2 k_0^2 \Delta} \right]^{(g+2)/g} \qquad (11.6\text{-}13)$$

Solving (11.6-13) for β yields

$$\beta = n_1 k_0 \left\{ 1 - 2\Delta \left[\frac{M(\beta)}{N} \right]^{g/(g+2)} \right\}^{1/2} \tag{11.6-14}$$

with

$$N = \frac{g}{g+2} (n_1 k_0 a)^2 \Delta \tag{11.6-15}$$

We have now derived the desired expression for the propagation constant β in terms of the cumulative mode number $M(\beta)$. The expression (11.6-14) is not a solution of the eigenvalue equation (11.5-45) because β appears implicitly on its right hand side as the argument of $M(\beta)$. But (11.6-14) is very useful for other purposes. For example, it immediately allows us to identify N of (11.6-15) as the total number of modes supported by the fiber. To see this, let $M = N$ so that we obtain from (11.6-14) with the help of (8.7-11)

$$\beta = n_1 k_0 (1 - 2\Delta)^{1/2} = n_2 k_0 \tag{11.6-16}$$

This shows that N is the total number of modes with propagation constants larger that $n_2 k_0$, which, of course, includes all the guided modes. Using (8.7-15) we can express N in terms of the important normalized frequency parameter V

$$N = \frac{g}{2(g+2)} V^2 \tag{11.6-17}$$

For the step-index fiber with $g = \infty$ we thus have

$$N = \tfrac{1}{2} V^2 \qquad \text{for } g = \infty \tag{11.6-18}$$

and for the square law fiber

$$N = \tfrac{1}{4} V^2 \qquad \text{for } g = 2 \tag{11.6-19}$$

Comparing (11.6-14) with (11.6-5) we find for the square law profile fiber ($g = 2$)

$$M(\beta) = (2\mu + \nu + 1)^2 \tag{11.6-20}$$

We can also use (11.6-14) to compute the impulse response of the multimode fiber with power law profile. To accomplish this we remind the reader that the group velocity of a wave is defined as

$$v = \frac{d\omega}{d\beta} \tag{11.6-21}$$

The specific group delay (group delay per unit length) is thus

$$\hat{\tau} = \frac{1}{v} = \frac{d\beta}{d\omega} = \frac{1}{c} \frac{d\beta}{dk_0} \tag{11.6-22}$$

To evaluate (11.6-22) we write (11.6-14) in the following form

$$\beta = n_1 k_0 \left\{ 1 - 2\Delta \left[\frac{(g+2)M}{gn_1^2 k_0^2 a^2 \Delta} \right]^{g/(g+2)} \right\}^{1/2} \qquad (11.6\text{-}23)$$

It seems that this expression is still not very useful for computing the specific group delay since $M = M(\beta)$ depends on β, and through β on k_0. However, the example (11.6-20) for the square law profile shows that M depends only on the integral values μ and ν and not on k_0. For a given mode with fixed μ and ν, $dM/dk = 0$, at least for the square law index profile. However, $dM/dk = 0$ actually holds in general. When we derived (11.6-23) from the eigenvalue equation (11.5-45) we treated μ and ν as continuous variables. But since we want to compute the specific group delay for a given, definite mode we must asign fixed, integral values to μ and ν. The number of guided modes $M(\beta)$ with propagation constants below a certain value of β does not change when β is varied by an infinitesimal amount. For the purpose of computing $d\beta/dk_0$ the mode number M must be regarded as constant. With constant M we obtain the following expression for the group delay from (11.6-22) and (11.6-23).

$$\tau = L\hat{\tau} = \frac{n_1 L}{c} \frac{1 - \dfrac{4\Delta}{g+2}\left(\dfrac{M}{N}\right)^{g/(g+2)}}{\left[1 - 2\Delta\left(\dfrac{M}{N}\right)^{g/(g+2)} \right]^{1/2}} \qquad (11.6\text{-}24)$$

L is the distance traveled by the pulse (fiber length).

For small values of Δ (11.6-24) can be approximated as

$$\tau = \frac{n_1 L}{c}\left\{ 1 + \Delta\frac{g-2}{g+2}\left(\frac{M}{N}\right)^{g/(g+2)} + \Delta^2 \frac{3g-2}{2(g+2)}\left(\frac{M}{N}\right)^{2g/(g+2)} \right\} \qquad (11.6\text{-}25)$$

This approximation shows clearly that, to first order in Δ, τ becomes independent of M/N for $g = 2$. No value of g can make τ exactly independent of M/N, thus there is no power law profile that equalizes the group velocity of all the modes exactly. For $M = 0$ the group delay becomes $\tau = n_1 L/c$. Generally, τ deviates from this value for $0 < M < N$, but if we let

$$g = 2(1 - \Delta) \qquad (11.6\text{-}26)$$

then to second order in Δ $\tau = n_1 L/c$ for $M = 0$ and $M = N$. This choice of the exponent g of the power law profile results in a good equalization compromise. Olshansky[134] has shown that the rms width of the pulse carried by all equally excited modes of a multimode fiber becomes a minimum if

$$g = 2 - \frac{12}{5}\Delta \qquad (11.6\text{-}27)$$

Next we compute the impulse response of a power law, multimode fiber. When a short pulse, an impulse, is launched into a multimode fiber such that all of its modes are equally excited, the length of the output pulse will depend on the group delay differences among the modes. Each mode still carries a short impulse, but the actual power arriving during a given time interval $d\tau$ is proportional to the number dM of modes arriving during this interval. Thus, if we ignore the possibility of differential mode losses we conclude that the output pulse $P(\tau)$ must be proportional to[129]

$$P(\tau) = \frac{dM}{d\tau} \tag{11.6-28}$$

Since $P(\tau)$ results from an impulse, launched into the fiber at its input end, it is called the impulse response.

It is hard to give a general expression for $M(\tau)$, and hence for $P(\tau)$, for all values of g. But for g values that are not too close to $g = 2$ we find from (11.6-25)

$$M = N \left[\frac{1}{\Delta} \frac{g+2}{g-2} \left(\frac{c\tau}{n_1 L} - 1 \right) \right]^{(g+2)/g} \qquad \text{for } g \neq 2 \tag{11.6-29}$$

For $g = 2$ we have

$$M = 2\frac{N}{\Delta^2} \left(\frac{c\tau}{n_1 L} - 1 \right) \qquad \text{for } g = 2 \tag{11.6-30}$$

The impulse response can now be compared from (11.6-28) and (11.6-29) for values $g \neq 2$

$$P(\tau) = N\frac{c}{n_1} \frac{g+2}{gL} \left(\frac{g+2}{g-2} \frac{1}{\Delta} \right)^{(g+2)/g} \left(\frac{c\tau}{n_1 L} - 1 \right)^{2/g}$$

$$\text{for} \qquad \frac{n_1 L}{c} \leqslant \tau \leqslant \frac{n_1 L}{c} \left(1 + \frac{g-2}{g+2} \Delta \right) \tag{11.6-31}$$

and from (11.6-30) for $g = 2$

$$P(\tau) = 2 \frac{c}{n_1 L} \frac{N}{\Delta^2}$$

$$\text{for} \qquad \frac{n_1 L}{c} \leqslant \tau \leqslant \frac{n_1 L}{c} \left(1 + \frac{\Delta^2}{2} \right) \tag{11.6-32}$$

The shapes of these functions are shown in Figure 11.6.2.[129]

From (11.6-31) we see that the width of the pulse is proportional to Δ,

$$\delta\tau = \frac{g-2}{g+2} \frac{n_1 L}{c} \Delta \qquad \text{for } g \neq 2 \tag{11.6-33}$$

For $g = 2$ the impulse response is much narrower because it is proportional to Δ^2

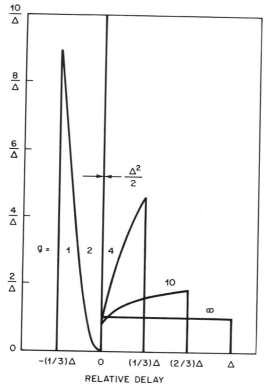

Figure 11.6.2 Impulse responses for power law profile fibers for several different exponent values g as functions of $\dfrac{c\tau}{n_1 L} - 1$. Copyright 1973, American Telephone and Telegraph Company. Reprinted with permission (Ref. 122).

$$\delta\tau = \frac{1}{2}\frac{n_1 L}{c}\Delta^2 \qquad \text{for } g = 2 \qquad (11.6\text{-}34)$$

For $g = 2(1 - \Delta)$ the factor $1/2$ in (11.6-34) is further reduced to $7/32$.

We conclude this section by deriving from the WKB eigenvalue equation (11.5-45) a general expression for the specific group delay of the modes of a general graded index fiber. We take the derivative of (11.5-45) with respect to k_0, keeping μ, ν and n (no material dispersion) constant. Neglecting $1/2$ relative to μ and solving for $d\beta/dk_0$, we obtain with the help of (11.6-22)

$$\hat{\tau} = \frac{k_0}{c\beta} \frac{\displaystyle\int_{r_0}^{r_1} \frac{n^2(r)\,dr}{[n^2(r)k_0^2 - \beta^2 - \nu^2/r^2]^{1/2}}}{\displaystyle\int_{r_0}^{r_1} \frac{dr}{[n^2(r)k_0^2 - \beta^2 - \nu^2/r^2]^{1/2}}} \qquad (11.6\text{-}35)$$

The limits r_0 and r_1 of the integral in (11.5-45) are dependent on k_0, but differentiation of the integral with respect to its limits does not make a contribution to (11.6-35) because the integrand vanishes at $r = r_0$ and $r = r_1$. It is remarkable that (11.6-35) is in exact agreement with (11.2-44), which was derived from ray optics. This comparison shows that the expression for the group delay derived from the WKB method is no more accurate than the corresponding ray optics result.

Throughout this chapter we have ignored material dispersion, that is we have assumed that $dn/dk_0 = 0$. Important additional effects occur when material dispersion is included. Of particular importance is profile dispersion, which is caused by the wavelength dependence of the relative refractive index difference Δ.

A more detailed discussion of modal dispersion in multimode fibers, including material and profile dispersion, is given in Section 12.5 of the next chapter.

11.7 LEAKY WAVES

Light propagation in graded-index fibers can be described either in terms of ray or wave optics. In the language of geometrical optics light travels along rays paths. A trapped ray follows an undulating path, spiraling around the fiber axis. At the lowest point of its trajectory its distance from the fiber axis is r_0, at its highest point its largest distance from the axis is r_1. The ray seems to be securely guided as long as its upper turning point, r_1 remains inside the fiber core.

Wave optics describes light propagation differently. It too knows of turning points. However, in terms of wave optics, a turning point occurs where the character of the wave changes from oscillatory to evanescent behavior. This is best seen with the help of the WKB solution. In Section 11.5, the solution (11.5-2) of the wave equation was expressed as a product of three terms, the radial, azimuthal, and z dependent parts. According to (11.5-4) and (11.5-7) the radial wave function can be expressed as

$$F(r) = A(r)e^{iS(r)} \tag{11.7-1}$$

The exponent of the exponential function is defined as

$$S(r) = \int_{r_0}^{r} \left[n^2(r)k_0^2 - \beta^2 - \frac{\nu^2}{r^2} \right]^{1/2} dr \tag{11.7-2}$$

If the square root expression under the integration sign is real the wave is oscillatory. This can best be seen by combining (11.7-1) with its complex conjugate counterpart and expressing the solution in the form (11.5-17) where the radial function assumes the form

$$F(r) = \hat{A}(r)\cos\left[S(r) + \Psi\right] \tag{11.7-3}$$

The function (11.7-1) remains oscillatory between the two turning points $r = r_0$ and $r = r_1$ that are defined as solutions of Equation (11.2-36)

$$n^2(r)k_0^2 - \beta^2 - \frac{\nu^2}{r^2} = 0 \tag{11.7-4}$$

Outside of the turning points the square root in (11.7-2) becomes imaginary, consequently, $iS(r)$ in (11.7-1) is now real so that the wave function assumes an evanescent behavior, that is it decays exponentially. Exponential growth is forbidden by the boundary condition that requires the field to remain finite at infinity. The WKB solution (11.7-1) does not work in the immediate vicinity of the turning points, but it is valid on either side of them, provided they are not approached too closely.

The solution of the guided wave problem differs fundamentally from the ray optics solution by postulating that the field extends outside of the region between the two turning points. This difference has important consequences for the question: what defines a guided wave? In terms of ray optics a wave remains guided as long as the upper turning point r_1 remains inside the fiber core, $r_1 < a$. The answer provided by wave optics is not quite that simple. To understand the difference, Figure 11.7.1 shows a plot of the two functions, $n^2(r)k_0^2 - \beta^2$ and $(\nu/r)^2$ that appear in (11.7-2). The exponent $S(r)$ of (11.7-1) remains real as long as $n^2(r)k_0^2 - \beta^2$, the solid curve in Figure 11.7.1, lies

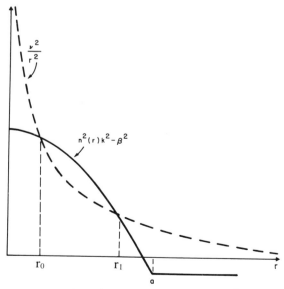

Figure 11.7.1   Graphical representation of the WKB solution. The solid line represents the curve $n^2(r)k_0^2 - \beta^2$, the dotted curve represents $(\nu/r)^2$. An oscillatory wave exists in regions where the solid curve lies above the dotted curve.

above the dotted curve representing $(\nu/r)^2$. As long as the solid curve becomes negative for $r > a$ (a = core radius) the two curves cannot cross again and the wave remains evanescent for all values $r > r_1$. The requirement that $n^2(r)k_0^2 - \beta^2 < 0$ for $r > a$ can be expressed as

$$\beta > n_2 k_0 \qquad (11.7\text{-}5)$$

with $n_2 = n(r)$ for $r > a$.

Figure 11.7.2 shows what happens if $\beta < n_2 k_0$. Now, the solid curve remains positive even for $r > a$ and, inevitably, the solid and dotted curves cross again. But we know that the field is oscillatory whenever the solid curve lies above the dotted curve. Figure 11.7.2 shows clearly that the field resumes its oscillatory behavior outside of a third turning point $r = r_2$. For $r > r_2$ we have an oscillatory traveling wave carrying energy away to infinity. Once $\beta < n_2 k_0$ the wave is no longer truly guided but loses power by radiation. The dividing line between a wave that is truly guided and one that loses power by radiation is given by the cutoff condition

$$\beta = n_2 k_0 \qquad (11.7\text{-}6)$$

A. W. Snyder coined the phrase "tunneling leaky wave" to characterize a wave described by the situation illustrated in Figure 11.7.2[135,136] This term is borrowed from quantum mechanics where a particle can tunnel through a potential barrier that is too high for it to pass over. The light wave tunnels through the region between the turning points r_1 and r_2 where it behaves as an evanescent wave. Its leakage loss depends on the length of the tunneling

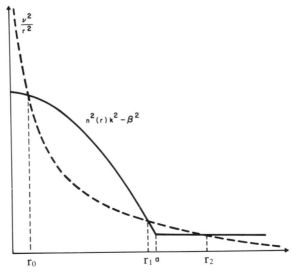

Figure 11.7.2   This figure is similar to Figure 11.7.1 but applies to leaky rays with $\beta < n_2 k_0$ and three turning points.

region. For large values of ν this region can be relatively long so that the leakage loss remains infinitesimally small. Some tunneling leaky waves are almost indistinguishable from truly guided waves. It is apparent from Figure 11.7.2 that, for constant β, the length of the leakage path increases with increasing values of the azimuthal mode number ν. For this reason, low loss leaky tunneling modes have large values of ν. On the other hand, as the distance between r_1 and r_2 becomes small, the leakage losses become high and the wave cannot be regarded as being guided.

When $r_1 = r_2 = a$ the tunneling region shrinks to zero and we have leaky modes of a different kind. Snyder coined the term refracting leaky mode[135,136] to describe highly lossy leaky waves that do not even have a tunneling region. The term refracting leaky modes comes from step-index fibers where the guided modes are trapped by total internal reflection at the refractive index discontinuity at the core-cladding boundary. Rays impinging on the core-cladding boundary at angles too large for total internal reflection are refracted out of the core—hence the name refracting leaky rays (or modes). For refracting leaky rays, the solid and dotted curves in Figure 11.7.2 intersect only once. The dotted curve remains below the solid curve for all $r > r_0$.

It is interesting to trace the boundaries in mode number space that delineate the region between guided modes and tunneling leaky modes. For simplicity we limit ourselves to fibers with square law refractive index cores and constant refractive index cladding. Mode number space was discussed in Section 11.6 (see Figure 11.6.1) where we pointed out that, for square law fibers, the loci for modes with the same value of the propagation constant are straight lines. All guided modes are contained in a triangular region whose boundary is defined by the condition $\beta = n_2 k_0$ as shown in Figure 11.7.3. Using the solution (11.6-5) for the propagation constant of square law fibers

$$\beta^2 = n_1^2 k_0^2 - 2n_1 k_0 \frac{\sqrt{2\Delta}}{a} (2\mu + \nu) \qquad (11.7-7)$$

(the "1" was neglected inside the parenthesis), we obtain the equation for the guided mode boundary by equating (11.7-6) and (11.7-7)

$$\mu = \tfrac{1}{2} (\tfrac{1}{2} n_1 k_0 a \sqrt{2\Delta} - \nu) \qquad (11.7-8)$$

We return to Figure 11.7.2 to find the boundaries for tunneling leaky modes. The diagram applies to leaky modes if $\beta < n_2 k_0$ so that the straight portion of the solid curve lies above the r axis. For a tunneling leaky mode there must be two turning points between $r = 0$ and $r = a$ so that there must be two crossings of the dotted and solid curves in this region. The boundary of the tunneling leaky mode domain is defined by the condition $r_1 = a$. The upper turning point is defined by (11.3-6),

$$r_1 = \frac{a}{n_1 k_0 \sqrt{4\Delta}} \left\{ \kappa^2 + \left[\kappa^4 - \left(\frac{n_1 k_0}{a} \right)^2 (8\Delta) \nu^2 \right]^{1/2} \right\}^{1/2} \qquad (11.7-9)$$

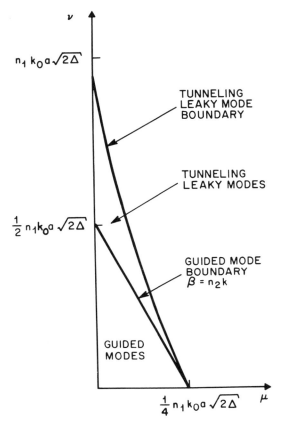

Figure 11.7.3   This figure shows the boundaries for guided and tunneling leaky modes in mode number space.

with

$$\kappa^2 = n_1^2 k_0^2 - \beta^2 \tag{11.7-10}$$

Setting $r_1 = a$ and using (11.7-7) and (11.7-10) we obtain from (11.7-9) the equation of the upper leaky mode boundary

$$\mu = \frac{(n_1 k_0 a \sqrt{2\Delta} - \nu)^2}{4 n_1 k_0 a \sqrt{2\Delta}} \tag{11.7-11}$$

which can be seen in Figure 11.7.3. The figure shows that the majority of leaky modes have large ν values, and we know that their losses decrease with increasing values of ν.

We can easily compute the total number of leaky and guided modes by solving the integral

$$N_g + N_t = 4 \int_0^{n_1 k_0 a \sqrt{2\Delta}} \mu \; d\nu \tag{11.7-12}$$

with the result

$$N_g + N_t = \frac{2}{3} (n_1 k_0 a)^2 \Delta \tag{11.7-13}$$

The factor 4 in front of (11.7-12) accounts for the fourfold degeneracy of the modes. The total number of guided modes is obtained from (11.6-15) with $g = 2$

$$N_g = \frac{1}{2} (n_1 k_0 a)^2 \Delta \tag{11.7-14}$$

so that the total number of tunneling leaky modes is

$$N_t = \frac{1}{6} (n_1 k_0 a)^2 \Delta = \frac{1}{3} N_g \tag{11.7-15}$$

Not all of these modes have low losses. The tunneling leaky modes with lowest losses lie near the guided mode boundary, the losses decrease with increasing values of the azimuthal mode number ν, provided $\beta \approx n_2 k_0$.

In Section 8.4, we pointed out that the complete set of modes of an optical fiber consists of guided and radiation modes. Thus, leaky modes are not members of the complete set of fiber modes. Strictly speaking, all modes lying above the guided mode boundary in Figure 11.7.3 belong to the continuum of radiation modes (provided we assume an infinitely extended cladding). The leaky modes are not truly legitimate solutions of Maxwell's equations because they cannot be normalized. After an initial decrease, the power density of leaky modes increases indefinitely with increasing radial distance from the fiber core. But, even though leaky modes cannot be generated in practice, they can be approximated by transient fields whose losses are very similar to the corresponding leaky modes. The transient fields that approximate leaky modes can be excited by shining light into a fiber at such an angle that the z component of its propagation constant satisfies the condition $\beta < n_2 k_0$. Such fields cannot be represented by a series expansion in terms of guided modes but must be expressed as a superposition of an infinite number of radiation modes. Such a superposition can resemble the field of a leaky mode very closely in and near the fiber core. If the corresponding leaky mode has low loss, the transient field can persist for long distances in the fiber core and propagate almost as a guided mode. In fact, its behavior is well approximated by the leaky mode solution. In mathematical terms, leaky mode solutions occur when the integral, describing the superposition of radiation modes, is transformed by moving the integration path into the complex plane. When the integration path is shifted off the real axis it crosses poles of the integrand whose residue contributions correspond to the leaky modes we have been discussing in this section.

12

DISPERSION IN FIBERS

12.1 INTRODUCTION

As a transmission medium for wide band communications, optical fibers are characterized by two important parameters: loss and dispersion. Loss determines how far a signal can travel in the fiber before it must be regenerated. However, the permissable repeater span is not only governed by fiber loss but is also limited by dispersion. The word dispersion is used to describe the tendency of the fiber to distort a light pulse (or light modulated with an analog signal) traveling through it. Since dispersion limits the spectral width of the modulation that a light wave can carry through the fiber, we also use the term "fiber bandwidth" (or simply bandwidth) to describe the dispersive property of the fiber. A fiber with high bandwidth can be used to transmit short pulses—it has high channel capacity. All optical fibers distort pulses (or analog signals) passing through them. If its length is short, the signal bandwidth that can be transmitted through the fiber is very large. But with increasing length signal distortion becomes more severe so that the desired channel capacity limits the length over which a signal can be transmitted through the fiber. If long distances between repeaters are to be achieved, the fiber must simultaneously have low loss and high bandwidth.

Dispersion in single mode fibers is governed by a different mechanism than in multimode fibers. In single mode fibers, dispersion is caused mainly by the dispersive properties of the fiber material and to a lesser extent by the waveguiding mechanism. The light energy of a pulse traveling in a single mode fiber is spread over a certain spectral range, partly because of the Fourier spectrum associated with the signal pulse itself, but also because the light source usually has a significant spectral width even before it is modulated. Due to chromatic dispersion, different parts of the spectrum travel with different velocity causing the pulse to spread out as it advances in the fiber.

The same mechanism exists, of course, also in the multimode fiber. Each mode carries a portion of the pulse and spreads due to chromatic dispersion. However, unless the multimode fiber is very carefully designed to achieve nearly optimum delay equalization of its modes, the major contribution to pulse distortion in multimode fibers stems from the difference in the group delays of the many modes. This contribution to pulse distortion is called intermodal dispersion, or just modal dispersion for short. However, there is an interesting effect in multimode fibers that is related to material dispersion. It stems from the fact that the dispersive properties of dopant and host glasses are not the same. The undoped cladding and lightly doped outer regions of the fiber core have slightly different chromatic dispersion characteristics than the more heavily doped core center. As a consequence, the group delays of rays that travel near the core axis are different than those of rays that stay farther away from the axis. This effect would cause intermodal dispersion even if the group delays of all modes were perfectly equalized in the absence of chromatic dispersion or for a uniformly dispersive fiber material. This profile dispersion effect[137] must be taken into account if high bandwidth is to be achieved in practical multimode fibers.

The fact that optical fibers have finite bandwidth can easily be explained. Consider a multimode fiber that carries a light signal whose power is sinusoidally modulated. A given component of the signal in the detection current, contributed by a given mode, has a phase that depends on the phase of the modulating signal and on the group delay of the mode that carried that element of the signal to the receiver. Since each mode has a slightly different phase delay the elements of the signal contributions of all the modes are not all strictly in-phase. In fact, if the modulation frequency is sufficiently high, destructive interference must occur, while there is no destructive interference if the modulation frequency is sufficiently low. The frequency that causes the amplitude of the detected signal to drop to one half of the amplitude of a low frequency signal is defined as the fiber bandwidth. Chromatic dispersion has a similar, if lesser, influence on signal transmission through single mode fibers.

This chapter is devoted to a discussion of dispersion in single mode and multimode fibers. We begin with a detailed description of pulse propagation in single mode fibers and discuss the interplay of material and waveguide dispersion. Finally, we return to a more detailed discussion of intermodal dispersion in multimode fibers. Pulse broadening in multimode fibers was already discussed briefly in Section 11.6 without including important material and profile dispersion effects.

12.2 MATERIAL DISPERSION

Light in material media is slowed relative to its vacuum velocity c by the factor $1/n$, the inverse of the refractive index of the material, because of the interaction of the electromagnetic wave with bound electrons. Since elastically bound particles are always able to oscillate at characteristic resonance frequencies, the interaction between light waves and bound electrons is frequency dependent. The dependence of the refractive index on the light frequency, $n = n(f)$, and hence also on its vacuum wavelength, $n = n(\lambda)$, gives rise to the phenomenon of chromatic dispersion that is responsible for the colors of the rainbow and for the distortion of light pulses in optical fibers.

An exact treatment of dispersion must be based on quantum mechanics.[138] In this section, we present a simplified, classical theory based on a model that regards the bound electrons in gases, solids, or liquids as harmonic oscillators.

We consider an electron of mass m that is bound to its equilibrium position by a restoring force that is proportional to the displacement x from the equilibrium position. If we also include damping, the motion of the electron with charge e, subject to the force eE exerted on it by the electric field E of the light wave, is described by the differential equation

$$m\left(\frac{d^2x}{dt^2} + \frac{1}{\tau}\frac{dx}{dt}\right) + Kx = eE \qquad (12.2\text{-}1)$$

K is the proportionality constant of the elastic restoring force and τ is the decay time of the free oscillation due to damping. Using complex notation, the time dependence of the electric field can be expressed as

$$E = E_0 e^{i\omega t} \qquad (12.2\text{-}2)$$

The forced motion of the electron is obtained as the solution of (12.2-1) after transient oscillations at the resonance radian frequency

$$\omega_0 = (K/m)^{1/2} \qquad (12.2\text{-}3)$$

have died out,

$$x = \frac{(e/m)E}{\omega_0^2 - \omega^2 + i\omega/\tau} \qquad (12.2\text{-}4)$$

We now extend our model assuming that there are N bound electrons per unit volume, each oscillating according to (12.2-4). The combined motion of these electrons gives rise to an induced polarization P of the material,

$$P = exN \qquad (12.2\text{-}5)$$

The polarization P is related to the electric displacement by the relation

$$D = \epsilon_0 E + P \qquad (12.2\text{-}6)$$

while, according to 1.2-3), D can also be expressed in terms of E as

$$D = \epsilon_0 n^2 E \tag{12.2-7}$$

Combining (12.2-6) and 12.2-7) yields the following expression for the square of the refractive index

$$n^2 = 1 + \frac{1}{\epsilon_0} \frac{P}{E} \tag{12.2-8}$$

Substitution of (12.2-4) and (12.2-5) into (12.2-8) finally results in

$$n^2 = 1 + \frac{e^2 N}{\epsilon_0 m (\omega_0^2 - \omega^2 + i\omega/\tau)} \tag{12.2-9}$$

This formula for the refractive index of our model of a material composed of damped, charged harmonic oscillators contains the most important features of the refractive index of a real dispersive medium. We see that the refractive index depends on the radian frequency ω of the light wave and that it is a complex quantity. We have seen in Section 2.6, Eqs. (2.6-18) and (2.6-19), that a wave in a medium with complex refractive index suffers loss; our model thus includes loss and dispersion. Dispersion is caused by the presence of a resonance of the charged particles in the material and loss by their damping. In real materials more than one resonance affects its dispersive properties, there are electronic resonances at ultraviolet frequencies and vibrational resonances of atoms or molecules at infrared frequencies. Neglecting the effect of damping we extend the formula (12.2-9) by including p resonances and obtain the important Sellmeier equation

$$n^2 - 1 = \sum_{j=1}^{p} \frac{e^2 N_j}{\epsilon_0 m_j (\omega_j^2 - \omega^2)} \tag{12.2-10}$$

The indices j label the corresponding quantities associated with each resonance. Since the parameters appearing in (12.2-10) are not easily computed from first principles (they must actually be replaced with the correct quantities obtained from a quantum mechanical calculation), it is more convenient to write (12.2-10) in a different form, replacing the frequencies with wavelengths according to

$$\omega = 2\pi c/\lambda \tag{12.2-11}$$

so that we obtain

$$n^2 - 1 = \sum_{j=1}^{p} \frac{\lambda^2 B_j}{\lambda^2 - \lambda_j^2} \tag{12.2-12}$$

The λ_j indicate the wavelengths at which resonances occur (they coincide with peaks of material absorption) and the B_j are constants that must be found from experimentally obtained dispersion curves.

In Section 11.6, we introduced the specific group delay (11.6-22)

$$\hat{\tau} = \frac{1}{c} \frac{d\beta}{dk_0} \tag{12.2-13}$$

The propagation constant of a plane wave is

$$\beta = nk_0 \tag{12.2-14}$$

If we regard n as an effective refractive index (12.2-14) assumes universal validity. If we express λ and its derivative with respect to k_0 as

$$\lambda = 2\pi/k_0 \quad \text{and} \quad \frac{d\lambda}{dk_0} = -\lambda^2/(2\pi) \tag{12.2-15}$$

we can convert the k_0 derivative of n to a derivative with respect to λ

$$\frac{dn}{dk_0} = \frac{dn}{d\lambda} \frac{d\lambda}{dk_0} = -\frac{\lambda^2}{2\pi} \frac{dn}{d\lambda} \tag{12.2-16}$$

These formulas allow us to express the specific group delay (12.2-13) in terms of n and its λ derivative

$$\hat{\tau} = \frac{1}{c}\left(n - \lambda \frac{dn}{d\lambda}\right) \tag{12.2-17}$$

Pulse distortion is caused by the dependence of the specific group delay $\hat{\tau}$ on the wavelength. Typically, light pulses are produced by intensity modulation of a light source that is not monochromatic but has a spectrum of finite width $\Delta\lambda$ even if it is not modulated. The light pulse may then be regarded as a superposition of pulses each supported by a light carrier of slightly different wavelength. Pulse spreading occurs because the component pulses arrive at the receiver at slightly different times depending on the group delay of their carrier wavelenghts. The increase $\Delta\hat{\tau}$ of the pulse width thus may be expressed as

$$\Delta\hat{\tau} = \frac{d\hat{\tau}}{d\lambda} \Delta\lambda \tag{12.2-18}$$

This consideration makes it clear that it is the derivative of $\hat{\tau}$ with respect to λ that is responsible for pulse dispersion. By differentiation of (12.2-17) we obtain

$$\frac{d\hat{\tau}}{d\lambda} = -\frac{\lambda}{c} \frac{d^2n}{d\lambda^2} \tag{12.2-19}$$

As an example, the broken curve in Figure 12.2.1 shows the refractive index of fused silica (SiO$_2$ is an important constituent of fibers) as a function of the wavelength.[139] Also shown in the figure is the derivative $d\hat{\tau}/d\lambda$ in units of nanoseconds per nanometer of source bandwidth and per kilometer of (fiber)

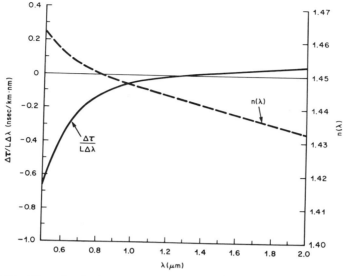

Figure 12.2.1   The refractive index and dispersion are shown for SiO_2 as functions of the wavelength λ.

length. The curves in this figure were computed from the Sellmeier equation (12.2-12) containing two terms, one resonance in the ultraviolet region at $\lambda_1 = 0.1\,\mu m$ with $B_1 = 1.0955$ and a second resonance in the infrared region at $\lambda_2 = 9\,\mu m$ with $B_2 = 0.9$. Figure 12.2.1 reveals an important property of fused silica,[140] the dispersion vanishes at a wavelength of $1.27\,\mu m$! Since optical fibers are typically made of fused silica, it is possible to make use of this fact to design a communications system with no first order dispersion. However, this does not mean that all dispersion vanishes when $d\hat{\tau}/d\lambda = 0$. In the next section, we treat pulse distortion in single mode fibers more rigorously and show that second order dispersion dominates when first order dispersion vanishes. The definitions of first and second order dispersion are given in Section 12.3.

12.3 PULSE DISTORTION IN SINGLE MODE FIBERS

In this section, we study the propagation of a Gaussian pulse in a dispersive medium, for example a single mode fiber.[141-143] The problem will be addressed in two stages. First we consider a monochromatic light source that is modulated by a Gaussian pulse. In this case we limit ourselves to first order dispersion. Next, we address the more realistic problem of a light source with finite spectral width that is again modulated by a Gaussian pulse. In this case, we allow the transmission medium to exhibit first and second order dispersion.

Beginning with the simpler problem, we assume that the electric field of the unmodulated monochromatic light source is described by the complex function

$$\psi_0(t) = Ae^{i\omega_0 t} \tag{12.3-1}$$

with amplitude A and radian frequency ω_0. The Gaussian signal pulse is represented by the function

$$s(t) = S \exp[-(t/T)^2] \tag{12.3-2}$$

with amplitude S and half width T at the $1/e$ points. The signal is used to modulate the power of the light source—not its amplitude. The electric field of the modulated light source is thus described by the product of (12.3-1) with the square root of (12.3-2). The spectrum of the modulated light field is

$$\phi(\omega) = \frac{1}{2\pi} \int_{-\infty}^{\infty} s^{1/2}(t)\, \psi_0(t) e^{-i\omega t}\, dt$$

$$= AT\left(\frac{S}{2\pi}\right)^{1/2} \exp\left[-\frac{1}{2}T^2(\omega - \omega_0)^2\right] \tag{12.3-3}$$

If we place the light source at $z = 0$, the field at any point z along the transmission path is given by the Fourier integral

$$\psi(z,t) = \int_{-\infty}^{\infty} \phi(\omega) \exp[i(\omega t - \beta z)] d\omega \tag{12.3-4}$$

The propagation constant β of the wave in the transmission medium is usually not a linear function of frequency, it is more accurately described by (12.2-14) with n given by (12.2-12). However, with such a complicated functional dependence for β, the integral in (12.3-4) cannot be solved analytically. Fortunately, it is not necessary to use the exact functional form of β. For all signal pulses of practical interest, (12.3-3) is sufficiently narrow that β deviates very little from a linear function over the spectral range that contributes to the integral in (12.3-4). For this reason, we approximate β by the first three terms of a Taylor series expansion

$$\beta = \beta_0 + \dot{\beta}_0(\omega - \omega_0) + \frac{1}{2}\ddot{\beta}_0(\omega - \omega_0)^2 \tag{12.3-5}$$

The subscript 0 indicates that β and its ω derivatives (indicated by dots) are evaluated at the (carrier) frequency of the light wave $\omega = \omega_0$. The modulated light field at point z and time t now follows from (12.3-3) through (12.3-5)

$$\psi(z,t) = AT\left(\frac{S}{2\pi}\right)^{1/2} \exp[i(\omega_0 t - \beta_0 z)]$$

$$\times \int_{-\infty}^{\infty} \exp\left[-\frac{1}{2}(T^2 + i\ddot{\beta}_0 z)(\omega - \omega_0)^2\right] \exp[i(t - \dot{\beta}_0 z)(\omega - \omega_0)] d\omega \tag{12.3-6}$$

The integral can be found in tables[61] so that we obtain the explicit analytical solution

$$\psi(z,t) = \frac{A\,T S^{1/2}}{(T^2 + i\ddot{\beta}_0 z)^{1/2}} \exp\left\{\frac{-(t - \dot{\beta}_0 z)^2\, T^2}{2[T^4 + (\ddot{\beta}_0 z)^2]}\right\}$$

$$\cdot \exp\left\{\frac{i\ddot{\beta}_0 z\,(t - \dot{\beta}_0 z)^2}{2[T^4 + (\ddot{\beta}_0 z)^2]}\right\} \exp\left[i(\omega_0 t - \beta_0 z)\right] \quad (12.3\text{-}7)$$

The solution (12.3-7) shows that the Light pulse has a Gaussian envelope that is multiplied by the propagation factor $\exp[i(\omega_0 t - \beta_0 z)]$ and by a second oscillatory function of more complicated form. The time averaged power of the light pulse is a simple Gaussian function

$$P(z,t) = |\psi(z,t)|^2 = \frac{A^2\,T^2\,S}{[T^4 + (\ddot{\beta}_0 z)^2]^{1/2}} \exp\left[\frac{-(t - \dot{\beta}_0 z)^2\,T^2}{T^4 + (\ddot{\beta}_0 z)^2}\right] \quad (12.3\text{-}8)$$

We defer a discussion of the properties of the spreading light pulse to combine it with a discussion of a pulsed non-monochromatic source.

Next, we generalize the problem by assuming that the field intensity of the light source has the following form[141]

$$\psi_0(t) = A(t)e^{i\omega_0 t} \quad (12.3\text{-}9)$$

The only difference between (12.3-1) and (12.3-9) is the time dependence of the complex amplitude $A(t)$ which is assumed to be a random function. The light source described by (12.3-9) is no longer monochromatic but consists of a carrier wave at angular frequency ω_0 with randomly fluctuating amplitude and phase. Such behavior is typical of partially coherent light sources such as light emitting diodes and even of injection lasers provided that they oscillate at a single carrier frequency. The degree of coherence of the light source depends on the nature of $A(t)$.

Because of its random fluctuations it is no longer possible to compute the propagation of the field intensity of the light pulse in a dispersive medium. This could only be done if we assumed a particular function $A(t)$ and even then it would be difficult. Instead we resort to a statistical treatment, computing an ensemble average of the light power. This means that instead of $\psi(z,t)$ we consider the average power

$$\langle P(z,t)\rangle = \langle |\psi(z,t)|^2\rangle \quad (12.3\text{-}10)$$

The bracket symbol indicates an ensemble average; that is we consider an average taken over many similar pulses that differ from each other only in the particular shape of the amplitude and phase of $A(t)$.

The ensemble average of the product of $A(t)$ with its complex conjugate counterpart taken at a different time,

$$R(t - t') = \langle A(t)A^*(t')\rangle \quad (12.3\text{-}11)$$

is an important quantity called the autocorrelation function of $A(t)$. As the notation indicates, $R(t - t')$ is assumed to be a function of the difference of $t - t'$ but not of t or t' themselves. A random function with this property is called a stationary random process. We encountered examples of stationary random processes and autocorrelation functions in Sections 5.5 and 9.4.

For the further development of the theory we need the autocorrelation function of the spectrum of the unmodulated source (12.3-9). The spectrum is defined as

$$\phi_0(\omega) = \frac{1}{2\pi} \int_{-\infty}^{\infty} \psi_0(t)e^{-i\omega t} \, dt \qquad (12.3\text{-}12)$$

so that its autocorrelation function follows from (12.3-9) and (12.3-11)

$$\langle \phi_0(\omega) \, \phi_0^*(\omega') \rangle = \frac{1}{(2\pi)^2} \int_{-\infty}^{\infty} \exp[i(\omega' - \omega)t]$$

$$\cdot \int_{-\infty}^{\infty} R(t - t') \exp[i(\omega_0 - \omega')(t - t')]dt' \, dt \qquad (12.3\text{-}13)$$

By introducing the new variables $u = t - t'$ and $t = v$, we see immediately that the double integral decomposes into the product of two integrals. The first integral is Dirac's delta function

$$\delta(\omega - \omega') = \frac{1}{2\pi} \int_{-\infty}^{\infty} \exp[i(\omega' - \omega)v] \, dv \qquad (12.3\text{-}14)$$

and the second integral is the power spectrum of the random source amplitude

$$|\phi_0(\omega - \omega_0)|^2 = \frac{1}{2\pi} \int_{-\infty}^{\infty} R(u) \exp[i(\omega_0 - \omega)u] \, du \qquad (12.3\text{-}15)$$

With these definitions we can write the autocorrelation function of the source spectrum as

$$\langle \phi_0(\omega) \, \phi_0^*(\omega') \rangle = |\hat{\phi}_0(\omega - \omega_0)|^2 \, \delta(\omega - \omega') \qquad (12.3\text{-}16)$$

It is remarkable that the autocorrelation function of the power spectrum of the source is proportional to a delta function. The only assumption needed to obtain this result was to require that the amplitude $A(t)$ of the unmodulated source is a stationary random process.

As in the case of the monochromatic source, we obtain the spectrum of the modulated, partially coherent source as the Fourier transform of the product of the unmodulated source function $\psi_0(t)$ and the square root of the modulation function $s(t)$

$$\phi(\omega) = \frac{1}{2\pi} \int_{-\infty}^{\infty} s^{1/2}(t) \, \psi_0(t)e^{-i\omega t} \, dt$$

$$= \int_{-\infty}^{\infty} \phi_0(\omega') F(\omega - \omega') \, d\omega' \qquad (12.3\text{-}17)$$

where $F(\omega)$ is the Fourier transform of the square root of the modulation function (12.3-2)

$$F(\omega) = \frac{1}{2\pi} \int_{-\infty}^{\infty} s^{1/2}(t) \, e^{-i\omega t} \, dt$$

$$= \left(\frac{S}{2\pi}\right)^{1/2} T \exp\left[-\frac{1}{2} T^2 \omega^2\right] \qquad (12.3\text{-}18)$$

The propagation of the pulse along the fiber is again given by (12.3-4) so that the ensemble average of the power (12.3-10) becomes

$$\langle P(z,t) \rangle \cdot = \iint_{-\infty}^{\infty} \langle \phi(\omega) \, \phi^*(\omega') \rangle$$

$$\cdot \exp\{i[(\omega - \omega') t - (\beta - \beta')z]\} \, d\omega \, d\omega' \qquad (12.3\text{-}19)$$

The autocorrelation function of the spectral function does automatically occur when we form the expression for the ensemble average of the light power. With the help of (12.3-16) and (12.3-17) we obtain

$$\langle \phi(\omega) \, \phi^*(\omega') \rangle = \int_{-\infty}^{\infty} |\hat{\phi}_0(\omega_0 - \omega'')|^2 F(\omega - \omega'') F^*(\omega' - \omega'') d\omega'' \qquad (12.3\text{-}20)$$

Substitution of (12.3-20) into (12.3-19) and rearranging of terms yields

$$\langle P(z,t) \rangle = \int_{-\infty}^{\infty} |\hat{\phi}_0(\omega_0 - \omega')|^2$$

$$\cdot \left| \int_{-\infty}^{\infty} F(\omega - \omega') \exp\left\{i\left[(\omega - \omega') t - (\beta - \beta')z\right]\right\} d\omega \right|^2 d\omega' \qquad (12.3\text{-}21)$$

Equation (12.3-21) is exact and holds for a partially coherent source whose amplitude and phase fluctuate like stationary random processes. At this point, we must again introduce approximations to be able to make further progress. However, we improve the approximation by including in the Taylor expansion of the propagation constant one more term

$$\beta = \beta_0 + \dot{\beta}_0 (\omega - \omega_0) + \frac{1}{2} \ddot{\beta}_0 (\omega - \omega_0)^2 + \frac{1}{6} \dddot{\beta}_0 (\omega - \omega_0)^3 \qquad (12.3\text{-}22)$$

The difference of the propagation constants taken at two neighboring frequencies occuring in (12.3-21) can be written in the form

$$\beta - \beta' = \left\{ \ddot{\beta}_0 + \frac{1}{2} \dddot{\beta}_0 [(\omega - \omega_0) + (\omega' - \omega_0)] + \frac{1}{6} \dddot{\beta}_0 [(\omega - \omega_0)^2 \right.$$

$$\left. + (\omega' - \omega_0)^2 + (\omega - \omega_0)(\omega' - \omega_0)] \right\} (\omega - \omega') \quad (12.3\text{-}23)$$

The power spectrum of the source fluctuations is also assumed to be a Gaussian function,

$$|\hat{\phi}_0(\omega_0 - \omega)|^2 = \frac{P_0}{\pi^{1/2} W} \exp \left[-(\omega - \omega_0)^2 / W^2 \right] \quad (12.3\text{-}24)$$

The normalization of this function ensures that integration over the radian frequency results in the power P_0 of the unmodulated light source. The parameter W is the half width at the $1/e$ points of the spectral function of the source.

Because of the improved accuracy of the expansion (12.3-22) the integrals in (12.3-21) can no longer be solved exactly. However, we can express the average power as a Fourier transform (the 0 of β_0 is omitted.)

$$\langle P(z,t) \rangle = \int_{-\infty}^{\infty} G(z,x) \exp \left(i \frac{t - \dot{\beta} z}{T} x \right) dx \quad (12.3\text{-}25)$$

with the spectral function

$$G(z,x) = \frac{P_0 S}{2\sqrt{\pi}} \frac{\exp(-x^2/4) \exp(-iBx^3/4)}{[1 + 3iBx(1 + V^2)]^{1/2}}$$

$$\cdot \exp \left[\frac{-Dx^2(1 + V^2)}{1 + 3iBx(1 + V^2)} \right] \quad (12.3\text{-}26)$$

We used a number of abbreviations. The parameter V is the normalized spectral width of the light source

$$V = TW \quad (12.3\text{-}27)$$

The parameter D is proportional to the second derivative of the propagation constant and the fiber length z

$$D = \ddot{\beta} z / (2T^2) \quad (12.3\text{-}28)$$

Finally, B is proportional to the third derivative of the propagation constant and also to z

$$B = \dddot{\beta} z / (6T^3) \quad (12.3\text{-}29)$$

D is the parameter describing first order dispersion while B is the second order dispersion parameter. If D and B are both zero there is no dispersion, the pulse propagates without changing its shape. The second derivative of β is responsible for a change of the pulse width. Since $\ddot{\beta}$ is the first term in the series expansion that causes any pulse distortion, we say that it causes first order dispersion. The next term in the expansion of β is responsible for second order dispersion. It is totally insignificant unless $\ddot{\beta} \approx 0$ and hence $D \approx 0$.

The derivation of (12.3-25) and (12.3-26) from (12.3-21) involves some cumbersome integrations which we do not demonstrate in detail.

If second order dispersion can be neglected, $B = 0$, the Fourier integral (12.3-25) can be solved, with the result

$$\langle P(z,t) \rangle = \frac{SP_0}{[1 + 4D^2(1 + V^2)]^{1/2}} \exp \left\{ - \frac{(t - \dot{\beta}_0 z)^2 / T^2}{1 + 4 D^2(1 + V^2)} \right\} \qquad (12.3\text{-}30)$$

Considering that $A^2 = P_0$, we see that for $W = 0$ (and hence $V = 0$) this expression becomes identical with (12.3-8).

With Gaussian input pulses, the output pulse retains its Gaussian shape provided second order dispersion is negligible ($B = 0$). Figure 12.3.1 shows the input pulse ($z = 0$ and hence $D = 0$) and the average power of the output pulse for a combination of dispersion and fiber length such that $D = 3$. These curves were computed from (12.3-30) with $V = 0$. However, since V appears in (12.3-30) only in the combination $D^2 (1 + V^2)$ the pulse shape for nonzero V would not change provided $D^2(1 + V^2) = 9$.

Looking at the input and output pulses shown in Figure 12.3.1, it appears as though the Fourier spectrum of the pulse should have become narrower as the pulse itself widens as it travels along the fiber. However, since the fiber is a linear device and since losses are being ignored, the spectral distribution of the pulse cannot change. This seeming paradox is resolved if we consider the electric field strength (12.3-7) of the light pulse instead of its time averaged

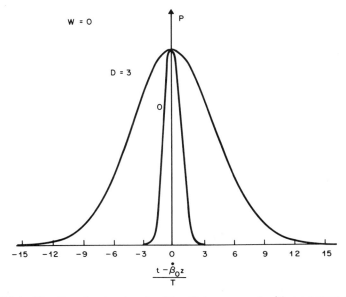

Figure 12.3.1 The Gaussian input pulse ($D = 0$) is compared with an output pulse corresponding to $D = 3$.

power. (The time average is automatically performed when we compute the power according to (12.3-10).) We remove the effect of the rapid oscillations of the pulse at the carrier frequency by multiplying $\psi\,(z,t)$ of (12.3-7) with exp $(-\omega_0 t)$. The square of the real part of this product is plotted in Fig. 12.3.2 as the solid line. This figure was also computed for $D = 3$. The corresponding average pulse from Figure 12.3.1 is reproduced as the dotted curve in Figure 12.3.2. It is now apparent that there is actually no bandwidth reduction as the pulse travels in the dispersive medium. Inside the pulse envelop the field performs rapid oscillations (which have nothing to do with the even more rapid field oscillation at the carrier frequency). The time averaged power is representative of a received pulse where the time constant of the receiver (its narrow bandwidth) suppresses the more rapid oscillations inside the pulse envelope, producing the pulse shown in Figure 12.3.1.

The influence of the third derivative of the propagation constant with respect to the radian frequency becomes noticeable only if D is very nearly zero. Once first order dispersion vanishes ($D = 0$) second order dispersion can cause a distortion of the average pulse from its initial Gaussian shape. Figure 12.3.3 shows how the pulse (shown in Figure 12.3.1 as the curve with $D = 0$) would look if second order dispersion became strong enough to make $B = 1$. This and the following pulse figures were obtained by numerical computation of the Fourier integral (12.3-25) with the help of the fast Fourier transform algorism. The pulse shown in Figure 12.3.3. is no longer symmetrical relative

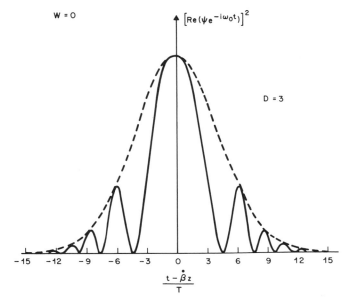

Figure 12.3.2. The dotted curve repeats the curve for $D = 3$ of Figure 12.3.1. The solid curve shows how the instantaneous pulse power fluctuates in the absence of narrow band filtering.

to its peak. The oscillation on the trailing edge of the pulse would be seen on its leading edge if B changed its sign. It is interesting that the theory now predicts fluctuations of the average pulse power not unlike those shown in Figure 12.3.2 for a pulse which is not time averaged (except for removing the carrier frequency). A receiver with insufficient bandwidth would, of course, not be able to reproduce the rapid fluctuations of the pulse shown in Figure 12.3.3. Thus, the averaging process inherent in the analysis is, after all, not fully equivalent to a narrow band filter.

In Figure 12.3.4, the value of B has been increased to $B = 5$ causing even more rapid pulse fluctuations and more pulse distortion. When second order dispersion predominates, an increase in source bandwidth is not equivalent to increased dispersion. The influence of an increase of the value of V on the pulse shape cannot be reproduced by leaving $V = 0$ and increasing the value of B. Figure 12.3.5 shows a pulse with $B = 5$ (as before) and $V = 5$. Instead of more rapid oscillations we see a lifting of the pulse tail. The distorting influence of second order dispersion can be removed by increasing D even if B is held constant. However, the return of the pulse to its Gaussian shape is paid

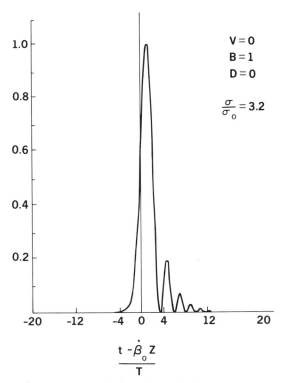

Figure 12.3.3.   Pulse distortion caused by second order dispersion with $V = 0$, $D = 0$, and $B = 1$. (From D. Marcuse, Pulse Distortion in Single-Mode Fibers, *Applied Optics*, Vol. 19, No. 10, May 1980, pp. 1653–60. Reprinted with permission.)

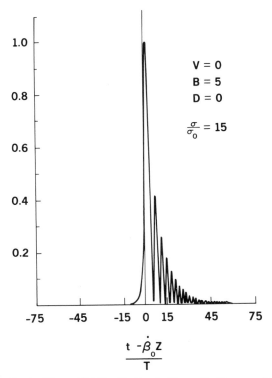

Figure 12.3.4   Similar to Figure 12.3.3 with $B = 5$. (From D. Marcuse, Pulse Distortion in Single-Mode Fibers, *Applied Optics*, Vol. 19, No. 10, May 1980, pp. 1653–60. Reprinted with permission.)

for in increased pulse width. Figure 12.3.6 shows a pulse with $B = 5$ and $D = 30$. Even though the tendency for a return to its Gaussian shape is apparent, it is not yet fully accomplished.

The values of D and B used in the preceding figures were chosen arbitrarily without regard for the question if such values are indeed achievable in a practical optical fiber.

Whenever second order dispersion becomes important the pulse shape must be computed by solving the Fourier integral (12.3-25) with the help of the spectral function (12.3-26). However, the spectral function can also be used directly to find the rms width σ of the pulse that is defined as

$$\sigma^2 = \frac{\displaystyle\int_{-\infty}^{\infty} t^2 \langle P(z,t) \rangle \; dt}{\displaystyle\int_{-\infty}^{\infty} \langle P(z,t) \rangle \; dt} - \left\{ \frac{\displaystyle\int_{-\infty}^{\infty} t \langle P(z,t) \rangle \; dt}{\displaystyle\int_{-\infty}^{\infty} \langle P(z,t) \rangle \; dt} \right\}^2 \qquad (12.3\text{-}31)$$

Fortunately, the integrals in this expression need not be solved! Instead, they

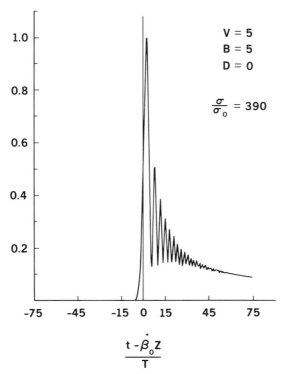

Figure 12.3.5   Similar to Figure 12.3.3 with $B = 5$ and $V = 5$. (From D. Marcuse, Pulse Distortion in Single-Mode Fibers, *Applied Optics*, Vol. 19, No. 10, May 1980, pp. 1653–60. Reprinted with permission.)

can be evaluated by differentiation of the spectral function. To see this, we consider the definition of the spectral function in terms of the Fourier transformation

$$\exp\left(-i\frac{\dot{\beta}z}{T}\right)G(z,x) = \frac{1}{2\pi T}\int_{-\infty}^{\infty} \langle P(z,t) \rangle \; e^{-i(t/T)x} \, dt \qquad (12.3\text{-}32)$$

This definition allows us to express the square of the rms pulse width in the following way

$$\sigma^2 = -T^2\left[\frac{d^2 G}{dx^2}\Big/ G\right]_{x=0} - \left\{iT\left[\frac{dG}{dx}\Big/ G\right]_{x=0}\right\}^2 \qquad (12.3\text{-}33)$$

When the differentiations are carried out we obtain[144]

$$\sigma = \frac{T}{\sqrt{2}}\left[1 + \left(\frac{\ddot{\beta}_0 z}{T^2}\right)^2 (1 + T^2 W^2) + \left(\frac{\ddot{\beta}_0 z}{2T^3}\right)^2 (1 + T^2 W^2)^2\right]^{1/2} \qquad (12.3\text{-}34)$$

The rms width of the input pulse follows from this expression with $z = 0$

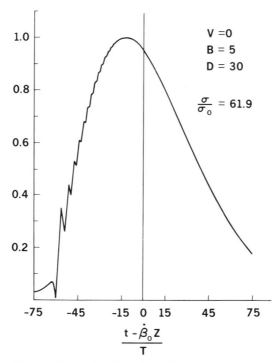

Figure 12.3.6   In this curve first order dispersion with $D = 30$ tends to restore the Gaussian shape of the pulse; $B = 5$, $V = 0$. (From D. Marcuse, Pulse Distortion in Single-Mode Fibers, *Applied Optics*, Vol. 19, No. 10, May 1980, pp. 1653–60. Reprinted with permission.)

$$\sigma_0 = \frac{T}{\sqrt{2}} \qquad (12.3\text{-}35)$$

If the fiber length z is so large that the "1" in (12.3-34) becomes negligible compared to the other two terms, the rms pulse width becomes directly proportional to z. This means that the rms pulse width is proportional to the fiber length for sufficiently strong first and/or second order dispersion. If second order dispersion is negligible, $\ddot{\beta}_0 = 0$, the rms pulse width becomes

$$\sigma = \frac{1}{\sqrt{2}} \frac{\ddot{\beta}_0 z}{T} (1 + T^2 W^2)^{1/2} \qquad (12.3\text{-}36)$$

For large source spectral width, $T \cdot W \gg 1$, (12.3-36) assumes the form

$$\sigma = \frac{1}{\sqrt{2}} W \ddot{\beta}_0 z \qquad (12.3\text{-}37)$$

It is interesting to compare this expression for the rms pulse width with the simple expression (12.2-18) for the increase in pulse width that we derived by heuristic arguments. To achieve this comparison we must express the spectral half width of the source W in terms of the (full) source width $\Delta\lambda$

$$W = -\pi c \frac{\Delta\lambda}{\lambda^2} \qquad (12.3\text{-}38)$$

Next we convert the ω derivative of the propagation constant into a derivative with respect to wavelength and obtain with the help of (11.6-22)

$$\ddot{\beta}_0 = \frac{d^2\beta_0}{d\omega^2} = \frac{d\hat{\tau}}{d\omega} = -\frac{\lambda^2}{2\pi c}\frac{d\hat{\tau}}{d\lambda} \qquad (12.3\text{-}39)$$

With these equations we obtain from (12.3-37)

$$\sigma/z = 2^{-3/2}\frac{d\hat{\tau}}{d\lambda}\,\Delta\lambda \qquad (12.3\text{-}40)$$

This formula does indeed agree with the heuristic formula (12.2-18). The factor $2^{-3/2}$ occurs because we are now dealing with the rms pulse width. Even though (12.2.-18) was supposed to represent the increase of the width of the input pulse, this comparison shows that it actually represents the width of the output pulse and not just its increase.

We have based the comparison between (12.2-18) and (12.3-40) on the assumption that the spectral width of the source is much larger than the width of the Fourier spectrum of the input pulse. However, if we let $W = 0$ in (12.3-36) we see that $1/T$ replaces W in (12.3-37). Since $1/T$ is proportional to the half width of the Fourier spectrum of the pulse, we see that (12.3-40) still applies if $\Delta\lambda$ is now interpreted as the spectral width of the pulse. This shows that the heuristic formula is correct (at least to order of magnitude) regardless of the spectral width of the unmodulated light source as long as $\ddot{\beta}_0$ is negligible.

To evaluate the rms pulse width we use the dispersion curves for $\ddot{\beta}_0$ and $\dddot{\beta}_0$ shown in Figures 12.3.7 and 12.3.8 that are representative[145,146] of fused silica, SiO_2. Even though these curves are plotted as functions of the wavelength λ they represent derivatives of the propagation constant taken with respect to the angular frequency ω. We see once more (compare Figure 12.2.1) that first order dispersion vanishes at a certain wavelength $\lambda = \lambda_0$ which, in fused silica, is $\lambda_0 = 1.27\ \mu m$. Figure 12.3.7 and several subsequent figures are drawn as functions of $\lambda - \lambda_0$ to make them slightly more general. The dispersion characteristics of high silica fibers are often similar to that of pure SiO_2 but the value of λ_0 may differ. For this reason we show wavelength plots relative to the minimum dispersion wavelength λ_0.

Figure 12.3.9 shows several curves of the output rms pulse width as functions of the input rms width σ_0. These curves are computed for a wavelength very close to λ_0, $\lambda - \lambda_0 = 0.02\ \mu m$, and for a fiber of 10 km length using the dispersion curves Figures 12.3.7 and 12.3.8 and several different values for the full width $2W$ of the unmodulated source spectral width.[147] Corresponding values of the source width expressed in wavelength units $\Delta\lambda$ at $\lambda = 1.3\ \mu m$ are also indicated at each curve. The most conspicuous feature of

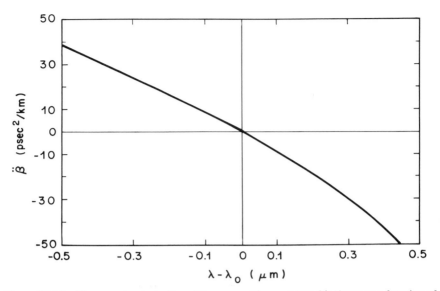

Figure 12.3.7   The second ω derivative of the propagation constant β is shown as a function of wavelength (relative to λ_0) for SiO_2. (From D. Marcuse and C. Lin, Low Dispersion Single-Mode Fiber Transmission—The Question of Practical vs. Theoretical Maximum Transmission Bandwidth. *IEEE J. Quantum Electronics* © 1981 IEEE. Reprinted with permission.)

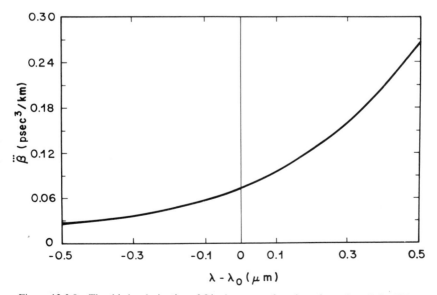

Figure 12.3.8   The third ω derivative of β is shown as a function of wavelength for SiO_2. (From D. Marcuse and C. Lin, Low Dispersion Single-Mode Fiber Transmission—The Question of Practical vs. Theoretical Maximum Transmission Bandwidth. *IEEE J. Quantum Electronics* © 1981 IEEE. Reprinted with permission.)

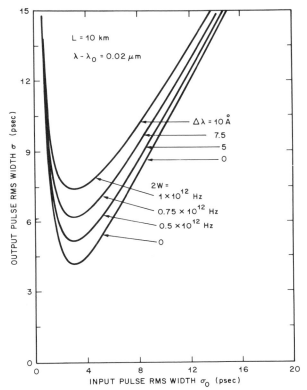

Figure 12.3.9   Output rms pulse width as function of the input rms pulse width for several values of the source spectral width for $\lambda - \lambda_0 = 0.02$ μm. (From D. Marcuse and C. Lin, Low Dispersion Single-Mode Fiber Transmission—The Question of Practical vs. Theoretical Maximum Transmission Bandwidth. *IEEE J. Quantum Electronics* © 1981 IEEE. Reprinted with permission.)

these curves is a pronounced minimum occuring at the same value of σ_0. This means that there is an optimum input width that minimizes the width of the output pulse. Figure 12.3.10 shows similar curves also for a fiber of 10 km length but for a wavelength that is farther removed from λ_0, $\lambda - \lambda_0 = 0.28$ μm. The mimima of the rms output pulse width are now much shallower, but the curves still rise very steeply for rms input pulse widths that are shorter than the optimum width.

To determine the value of T, or equivalently of σ_0, at which the minimum occurs, we differentiate the expression (12.3-34) for the rms output pulse width with respect to T and find the following condition for a minimum

$$\left(\frac{\ddot{\beta}_0 z}{T^2}\right)^2 + \frac{1}{2}\left(\frac{\ddot{\beta}_0 z}{T^3}\right)^2 (1 + T^2 W^2) = 1 \qquad (12.3\text{-}41)$$

Instead of presenting a cumbersome general solution for this third order equation in T^2 we list several special cases. Using (12.3-35) to express T in terms of the input rms pulse width σ_0 we find the following optimum input rms

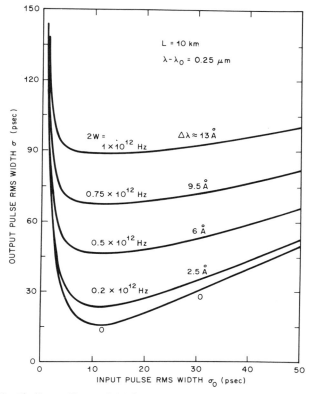

Figure 12.3.10   Similar to Figure 12.3.9 for a wavelength farther removed from the point of minimum dispersion, $\lambda - \lambda_0 = 0.25 \, \mu$m. (From D. Marcuse and C. Lin, Low Dispersion Single-Mode Fiber Transmission—The Question of Practical vs. Theoretical Maximum Transmission Bandwidth. *IEEE J. Quantum Electronics* © 1981 IEEE. Reprinted with permission.)

pulse width in case of negligible second order dispersion

$$(\sigma_0)_{opt} = \left(\frac{1}{2} \left| \ddot{\beta}_0 \right| z \right)^{1/2} \tag{12.3-42}$$

This formula shows that the optimum input pulse width increases only proportionally to the square root of the fiber length z. It is also noteworthy that this optimum value of the input pulse width is independent of the source bandwidth 2W.

If $\ddot{\beta}_0 = 0$ we can again obtain simple solutions of (12.3-41) for two special cases. For $W = 0$ and $\ddot{\beta}_0 = 0$ we have

$$(\sigma_0)_{opt} = \left(\frac{1}{4} \left| \dddot{\beta}_0 \right| z \right)^{1/3} \tag{12.3-43}$$

indicating that the optimum rms input pulse width is proportional to the third root of the fiber length, if second order dispersion predominates and a monochromatic source is used. In the opposite case, $T \cdot W >> 1$ and $\ddot{\beta}_0 = 0$, the optimum input pulse width is again proportional to the square root of fiber length

$$(\sigma_0)_{opt} = \left(\frac{W}{\sqrt{8}} \left| \ddot{\beta}_0 \right| z \right)^{1/2}$$

(12.3-44)

The optimum rms input pulse width is shown in Figure 12.3.11 as a function of wavelength (relative to the minimum dispersion wavelength λ_0) for several fiber lengths. Second order dispersion affects these curves only in the immediate vicinity of the point $\lambda = \lambda_0$ and is not discernible on the scale of Figure 12.3.11. Figure 12.3.12 shows the optimum rms input pulse width in the immediate vicinity of λ_0 for a fiber of 100 km length. The extremely low fiber losses already achieved in practice make it feasible to consider fibers of such length. Figure 12.3.12 shows the optimum rms pulse width for a source of spectral width $\Delta \lambda = 100$ Å and for a monochromatic source. Second order dispersion prevents these curves to go down to $(\sigma_0) = 0$ for $\lambda = \lambda_0$. However, when the two curves merge into one, the influence of second order dispersion is already lost. Thus it is clear that second order dispersion is noticeable only as long as $\lambda - \lambda_0 < 0.01$ Å. These curves show also that the optimum rms input pulse width is extremely short and becomes of interest only for single mode fiber systems that use picosend (1 psec = 10^{-12} sec) pulses.

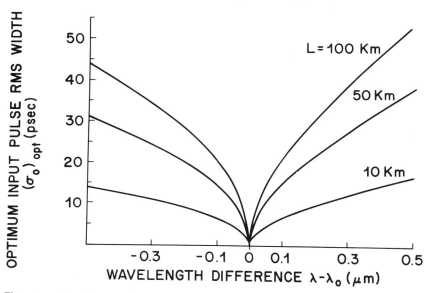

Figure 12.3.11   Optimum input pulse as a function of the wavelength difference $\lambda - \lambda_0$ for several different fiber lengths. (From D. Marcuse and C. Lin, Low Dispersion Single-Mode Fiber Transmission—The Question of Practical vs. Theoretical Maximum Transmission Bandwidth. *IEEE J. Quantum Electronics* © 1981 IEEE. Reprinted with permission.)

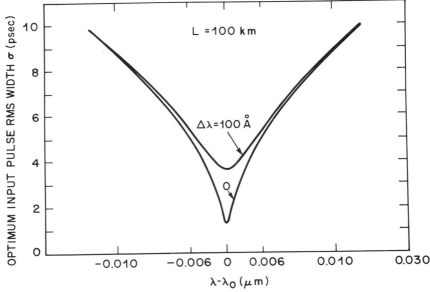

Figure 12.3.12   This figure shows one of the curves of Figure 12.3.11 in the immediate vicinity of the point of minimum dispersion for two different values of the source spectral width. (From D. Marcuse and C. Lin, Low Dispersion Single-Mode Fiber Transmission—The Question of Practical vs. Theoretical Maximum Transmission Bandwidth. *IEEE J. Quantum Electronics* © 1981 IEEE. Reprinted with permission.)

12.4 DISPERSION IN SINGLE MODE FIBERS

In Section 12.3, we discussed pulse dispersion in single mode fibers in general terms and gave examples that are representative of plane waves in fused silica. As long as the core radius of a single mode fiber is large (requiring a small index difference between core and cladding) the dispersion of the guided mode differs only slightly from that of a plane wave. But when the core radius becomes small and the index difference large the dispersion of the guided mode is significantly influenced by the fiber waveguide.[148,149]

To show this effect we compute $d\hat{\tau}/d\lambda$ [see (12.2-18) or (12.3-40)] from the approximate eigenvalue equation (8.6-32) for the HE_{11} mode of a step index fiber with a core of radius a and constant refractive index n_1. The core is surrounded by an infinitely extended cladding with refractive index n_2. Setting $\nu = 1$ in (8.6-32) we obtain

$$\kappa \frac{J_1(\kappa a)}{J_0(\kappa a)} = \gamma \frac{K_1(\gamma a)}{K_0(\gamma a)} \tag{12.4-1}$$

with the modified Hankel functions defined as[8]

$$K_\nu(x) = i \frac{\pi}{2} e^{i\nu\pi/2} H_\nu^{(1)}(ix) \tag{12.4-2}$$

The transverse propagation constant κ (in the core) is related to the longitudinal propagation constant β by the equation

$$\kappa = (n_1^2 k_0^2 - \beta^2)^{1/2} \tag{12.4-3}$$

and the transverse decay parameter γ (in the cladding) is related to β by

$$\gamma = (\beta^2 - n_2^2 k_0^2)^{1/2} \tag{12.4-4}$$

The sum of the squares of κ and γ is related to the normalized frequency parameter V

$$V = k_0 a (n_1^2 - n_2^2)^{1/2} = [(\kappa a)^2 + (\gamma a)^2]^{1/2} \tag{12.4-5}$$

The parameter κ, and hence all the other parameters, can be computed as solutions of the simultaneous equations (12.4-1) and (12.4-5).

We incorporate material dispersion into the theory by letting n_2 depend on the wavelength λ via a two-term Sellmeier equation (12.2-12) with the coefficients chosen as in Figure 12.2.1. The ratio n_1/n_2 is held constant.

The specific group delay (group delay per unit length) $\hat{\tau}$ is obtained as the derivative of

$$\hat{\tau} = \frac{d\beta}{d\omega} = -\frac{\lambda^2}{2\pi c}\frac{d\beta}{d\lambda} \tag{12.4-6}$$

According to (12.2-18) or (12.3-40) first order dispersion is the derivative of $\hat{\tau}$ so that we define the dispersion parameter D as

$$D = \frac{d\hat{\tau}}{d\lambda} \tag{12.4-7}$$

Since the relative variation of β is very small, higher computational accuracy can be achieved by expressing D in terms of derivatives of n_1 and κ

$$D = \frac{k_0}{c\lambda\beta^3}\left\{\lambda^2\beta^2\left(\frac{\kappa}{k_0^2}\frac{d^2\kappa}{d\lambda^2} - n_1\frac{d^2n_1}{d\lambda^2}\right) + \kappa^2\left(n_1 - \lambda\frac{dn_1}{d\lambda}\right)^2\right.$$
$$\left. + \lambda\frac{d\kappa}{d\lambda}\left[2n_1\kappa\left(n_1 - \lambda\frac{dn_1}{d\lambda}\right) + 2\frac{\kappa\beta^2}{k_0^2} + n_1^2\lambda\frac{d\kappa}{d\lambda}\right]\right\} \tag{12.4-8}$$

The expression for material dispersion is obtained by setting κ and its derivatives equal to zero and using $\beta = n_1 k_0$

$$D = -\frac{\lambda}{c}\frac{d^2n_1}{d\lambda^2} \tag{12.4-9}$$

The dispersion equation (12.4-8) cannot be solved analytically. Figure 12.4.1 shows numerical solutions that were obtained by solving the eigenvalue equation (12.4-1) for κ at ten different wavelengths.[150] These ten solutions were used to determine the coefficients of a fourth order polynomial by least mean square fitting. The derivatives of κ were then computed as derivatives of

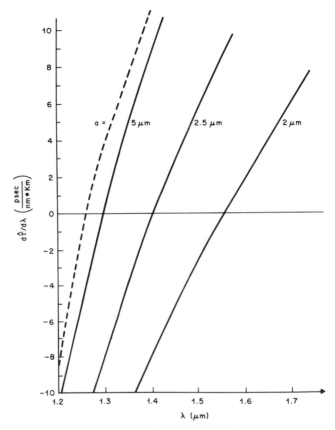

Figure 12.4.i   Dispersion curves of three different step-inex fibers. The dotted curve represents pure material dispersion, the shifts of the curves are caused by waveguide effects.

the least mean square polynomial. The derivatives of n_1 were obtained by differentiating the Sellmeier equation.

The dotted curve in Figure 12.4.1 represents the material dispersion (12.4-9). The solid curves represent the dispersion of the guided HE_{11} mode for several core radii a (in–μm) with $V = 1.8$ at the zero crossings. For radii larger than $a = 5\ \mu m$ wave guidance has little influence on dispersion. For large values of "a" the dispersion curves merge with the dotted (material dispersion) curve.

The dispersion curves shown in Figure 12.4.1 are of great significance for light communications. Optical fibers made of germanium doped fused silica have a loss minimum near $\lambda = 1.55\ \mu m$. It is, of course, desirable to shift the point of vanishing (first order) dispersion to the point where the losses reach a minimum. Since the zero dispersion point for fused silica lies at $\lambda = 1.27\ \mu m$, waveguide dispersion can be used to shift it to the 1.5 μm region.

Practical realization of this scheme is complicated by the required small core radii. Ease of handling and splicing makes fibers with larger core radii more desirable.

Our discussion applies to step index fibers. The shape of the refractive index profile influences the dispersion curve. Other index profiles may lead to even more desirable dispersion curves. For example, it has been shown that the so called W fiber has a dispersion curve with two zero crossings in the wavelength range shown in Figure 12.4.1.[151]

12.5 DISPERSION IN MULTIMODE FIBERS

The principal contribution to dispersion in multimode fibers was identified in Section 11.6 as intermodal dispersion caused by the group delay differences of the modes. In this section, we return to intermodal dispersion effects.

Olshansky and Keck[134] made a significant contribution to the understanding of intermodal disersion by incorporating material and profile dispersion into the theory of Gloge and Marcatili.[129] Material dispersion is present whenever $dn_1/d\lambda \neq 0$, profile dispersion makes itself felt when $d\Delta/d\lambda \neq 0$. When we derived the expression (11.6-24) for the group delay and its approximation (11.6-25) we asumed that $dn_1/d\lambda = 0$ and $d\Delta/d\lambda = 0$. If these restrictions are dropped the expression (11.6-25) assumes the more general form

$$\hat{\tau} = \frac{\bar{n}_1}{c} \left[1 + \Delta \frac{g - 2 - \epsilon}{g + 2} \left(\frac{M}{N} \right)^{g/(g+2)} \right.$$

$$\left. + \frac{\Delta^2}{2} \frac{3g - 2 - 2\epsilon}{g + 2} \left(\frac{M}{N} \right)^{2g/(g+2)} \right] \quad (12.5\text{-}1)$$

Two new quantities appear in this equation, the group index

$$\bar{n}_1 = n_1 - \lambda \frac{dn_1}{d\lambda} \quad (12.5\text{-}2)$$

and the propfile dispersion parameter

$$\epsilon = - \frac{2n_1\lambda}{\bar{n}_1\Delta} \frac{d\Delta}{d\lambda} \quad (12.5\text{-}3)$$

Equation (12.5-1) is an approximation of $\hat{\tau}$ valid to second order in Δ.

In Section 11.6, we derived the impulse response $P(\tau)$ of power law, graded index fibers from the group delay [see Eq. (11.6-28)] under the assumption that all modes carry equal amounts of power. Much useful information can also be gained from the rms width of the pulse defined by (12.3-31). If we identify the integration variable t in (12.3-31) with the group delay τ and the (average) pulse power $\langle P(z,t) \rangle$ with the impulse response $P(\tau)$, we can use the relation between τ and the cumulative mode number M to write

$$F = \int_{-\infty}^{\infty} f(\tau) P(\tau) \, d\tau = \int_0^N f[\tau(M)] P(\tau) \, \frac{d\tau}{dM} \, dM \qquad (12.5\text{-}4)$$

where M is defined by (11.6-10) and N is the total number of modes, (11.6-15). The function $f(\tau)$ stands for either 1, τ, or τ^2 in (12.3-31). According to (11.6-28), the impulse response $P(\tau)$ is the derivative of M with respect to τ so that we can simplify (12.5-4) to read

$$F = \int_0^N f[\tau(M)] \, dM \qquad (12.5\text{-}5)$$

The formula (12.3-31) now assumes the simpler form

$$\sigma^2 = \frac{1}{N} \int_0^N \tau^2(M) \, dM - \left[\frac{1}{N} \int_0^N \tau(M) \, dM \right]^2 \qquad (12.5\text{-}6)$$

The detailed derivation of the rms pulse width from (12.5-6) requires an approximation of τ that includes terms up to the fourth power of Δ. Instead of retracing the necessary steps we immediately quote the result derived by Olshansky and Keck[134]

$$\sigma = \frac{L \bar{n}_1 \Delta}{2c} \frac{g}{g+1} \left(\frac{g+2}{3g+2} \right)^{1/2}$$

$$\cdot \left\{ C_1^2 + \frac{4 C_1 C_2 \Delta (g+1)}{2g+1} + \frac{16 \Delta^2 C_2^2 (g+1)^2}{(5g+2)(3g+2)} \right\}^{1/2} \qquad (12.5\text{-}7)$$

with the abbreviations

$$C_1 = \frac{g - 2 - \epsilon}{g + 2} \qquad (12.5\text{-}8)$$

and

$$C_2 = \frac{3g - 2 - 2\epsilon}{2(g + 2)} \qquad (12.5\text{-}9)$$

Olshansky and Keck have shown that the rms pulse width assumes a minimum for the following optimum value of the power law exponent

$$g_{opt} = 2 + \epsilon - \frac{(4 + \epsilon)(3 + \epsilon)}{5 + 2\epsilon} \Delta \qquad (12.5\text{-}10)$$

The profile dispersion parameter ϵ is defined by (12.5-3). For vanishing profile dispersion (12.5-10) becomes simply

$$g_{opt} = 2 - \frac{12}{5} \Delta \qquad (12.5\text{-}11)$$

The expression (12.5-10) for the optimum value of the power law exponent is extremely important for designing multimode fibers with maximum bandwidth. The graded index profiles of multimode fibers are produced by doping fused silica with suitable low-loss materials. The wavelength dependence of the dopant materials is usually not exactly the same as that of pure fused silica so that ϵ of (12.5-3) is not equal to zero. The dispersive properties of the core material depends on dopant concentration and hence varies as a function of the radial coordinate. This phenomenon is called profile dispersion. The shape of the optimum refractive index profile thus depends on the type of dopant material and on the wavelength of light at which the fiber is to be operated. However, the formulas (12.5-7) and (12.5-10) apply only to fibers whose cores contain only a single dopant species. Fibers doped with a mixture of dopant materials have more complicated dispersion behavior.

We saw in Section 11.6 that the WKB method can be used to derive a formula for the group delay of the modes of a multimode fiber. By allowing $n(r)$ to be a function of the wavelength, (11.6-35) generalizes to read

$$\hat{\tau} = \frac{k_0}{c\beta} \frac{\int_{r_0}^{r_1} \dfrac{n(r)[n(r) + k_0 \, dn(r)/dk_0]}{[n^2(r) k_0^2 - \beta^2 - \nu^2/r^2]^{1/2}} \, dr}{\int_{r_0}^{r_1} \dfrac{dr}{[n^2(r) k_0^2 - \beta^2 - \nu^2/r^2]^{1/2}}} \qquad (12.5\text{-}12)$$

As always, the integral extends from the lower turning point r_0 to the upper turning point r_1 where r_0 and r_1 are defined as the points where the square root expression in the denominators of the integrands vanish.

To evaluate the derivative $dn(r)/dk_0$ we assume that the refractive index profile can be expressed as

$$n^2(r) = n_1^2[1 - 2f(r)\Delta] \qquad (12.5\text{-}13)$$

In addition, we assume that, whereas n_1 and Δ are functions of λ, $f(r)$ is independent of the wavelength. This assumption restricts (12.5-13) to a fiber whose core contains only one species of dopant. The equation may still be valid if the core consists of a composite of dopant materials if one of them clearly predominates. In cores consisting of a mixture of several different dopants, that all contribute significantly to the increase (or decrease) of the refractive index, $f(r)$ would also be wavelength dependent. If we express $f(r)$ in terms of $n(r)$, using (12.5-13), we obtain the expression

$$n(r) \frac{dn(r)}{dk_0} = n^2(r) \left(\frac{1}{n_1} \frac{dn_1}{dk_0} + \frac{1}{2\Delta} \frac{d\Delta}{dk_0} \right) - \frac{n_1^2}{2\Delta} \frac{d\Delta}{dk_0} \qquad (12.5\text{-}14)$$

Converting the k_0 derivative to a λ derivative results in

$$k_0 n(r) \frac{dn(r)}{dk_0} = n^2(r)\ D_1 - D_2 \qquad (12.5\text{-}15)$$

with the abbreviations

$$D_1 = -\frac{\lambda}{n_1} \frac{dn_1}{d\lambda} - \frac{\lambda}{2\Delta} \frac{d\Delta}{d\lambda} \qquad (12.5\text{-}16)$$

and

$$D_2 = -\frac{n_1^2 \lambda}{2\Delta} \frac{d\Delta}{d\lambda} \qquad (12.5\text{-}17)$$

The specific group delays of all the modes can now be computed with the help of (12.5-12) and (12.5-15) provided $n(r)$ is given either by an analytical expression or as a result of a measurements. The derivatives $dn_1/d\lambda$ and $d\Delta/d\lambda$ must be obtained from suitable measurements. The integrals in (12.5-12) can usually not be solved analytically, requiring numerical methods for their evaluation. The computation of $\hat{\tau}$ from (12.5-12) seems to require that β be obtained as a solution of the eigenvalue equation. However, the following approximate procedure can save considerable computing time. Clearly, $\hat{\tau}$ is a continuous function of β. Instead of computing $\hat{\tau}$ for the correct eigenvalues it is simpler to compute it for a number of descrete values of β in the range $n_2 k_0 < \beta < n_1 k_0$. Simultaneously, we compute a number of μ values for the same set of β values using the eigenvalue equation (11.5-45). In general, these μ values are not integers. However, we now find approximate solutions for β by linear interpolation by requiring that μ assumes all the integral values that correspond to β values in the range $n_2 k_0 < \beta < n_1 k_0$. Simultaneously, a linear interpolation is used to find the corresponding $\hat{\tau}$ values.

The numerical integration can be performed by any of the available methods. However, there is one complication, the integrands become infinite at the lower and upper limits of the integrals! Denoting the expression under the square root sign in the denominator by $F(r)$, with $F(r_0) = F(r_1) = 0$, and assuming that $g(r)$ is an arbitrary continuous function, we have

$$\int_{r_0}^{r} [F(r)]^{-1/2}\ dr = 2(r - r_0)^{1/2}\ [1 - (F''/12F')\ (r - r_0)](F')^{-1/2} \qquad (12.5\text{-}18)$$

and

$$\int_{r_0}^{r} g(r)\ [F(r)]^{-1/2}\ dr = 2(r - r_0)^{1/2}\ g \left\{ 1 + \frac{1}{3}\ [(g'/g) \right.$$

$$\left. - (F''/4F')]\ (r - r_0) \right\} (F')^{-1/2} \qquad (12.5\text{-}19)$$

These expressions were obtained by approximating $F(r)$ and $g(r)$ by the first two terms of their Taylor series expansion at $r = r_0$. Since $F(r_0) = 0$ the second nonvanishing term of its expansion contains the second derivative F''. The interval $r - r_0$ is kept small but it is made larger than the steps used in the remainder of the integration range. The poles at the upper turning point can be handled by formulas similar to (12.5-18) and (12.5-19). To increase speed the trapezoidal rule was used for numerical integration.

We can now calculate the specific group delays for all the modes from (12.5-12). The impulse response can be obtained as follows. We divide the interval between the arrival times of the fastest and slowest modes into 15 time slots (or any other suitable number) and count the number of modes arriving in each slot. It it is assumed that all modes carry equal amounts of power, the total light power arriving in a given time slot is directly proportional to the number of modes counted in this slot. It the modes carry unequal amounts of power, each is counted with an appropriate weighting factor. By this method we obtain staircase approximations to the true impulse response function.

Figures 12.5.1 through 12.5.4 show staircase approximations of impulse response functions for fibers with power law refractive index profiles with $n_2 = 1.457$ and $n_1 - n_2 = 0.02$ so that $\Delta = 0.01345$.[152] Material and profile dispersion were excluded from the calculation of these figures. The power law exponent g is increasing from Figure 12.5.1 through 12.5.4. It should be noted that the horizontal scales are different in each figure because of the widely differing widths of the impulse response functions. The origin of the horizontal time axis coincides with the arrival time of the mode of lowest order. A comparison with Figure 11.6.2 shows that the general shapes of the

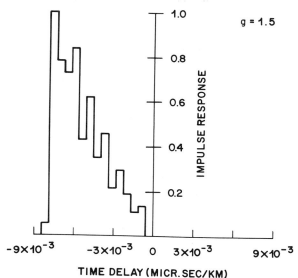

TIME DELAY (MICR. SEC/KM)

Figure 12.5.1   Staircase approximation of the impulse response of a multimode power law index fiber with $n_2 = 1.457$, $n_1 - n_2 = 0.02$, $\Delta = 0.01345$, core radius a $= 30$ μm and power law exponent $g = 1.5$. (From D. Marcuse, Calculation of Bandwidth from Index Profiles of Optical Fibers, Part I: Theory, *Applied Optics*, Vol. 18, No. 12, June 1979, pp. 2073–80. Reprinted with permission.)

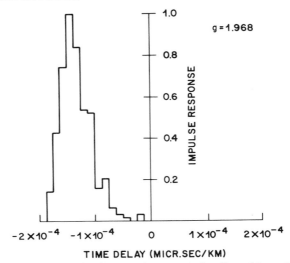

TIME DELAY (MICR.SEC/KM)

Figure 12.5.2   Same as Figure 12.5.1 but for an optimum power law fiber with g = 1.968. (From D. Marcuse, Calculation of Bandwidth from Index Profiles of Optical Fibers, Part I: Theory, *Applied Optics*, Vol. 18, No. 12, June 1979, pp. 2073–80. Reprinted with permission.)

staircase curves agree with the predictions of the theory. For $g < 2$ the peak of the pulse occurs before the arrival of the mode of lowest order, for $g > 2$ the peak arrives later. Figure 12.5.2 shows the optimum case for $g = 2 - (12/5)\Delta = 1.968$. Figure 12.5.4 illustrates that the impulse response of a fiber with $g = 2$ is square shaped.

Once the shape of the impulse response is known its rms width can be computed from (12.3-31).

Figure 12.5.5 shows plots of the rms width of the impulse response as a

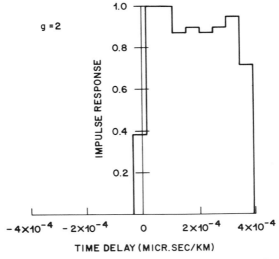

TIME DELAY (MICR.SEC/KM)

Figure 12.5.3   Impulse response of a fiber with square law index distribution, $g = 2$. (From D. Marcuse, Calculation of Bandwidth from Index Profiles of Optical Fibers, Part I: Theory, *Applied Optics*, Vol. 18, No. 12, June 1979, pp. 2073–80. Reprinted with permission.)

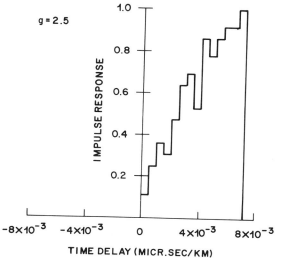

g = 2.5

IMPULSE RESPONSE

TIME DELAY (MICR.SEC/KM)

Figure 12.5.4 Same as the preceeding three figures but with $g = 2.5$. (From D. Marcuse, Calculation of Bandwidth from Index Profiles of Optical Fibers, Part I: Theory, *Applied Optics*, Vol. 18, No. 12, June 1979, pp. 2073–80. Reprinted with permission.)

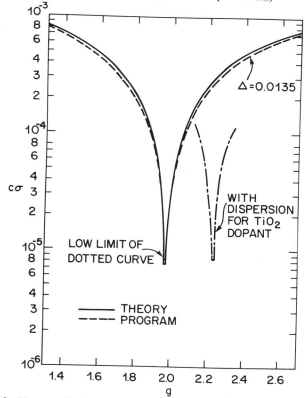

$c\sigma$

$\Delta = 0.0135$

WITH DISPERSION FOR TiO_2 DOPANT

LOW LIMIT OF DOTTED CURVE

——— THEORY
- - - - PROGRAM

g

Figure 12.5.5 The normalized rms pulse width $c\sigma/L$ as a function of the exponent g of the power law profile. The solid curve represents (12.5-7) with $\epsilon = 0$, the dotted curve is computed from (12.5-12). The dashed–dotted curve is computed from (12.5-12) for a titanium doped fiber. (From D. Marcuse, Calculation of Bandwidth from Index Profiles of Optical Fibers, Part I: Theory, *Applied Optics*, Vol. 18, No. 12, June 1979, pp. 2073–80. Reprinted with permission.)

function of the exponent g of the power law profile. The solid curve represents (12.5.7) with $\epsilon = 0$, the dotted curve was computed with the help of the numerical analysis described above.[152] The excellent agreement between the two curves establishes the validity of the numerical computation. The dashed-dotted curve in Figure 12.5.5 provides an illustration of the effect of profile dispersion. It was computed for a fused silica fiber with titanium doped core using information about its dispersive properties provided by Olshansky and Keck.[134] The shift of the curve caused by profile dispersion is again in perfect agreement with Keck and Olshansky's theory. Furthermore, it should be noted that profile dispersion does not change the minimum value assumed by the rms pulse width, it only shifts its position.

Titanium is not a typical dopant material for low loss optical fibers. This example was chosen in order to compare the numerical computation with a well known published result, because Olshansky and Keck demonstrate the effect of a finite source bandwidth on the rsm pulse width by using a Ti doped fiber as an example. We reproduce their curve[134] in Figure 12.5.6, which shows the rms pulse width plotted as a function of g for three different light sources: a

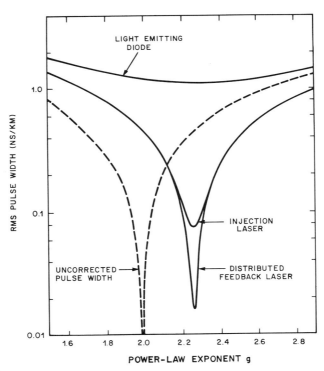

Figure 12.5.6   The rms of the impulse response for sources with different spectral width for a titanium doped fiber. The dotted curve represents a fiber without material or profile dispersion excited with a monochromatic source. (From R. Olshansky and D. B. Keck, Pulse Broadening in Graded-Index Optical Fibers, *Applied Optics*, Vol. 15, No. 2, February 1976, pp. 483–91. Reprinted with permission.)

light emitting diode of spectral width $\Delta\lambda = 150\text{Å}$, an injection laser with $\Delta\lambda = 10\text{Å}$ and a distributed feedback injection laser with $\Delta\lambda = 2\text{Å}$. The dotted curve in Figure 12.5.6 represents the rms pulse width that would result if material and profile dispersion were absent. This example shows that the source bandwidth may have a considerable influence on fiber performance and that material dispersion may not be negligible relative to intermodal dispersion effects. Operation at the dispersion minimum near $\lambda = 1.3\,\mu m$ alleviates the influence of material dispersion.

The Fourier transform of the impulse response function represents the spectral response of the fiber system. It may seem surprising that a system with power modulation and square law detection can be regarded as linear. This question has been answered affirmatively by Personick.[153] The frequency at which the amplitude of the spectrum decreases to one half of its peak value is defined as the fiber bandwidth. The staircase approximation of the impulse response may be used for computing the Fourier transform because its abrupt changes contribute only to the high frequency portion of the spectrum but do not affect the bandwidth.

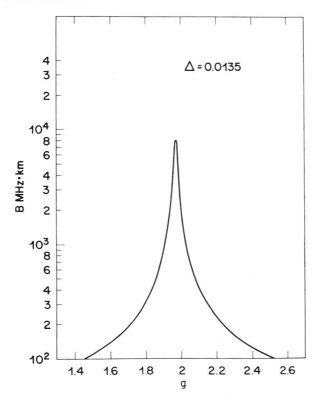

Figure 12.5.7   Bandwidth of power law profile fibers plotted as a function of the power law exponent g for $n_2 = 1.457$, $n_1 - n_2 = 0.02$, $\Delta = 0.01345$. (From D. Marcuse, Calculation of Bandwidth from Index Profiles of Optical Fibers, Part I: Theory, *Applied Optics*, Vol. 18, No. 12, June 1979, pp. 2073–80. Reprinted with permission.)

Figure 12.5.7 shows a plot of bandwidth versus the power law exponent g for a fiber with $\Delta = 1.345$ in the absence of profile dispersion. This curve makes it clear how strongly the bandwidth depends on g. A fiber designed to achieve maximum bandwidth must be built to extremely close tolerances.

Once a computer program exists for evaluating the impulse response and bandwidth of fibers with arbitrary refractive index profiles it can be used to study the effect of typical profile deformations. Fibers made by the process of chemical vapor deposition often have a refractive index depression (central dip) on-axis. Such a fault is simulated in Figure 12.5.8. Figure 12.5.9. shows the bandwidth of a fiber as a function of the relative width w/a of the central dip (a is the core radius). It is striking how much even a relatively narrow dip can reduce the fiber bandwidth. The dotted curve in the figure illustrates what would happen if modes with azimuthal mode numbers $\nu = 0$ and 1 did not carry any power. In this case the fiber would be more tolerant of the central dip.

Another important and typical flaw is a radial ripple on the refractive index profile is illustrated in Figure 12.5.10. The bandwidth reduction attributable to this flaw depends on the frequency of the periodic ripple. Figure 12.5.11 shows a plot of bandwidth as a function of the number N of periods that fit into the range $0 < r < a$. Again, there is a dramatic decrease of bandwidth with increasing ripple frequency. This example provides a test for the WKB theory. According to our result, the bandwidth would decrease indefinitely with increasing ripple frequency. However, independent calculations based on perturbation theory have shown that the bandwidth increases again after having reached a minimum.[154] This behavior is to be expected when we

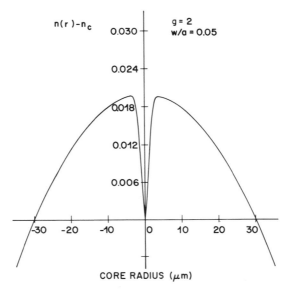

Figure 12.5.8 "Ideal" refractive index distribution distorted by a central dip. (From D. Marcuse, Calculation of Bandwidth from Index Profiles of Optical Fibers, Part I: Theory, *Applied Optics*, Vol. 18, No. 12, June 1979, pp. 2073–80. Reprinted with permission.)

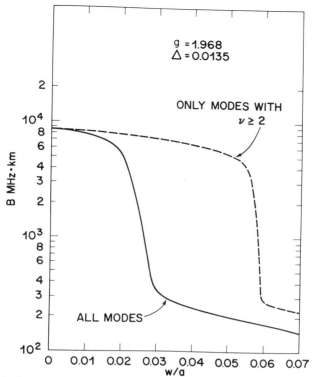

Figure 12.5.9 Bandwidth reduction caused by a central dip plotted as a function of the relative width of the dip. The dotted curve shows the bandwidth that would result if modes with $\nu = 0$ and 1 were absent. (From D. Marcuse, Calculation of Bandwidth from Index Profiles of Optical Fibers, Part I: Theory, *Applied Optics*, Vol. 18, No. 12, June 1979, pp. 2073–80. Reprinted with permission.)

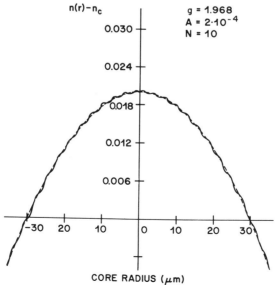

Figure 12.5.10 "Ideal" power law profile distorted by a sinusoidal ripple. (From D. Marcuse, Calculation of Bandwidth from Index Profiles of Optical Fibers, Part I: Theory, *Applied Optics*, Vol. 18, No. 12, June 1979, pp. 2073–80. Reprinted with permission.)

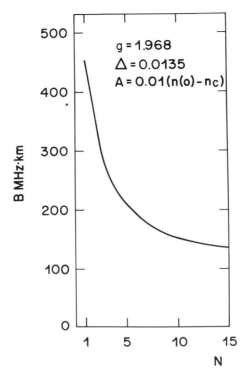

Figure 12.5.11   Bandwidth as a function of the number N of ripple periods (spatial frequency). This curve is not quite correct. For $N > 10$ the bandwidth should rise again. (From D. Marcuse, Calculation of Bandwidth from Index Profiles of Optical Fibers, Part I: Theory, *Applied Optics*, Vol. 18, No. 12, June 1979, pp. 2073–80. Reprinted with permission.)

consider that a wave cannot "sense" the ripple once its spatial period has become substantially shorter than the wavelength. The WKB theory, on which our calculation is based, is only as accurate as geometrical optics. Indeed, we have seen that the formula (11.6-35) can also be obtained from geometrical optics [see (11.2-44)]. Geometrical optics assumes that the wavelength of light vanishes and hence can not accomodate effects attributable to finite wavelength.

The properties of multimode fibers have been studied in great detail and fill a considerable literature. In this book we could only touch on some of their many interesting properties.

REFERENCES

1. M. Born and E. Wolf, *Principles of Optics*, 3rd edition, Pergamon Press, New York, 1964.
2. J. A. Stratton, *Electromagnetic Theory*, McGraw-Hill Book Company, New York, 1941.
3. C. H. Durney and C. C. Johnson, *Introduction to Modern Electromagnetics*, McGraw-Hill Book Company, New York, 1969.
4. R. Courant and D. Hilbert, *Methods of Mathematical Physics*, Vol. II, *Partial Differential Equations*, Interscience Publishers, New York, 1962.
5. P. A. M. Dirac, *The Principles of Quantum Mechanics*, 4th edition, New York, Oxford, 1958.
6. R. Courant and D. Hilbert, *Methods of Mathematical Physics*, Vol. I, Interscience Publishers, New York, 1953.
7. J. W. Goodman, *Introduction to Fourier Optics*, McGraw-Hill Book Company, New York, 1968.
8. M. Abromovitz and I. A. Stegun, *Handbook of Mathematical Functions with Formulas, Graphs and Mathematical Tables*, National Bureau of Standards Applied Mathematics Series, Vol. 55, National Bureau of Standards, Washington, D. C., 1965.
9. J. Mathews and R. L. Walker, *Mathematical Methods of Physics*, Benjamin, New York, 1965.
10. D. Marcuse, *Engineering Quantum Electrodynamics*, Harcourt, Brace & World, New York, 1970.
11. E. Jahnke and P. Emde, *Tables of Functions with Formulas and Curves*, 4th edition, Dover Publications, New York, 1945.
12. W. Heitler, *The Quantum Theory of Radiation*, 3rd edition, New York, Oxford, 1957.
13. H. Kogelnik, "Coupled Wave Theory for Thick Hologram Gratings," *B.S.T.J.*, Vol. 48, No. 9, November, 1969, pp. 2909–2947.
14. J. B. Keller and W. Streifer, "Complex Rays with an Application to Gaussian Beams," *J. Opt. Soc. Am.*, Vol. 61, No. 1, January, 1971, pp. 40–43.
15. M. Herzberger, *Modern Geometrical Optics*, Interscience Publishers, New York, 1958.
16. R. K. Luneberg, *Mathematical Theory of Optics*, University of California Press, Berkeley and Los Angeles, 1964.
17. W. Kaplan, *Advanced Calculus*, Addison-Wesley Publishing Company, Reading, Mass., 1953.
18. M. Kline and I. W. Kay, *Electromagnetic Theory and Geometrical Optics*, Interscience Publishers, New York, 1965.
19. H. Goldstein, *Classical Mechanics*, Addison-Wesley Publishing Company, Reading, Mass., 1950.
20. I. M. Gelfand and S. V. Fomin, *Calculus of Variations*, Prentice-Hall, Englewood Cliffe, N.J., 1963.
21. V. Flock, *The Theory of Space Time and Gravitation*, Pergamon Press, New York, 1959.

22. L. I. Schiff, *Quantum Mechanics*, 2nd edition, McGraw-Hill Book Company, New York, 1955.
23. I. Kitano, K. Koizumi, H. Matsumura, T. Uchida, and M. Furukawa, "A Light-Focusing Fibre Guide Prepared by Ion Exchange Techniques," *J. Japan Soc. Appl. Phys.* (Supplement), Vol. 39, January, 1970, pp. 63–70.
24. D. Marcuse, "Physical Limitations on Ray Oscillation Suppressors," *B.S.T.J.*, Vol. 45, No. 5, May-June, 1966, pp. 743–751.
25. R. C. Tolman, *The Principles of Statistical Mechanics*, Oxford University Press, New York, 1938.
26. R. Courant, *Differential and Integral Calculus*, Vol. 2, John Wiley, New York, 1936.
27. D. W. Berreman, "A Lens or Light Guide Using Convectively Distorted Thermal Gradients in Gases," *B.S.T.J.*, Vol. 43, No. 4, July, 1964, pp. 1469–1475.
28. P. Kaiser, "Measured Beam Deformation in a Guide Made of Tubular Gas Lenses," *B.S.T.J.*, Vol. 47, No. 2, February, 1968, pp. 179–194.
29. P. Kaiser, "An Optical Beam Waveguide Made of Counterflow Gas Lenses," to be published.
30. A. C. Beck, "An Experimental Gas Lens Optical Transmission Line," *IEEE Trans. Microwave Theory and Techniques*, MTT-15, No. 7, July, 1967, pp. 433–434.
31. D. Marcuse and S. E. Miller, "Analysis of a Tubular Gas Lens," *B.S.T.J.*, Vol. 43, No. 4, July, 1964, pp. 1759–1782.
32. M. Jacob, *Heat Transfer*, Vol. 1, John Wiley, New York, 1949.
33. D. Gloge, "Deformation of Gas Lenses by Gravity," *B.S.T.J.*, Vol. 46, No. 2, February, 1967, pp. 357–365.
34. D. Marcuse, "Theory of a Thermal Gradient Gas Lens," *IEEE Trans. Microwave Theory and Techniques*, MTT-13, No. 6, November, 1965, pp. 734–739.
35. W. H. Steier, "Measurements on a Thermal Gradient Gas Lens," *IEEE Trans. Microwave Theory and Techniques*, MTT-13, No. 6, November, 1965, pp. 740–748.
36. D. Marcuse, "Comparison between a Gas Lens and Its Equivalent Thin Lens," *B.S.T.J.*, Vol. 45, No. 8, October, 1966, pp. 1339–1344.
37. G. Toraldo di Francia, "Degrees of Freedom of an Image," *J. Opt. Soc, Am.,* Vol. 59, No. 7, July, 1969, pp. 799–804.
38. M. R. Spiegel, *Theory and Problems of Complex Variables with an Introduction to Conformal Mapping and Its Applications*, Schaum Publishing Company, New York, 1964.
39. C. E. Shannon, "A Mathematical Theory of Communications," *B.S.T.J.*, Vol. 27, 1948, pp. 379–423 and 623–656.
40. D. Slepian, "Prolate Spheroidal Wave Functions, Fourier Analysis and Uncertainty-IV: Extension to Many Dimensions, Generalized Prolate Spheroidal Functions," *B.S.T.J.*, Vol. 43, No. 6, November, 1964, pp. 3009–3057.
41. D. Slepian and E. Sonnenblick, "Eigenvalues Associated with Prolate Spheroidal Wave Functions of Zero Order," *B.S.T.J.*, Vol. 44, No. 8, October, 1965, pp. 1745–1759.
42. D. Gloge and D. Marcuse, "Formal Quantum Theory of Light Rays," *J. Opt. Soc. Am.*, Vol. 59, No. 12, December, 1969, pp. 1629–1631.

43. G. Goubau and F. Schwering, "On the Guided Propagation of Electromagnetic Beam Waves," *IRE Trans. on Antennas and Propagation*, Vol. AP-9, May, 1961, pp. 248–256.

44. J. R. Pierce, *Theory and Design of Electron Beams*, Van Nostrand Reinhold, New York, 1954.

45. S. E. Miller, "Alternating Gradient Focusing and Related Properties of Conventional Convergent Lens Focusing," *B.S.T.J.*, Vol. 43, No. 4, Part II, July, 1964, pp. 1741–1758.

46. G. D. Boyd and H. Kogelnik, "Generalized Confocal Resonator Theory," *B.S.T.J.*, Vol. 41, No. 4, July, 1962, pp. 1347–1369.

47. G. D. Boyd and J. P. Gordon, "Confocal Multimode Resonator for Millimeter through Optical Wavelength Masers," *B.S.T.J.*, Vol. 40, No. 2, March, 1961, pp. 489–508.

48. A. G. Fox and T. Li, "Resonant Modes in a Maser Interferometer," *B.S.T.J.*, Vol. 40, No. 2, March, 1961, pp. 453–488.

49. H. Kogelnik and T. Li, "Laser Beams and Resonators," *Applied Optics*, Vol. 5, No. 10, October, 1966, pp. 1550–1567.

50. J. Hirano and Y. Fukatsu, "Stability of Light Beams in a Beam Waveguide," *Proc. IEEE*, Vol. 52, No. 11, November, 1964, pp. 1284–1292.

51. D. Marcuse, "Propagation of Light Rays through a Lens Waveguide with Curved Axis," *B.S.T.J.*, Vol. 43, No. 2, March, 1964, pp. 741–753.

52. D. Marcuse, "Statistical Treatment of Light-Ray Propagation in Beam-Waveguides," *B.S.T.J.*, Vol. 44, No. 9, November, 1965, pp. 2065–2081.

53. D. Marcuse, "Probability of Ray Position in Beam Waveguides," *IEEE Trans. on Microwave Theory and Techniques*, Vol. MTT-15, No. 3, March, 1967, pp. 167–171.

54. A. Papoulis, *Probability, Random Variables and Stochastic Processes*, McGraw-Hill Book Company, New York, 1965.

55. P. Beckmann, *Probability in Communication Engineering*, Harcourt, Brace & World, New York, 1967.

56. E. A. J. Marcatili, "Ray Propagation in Beam-Waveguides with Redirectors, *B.S.T.J.*, Vol. 45, No. 1, January, 1966, pp. 105–115.

57. E. A. J. Marcatili, "Effect of Redirectors, Refocusers, and Mode Filters on Light Transmission through Aberrated and Misaligned Lenses," *B.S.T.J.*, Vol. 46, No. 8, October, 1967, pp. 1733–1752.

58. C. C. Johnson, *Field and Wave Electrodynamics*, McGraw-Hill Book Company, New York, 1965.

59. V. Heine, *Group Theory in Quantum Mechanics*, Pergamon Press, New York, 1960.

60. J. M. Ziman, *Electrons and Phonons*, Oxford University Press, New York, 1960.

61. I. S. Gradshteyn and I. M. Ryzhik, *Tables of Integrals, Series, and Products*, 4th edition, Academic Press, New York, 1965.

62. D. Slepian and H. O. Pollak, "Prolate Spheroidal Wave Functions, Fourier Analysis and Uncertainty" — I, *B.S.T.J.*, Vol. 40, No. 1, January, 1961, pp. 43–63.

63. H. J. Landau and H. O. Pollak, "Prolate Spheroidal Wave Functions, Fourier Analysis and Uncertainty" — II, *B.S.T.J.*, Vol. 40, No. 1, January, 1961, pp. 65–84.

64. D. Marcuse, "Deformation of Fields Propagating through Gas Lenses," *B.S.T.J.*, Vol. 45, No. 8, October, 1966, pp. 1345–1368.
65. E. A. J. Marcatili, "Off-Axis Wave-Optics Transmission in a Lens-Like Medium with Aberrations," *B.S.T.J.*, Vol. 46, No. 1, January, 1967, pp. 149–166.
66. G. Eichmann, "Quasi-Geometric Optics of Media with Inhomogeneous Index of Refraction," *J. Opt. Soc. Am.*, Vol. 61, No. 2, February, 1971, pp. 161–168.
67. H. Kogelnik, "Imaging of Optical Mode Resonators with Internal Lenses," *B.S.T.J.*, Vol. 44, No. 3, March, 1965, pp. 455–494.
68. H. Kogelnik, "On the Propagation of Gaussian Beams of Light through Lenslike Media Including Those with Loss or Gain Variation," *Appl. Optics*, Vol. 4, No. 12, December, 1965, pp. 1562–1569.
69. P. K. Tien, J. P. Gordon, and J. R. Whinnery, "Focusing of a Light Beam of Gaussian Field Distribution in Continuous and Periodic Lens-Like Media," *Proc. IEEE*, Vol. 53, No. 2, February, 1965, pp. 129–136.
70. R. L. Fork, D. R. Herriot, and H. Kogelnik, "A Scanning Spherical Mirror Interferometer for Spectral Analysis of Laser Radiation," *Appl. Optics*, Vol. 3, No. 12, December, 1964, pp. 1471–1484.
71. P. Smith, "Stabilized Single Frequency Output from a Long Laser Cavity," *IEEE J. of Quantum Electronics*, Vol. QE-1, No. 8, November, 1965, pp. 343–348.
72. G. D. Boyd and H. Kogelnik, "Generalized Confocal Resonator Theory," *B.S.T.J.*, Vol. 41, No. 4, July, 1962, pp. 1347–1369.
73. A Erdelyi, W. Magnus, F. Oberhettinger, and F. G. Tricomi, *Higher Transcendental Functions*, Vol. II, McGraw-Hill Book Company, New York, 1953.
74. D. Marcuse, "Modes and Pseudomodes in Dielectric Waveguides," *IEEE Trans. on Microwave Theory and Techniques*, Vol. MTT-18, No. 1, January, 1970, pp. 62–63.
75. P. Hirsch and A. H. Carter, "Mathematical Models for the Prediction of SOFAR Propagation Effects," *J. Acoustical Soc. Am.*, Vol. 37, No. 1, January, 1965, pp. 90–94.
76. S. E. Miller, "Light Propagation in Generalized Lenslike Media," *B.S.T.J.*, Vol. 44, No. 9, November, 1965, pp. 2017–2064.
77. J. P. Gordon, "Optics of General Guiding Media," *B.S.T.J.*, Vol. 45, No. 2, February, 1966, pp. 321–332.
78. E. A. J. Marcatili, "Modes in a Sequence of Thick Astigmatic Lens-Like Focusers," *B.S.T.J.*, Vol. 43, No. 6, November, 1964, pp. 2887–2904.
79. N. S. Kapany, *Fiber Optics*, Academic Press, New York, 1967.
80. E. Snitzer, "Cylindrical Dielectric Waveguide Modes," *J. Opt. Soc. Am.*, Vol. 51, No. 5, May, 1961, pp. 491–498.
81. R. E. Collin, *Field Theory of Guided Waves*, McGraw-Hill Book Company, New York, 1960.
82. H. E. Rowe and W. D. Warters, "Transmission in Multimode Waveguide with Random Imperfections," *B.S.T.J.*, Vol. 41, No. 3, May, 1962, pp. 1031–1170.
83. S. P. Schlesinger, P. Diament, and A. Vigants, "On Higher Order Hybrid Modes of Dielectric Cylinders," *IEEE Trans. on Microwave Theory and Techniques*, Vol. MTT-8, No. 2, March, 1960, pp. 252–253.

84. A. L. Jones, "Coupling of Optical Fibers and Scattering in Fibers," *J. Opt. Soc. Am.*, Vol. 55, No. 3, March, 1965, pp. 261–271.

85. R. Vanclooster and P. Phariseau, "The Coupling of Two Parallel Dielectric Fibers I- Basic Equations," *Physica*, Vol. 47, No. 4, June, 1970, pp. 485–500.

86. R. Vanclooster and P. Phariseau, "The Coupling of Two Parallel Dielectric Fibers II- Characteristic of Coupling in Two Fibers," *Physica*, Vol. 47, No. 4, June, 1970, pp. 501–514.

87. R. Vanclooster and P. Phariseau, "Light Propagation in Fiber Bundles," *Physica*, Vol. 49, No. 4, November, 1970, pp. 493–501.

88. S. E. Miller, "Coupled Wave Theory and Waveguide Applications," *B.S.T.J.*, Vol. 33, No. 3, May, 1954, pp. 661–720.

89. A. W. Snyder, "Radiation Loss Due to Variations of Radius on Dielectric or Optical Fibers," *IEEE Trans. on Microwave Theory and Techniques*, Vol. MTT-18, No. 9, September, 1970, pp. 608–615.

90. A. W. Snyder and R. De La Rue, "Asymptotic Solutions of Eigenvalue Equations for Surface Waveguide Structures," *IEEE Trans. on Microwave Theory and Techniques*, Vol. MTT-18, No. 9, September, 1970, pp. 650–651.

91. A. W. Snyder, "Coupling of Modes on a Tapered Dielectric Cylinder," *IEEE Trans. on Microwave Theory and Techniques*, Vol. MTT-18, No. 7, July, 1970, pp. 383–392.

92. A. W. Snyder, "Asymptotic Expressions for Eigenfunctions and Eigenvalues of a Dielectric or Optical Waveguide, *IEEE Trans. on Microwave Theory and Techniques*, Vol. MTT-17, No. 12, December, 1969, pp. 1130–1138.

93. A. W. Snyder, "Excitation and Scattering of Modes on a Dielectric or Optical Fiber," *IEEE Trans. on Microwave Theory and Techniques*, Vol. MTT-17, December, 1969, pp. 1138–1144.

94. R. B. Dyott and J. R. Stern, *Group Delay in Glass Fibre Waveguide*, pp. 176–181, Conference on Trunk Telecommunications by Guided Waves, Conference Publication No. 71, Institute of Electrical Engineers, London, 1970.

95. F. P. Kapron, D. B. Keck, and R. D. Maurer, "Radiation Losses in Glass Optical Waveguides," *Appl. Phys. Letters*, Vol. 17, No. 10, November 15, 1970, pp. 423–425.

96. D. Marcuse, "Radiation Losses of the Dominant Mode in Round Dielectric Waveguides," *B.S.T.J.*, Vol. 49, No. 8, October, 1970, pp. 1665–1693.

97. R. B. Adler, "Waves on Inhomogeneous Cylindrical Structures," *Proc. IRE*, Vol. 40, No. 3, March, 1952, pp. 339–348.

98. D. Marcuse, "Mode Conversion Caused by Surface Imperfections of a Dielectric Slab Waveguide," *B.S.T.J.*, Vol. 48, No. 10, December, 1969, pp. 3187–3215.

99. D. Marcuse and R. M. Derosier, "Mode Conversion Caused by Diameter Changes of a Round Dielectric Waveguide," *B.S.T.J.*, Vol. 48, No. 10, December, 1969, pp. 3217–3232.

100. D. Marcuse, "Radiation Losses of Dielectric Waveguides in Terms of the Power Spectrum of the Wall Distortion Function," *B.S.T.J.*, Vol. 48, No. 10, December, 1969, pp. 3233–3242.

101. D. Marcuse, "Radiation Losses of Tapered Dielectric Slab Waveguides," *B.S.T.J.*, Vol. 49, No. 2, February, 1970, pp. 273–290; Erratum, *B.S.T.J.*, Vol. 49, No. 5, May-June, 1970, p. 919.

102. D. Marcuse, "Compression of a Bundle of Light Rays," *Appl. Optics*, Vol. 10, No. 3, March, 1971, pp. 494–497.
103. D. Marcuse, "The Coupling of Degenerate Modes in Two Parallel Dielectric Waveguides," *B.S.T.J.*, Vol. 50, No. 6, July-August, 1971, pp. 1791–1816.
104. D. Marcuse, "Crosstalk Caused by Scattering in Slab Waveguides," *B.S.T.J.*, Vol. 50, No. 6, July-August, 1971, pp. 1817–1831.
105. A. R. Tynes, A. D. Pearson, and D. L. Bisbee, "Loss Mechanisms and Measurements in Clad Glass Fibers and Bulk Glass," *J. Opt. Soc. Am.*, Vol. 61, No. 2, February, 1971, pp. 143–153.
106. E. A. J. Marcatili, "Bends in Optical Dielectric Guides," *B.S.T.J.*, Vol. 48, No. 7, September, 1969, pp. 2103–2132.
107. J. H. Harris, R. Shubert, and J. N. Polky, "Beam Coupling to Films," *J. Opt. Soc. Am.*, Vol. 60, No. 8, August, 1970, pp. 1007–1016.
108. P. K. Tien, R. Ulrich, and R. J. Martin, "Modes of Propagating Light Waves in Thin Deposited Semiconductor Films," *Appl. Phys. Letters*, Vol. 14, No. 9, May, 1969, pp. 291–294.
109. R. Ulrich, "Optimum Excitation of Optical Surface Waves," *J. Opt. Soc. Am.*, Vol. 61, No. 11, November, 1971 pp. 1467–1477.
110. M. F. Bracey, A. L. Cullen, E. F. T. Gillespie, and J. A. Staniforth, "Surface-Wave Research in Sheffield," *IRE Trans. on Antennas and Propagation*, Vol. AP-7, Supplement, December, 1959, p. S 219.
111. S. S. Schweber, *An Introduction to Relativistic Quantum Field Theory*, Harper and Row, New York, 1961.
112. A. D. Pearson, W. G. French, and E. G. Rawson, "Preparation of a Light Focusing Glass Rod by Ion Exchange Techniques," *Appl. Phys. Letters*, Vol. 15, No. 2, July 15, 1969, pp. 76–77.
113. V. V. Shevchenko, *Continuous Transitions in Open Waveguides*, Golem Press, Boulder, Colorado, 1971.
114. M. A. Miller and V. I. Talanov, "Electromagnetic Surface Waves Guided by a Boundary with Small Curvature," *Zh.T.F.*, Vol. 26, No. 12, 1956, p. 2755.
115. L. Bergstein and E. Marom, "Angular Spectra of Optics Cavities," *J. Opt. Soc. Am.*, Vol. 56, No. 1, January, 1966, pp. 16–32.
116. R. L. Sanderson and W. Streifer, "Comparison of Laser Mode Calculations," *Appl. Optics*, Vol. 8, No. 1, January, 1969, pp. 131–136.
117. A. E. Siegman, "Unstable Optical Resonator for Laser Applications," *Proc. IEEE*, Vol. 53, No. 3, March, 1965, pp. 277–287.
118. S. Kawakami and J. Nishizawa, "An Optical Waveguide with the Optimum Distribution of the Refractive Index with Reference to Waveform Distortion," *IEEE Trans. Microwave Theory and Techniques*, Vol. 16, No. 10, October, 1968, pp. 814–818.
119. E. G. Rawson, D. R. Herriott, and J. McKenna, "Analysis of Refractive Index Distributions in Cylindrical, Graded Index Glass Rods (Grin Rods) Used as Image Relays," *Appl. Optics*, Vol. 9, No. 3, March, 1970, pp. 753–759.
120. J. C. Heurtley, "Hyperspheroidal Functions—Optical Resonators with Circular Mirrors," *Proceedings of the Symposium on Quasi-Optics*, Polytechnic Press of the Polytechnic Institute of Brooklyn, 1964, pp. 367–375.
121. J. H. Harris and R. Shubert, "Variable Tunneling Excitation of Optical Surface Waves," *IEEE Trans. Microwave Theory and Techniques*, Vol. 19, No. 3, March, 1971, pp. 269–276.

122. D. Marcuse, "Gaussian Approximation of the Fundamental Modes of Graded-Index Fibers," *J. Opt. Soc. Am.*, Vol. 68, No. 1, January, 1978, pp. 103–109.

123. G. A. Bliss, *Lectures on the Calculus of Variations*, University of Chicago Press, Chicago, 1946.

124. G. A. Korn and T. M. Korn, *Mathematical Handbook for Scientists and Engineers*, McGraw-Hill, New York, 1961.

125. A. W. Snyder and R. A. Sammut, Fundamental (HE_{11}) Mode of Graded Optical Fibers, *J. Opt. Soc. Am.*, Vol. 69, No. 2, December, 1979, pp. 1663–1671.

126. S. E. Miller and A. G. Chynoweth, *Optical Fiber Telecommunications*, Academic Press, New York, 1979.

127. J. A. Arnaud, *Beam and Fiber Optics*, Academic Press, New York, 1976.

128. K. Petermann, "The Mode Attenuation in General Graded Core Multimode Fibers," *Arch. Elect. Uebertr.* Vol. 29, No. 7/8, July/August, 1975, pp. 345–348.

129. D. Gloge and E. A. J. Marcatili, "Multimode Theory of Graded-Core Fibers," *Bell Syst. Tech. J.*, Vol. 52, No. 9, November, 1973, pp. 1563–1578.

130. D. Gloge, "Propagation Effects in Optical Fibers," *IEEE Trans. Microwave Theory and Tech.*, MTT-23, No. 1, January, 1975, pp. 106–120.

131. H. G. Unger, *Planar Optical Waveguides and Fibers*, Clarendon Press, Oxford, 1977.

132. J. E. Midwinter, *Optical Fibers for Transmission*, John Wiley and Sons, New York, 1979.

133. P. M. Morse and H. Feshbach, *Methods of Theoretical Physics*, Vol. II, McGraw-Hill, New York, 1953.

134. R. Olshansky and D. B. Keck, "Pulse Broadening in Graded-Index Optical Fibers," *Applied Optics*, Vol. 15, No. 2, February, 1976, pp. 483–491.

135. A. W. Snyder and D. J. Mitchell, "Leaky Rays on Circular Fibers," *J. Opt. Soc.*, Vol. 64, No. 5, May, 1974, pp. 599–607.

136. A. W. Snyder and C. Pask, "*Optical Fibre: Spatial Transient and Steady State, Opt. Comm.*," Vol. 15, No., 2, October, 1975, pp. 314–316.

137. D. Gloge, I. P. Kaminow and H. M. Presby, "Profile Dispersion in Multimode Fibers: Measurement and Analysis," *Electron. Lett.*, Vol. 11, No. 19, September, 1975, pp. 469–470.

138. D. Marcuse, *Principles of Quantum Electronics*, Academic Press, New York, 1980.

139. D. Marcuse, *Principles of Optical Fiber Measurements*, Academic Press, New York, 1981.

140. D. N. Payne and W. A. Gambling, "Zero Material Dispersion in Optical Fibres," *Electron. Lett.*, Vol. 11, No. 8, April, 1975, pp. 176–178.

141. D. Marcuse, "Pulse Distortion in Single-Mode Fibers," *Applied Optics*, Vol. 19, No. 10, May, 1980, pp. 1653–1660.

142. H. G. Unger, "Optical Pulse Distortion in Glass Fibres at the Wavelength of Minimum Dispersion," *Arch. Electr, Uebertr.*, Vol. 31, No. 12, December, 1977, pp. 518–519.

143. F. P. Kapron, "Maximum Information Capacity of Fibre-Optic Waveguides," *Electron. Lett.*, Vol. 13, No. 4, February, 1977, pp. 96–97.

144. D. Gloge, "Effect of Chromatic Dispersion on Pulses of Arbitrary Coherence," *Electron. Lett.*, Vol. 15, No. 21, October, 1979, pp. 686–687.

145. S. Kobayashi, S. Shibata, N. Shibata, and T. Izawa, "Refractive Index Dispersion of Doped Fused Silica," *IOOC '77*, Tokyo, July, 1977, paper B8-3.

146. J. W. Fleming, "Material and Mode dispersion in GeO_2-B_2O_3-SiO_2 Glasses," *J. Am. Ceram. Soc.*, Vol. 59, No. 11-12, November–December, 1976, pp. 503–507.

147. D. Marcuse and C. Lin, "Low Dispersion Single-Mode Fiber Transmission—The Question of Practical vs. Theoretical Maximum Transmission Bandwidth," IEEE *J. Quant. Electr.* QE-17, No. 6, June 1981, pp. 869–877.

148. L. G. Cohen, W. L. Mammel and H. M. Presby, "Correlation Between Numerical Predictions and Measurements of Single-Mode Fiber Dispersion Characteristics," *Applied Optics*, Vol. 19, No. 12, June, 1980, pp. 2007–2010.

149. C. Lin and L. G. Cohen, "Pulse Delay Measurements in the Zero-Material Dispersion Region for Germanium and Phosphorus-Doped Silica Fibers," *Electron. Lett.*, Vol. 14, No. 6, March, 1978, pp. 170–172.

150. D. Marcuse, "Interdependence of Waveguide and Material Dispersion," *Applied Optics*, Vol. 18, No. 17, September, 1979, pp. 2930–2932.

151. K. Okamoto, T. Edahiro, A. Kawana, and T. Miya, "Dispersion Minimization in Single-Mode Fibres over a Wide Spectral Range," *Electron. Lett.*, Vol. 15, No. 22, October, 1979, pp. 729–731.

152. D. Marcuse, "Calculation of Bandwidth from Index Profiles of Optical Fibers. Part 1: Theory," *Applied Optics*, Vol. 18, No. 12, June, 1979, pp. 2073–2080.

153. S. D. Personick, "Baseband Linearity and Equalization in Fiber Optic Digital Communication Systems," *Bell Syst. Tech. J.*, Vol. 53, No. 7, September, 1973, pp. 1175–1194.

154. R. Olshansky, "Pulse Broadening Caused by Deviations from the Ideal Optimal Index Profile," *Applied Optics*, Vol. 15, No. 3, March, 1976, pp. 782–788.

INDEX